100 Years of Swiss Design

Museum
für Gestaltung
Zürich

100 Years of Swiss Design

Lars Müller
Publishers

Edited by
Christian Brändle
Renate Menzi
Arthur Rüegg

Contents

		7	Christian Brändle	Foreword
		8	Renate Menzi Arthur Rüegg	Introduction
		10	Stanislaus von Moos	Tools and Tidbits
		22	Claude Enderle	1850–1900: Where Does Swiss Design History Begin?
28	1910—Reform	36	Key Objects	
		46	Claude Lichtenstein	Reform and New Form: The 1918 Swiss Werkbund Exhibition in Zurich
		52	Arthur Rüegg	Arts and Crafts, Industrial Art, Machine Art: From Furniture Type to Type Furniture
58	1920—Mission	66	Key Objects	
		76	Meret Ernst	Small and Precise
		82	Hans Ulrich Gumbrecht	Rational Impatience with the Past: Hannes Meyer's *Co-op. Interieur,* 1926
		88	Arthur Rüegg	Modern and Inhospitable? Bridges from Tradition to Modernity
94	1930—Pragmatics	102	Key Objects	
		112	Anne-Claire Schumacher Sophie Wirth Brentini	The Shifting Contours of the Arts of the Table: Ceramic Design in Switzerland
		118	Arthur Rüegg	The "New" Light: 100 Years of Swiss Lighting Design
		124	Christina Sonderegger	Mobile and Lightweight: Designing with Aluminum
		130	Peter Lepel	The Aluminum Industry Competition of 1933
		132	Claude Lichtenstein	Furniture for All Seasons: The Early Days of Wohnbedarf
138	1940—Réduit	146	Key Objects	
		156	Claude Lichtenstein	Retranslation of Modernism: Design in the Forties
		162	Arthur Rüegg	"Solutions d'urgence": The Economy of Means and the Aesthetics of Necessity
		168	Renate Menzi	Bones, Splints, and Shells: Ergonomics and Organic Form
174	1950—Ethics	182	Key Objects	
		192	Renate Menzi	Designing "Die gute Form"
		198	Christof Kübler	If ... a Traveler
		204	Arthur Rüegg	Nomadic Living: The "New Wood" in Swiss Design
210	1960—Rationality	218	Key Objects	
		228	Arthur Rüegg	A Place for Everything and Everything in Its Place: Systems, Modules, Formats
		234	Klaus Spechtenhauser	Individual Piece, Modular System, Compact Unit: The Evolution of Swiss Kitchen Design
		240	Claude Lichtenstein	The Hidden Dimension: Industrial Design, 1960–1980
246	1970—Pop	254	Key Objects	
		264	Sebastian Hackenschmidt	Forming Swiss Taste: From *Gute Form* to Pop Culture
		270	Sebastian Hackenschmidt	"Inspiration for Playful Experimentation": Do-It-Yourself Design
276	1980—History	284	Key Objects	
		294	Fredi Fischli	Design Tools circa 1980
		300	Arthur Rüegg	The Beginning—or the End—of History? Neo-Historicism, Reeditions, and Tributes
306	1990—Statement	314	Key Objects	
		324	Meret Ernst	Swiss Made, Swiss Design?
		330	Rolf Fehlbaum	What Makes a Good Producer?
		331	Daniel Freitag	What Makes a Good Designer?
		332	Renate Menzi	The New Simplicity in Form and Production
338	2000—Perspectives	346	Key Objects	
		356	Renate Menzi	Fields of Activity in Current Design Practice: When Does Good Design Pay Off?
		358	Meret Ernst	Teaching, Research, Promotion
		360	Christian Brändle	Open Collections for Tomorrow
		362	Bibliography	
		370	Index of Names	
		373	Authors	
		374	Credits and Copyrights	

Christian Brändle

Preface

Relative to the long history of the Museum für Gestaltung Zürich, its Design Collection is quite a recent addition. Founded in 1987, it joins the Applied Art, Graphics, and Poster collections to create overall holdings that aspire to comprehensively cover Swiss design history since the advent of serial production. The museum began its collecting activity with its Applied Art Collection (founded together with the Graphics Collection in 1875), which focuses on international arts and crafts and unique handcrafted pieces, while the Design Collection instead targets industrially produced objects of daily use with a clear link to Switzerland.

Design unfolds its meaning only when placed in context. A chair, for example, might on its own convey insight into a particular design attitude or a structural solution, but its significance and purpose will usually only reveal itself to us once we know something about its historical background and setting. Accordingly, this publication aims not only to provide a historical overview of outstanding design achievements both unknown and familiar but also to embed the individual objects in the aesthetic and discursive context of their day.

The museum is pursuing a similar goal with the establishment of its new Schaudepot (Open Collections), which in 2014 brings together all 500,000 collection objects at the new campus of the Zurich University of the Arts (ZHdK). With its university setting, and above all thanks to a collection strategy that is devoted not only to the material object itself but also to its design history, marketing, and reception, the future Schaudepot will form a unique research resource for scholars, students, and the interested public.

In parallel, the museum is also working on the global publication of its collections: 80,000 objects along with exhibition documentation are already available today on the Internet portal eMuseum.ch. The collections of the Museum für Gestaltung Zürich thus constitute a one-of-a-kind real and virtual archive of design history, with a focus on Switzerland.

Finally, the present publication and its companion, *100 Years of Swiss Graphic Design,* serve to make the collections accessible to a broad public in varying contexts. The cooperative endeavor that went into their making was extremely rewarding and further strengthened the museum's diverse relationships with designers, businesses, and institutions.

On behalf of the museum and the editors, I would like to express my heartfelt thanks to the authors, advisors, and designers who contributed to this ambitious project, as well as to the Lottery Fund of the Canton of Zurich, the "Friends of the Museum für Gestaltung Zürich," and all other partners for their trust and their generous financial support.

Renate Menzi
Arthur Rüegg

Introduction

What do Glarus handkerchiefs circa 1900 have to do with Sophie Taeuber's marionettes from 1918, the Patent Ochsner garbage can from 1926, the *Landi Chair* from 1939, the Swiss Railway station clock from 1944, the Arwa Therm single-lever faucet from 1952, a Linck vase from the 1960s, the Wisa-Gloria tricycle from 1970, the Logitech computer mouse from 1982, or Freitag bags from 1995? Many of these are ordinary everyday items produced in series, while others are spectacular handcrafted unique pieces. But what they all have in common is a carefully thought-out concept, absolute credibility in the implementation, and high-quality fabrication. In today's view of cultural history, we classify all of them—as we do any deliberately conceived and crafted real or virtual object, service, or brand—under the generic term "design." Borrowing from Joseph Beuys's extended concept of what constitutes art, this term implies that social changes can also be understood as design processes; in 1980, the Basel sociologist Lucius Burckhardt even went so far as to call for a concept of design integrating invisible personal relationships.

Just half a century ago, Swiss *Industrie-Entwerfer* or *Gestalter* like Max Bill or Hans Bellmann were still wary of the American catchword. They distanced themselves from everything "fashionable about an object, and from the idea of a fashionable object in the first place"—something the term "design" evidently connoted for them. Doubts about whether a productive partnership between idealistic formal considerations and commercial exploitation could even be possible had nagged at the Swiss Werkbund (SWB) ever since its inception in 1913. Despite all misgivings, however, the English terms "industrial design" and, later, "design" and "designer" inexorably took root in Switzerland. Today, design refers not only to the work involved but also to the entire field of endeavor, and not only to the form but also to its authorship (such as in the term "designer furniture"). As a lifestyle buzzword, the English term is almost overused in the media—not least because objects that were once mere utilitarian necessities have long since become brand vehicles and status symbols. In the introductory essay to this publication, Stanislaus von Moos analyzes the related dilemma in visual culture under the title "Tools and Tidbits."

Swiss design may very well owe its good reputation to the way it ignores expeditious paradigm shifts and the vagaries of public taste to adhere firmly to the modernist concept of quality. Numerous publications on "Swissness" and "Swiss Made" perpetuate the myth of Swiss provenance. Perhaps this is a way to compensate for the fact that Switzerland's importance as a production location has steadily dwindled in the face of globalization. "Designed in Switzerland" is thus no longer necessarily tied to the coherent technical realization that was once associated with Swiss quality. The present publication also participates in this inquiring as to whether industrial design in Switzerland might nonetheless have its own unique features or special qualities. The term "Swiss design" is, however, defined relatively openly here: as a description of products, concepts, and contemporaneous historical currents that emerged in the field of furniture and product design in Switzerland, from Switzerland, or by Swiss designers living abroad. In addition to presenting outstanding objects, the book illuminates products and processes that have shaped the ideological history and material culture of the past hundred years.

The first stirrings of the discipline can already be discerned in the manufactories of the eighteenth and nineteenth centuries. Examples are the porcelain manufactories in Kilchberg-Schooren and Nyon, the watchmakers of La Chaux-de-Fonds, and, of course, the textile, food, and mechanical engineering industries. Their products were presented in the first half of the nineteenth century—barely separated from the fine arts—at "art and industry exhibitions" and later on the international stage at the international expositions. The second source is to be found in the decorative and applied arts with their focus on precious unique pieces. Here, the industrial revolution was accompanied by growing professionalism; between 1873 and 1884 alone, no less than six institutions dedicated to aesthetic education opened their doors to students. As late as 1918, however, the first major show presented by the Swiss Werkbund was still essentially an exhibition of decorative arts and interior design. Nevertheless, Sophie Taeuber's marionettes for the production of *König Hirsch* (*The Stag King*) at that show still haunt collective memory today as powerful images of the artistic inclination to abstraction.

Swiss design's international profile began to take shape in 1926 with a provocation: Hannes Meyer advocated in his *Co-op. Interieur* the exclusive use in interior design of standardized, serially made "type" products. In 1927, the first *Typenmöbel,* or "type furniture," designed by the Schweizer Kollektivgruppe commissioned by the SWB attracted attention from both critics and the public at the exhibition *Die Wohnung* (The Dwelling) in Stuttgart. The construction of the Werkbund's Neubühl housing development in Zürich-Wollishofen in 1931 provided a vital impetus for furniture, ceramic, and textile design geared toward maximum multifunctionality. In the push and pull between traditional and progressive forces, this period evokes a specifically Swiss stance that made its own unique contribution by innovatively responding to international developments with "pretentious humility." Resource scarcity during World War II and the reconstruction of Europe triggered trend-setting developments guided by Swiss pragmatism. With a 1949 traveling exhibition organized by Max Bill on behalf of the SWB, the Swiss idea of *Gute Form* ("good form," or good design) was propagated throughout postwar Europe. People cherished the hope that economically and beautifully crafted everyday objects might help make the new world a better place. In 1952, the SWB inaugurated an annual award for well-designed products. With time, however, the idealistic aspiration toward a normative Swiss design practice flagged, and during the social upheavals of the seventies efforts were made to expose the "deceitful nature of commodity aesthetics." The furniture design of the ensuing decades would be shaped by outstanding accomplishments that took off in divergent directions, such as the USM Haller furniture system and the ironically critical furniture pieces by Trix and Robert Haussmann or Susi and Ueli Berger. Technological innovation dominated industrial design, with solutions such as the Logitech computer mouse, the Phonak hearing aid, and the Swatch watch leaving a lasting mark on the international product world.

Design in Switzerland receives support at the national level from the Federal Office of Culture (BAK) and the Pro Helvetia Cultural Foundation. The Federal Government already initiated a design promotion program in 1917 as a way of stimulating industry after World War I. And in 1991, the Design Center Langenthal inaugurated the Swiss Design Award in an effort to influence the market, recognize distinguished achievements, and promote dialogue between disciplines, designers, and businesses. Changing framework conditions now call for a new understanding of "Swissness." Large companies—as well as young labels such as INCH Furniture and Aekae—have long since begun to shift the manufacture of utilitarian objects to countries with specialized resources. In a globalizing world, most Swiss designers work with international manufacturers and for international producers, while a few—such as the Freitag brothers—have instead successfully invested in their own labels with proprietary production and sales. Whether they like it or not, though, all of them are part of a global network, one that they know how to enable for specific projects.

The publication *100 Years of Swiss Design* is in principle arranged chronologically. Each decade is introduced by a photo spread that helps to immerse the reader in the atmosphere of that particular period. We then zero in as if through a telescope on ten exemplary products, examining both their significance in design history and their intriguing details. These key objects also serve to provide deeper insight into the collections of the Museum für Gestaltung. The essays then tie together the temporal and thematic axes. They place different generational successions and paradigm shifts in a wider context, often exceeding the confines of the respective decade. It is more a case here of looking through a reverse telescope to see the bigger picture. More than ten authors contributed essays for the book. Their theses may be contradictory depending on the issue at hand and the angle from which it is viewed, but this only makes them all the more fruitful for future discussion. The visual design of the book captures the different characters of the photo spreads, object series, and essays, highlighting the aesthetic standing as well as the expressive power of outstanding Swiss design achievements.

1

1 Herzog & de Meuron, *Sweet Dreams,* "Sugar Object," presented at the Herzog & de Meuron retrospective exhibition, Schaulager, Basel, 2004

"Good design" has often been defined as a way of serving material function in the most straightforward way while also giving form to the logic of the production process. In fact, modernism's preoccupation with manufacture and industrial production has been something like the alpha and omega of design reform from William Morris up to Max Bill and beyond. At first sight, by placing a "Zuckerobjekt" (Sugar Object) near the entrance to their 2004 retrospective exhibition in Basel, Herzog & de Meuron seem to have placed their work within this tradition[fig. 1]. Though hardly functional in any usual way, the form of the object is patently about "process"—in other words, about "dripping" as found in nature, resulting in a myriad of small and large stalagmites rising from a tub of hard, glass-clear molasses like a fountain reminiscent of certain Art Nouveau forms. However, its title, its juicy color, and its apparent stickiness seem to point beyond the traditional territory of design toward the titillations of candy and perfume. Indeed, what is here suggested are pleasures of consumption as well as labors of production.

2

Edible Buildings

It is difficult not to be reminded of Salvador Dalí, who, at the heyday of architectural functionalism, turned to *confiserie* as the ultimate paragon of aesthetic experience. The examples Dalí used were buildings by Guimard and Gaudí as photographed by Brassaï[fig. 2] and Man Ray. Art Nouveau is the most concrete and the craziest form of hypermaterialism; Dalí writes, "An illustration of the apparent paradox can be found ... in comparing an art nouveau building with a cake, an exhibitionistic and ornamental piece of *confiserie.*"[1]

No longer "convulsive" (as in Breton's phrase[2]), aesthetic perfection and consumption in the most literal sense become one in these examples. According to Dalí, to declare buildings as intrinsically "edible" is as "lucid as it is intelligent," "not just because it is attacking the violently materialistic prosaic quality of the 'immediate needs' on which ideal needs are based, but because in this way ... the nutritive, edible character of buildings of this kind is pointed out." Such buildings, Dalí adds, "are nothing other than the first edible buildings, and so also the first and only buildings that can be eroticized, whose mere existence proves the amorous notion of this function that is as urgent as it is essential, which is really being able to eat the object of desire."[3] Dalí's indulgence in the "psychic automatism" of chatter and fantasy may come over as mere nonsense, but while his philosophical nonsense resonates with memories from fairytales—think of the gingerbread house in *Hansel and Gretel*—it also has a philosophical prehistory. In *De re aedificatoria,* first published in 1485, Leon Battista Alberti describes the apparently infinite passion that forces the eye to devour its object, provided it obeys the laws of beauty.[4] Hegel, in turn, speaks of the "satisfying, as it were satiating appearance" offered by Greek temples.[5]

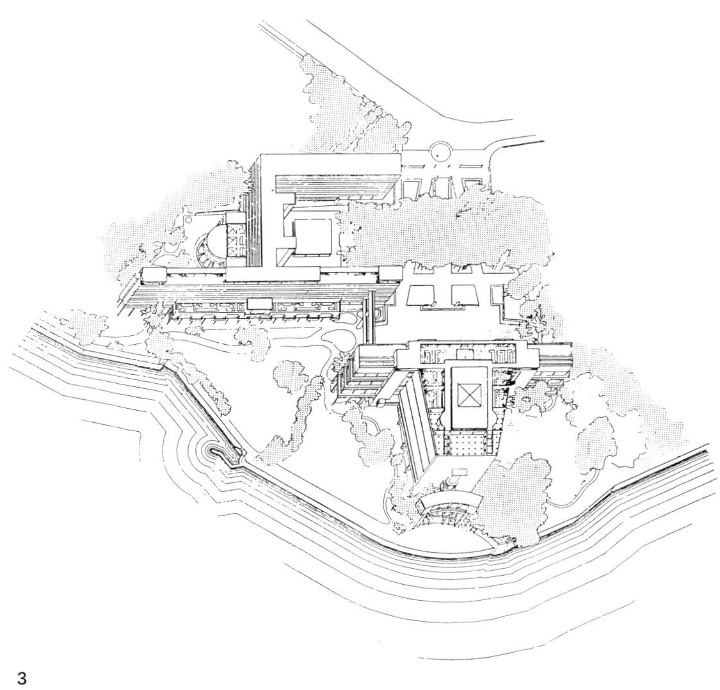

3

What interests in Dalí's text is yet another aspect. By associating the pleasures of *confiserie* to the enjoyment of buildings, if not of design altogether, and by so blatantly evoking the issue of obsolescence (*confiserie,* after all, is produced not in order to stay but in order to disappear—to be literally *consumed*) the artist laid his finger on what is or has been a distinctive mark of modernist design, that is, its prejudice in favor of defining function in terms of biotechnical *utilitas,* or rather, in Dalí's words, "the violently materialistic prosaic quality of the 'immediate needs'" on which, allegedly, if Dalí is right, "ideal needs are based."

As is well known, the film, fashion, and advertising worlds of America turned out to be phenomenally receptive to Dalí's extravagant clowneries from his first visit to New York—from the scandal of the Bonwit Teller displays on Fifth Avenue to his later covers for *Vogue* and *Dalí News.* In fact, Dalí's provocative capacity of transforming Surrealism into a commercial proposition had resulted early on in the proverbial mutation of "Salvador Dalí" into "Avida Dollars" and in the artist's de facto expulsion from the movement's core group.[6] Yet in the aftermath of these tribulations (though hardly triggered by them), the borderlines between "High" and "Low," between the museum and the department store, were permanently blurred, and surfing on the waves of psychoanalysis, mass taste, and product marketing became a legitimate artistic pursuit. Twenty-five years after Dalí's essay on the "terrific beauty of edible

2 Salvador Dalí, "Avez-vous vous déjà vu l'entrée du métro de Paris?" (Have you seen the entrance to the Paris Métro yet?). Photographs by Brassaï illustrating Dalí's essay "De la beauté terrifiante et comestible du Modern Style," *Minotaure,* no. 3–4 (1933)

3 Le Corbusier and Pierre Jeanneret, League of Nations Palace, Geneva. Competition design, 1927

4

5

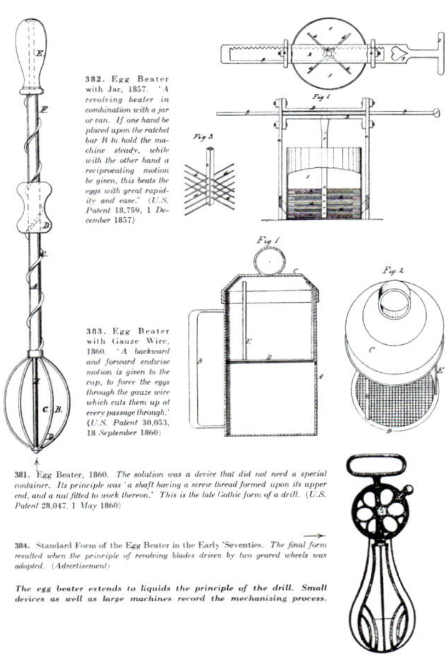
382. Egg Beater with Jar, 1857. 'A revolving beater in combination with a jar or can. If one hand be placed upon the ratchet bar B to hold the machine steady, while with the other hand a reciprocating motion be given, this beats the eggs with great rapidity and ease.' (U.S. Patent 18,759, 1 December 1857)

383. Egg Beater with Gauze Wire, 1860. 'A backward and forward endwise motion is given to the cup, to force the eggs through the gauze wire which cuts them up at every passage through.' (U.S. Patent 30,053, 18 September 1860)

381. Egg Beater, 1860. The solution was a device that did not need a special container. Its principle was 'a shaft having a screw thread formed upon its upper end, and a nut fitted to work thereon.' This is the late Gothic form of a drill. (U.S. Patent 28,047, 1 May 1860)

384. Standard Form of the Egg Beater in the Early 'Seventies. The final form resulted when the principle of revolving blades driven by two geared wheels was adopted. (Advertisement)

The egg beater extends to liquids the principle of the drill. Small devices as well as large machines record the mechanizing process.

6

4 Richard Hamilton, *Just what is it that makes today's homes so different, so appealing?,* 1956. Collage
Kunsthalle Tübingen, Sammlung Zundel

5 "Poultry on the Production Line, 1944."
From: Sigfried Giedion, *Mechanization Takes Command,* 1948

6 US patent drawings for egg beaters, ca. 1860.
Page from Sigfried Giedion, *Mechanization Takes Command,* 1948

architecture"—it is not far-fetched to suspect that the text was written as a deliberate polemic against Le Corbusier[7] / fig. 3 —Andy Warhol brought the Campbell Soup can and Coca-Cola to the museum and declared "Business Art" to be identical with "Art Business."

In the meantime, post-World War II consumer culture—cars, movies, household appliances, scooters, television, as well as, not to forget, canned foods and soft drinks—had done its own thorough job at undermining functionalist morals in design. Art turned out to be an acute seismograph of the situation, and once again the palate, if not more precisely the tongue, appears to be at the focus of the onslaught of merchandise that is pictured in these works. Note that Richard Hamilton's famous collage, *Just what is it that makes today's homes so different, so appealing?,* not only offers an almost comprehensive catalogue of that moment's consumer gadgetry—by placing a colossal, red lollipop at the center of the image it highlights the joy of candy as the ultimate consumer experience fig. 4—but some see Hamilton's lollipop as the initiatory moment of "Pop Art."[8]

Modernism and the Five Senses

Distrust in the dynamics of the market and its tendency to lure its publics into "consuming" goods for reasons that were often contrary to or at least independent from the proposed service has fueled modernist design ideology ever since Augustus Welby Pugin, Henry Cole, Gottfried Semper, and many others.[9] More specifically, it is a key theme in twentieth-century design theory: designers who played to the market "prostituted our fundamental precepts into modish trivialities," Walter Gropius argued as he witnessed the unstoppable trickling down of Bauhaus taste into 1950s supermarket aesthetics.[10] More recently, Kenneth Frampton insisted that it is one of the architect's duties to serve "living needs" but never the seamy side of "desire," which, as he points out, is only semiconscious.[11] Intriguingly—and here lies the "provocation" of Herzog & de Meuron's "Sugar Object" as well as of Dalí's amoral diatribe (not to mention Hamilton's *Just what is it that makes today's homes so different, so appealing?*)—this distrust has often been accompanied by sarcasm for gustatory pleasure, especially so if induced with the help of natural (or artificial, or metaphorical) "sweetness."

Le Corbusier, for example, saw the opulent design of the "Palais des Nations" in Geneva that was about to be built instead of his "rational" winning design of 1927 fig. 5 as equivalent to the self-indulgent table manners—if not the gluttony—in the world of diplomacy: "Brillat-Savarin: cooking for diplomatic lunches and dinners, wearing dinner jackets or tails (in the style of the generals of the Grande Armée). We take leeks, asparagus, potatoes, beef, butter, spices, fruit: and ... we denature everything and reduce everything to the same taste.... That's where architecture is at. The academic palaces of Geneva were the most inconceivable baldachins of red plush and golden ribbons imaginable." Straightforward service of function is what the League of Nations headquarters should have in common with a meal: "There was a purpose to the palace: to work for the welfare of the world, in the same way that there is a purpose to a meal: to nourish."[12]

... a Double Portrait

Indeed, modernism was tendentially suspicious of any lack of control in matters of indulgence (an attitude hardly to be reproached in an era in which obesity has become a worldwide health issue). Yet the pleasures of the senses was obviously not what modernism opposed. Modern art, after all, was all about "sensation": first, about the immediacy of sensorial perception; and second—and partly by implication—about "spectacle." Educating the public to restore its capacity of authentic sensorial experience and thereby to overthrow the alleged condition of the nineteenth century that had forced the "bourgeois" to live by surrogates was modernism's very raison d'être.

The curious lithograph by Eugen Zeller that shows Sigfried Giedion together with Hans Finsler in a rather traditional living room is a case in point fig. 7. Giedion, as one knows, was the general secretary and the ideologue of an organization that had been founded in the aftermath of the "scandal" of the League of Nations Palace in order to help modernism to get access to public commissions—

the CIAM (Congrès internationaux d'architecture moderne).[13] Hans Finsler, in turn, was the founder of the Fotoklasse at Zurich's Kunstgewerbeschule and a pioneer of "objective" photography.[14] The two friends are holding a print showing a wide-open eye and an ear—attributes of the modern intellectual sworn to the supremacy of the arts as the key to reality. Their fingers are finely chiseled as if to highlight the tactile sensibility required of specialists engaged in such pursuits. Giedion, wearing his pair of old-fashioned glasses, is entirely focused on the graphic work before him. Finsler, in turn, is looking up, as if interrupted by the sound of a trumpet coming from the other end of the room. The picture recalls works by Max Ernst. Its imagery—hands, eye, ear, trumpet—suggests touch, vision, and audition as the key senses. The radio antenna rising behind the scene once again evokes sound. Finally, there is a copy of the magazine *L'Esprit nouveau* conspicuously placed on the table before the two men, and next to the magazine, a glass of pastis. The siphon sputters soda water into the glass, suggesting a sense of prickle in the palate—refreshing like the salutary beam of spring water shown in Giedion's *Mechanization Takes Command* as it is shot by a primitive shower [fig. 8]. Indeed, *Mechanization Takes Command* will place the interface between the organic and the mechanical worlds at the very core of the modern experience.

Not only can the picture thus be read as a modern allegory of the five senses: the pleasures of the senses are emphatically pictured as providing vivifying energy. Water is seen as a life-sustaining tonic. The sound of the trumpet played by a young girl who has just entered the room (in fact, Giedion's daughter Verena) plays a similar role. True, sound also helps to stay connected to the global information network, as is suggested by the antenna Zeller has planted behind the two men. This was in 1931, when radio meant information, not spurious pastime. To imagine Giedion in 1956 with a TV set as an accessory—think, say, of Hamilton's collage of that year [fig. 4]—would be a joke.

7

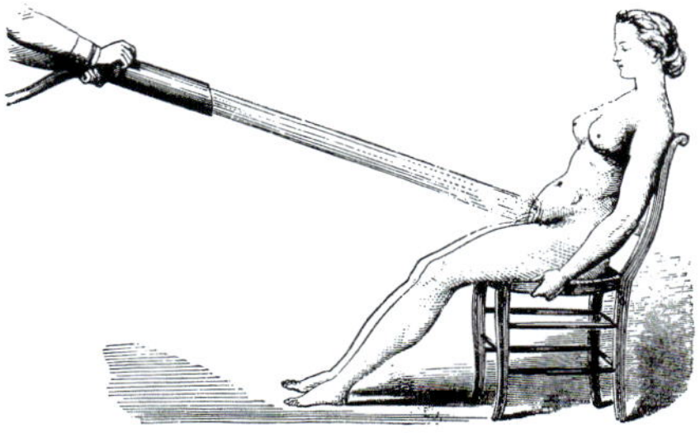

8

Brillat-Savarin and the Whirlpool of Consumption

Food and drink are essential to man's survival. Considering Le Corbusier's rumbling contempt of Brillat-Savarin and of the alleged perversions of nineteenth-century culinary art, it might be useful to take a second look at the book that established Brillat-Savarin's fame as one of the founders of the modern science of gastronomy, his *Physiologie du goût* (*Physiology of Taste*). Published in 1825, the *Physiologie du goût,* far from being a mere cookbook, has been described as the "deculpabilizing catechism of the moderate epicurean."[15] It would be no less appropriate to rank it among the reference texts in the physiology of the senses, side by side with its German counterpart, the famous art historian Carl Friedrich von Rumohr's treatise on the culinary arts *Geist der Kochkunst* (1822). Seen against the background of Brillat-Savarin's system, the Dalí / Le Corbusier diatribe in fact illustrates nothing so much than the dual elements of megalomania and culpability that characterizes the human response to orality:[16] Dalí representing the pole of narcissistic gluttony as opposed to what one might call Le Corbusier's "psychic masochism." For Giedion, who once declared gardening and cooking to be his preferred hobbies,[17] it may have been only natural that the production of meat, wheat, bread, etc., in other words, the inventions and the patents that contributed to the mechanization of slaughtering, reaping as well as of baking, egg beating, and eventually cooking, should be key aspects of modernity. In *Mechanization Takes Command* (1948), arguably the historian's most important work, he ventured to examine these techniques and procedures in great detail [figs. 5, 6]. Only rarely, however, does the viewpoint include the last phase in the lifecycle of foodstuffs: their scenographic presentation at the service of consumption (bread is among the exceptions).[18]

What I mean can be illustrated by the elaborately composed photographs of pastries and desserts that Giedion's friend Finsler contributed to Elisabeth Fülscher's famous *Kochbuch* (Cookbook) some ten years after Zeller's double portrait. Note that later editions of that book—a true vade mecum of postwar Swiss culinary arts—were amplified by rather noteworthy color photographs by Bernhard Moosbrugger, Finsler's assistant (from 1948 to 1951) [fig. 9]. Needless to add that there was no room for such things in *Mechanization Takes Command.*

9

7 Eugen Zeller, *Sigfried Giedion and Hans Finsler,* 1931. Etching
Private collection

8 "Shower for the Treatment of Abdominal Maladies. France, ca. 1860."
From: Sigfried Giedion, *Mechanization Takes Command,* 1948

9 Bernhard Moosbrugger, Color Plate 2, from Elisabeth Fülscher, *Kochbuch,* 8th ed., 1966

10

11

13

12

10 "Advertisement for Wrapped and Sliced Bread, 1944." From: Sigfried Giedion, *Mechanization Takes Command,* 1948

11 Herbert Matter, "Italian Bread," New York, 1944. From: Sigfried Giedion, *Mechanization Takes Command,* 1948

12 Advertisement for General Motors Switzerland, Biel, in *Neue Zürcher Zeitung,* February 20, 1959.

13 "Das erotische Genre" ("Erotic Mode"). Serving the "ruling taste" by way of "sweetness" (Süsslichkeit) in nineteenth-century painting. From: Sigfried Giedion, *Architektur und Gemeinschaft,* 1956

Is there a "morality" of sugar? Giedion deals at length with the temptations and the perils of "sweetness" in the production and the marketing of foods. Bread is crucial in this context. Sugar, he argues, played a key role in the way by which in the United States mass-manufactured bread began to eliminate small bakeries in the decades after 1900. It was used for sweetening the product, for giving it "a cakelike structure," and for stabilizing the crust fig. 10. As a result, so he continues, "the bread of full mechanization has the resiliency of a rubber sponge. When squeezed it returns to its former shape. The loaf becomes constantly whiter, more elastic, and frothier." Far from being only due to "mechanization," those qualities resulted from "new ways of pushing consumption," Giedion argues. It is the shortening used that "produce the 'soft velvet crumb,' a cakelike structure, so that the bread is half masticated, as it were, before reaching the mouth."[19]

And he goes on: "No king of France had bread of such white and silken quality on his table. But bread's position as the staff of life has been undermined. The complicated machinery of full mechanization has altered its structure and converted it into a body that is neither bread nor cake, but something half-way between the two. Whatever new enrichments may be devised, nothing can really ... [restore bread's character] ... as long as this sweetish softness continues to haunt its structure."[20]

Playing with the double entendre of the term "taste," Giedion then arrives at a curious blending of the physiopsychological risks of "sweetness" as he compares the process by which mechanization altered the taste of bread to "the ruling artistic taste in the nineteenth century" and its way of "exploiting certain dormant wishes of the public." Thus sugar in the production of bread turns out to play a role not so dissimilar from that played by the erotic stimulus in nineteenth-century "Salon painting" fig. 13: "The public loves what is sweet, smooth, and outwardly appealing.... The painters of the ruling taste set out to satisfy this public demand more and more, thus securing a market and rewards. The outcome was a blurring of the instincts in all classes of society, a disorientation that is doing its damage even today."[21] By thus associating sugar and sex with the "lower" human instincts, and coupling both with market needs, Giedion summarized what may be called a key stereotype in modernist discourse. Note that, once again, Guimard might have provided an eloquent precedent (as early as 1902 one critic blamed the lamps crowning the famous Paris Métro entrances as "unbearable public representations of the female sex"[22]).

Streamlining and Frosting: Sweet Mobility...

The American automobile and the way its body had been subjected to extravagant procedures of styling and frosting has often been seen as the ultimate symbol of a consumer object whose outer appearance was "sweetened" and "eroticized" in order to increase sales. One need only to remember that worldwide production of automobiles rocketed from two million in 1946 to eight million in 1955. By the time Giedion wrote *Mechanization Takes Command* and Max Bill, to take another example, began to launch his campaigns in favor of *Gute Form* ("good form," or good design), the automobile industry had long begun to market the car not merely as a practical tool for getting from one place to another but as a "motorized magic carpet on which social egos could ascend"[23].

Switzerland lived its own romance with American streamlined cars in the fifties (the sales were important enough as to justify a General Motors assembly plant for that country alone fig. 12).[24] In one of his short stories, Friedrich Dürrenmatt uses the Studebaker as an attribute of a textile salesman's all too shaky social ascension. "I don't want to sound like vaunting myself, but does anybody here own a Studebaker?" asks Alfredo Traps in self-defense, trying to escape the deadly logic of the fictive trial arranged by a group of retired attorneys and judges in the aftermath of the salesman's mishap with his beautiful red car on a highway (*Die Panne,* 1956).[25] Indeed, by streamlining American car-body baroque to European (or rather Italian) concepts of functionality and elegance, Raymond Loewy appears to have found a formula that "delivered" on both sides of the Atlantic. Walter Faber, too, the protagonist of Max Frisch's novel *Homo Faber* (1957), drives a Studebaker.[26] Once again, the color is red, though we don't know if the model is a Commander or a Starliner. By attributing the wish to own a Studebaker to Ivy, Faber's girlfriend, Frisch, incidentally, leaves no doubt as to the gendered nature of the consumer appetite involved.

... and the Jeep's revenge

For many Europeans, streamlined cars with Swiss license plates became a proverbial *accessoire* of the nouveau riche, a fact that may have contributed to the stiffening of the intellectual elite's contempt for that phenomenon. Design critics were notoriously dismissive. To add on "sales value" to products via body inflation or frosting, often by increasing the volume of a given "streamlined" car or household appliance, was seen as an aberration. Giedion devotes five pages to the issue of the streamlined automobile, citing the "logic of salesmanship" as the prime agent of form-giving in this context. With the frenzy of streamlining, the industrial designer gained in "influence on the shaping of public taste" that was comparable "only to that of the cinema," he argued.[27] He did not propose an alternative to the streamlined car, however. Remember that elsewhere in the book he proposed whole-wheat bread according to Sylvester Graham's recipe that secures "a full exercise of the teeth in mastication" as a cure against the perversion of bread into a malleable, spongelike substance. Would he have subscribed to Charles Eames's praise of the Willys Jeep?[28] "Beauty had to be thrust upon us," Eames had written in 1943, raving about the "naked truth" of the Jeep that had been flown by the thousands into Europe and Asia during World War II, and which, so he believed, represented all what a car should be: "Too bad it took a war to do away with the 'frosting,' but with that frosting gone we see the real side of many things for the first time."[29]

Note that the "real side" of the thing in this case involved a return to "simple" pleasures of bumpy rides on barely upholstered seats, exposed to wind and rain—not quite as healthy yet perhaps more fun than masticating Graham bread.

Max Bill shared Eames's fascination with the Jeep[fig. 17]. Nor could he help indulging in the gastronomic metaphor: the section that discusses streamline design in his programmatic book *Form* is entitled "Vom Werkzeug zum Schleckzeug" ("From Making to Faking"[figs. 14, 15]).[30] "Compared with producer goods, consumer goods are much more subject to recurrent changes in appearance than formerly," Bill laconically states: "As consumption is also more rapid, design is now being automatically misused as a device for increasing turnover. This dangerous tendency can be plainly seen in that flashy, spurious 'streamlining' which has usurped the place of ornament."[31]

As if to make clear that the provision of short-term gustatory pleasure lies outside the lofty tasks of design, Bill ends his chapter somewhat dramatically by associating such "business-motivated sugar icing" with the "final collapse of our civilization."[32] At the Hochschule für Gestaltung Ulm, Bill later institutionalized his politics of *Gute Form,* making the school a stronghold for the promotion of a Puritan neofunctionalism. While others returned to the bicycle or began to make the American streamlined baroque into a subject of anthropological curiosity, Bill concretized this opposition against "sugar icing" by conspicuously driving a Bentley Mark VI.

Austerity *versus* the Angry Young Men's Realism

War not merely universalized the Jeep. If Lewis Mumford was correct, it implanted a sense of what is essential for survival in people's minds. Mumford and other moralists and design reformers (American and non-American) thought that applying the war lessons of austerity might be the only way of regenerating true democracy and of preventing it from falling prey to a mindless consumerism. Bill's Hochschule für Gestaltung Ulm shared that dream of cultural regeneration via an ethic of curtailed consumption.[33] At the same time, in England, exploring the psychology of American car styling became the very subject of some quirky but extremely poignant paintings by Richard Hamilton. One of those, entitled *Homage to Chrysler Corp.*[fig. 16], shows the "distorted" face of an engine hood, with, hovering above, a pair of lips loosely connected with the curves of a nude vamp leaning over the car so as to almost merge with the bodywork. Another work, *Aahh!,* is basically about the rosy color of flesh and the mysterious fumes released by the drivers' subtle touch of the automobile's most erogenous zone: the gear

14

Entwicklung einer Gebrauchsform in England von 1900 bis 1933.

The evolution in the design of the household tap in England between the years 1900 and 1933.

Evolution d'une forme utilitaire en Angleterre entre 1900 et 1933.

(Doc.: Herbert Read «Art and Industry» 1934, Faber & Faber, London. Prod.: G. Jennings & Co.)

Die verchromte Blechdekoration der Vorderfront von Modell 1952 einer führenden und einflußreichen Automobilmarke.

The ornamental chromium-plated treatment of the radiator grille and buffer-bars in the 1952 model of one of the best known and most widely imitated makes of motor car.

La décoration de tôle chromée de l'avant du modèle 1952 d'une grande et influente marque d'automobile.

15

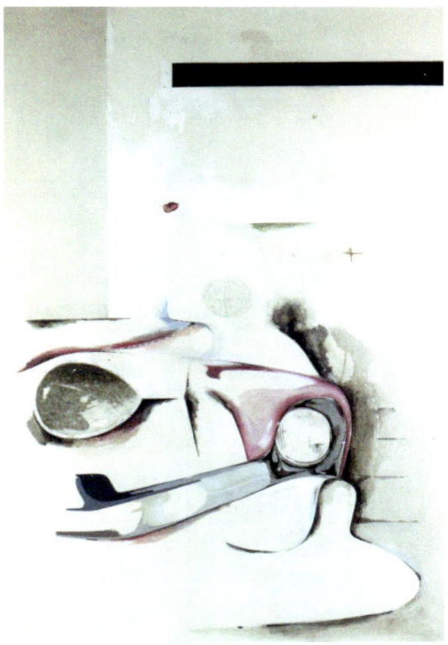

16

14 Willys Jeep on a technical mission in the Swiss Alps. From: Max Bill, *Form,* 1952

15 Evolution of the household tap in England, 1900–1933, compared to the chromium-plated radiator grille and buffer-bars in the 1952 model of an American motor car. Page from Max Bill, *Form,* 1951

16 Richard Hamilton, *Hommage à Chrysler Corp.,* 1957 Tate London

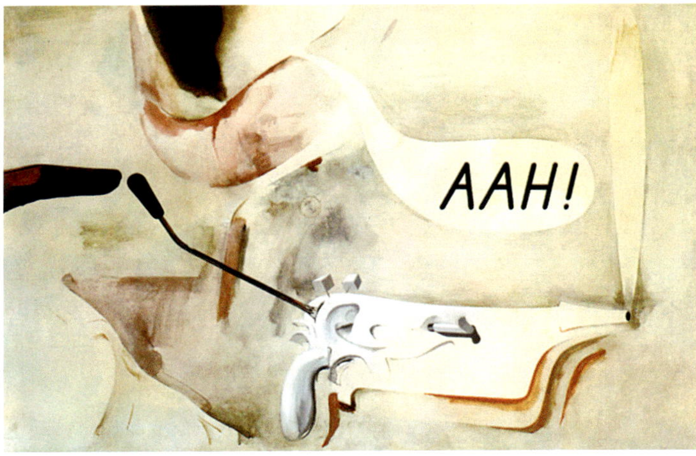

17

18

17 Richard Hamilton, *AAH!*, 1962
Tate London
18 Goldstripe Nylon Stockings, magazine advertisement. From: Marshall McLuhan, *The Mechanical Bride*, 1951

shift[fig. 17]. We may see more than a grain of satire here, even though the painter describes his intention as "a search for what is epic in everyday objects and everyday attitudes."[34]

Reyner Banham later provided the necessary underpinnings for these new interests. Once again, Banham's (as well as Hamilton's—or Marshall McLuhan's!) observations on the streamlined automobile's charge of eroticism, its implied significance as sex symbol, reverberate with Brillat-Savarin. According to the French physiologist, the function of "taste" is to help stabilize the individual in society, and ultimately to serve the continuity of the species. For, if Brillat-Savarin is correct (Sigmund Freud, for one, would not have disagreed), "physical desire had invaded the workings of all the other senses; no less strongly has it influenced all our sciences ... everything subtle and ingenious about them is due to this sixth sense, to the desire, the hope, the gratitude that spring from sexual union."[35]

The possible role of streamline design in the dynamics of species conservation may be more than what was at stake for the "Angry Young Men." For them, to take a "post-moralizing" attitude with regards to market aesthetics was one way of understanding the mechanisms that triggered it. The hope was that such an understanding might ultimately generate "knowing consumers"—like the sailor in Edgar Allen Poe's famous story who in order to free himself from the whirlpool into which he had been thrown understands that he has no choice but to go along with its deadly dynamic[fig. 18].[36]

Mechanization as Art Process

Marshall McLuhan admired Giedion's work. He saw it as being somehow caught up by the maelstrom while also keeping it at a critical distance. In *Mechanization Takes Command,* Giedion has taught us "how to see the entire process of mechanization as an art process," McLuhan writes.[37] "When printing was new, the Middle Ages became an art form. 'The Elizabethan world view' was a view of the Middle Ages. And the industrial age [the nineteenth century] turned the Renaissance into an art form as seen in the work of Jacob Burckhardt."[38]

Not by coincidence (given his formation as an art historian), Giedion comes closest to discussing broader socioeconomic issues in contemporary design, such as the market-driven dynamics of diversification, motivational research, obsolescence, etc., when he discusses art. Some drawings by Fernand Léger—or rather Giedion's own pictures made at the abandoned farm near the Canadian border Léger had chosen as his summer retreat during his American "exile"—thus became symbols of the more irrational sides of American industry[figs. 19, 20]: "In the shed was a reaper. 1892 model. I sat in its springy seat: the rusty break-lever still operated, the draw-pole was still strong, and it seemed that the horse need only to be harnessed.... It almost took one's breath away. The farm did not seem to have been abandoned many years before; it did not exhale the atmosphere of ruin—it seemed rather that some sudden catastrophe had arrested daily living, that we were in the midst of some kind of Pompeian excavation. And yet it had been an affair of only yesterday. Obsolescence. Artificial obsolescence, premature obsolescence."[39]

Among the most memorable sections of *Mechanization Takes Command* are those where Surrealism is used as a key to the psychology of the First Machine Age (in fact, despite Léger's fascination with the visuality of the modern metropolis and of industry, the painter plays only a marginal role in the book). Thanks to Surrealism, we have learned to discover the nineteenth century in its debris—a thought as crucial to Giedion as it was to Walter Benjamin.[40] For Giedion, what turned out to be useful in Surrealism is its capacity to unmask the perversity of the nineteenth century: "Such is the demonic side of the nineteenth century, ever scented behind the banal forms.... [W]hat the Surrealists mockingly portray is at bottom the same phenomenon that Henrik Ibsen, living within the period, attacked in deadly earnest and incarnated in his personages—the ceaseless roaming in search of one's soul: Nora's longing to escape from her domestic prison, the millstream *débâcle* in *Rosmersholm,* Oswald's madness in *Ghosts.* There, as here, it is the nineteenth century, never finding the way to its true self, devaluating symbols, without shaping new ones."[41 / fig. 21]

As any history writing, the practice of design history must negotiate its normative agenda with the logic of a captivating narrative. And Giedion knows how to tell his story and how to make us forget the profoundly "functionalist" bias that underlies the book. That prejudice implies that progress in design must result from the discovery of new uses and the ingeniousness of those who invented the respective forms. It also implies a moral contempt for the market instincts at work in the production of material culture. Quite predictably, perhaps, Reyner Banham as well as, more recently, Adrian Forty, both British (like Hamilton and Paolozzi), have followed the reverse strategy, insisting on the vital role of the market in determining success and failure in consumer culture.[42] More interested in treating design in terms of what it does than in terms of who did it, they argue that the consumer is as much an agent in design culture as the inventor or the designer (granted that, far from being a fixed datum, he or she needs to be reinvented with every new product).

Dalí, Bill, and the Dark Side of Pleasure

Seen in the general context of twentieth-century design, Dalí's manifesto on architecture as *confiserie* and Max Bill's promotion of *Gute Form* address the same issue, only they do so from opposite ends. What is at stake, ultimately, is the split between a modernist design culture that defines consumer pleasure in exclusively utilitarian terms (as if the house, and domesticity at large, were an extension of the factory and the office), and the reality of the marketplace where goods are merchandized with the help of clever salesmanship or by means of systematic psychoanalytical research into the subconscious desires that motivate consumer attitudes.[43]

When, after World War II, the Swiss Werkbund launched its campaign in favor of *Gute Form,* the chances for Swiss-made functional design to conquer international markets seemed better than ever—not least under the economic premises of the Marshall Plan.[44] However, for those willing to read the signs of the times, the limits of that project had already become clear almost two decades earlier.[45] By locating consumerism unbound in the nineteenth century, as Giedion had done in *Mechanization Takes Command,* he left the question of modernism's place in the contemporary whirlpool of consumption somewhat open. Max Bill's argumentation was at least more head-on in comparison. Note that it was during his collaboration with Wohnbedarf, the Zürich firm Giedion had cofounded in 1931 [fig. 22], that he had become personally involved in the politics of design, though his experience with industrial design had originated long before at the Kunstgewerbeschule Zürich and was later expanded during his studies at the Bauhaus in Dessau.[46] In a famous manifesto entitled "Schönheit aus Funktion und als Funktion," ("Beauty from Function and as Function"), published in 1949, one year after *Mechanization Takes Command,* Bill pleaded for a relative autonomy of aesthetic criteria in form-giving, thus updating a design approach that had been content to live by the proverbial (and proverbially misunderstood) Sullivanian "form follows function."[47] Of course, the market was quick to interpret this apparent departure from functionalist orthodoxy as an invitation toward a more playful take on the issue. As a result, Bill had no choice but to spend considerable energy in castigating what he saw as a fatal trend toward commercialism in modernist design.

Brazilian Dangers

If Giedion projected today's contradictions into the nineteenth century, Bill exported them to contemporary Brazil. He saw them epitomized in the formalistic "debauchery" of that country's recent architecture. With an important exhibition in the country's leading art institution to his credit (Museu de Arte de São Paulo, 1950), and the "Grand Prix" for sculpture awarded by the Bienal de São Paulo (1951), Bill was an authoritative figure in contemporary visual culture when the Brazilian authorities invited him to return to the country in order to study its principal sites of architectural innovation. "Architecture in your country stands in danger of falling into a perilous state of anti-social academism," he proclaimed in a lecture at the Museu de Arte Moderna do Rio de Janeiro ("The Architect, Architecture, and Society").[48]

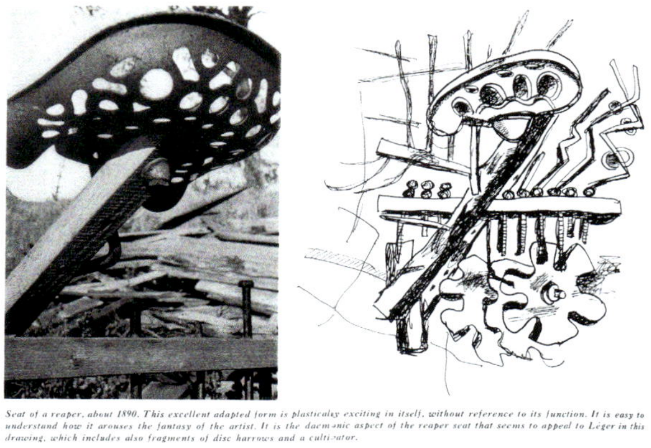

19

20

21

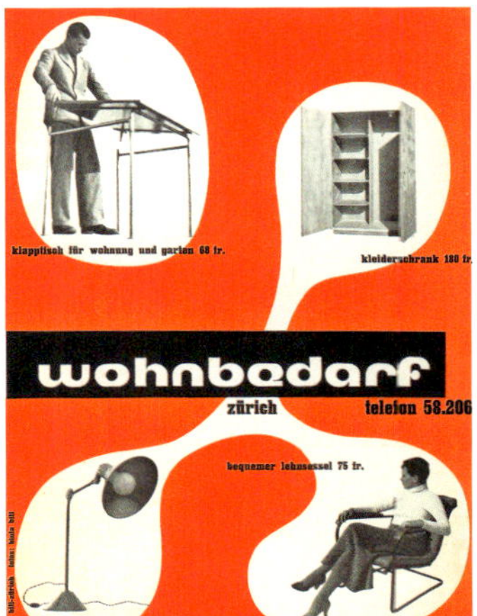

22

19 Seat of a reaper, ca. 1860, on an abandoned farm next to a drawing by Fernand Léger of the same motif. From: Sigfried Giedion, "Léger in America," *Magazine of Art* 38, no. 8 (1945)

20 Scrapheap of rusting and rotting farm implements. From: Sigfried Giedion, "Léger in America," *Magazine of Art* 38, no. 8 (1945)

21 "Surrealist Interpretation of the 19th-Century Interior, Max Ernst, 1929." From Sigfried Giedion, *Mechanization Takes Command,* 1948, after Max Ernst, *La femme 100 têtes,* 1929

22 Max Bill and Binia Bill (photo), Wohnbedarf Zürich, 1931–32. Poster

What he found particularly alarming appears to have been the Brazilian predilection for "the free form, the organic form or free planning." This disease, in his view, was nothing but the result of an illicit transfer from the visual arts to architecture: "Freedom of form ... owes its introduction into the art of today primarily to Kandinsky, in his pictures of around 1910." Applied to architecture, such forms, realized with the help of "pilotis, brise-soleils and walls of glass" may provide "spectacular" effects; in the end, however, they represent the opposite of free planning, Bill argues: "Has it to be as photogenic and as spectacular as all that? I don't believe it."[49]

For some, this verdict may have come as a surprise. Could not the lyricism of Oscar Niemeyer's Casa do Baile in Pampulha of 1950 be seen as resonating with the "calculated" but no less eroticized sensualism of Bill's sculpture *kontinuität* (Continuity) of 1947 (both illustrated, by the way, in Giedion's *A Decade of New Architecture* [figs. 23, 24])?[50] This kind of formal crossover with art is precisely what should be avoided in architecture, Bill argues. True, as he failed to single out particularly strident examples, the critique remained somewhat vague. Even so, Bill can't help sermonizing his Brazilian colleagues (as well as, indirectly, his own friends in Europe, let alone in Switzerland): "You argue, maybe, that architecture, too, is an art, an art moved by the urge to self expression and the urge to infuse buildings with the thoughts of artists. But such is *not* the function of architecture. The architect who so proceeds makes himself ridiculous." Indeed, Brazil may not have been the issue after all. The real worry was Le Corbusier. Note that the Ronchamp Chapel was under construction at that moment; Bill must have seen it as a betrayal of the very approach documented in the master's prewar production. As he had personally edited the recent volume of Le Corbusier's *Œuvre complète* that covered the respective period, the "turn" Le Corbusier had taken must have concerned him all the more directly:[51] "Firstly, an architect must above all be modest and clear. Architecture becomes an art when all its elements—function, construction, form—are in perfect harmony."[52]

It was with a building, not with a book, that Bill finally offered his most powerful answer to Niemeyer's as well as Le Corbusier's recent "formalisms": the Hochschule für Gestaltung Ulm. Despite if not precisely because of its programmatic cult of economy in the way the means are here related to the ends, the school is a powerful architectural statement in its own right: a tribute to design as a response to clearly defined need as opposed to vaguely imagined desire.

The "Sugar Object" and the New Hedonism

In what ways can an object made of sugar molasses be a paragon of design [fig. 1]? Returning to the work of Herzog & de Meuron that had been referred to earlier, the cranberry-flavored Ricola drops may at least be mentioned in passing (Ricola played a major role in Jacques Herzog and Pierre de Meuron's career). Granted the unambiguously gustatory charm of the work, compared to some of the projects imagined for Expo.01 (in other words, before the Swiss National Exhibition was rescheduled and renamed Expo.02)—think of the "pudding, fifteen meters high, designed to draw visitors toward the inside via a modern cockaigne made of slobbering and smacking fragments of vanilla-flavored pudding" that had been announced on the Expo website some time before the event—the Basel "Sugar Object" strikes one as a piece of abstract sculpture.[53]

Seen in a larger context, such works appear to reflect a shift of sensibility, both in design and in art. The ways by which produce of nature is turned into food, or the reverse process by which food turns into waste, but also the taste of food as it reaches the palate and odor as it reaches the nose, in short: the psychophysiological mechanism of taste and olfaction have become an object of intrigue and inquiry in art and design. The 1939 Swiss National Exhibition in Zurich was practically entirely about how objects of modern consumption were made and by whom. The show's 2002 successor, Expo.02, was practically entirely about the breaking down of former barriers between art and marketing, and about the psychophysiological sensations modern industries are able to provide—down to the nauseating thrill of the chemically induced smell of strawberry. The very temptation to interpret such an artifice in terms of an

23

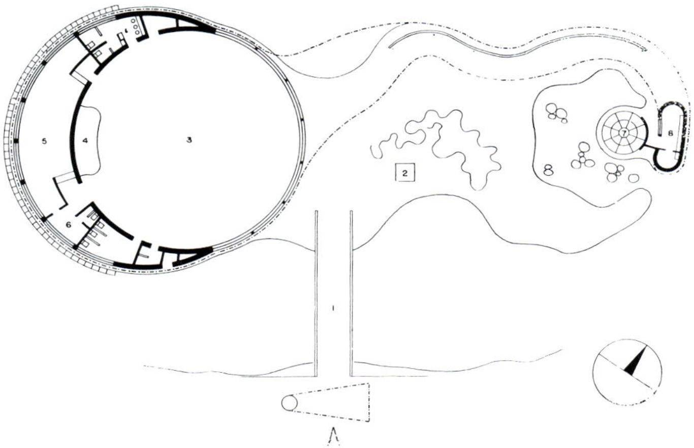

24

25

23 Max Bill, *kontinuität* (Continuity), n.d. Sculpture, shown at Züka, the Zurich Cantonal Agricultural and Industrial Exhibition, 1947
24 Oscar Niemeyer, Casa do Baile (Dancing Pavilion), Pampulha, Belo Horizonte, Brazil, 1942. Plan
25 Diller + Scofidio, *Blur*, animation, ca. 1999. Project for the Swiss national exhibition Expo.02, Arteplage, Yverdon-les-Bains
Courtesy of Diller Scofidio + Renfro

inherent contradiction between the logic of money and the logic of design literally evaporated in a big "Blur"[fig. 25].[54]

In fact, the "Blur"—a machine that produced a cloud of mist hovering above Lake Neuchâtel, and arguably the most memorable work of the show—has thrown us in the midst of a sensualistic renaissance. Art and architecture have rediscovered "experience" as their raison d'être. Such a shift of interest may be anything but unexpected at a moment when the gourmet sections in newspapers and the culinary programs on TV have become a prime platform of lifestyle marketing. Maybe we are even up for artistic recyclings of chocolate and cheese, now that the art world is no longer forced to define its endeavors in opposition to mainstream culture's definition of "Swissness" by ways of products such as these. After all, the first retrospective exhibition ever shown at the Schaulager in Basel, immediately preceding the Herzog & de Meuron show referred to above, had been dedicated to Dieter Roth, whose work includes chocolate objects in various states of putrefaction as well as sausage fabricated from literary texts. It is from here, perhaps, that Peter Fischli and David Weiss have taken off with *Die Wurstserie* of 1979 (literally "Sausage Series"—but can "Wurst" perhaps also be read as "worst"?).

In 1968, Dalí declared "Je suis fou du chocolat Lanvin." It was in a French television ad.[55] In fact, one may wonder why he had not been invited to design a Swatch. He almost lived long enough in order to accept such a commission: he died in 1989. One year earlier Swatch had celebrated the sale of the first fifty million Swatches in a large funfair on Lake Biel. As is well known, Swatch played a major role in helping the Swiss watch industry to get out of a near fatal slump in the seventies. In doing so, it ended up branding Switzerland more efficiently than chocolate and cheese had ever done in the past.

Meanwhile, the watch industry's traditional luxury brands continue to invest in the competition toward producing the most expensive watch ever built. Watches made of the most extravagant materials and functioning like miniaturized high-tech precision arms for James Bond movies continue to boom, as the front- or third-page newspaper ads demonstrate every day: "In this area, the Swiss are untouchable. No one comes close, and one can see Japanese buyers going to Lake Geneva to shop for the kind of watch that 'tells you something about yourself' (Patek); that is a 'precious and rare work of art' (Vacheron); an 'exclusive creation for the discerning few' (Audemars); 'the most expensive watch in the world' (Piaget)."[56]

Swatch did the opposite: it strove toward producing the cheapest watch. If the discerning few noticed, that was okay; in fact, the bourgeois-bohemians welcomed it as a ready-made alternative to the nouveau-riche fetish watch. However, the target group was the masses. Not perfection at the highest possible cost was what promised success, but diversification and obsolescence. Aggressively redefined as "gadget," the Swatch thus became a driving force within the economy of waste. Meanwhile, its second nature as fashion accessory restored it to its former liaison with jewelry, jokingly subverting it so as to respond to the visual expectations of a teenage clientele.

Dalí comes to mind once again as the most obvious precedent in art. His famous painting of 1931, *The Persistence of Memory,* ubiquitous in popular art books like Swatch Stores in shopping malls and airports (there even exists a tapestry version of the work[fig. 26]) describes nothing so much as the upcoming transfiguration of the watch from precision instrument to consumer gadget; in fact: visual cap bomb. By giving it the consistency of an omelet (or of a bedside rug), the artist not only indicates some directions as to the flow of psychic automatisms that may be unleashed by objects of everyday use (and as to the course of art history itself: think of Claes Oldenburg's "soft typewriters" and "soft plugs" realized in the seventies), he also offers a glimpse into a future of watchmaking as entertainment.

The asceticism of *Gute Form* may survive as an option among others within the current "system of objects," but given that we consume "signs" when we spend money on consumer items, as Baudrillard has argued long ago,[57] what counts today in marketing and perhaps also in design is the attention-getting gag, a not too esoteric form of metaphysics, in short, the readiness of the designer to enter into a dialogue with the public's longing for sentiment. That

26

27

26 Salvador Dalí, Untitled collage with *The Persistence of Memory* and other motifs. Tapestry
Courtesy of Gala-Salvador Dalí Foundation, Figueres

27 Advertisement for Omega showing men's wristwatch with luminous cifres, "English style."
From: *L'Illustration,* 1916; Stanislaus von Moos, *Industrieästhetik,* 1992

is not necessarily synonymous with the end of quality in design, but most likely with a fatal shift in the meaning of *Gute Form.* Max Bill's position as formulated in the eighties, toward the close of his career, could hardly be more consistent with its premises: "But now design has become so fashionable that form is no longer the 'unity of functions.' Instead it ranges between promotion and play. We have come full circle back to where it all began."[58]

1 Dalí 1933, p. 72 (translated by the author).
2 André Breton, "La beauté sera convulsive," *Minotaure* 5 (1934), pp. 9–16.
3 Dalí 1933, p. 72 (translated by the author).
4 Leon Battista Alberti, *De re aedificatoria,* 1485, book 9, ch. 8.
5 Those temples offer, Hegel writes, "einen befriedigenden, sozusagen sättigenden Anblick," in Hegel (1822) 1970, p. 318.
6 Scholz-Hänsel 2005, pp. 203–26. On Dalí in New York, see Koolhaas 1978, pp. 234–45.
7 Dalí had been familiar with Le Corbusier's ideas since the early 1920s. On Dalí, Federico García Lorca, and *L'Esprit nouveau,* see Lahuerta 1999, pp. 32–54.
8 Lucie-Smith 1969, p. 120.
9 See von Moos 1992 for the necessary references.
10 Gropius (1935) 1965, p. 92.
11 Frampton 2005, p. xviii.
12 Le Corbusier (1930) 1991, p. 10.
13 The story of the competition is summarized in Oechslin 1988, pp. 54–71 and passim. On Giedion in general, see Georgiadis 1993.
14 Koenig/Gasser 2006.
15 Revel 1982, p. 18.
16 Ibid., p. 19.
17 In *Who Is Who in Switzerland,* Zurich 1952, pp. 182–83. Like Le Corbusier (and many others), Giedion associated reactionary trends in modernism with gluttony, as in Giedion (1928) 1995, p. 190, where he characterizes a villa by Mallet-Stevens as having reinterpreted "Le Corbusier's astringency … for the *gourmand-élégant.*"
18 Giedion 1948, pp. 195–208.
19 Ibid., p. 198.
20 Ibid., p. 200.
21 Ibid., p. 201.
22 "La provocation pornographique de MM. Guimard et Bénard," *La Croix,* March 17, 1902, cited in Gamboni 2007, p. 108.
23 J. F. McCullough, "Design Review—Cars '59'," *Industrial Design* (1959), p. 79; cited in Whiteley 1987, p. 6.
24 On the promotion strategies by Ford and GM in Switzerland and on the GM assembly plant in Biel, built 1935–36, see Bignens 2003, pp. 161–65 and passim, as well as, on the building as such, Hildebrand et al. 2007, pp. 275–77.
25 Dürrenmatt 1956, p. 43.
26 Bignens 2003, pp. 149–51.
27 Giedion 1948, p. 610; see also Giedion 1946, pp. 155–62.
28 Note that Giedion did not object to streamlining as a way of improving airflow or of easing handling but only as a means of inflating the volume of a consumer object.
29 Eames 1943, p. 29.
30 Bill 1952, pp. 46–47. The literal translation of the German title is "From Tool to 'Sweet Tidbit.'"
31 Ibid., p. 47; what is described as "süss" (sweet) in the original version of the text (p. 46) is translated as "flashy" in its English version.
32 Once again, the explicit culinary implication of the original text is "neutralized" by the English translation given in the book, where "geschäftstüchtiger Zuckerguss" (ibid., p. 46) is translated as "worthless catchpenny trash" (p. 47). For more references to Bill's contempt of streamlining, see Bignens 2003, pp. 148, 163–65, and passim.
33 See Horowitz 2004, pp. 20–47 and passim.
34 "Irony has no place in it except in so far as irony is part of the ad man's repertoire." Hamilton (1962) 1982, p. 37. Hamilton's variations on the theme of streamline design had been preceded by Eduardo Paolozzi's collages made from American automobile advertisements; see Robbins 1990, pp. 94–108, 181–83, and passim, but within the Independent Group it was Reyner Banham who fueled the discussions on automobile design; see Craig-Martin/Hamilton (1990) 2010.
35 Brillat-Savarin (1825) 2009, p. 39.
36 Edgar Allen Poe's "A Descent into the Maelstrom" is referred to by Marshall McLuhan in *The Mechanical Bride,* McLuhan (1951) 2001, p. v. Like Giedion's *Mechanization Takes Command, The Mechanical Bride* didn't fail to enter the Independent Group's radar shortly after publication; see Craig-Martin/Hamilton (1990) 2010, p. 6. For Reyner Banham's ideas about automobile design, see Banham 1953 and Banham 1955.
37 McLuhan (1964) 1972, p. ix. Giedion's chapters on "Movement" and "Scientific Management" are key in this context; see Giedion 1948, pp. 14–32 and passim; and von Moos (1982) 2013, pp. 725–58, in particular pp. 745–53.
38 McLuhan (1964) 1972, p. ix.
39 Giedion 1945, pp. 295–99.
40 See Benjamin (1982) 1999, p. 391.
41 Giedion 1948, p. 388. What Giedion has in mind is best illustrated on pp. 361–62.
42 See, for example, Adrian Forty's critique of Giedion's approach in Forty 1986, pp. 92–93, 240.
43 Motivational research has been practiced in Switzerland since the 1930s. Modernist design discourse has been notoriously vocal in its rejection of such practices; see in particular Lohse 1947.
44 Erni 1983 is still key in this context, but among the more recent studies, note in particular Bignens 2001, as well as now Hünerwadel 2013.
45 For a summary of the discussions at the 1932 meeting of the Swiss Werkbund, and in particular the industrialist Ernst Wetter's and the art historian Georg Schmidt's contradictory views on the role of taste and fashion in design, see von Moos 1992, pp. 242–47, as well as Gnägi et al. 2004, pp. 231–32.
46 Hüttinger 1977 still offers the most authoritative overview of Bill's career and work. On the history of "Wohnbedarf," see Mehlau-Wiebking et al. 1989; and for the general context, von Moos 1992, pp. 247–57 (with references).
47 Bill (1948) 2010.
48 "L'architecte, l'architecture et la société," translated into English in Bill 1954b, pp. 38–39. Regarding the controversy on Brazil, see also Rogers 1954, pp. 1–3, and Bill 1954a, p. ii.
49 Bill 1954b, pp. 38–39.
50 Giedion 1951, p. 151.
51 Bill 1939.
52 Bill 1954b, pp. 38–39.
53 Spring 2000. See von Moos 2002, p. 127.
54 For a more careful if doubtlessly also biased discussion of the Expo.02, see von Moos 2004—but the legacy of that event is still in need of critical revision.
55 "I am crazy about Lanvin chocolate." Not a few among his viewers may have recalled his earlier claim: "The only difference between me and a madman is that I am not mad."
56 Landes 1983, p. 355, following a *New York Times* article of December 15, 1980; see von Moos 1992, pp. 44–48.
57 Baudrillard (1968) 1996.
58 Bill 1988, pp. 50, 52 (translation modified by the author).

* The present essay tries to update some research done around 1990 in view of my book *Industrieästhetik* (Disentis 1992), and in the context of seminars taught at the University of Zurich (1991–92, together with Paul Hugger) and the Yale School of Architecture (Spring 2014). For some thoughts on the gastronomic analogy in postmodern design discourse, see my "Rules of Fasting and Desire Derailed: Notes on Architecture and Gastronomy," in *The Architect, the Cook and Good Taste* (von Moos 2007); but for a broad discussion of gastronomy and design, see the essays by Petra Hagen Hodgson, Fritz Neumeyer, Akos Moravansky, and others in that book as well as the literature referred to there.

28

28 Lady's carved wooden secretary by cabinetmaker M. L. Wettly, exhibited in the Swiss section of the Great Exhibition in London, 1851
29 Swiss section of the Great Exhibition in London, 1851, steel engraving. At the center, a cardboard model of Strasbourg Cathedral; on the right, a model of the Säntis massif.

29

The question of the beginnings of Swiss design history is currently more relevant than ever in view of recent economic and political discussions of "Swissness" and "Swiss Made" that have inspired efforts to create a Swiss design mythology and to formulate a "grand narrative" of its history. Toblerone, the Rex vegetable peeler, and the Swiss Army knife are all essential milestones in this history. But where did it all begin, and to what extent is it categorically Swiss?

The retrospective view is especially interesting when looking not only at the chronology but at the constitutive conditions for these beginnings: the then original and still valid characteristics of what was created. Possibly, this interest is bound up with the search for a prototype, an object that crystallizes the central characteristics of Swiss design—if only to serve as a reference point for our postmodern discussions.

It is remarkable that the search for what is "typically Swiss" was already a central, recurring topic once before—in nineteenth-century discussions on the quality and style of everyday products. Beyond the consideration of the formal attributes and level of execution of individual products, the primary concern was how Swiss identity was conveyed and the nation represented through a consciously designed world of goods.

Heinrich Hübsch's question of the century—"In what style should we build?" (1828)—attained unexpected pertinence for the handcraftsmanship that predominated at the time when, toward mid-century, it was increasingly confronted with the new "aesthetic" of mass-produced goods. Industrially manufactured products generally had a bad reputation, regarded as cheap, poorly made copies of artisanal originals, with matters of style playing no part in their design. The art industry's attempts to copy as closely as possible the idiosyncrasies of Swiss handcraftsmanship—which celebrated its manual execution with regional specialties and echoes of folk art[1]—presented an existential threat.

How would artistic craftsmanship respond to the challenges of industrial mass production without losing its carefully cultivated and cherished national identity? Could art, craftsmanship, and industry find a way to work together? These questions were discussed in a different manner back then than they are today, if nothing else because the profession of "designer"—and the corresponding term "design" in the sense of industrial or product design—only took root in Switzerland after World War II.[2]

1851: The Products of Swiss Industriousness under Scrutiny

The Great Exhibition held in London's Hyde Park in 1851 would be the first touchstone for the products of the Swiss Confederation, founded just three years before. As part of the economic globalization then already emerging, the Swiss exhibits faced off directly with their foreign competitors in Joseph Paxton's Crystal Palace and were thus subjected to international scrutiny. Criticism came equally from the leading industrial nation, France, and its competitors England, Germany, and Austria, as well as from Switzerland itself.

But first: What exactly was presented there? A steel engraving of the fabric-draped "Swiss Section" prominently displays the Swiss export products par excellence: lace, printed fabrics, and silk ribbon. The display was representative in this sense. The model of the Säntis massif that can be seen at the right edge of the image not only highlighted the success story of Swiss cartography but also alludes to the future stocktaking of the country "en miniature." Whether the centrally placed cardboard model of Strasbourg Cathedral was meant as a tribute to stupendous Swiss craftsmanship or to humanist educational culture is immaterial in this context.[3] In any case, a lady's secretary, also displayed in London, featuring a naturalistic pastoral idyll and alpine decor was meant to embody the refinement of Swiss wood carving. This is unsurprising, considering that souvenirs and furniture adorned with such fantastical motifs glorifying alpine life, nowadays decried as kitsch,[4] had to meet the real demands of burgeoning "mass" tourism. In fact, it was hoped that such products would establish a new line of business that could help the rather barren Alpine region of the Bernese Highlands (or more precisely the wood-carving schools in Meiringen and Brienz and their offshoots reaching as far as Paris) to flourish.

The critique of Swiss exhibits mentioned above came in many different forms, most of them negative. While the relief of the Säntis massif was lauded as the most significant work of its kind at the Great Exhibition,[5] the carvings—seen as a symbol of simple "devotedness to a beloved nationality thus evidenced by a brave, industrious, and primitive people"[6]—were only able to gain favor in connection with the recent founding of the federation. The exhibition's staging, along with further exhibits—from jewelry and watches to music boxes, machines, and straw goods[7]—were rejected as being entirely unsatisfactory from an artistic point of view. The first inventory of Switzerland's material world demonstrates at least two things: first, the country's attempt to distinguish itself as an ambitious industrial nation by making use of its popular textile exports; and second, the stylistic cluelessness and lack of quality criteria with which Switzerland assembled its rather arbitrary selection of products.

Early Predecessors of Swiss Design

Beginning our analysis with the Great Exhibition certainly does not mean that Switzerland had not produced important applied arts before that time. We need only think here of the excellent quality of early nineteenth-century faience and earthenware from Kilchberg near Zurich, Matzendorf (est. 1798), Nyon, Carouge (est. 1803), and Schaffhausen (est. 1830). More extensive series were often produced in manufactories during the first half of the century—proto-industrially, one might say—as well as in art furniture workshops with up to fifty workers. Most products, however, were created in small series or as single items in small handicraft businesses with one master and a few artisans, for example Meubles d'Yverdon (1830–1900). As early as the 1830s, the cabinetmaker Jean-Pierre Guichard was already creating in Yverdon technically innovative, elegant armchairs from molded plywood, independently of the bentwood technique pioneered slightly later by the Thonet brothers. Edouard Wanner later developed this seating furniture into prefabricated models with variable decoration that could be individually modified to suit the client's wishes, anticipating the principle of "mass customization" that became popular in the 1970s.[8]

Even earlier than the faience manufactories, the watch industry had been establishing itself in the Jura region since the seventeenth century. Watchmaking was important economically, and in the early nineteenth century Swiss watches sold well abroad—namely in Paris. In terms of export, it was second only to the textile industry, which held a special status in Switzerland's primarily agricultural economy. Silk ribbons were produced in Basel, printed fabrics in Glarus, embroidery in Appenzell, and woven goods in St. Gallen and Zurich.

Throughout the first half of the nineteenth century, applied arts—as yet hardly separated from fine arts—were presented at "art and industry exhibitions" (for example in Geneva in 1789 or in Bern in 1804). These exhibitions were vital, as witnessed by the fact that in the 1820s art was considered an important "form of expression for a patriotic spirit and constitutional values."[9] Coupled with fine arts, applied arts thus became a main focus of the creation of national identity far prior to 1848. Even in these early days, the applied arts were the subject of criticism due to discrepancies between form and content, and because the items were often made from cheap substitute materials. In order to remedy these shortcomings, an award was instituted in 1839 to recognize "objects that stood out for their general usefulness, reasonable price, and good workmanship."[10] Apparently, the criteria applied to selecting examples and models of specifically Swiss products for the Great Exhibition were not yet substantial enough.

Nestlé, Maggi, and the Machine Industry

Those seeking out the earliest design icons that still retain their status to this day will no doubt come across the logos and logotypes of Maggi and Nestlé. Designed in the last third of the nineteenth century, Nestlé's essentially unaltered company logo—a scene picturing the feeding of two young birds—and the bold, yellow and red Maggi lettering on the enduring reddish-brown Maggi bottle and bouillon cube packaging are still being used today. The Maggi

30

31

32

33

34

30 Factory design, Handkerchief for export to Nigeria, 1924–28, cotton fabric with linen weave, Fröhlich Brunnschweiler & Cie., Ennenda
Museum für Gestaltung
31 Factory design, Pestemal/fouta (spa towel) for export to Turkey, 1910–14, cotton fabric with linen weave, Fröhlich Brunnschweiler & Cie., Ennenda
Museum für Gestaltung
32 Michael Thonet, Chairs made of bent and glued wood veneers, 1836–40, Thonet, Boppard (DE)
33 Jean-Pierre-Moïse Guichard, Chaise en gondole
34 and Canapé, ca. 1830, Yverdon. Two early examples of Yverdon furniture made of three-dimensionally molded and glued poplar and walnut veneers.

35

36

35 Factory design, Washstand pitcher with lid, ca. 1770,
36 Porzellanmanufaktur Kilchberg-Schooren
Private collection

advertising campaigns are noteworthy as precursors to the all-encompassing advertising strategies that became known in the 1970s as "corporate design." In all Maggi markets—from Paris to Vienna, from Milan to London—these revolutionary products were labeled with the same uniform logotype and additionally promoted through superbly designed advertisements. It is perhaps thanks to this omnipresence that one finds immortalized Maggi cubes in Pablo Picasso's Cubist *Paysage aux affiches* of 1912, and the Maggi lettering in Félix Vallotton's *Point du Jour* of 1901.

Around 1889, Switzerland's machine industry surpassed the textile industry. An increasing number of machine exhibits at the international expositions in 1873 and 1878 had already defined a more distinct image of Switzerland as an advanced export nation: for example, the steam-powered trams and mountain railways produced by the Winterthur engine works, as well as an omnibus, a generator, and a steam engine manufactured by the Sulzer brothers. Concurrently with industrialization, there was a move toward greater professionalism in the applied arts, including the founding of a number of specialized schools. In reaction to "pronouncements on good technology, bad style, and poor composition,"[11] institutes devoted to aesthetic education opened in La Chaux-de-Fonds (1873), Winterthur (1874–1915), Geneva (1876), Lucerne (1877), and Zurich (1878), as well as in Brienz (wood-carving school, 1884). However, throughout the latter half of the nineteenth century, compulsory historicism, and with it the artistic freedom to reinterpret and freely combine historical styles, held sway at these schools while also dominating applied arts practice. With a "Brienz Renaissance-style chamber," this historically eclectic form of applied art had been embedded even within the walls of the Bundeshaus (Federal Palace) and asserted itself against a chorus of critical voices that considered its backward-looking and heavily ornamental style utterly unsuitable and out of step with modern industrial design.

Until circa 1900 quite a few schools of applied art were able to establish themselves, with different regional focuses and often only succeeding after many comprehensive reforms. Of note in this context is the stylistic renewal brought about for the applied arts school in La Chaux-de-Fonds through Charles L'Eplattenier's *Style sapin* ("pine tree style"), which proved influential far beyond the local watch industry. At the Kunstgewerbeschule Zürich it was Jules de Praetere who from 1906 to 1912 sought to collaborate with partners in industry by means of a novel method of teaching. Both developments reach beyond the turn of the century and therefore beyond the scope of our analysis.

The Striking and Less Striking Aspects of Nineteenth-Century "Design"

Examining the well-documented Swiss submissions to the international expositions, it is apparent that the country wished to present itself primarily as an export nation with a strong industrial sector—from 1851 to 1900 with products from the wood-carving, watchmaking, and especially textile industries, and then from 1889 onward with machine works. The selected products served the purpose of national representation and shaped Switzerland's identity on the international stage. The products themselves were often harshly criticized based on stylistic, fashion, and taste criteria, as well as on principles of utility, manufacture, and price, or even on ethical and moral grounds. Within this open debate, the broad differences between regional cultures, local resources, and the forms of industry dependent on them rendered the exploration of what makes products typically Swiss quite complex, or even doomed it to failure.

What is less obvious is that, contrary to the ideas of the "pioneers of modern design"—such as John Ruskin and William Morris (with the Arts and Crafts movement in England) and Gottfried Semper in this country (1855–70 at the Federal Polytechnic in Zurich)—these German-Swiss attempts were decidedly aimed toward achieving "bourgeois mediocrity." One may belittle Georg Lasius's call for "artistically sound but simply made furniture ... for normal, middle-class households"[12] as an early form of Swiss, democratically trained impartiality, or even deride it as an unsophisticated lack of refinement. However, the dearth of normal, middle-class furniture lamented by Lasius in 1876 may also be read as pragmatism, which took into account not only the patchwork structure of Swiss consumer-goods markets but also the personal tastes of a small but prosperous clientele. In any case, the bourgeois mainstream became a very long-lived criterion of Swiss design, as well

37

38

39

40

37 Nestlé's Kindermehl ("children's flour"), Can with image and script logo, ca. 1890
38 Sepo (Severo Pozzati), Bouillon cube, before 1940. Poster
39 Pablo Picasso, *Paysage aux affiches,* 1912
40 Sulzer steam engine with valve control, Exposition Universelle, Paris, 1878
41 Bundeshaus in Bern, Antechamber of the assembly room of the Council of States with Brienz wood carvings in the Renaissance style.
42 Sample portfolio of the Eduard Binder company in Brienz, showing carved wood items, ca. 1910
43 Robert Belli and Henri Robert, Crematorium in La Chaux-de-Fonds, 1908–09. Ornament in the Style sapin, designed and executed by Léon Perrin, 1909

41

42

43

as an implicit guiding principle for award juries. After all, this approach focuses on the perfect interplay of functionality *and* manufacture *and* material.

That this very pointed pragmatism did not present a sufficient benchmark by which to judge the characterization of overall Swiss design achievements can be inferred from Gottlieb Ziegler's wish "that the German-Swiss achievements might be anointed with a drop of graceful western Swiss oil."[13] Was this desire for elegance inspired by the Meubles d'Yverdon? A brief look at the history of Guichard et Wanner (1830–1900) reveals a small business displaying both technical and design innovation that deftly maneuvered through the nineteenth century's changing tides of style. If this successful business can be said to be paradigmatic for Swiss design culture, the history of Swiss design could certainly be said to have begun before the founding of the federation in 1848.

1 The boundaries between folk art and applied arts are fluid and differ from region to region. See Oehler 2013.
2 The term "design" does not have a universal definition. This is due to thematic tensions between craftsmanship and industrial production in the nineteenth century that led to the different notions of design that still exist to this day. See Erlhoff/Tim Marshall 2008.
3 See Goethe's essay "On German Architecture," 1773.
4 The use of the term "kitsch" can be traced back to around 1870, when it was used to mean "daub" or "to hawk cheap wares." It was given a more positive spin in a programmatic postmodern essay by Alessandro Mendini; see Mendini 1984.
5 Carl Sartorius, *Die Industrieausstellung in London,* Darmstadt 1851, p. 69; cited in Müller Horn 2012, p. 40 n167.
6 Clark 1851, p. 309.
7 See von Moos 1992, pp. 5–15.
8 See Mager 2008.
9 Comment by Philipp Albert Stapfer; see Müller Horn 2012, p. 60.
10 Othmar Birkner, *Bauen und Wohnen in der Schweiz 1850–1920,* Zurich 1975, p. 169; cited in Schilder Bär/Wild (2001) 2003, p. 42 n40.
11 Wilhelm Meile, *Die Schweiz auf Weltausstellungen,* ed. Hans Schröer, Schweizerische Wirtschaftsstudien no. 2, Zurich 1914, p. 121; cited in von Moos 1992, p. 220.
12 Birkner 1969, p. 15.
13 Gottlieb Ziegler, *Brief von der Schweizerischen Landesausstellung in Zürich,* Winterthur 1883; cited in Birkner 1969, p. 20.

44

Swiss Werkbund Exhibition, Zurich, 1918. Main entrance with view of the lake and the Üetliberg hill. As World War I drew to a close, the still-young SWB managed to put the "Werkbund idea" in the limelight in a prominent location for an entire summer.

From: *Schweizer Illustrierte Zeitung* 7, no. 22 (1918)

1910 – Reform

With the advent of the mass production of consumer goods in the nineteenth century, many begin to take a critical view of the widespread practice within the applied arts of copying an eclectic range of historical styles. Does the true path lead back to the handcrafted original or rather toward the "period-style machine-made product"? After 1913, the Swiss Werkbund (SWB) becomes the authoritative interdisciplinary discussion forum for design issues in Switzerland. At applied arts schools in La Chaux-de-Fonds and Zurich, high-profile painters and sculptors eschew slavish copies of historical styles in favor of their own artistic designs. The Europe-wide penchant for neoclassicism along with the exemplary work of the Wiener Werkstätte now attune public taste to timeless forms and the quest for functional objects of daily use. In 1918, the SWB for the first time exercises its power by taking stock of the situation. In addition to precious originals, it exhibits artfully designed prototypes for series production; alongside elegant interiors for the bourgeois middle class, it puts a second focus on workers' housing.

The following key objects also include some significant precursors of Swiss design dating back to before 1910.

45

Having coffee on the balcony of a middle-class home on Holbeinstrasse, Basel, ca. 1910. While photos of the house's interior show historicist furniture and a few Thonet chairs, outdoors the spirit of Viennese modernism can be felt.

Complete photographic documentation of the house, Archive of Arthur Rüegg

46

Charles L'Eplattenier and the students in his Cours supérieur d'art et décoration (including Charles-Edouard Jeanneret, later known as Le Corbusier), music room in the Villa Albert Mathey-Doret, La Chaux-de-Fonds, 1906. The main attraction in this Gesamtkunstwerk rife with fir tree motifs was the square picture window framing the real fir trees in the Jura mountain forest.

47

Sophie Taeuber in Munich, ca. 1912. The young Swiss artist and designer already took part in masked balls while still a student. In 1916–17 she took courses in expressive dance at Rudolf von Laban's Schule für Bewegungskunst in Zurich and at his summer residence on Monte Verità in Ascona, where Mary Wigman, Suzanne Perrottet, and Katja Wulff also danced.

48

A classroom at the Schule für Holzbildhauerei in Brienz, 1905–08. Founded in 1894, the wood-carving school became an internationally known training center that played a key role in the success of the Brienz artists at the international expositions. This postcard promotes the school on the reverse (in English) by listing the medals won in Chicago (1893), Geneva (1896), Paris (1900), and Liège (1905), and citing the school subjects carving, modeling, drawing, and ornamental style. While the animal figurines made at the school were especially popular, richly decorated picture frames, spoons, and bowls were also highly successful Swiss export products.

Postcard: Schule für Holzbildhauerei, Brienz

49

Studio at the Kunstgewerbeschule Zürich (housed at the time in a side wing of the Swiss National Museum), ca. 1920. Under the directorship of Alfred Altherr, the school's renown rose steadily after 1912 as the result of practice-oriented lessons, a flexible curriculum, and the hiring of young freelance artists as teachers.

50

Sophie and Erika Taeuber, Zurich, ca. 1921, in costumes by Sophie Taeuber. Housed today in the Aargauer Kunsthaus in Aarau, the costumes with their abstract motifs are made of felt and other fabrics. The designs are based on the kachina dolls made by the Hopi of North America.

Manufacture de porcelaine
Factory design

Mille-fleurs Service
1781–1813

51

52

53

51 Anonymous, Mille-fleurs Service, cup and saucer,
52 1781–1813
Porcelain, 4.5 × 8 cm (cup), 3 × 13.5 cm (saucer),
Manufacture de porcelaine, Nyon
Collection of Arthur Rüegg

53 The manufactory moved into this building in Nyon in 1787; in 1788 it was expanded to double its size.

Almost a century before the Langenthal porcelain factory was founded in 1906, a chapter of industrial history came to an end in Nyon. Swiss porcelain production had begun in 1763 at the Zurich manufactory at Kilchberg-Schooren, which until 1790 produced figurines, table ornaments, tableware, and tea services in porcelain and faience for the wealthy bourgeoisie. Around twenty years later, the Frenchman Jacques Dortu and his German father-in-law, Ferdinand Müller, founded a second factory with about thirty employees in Nyon. Dortu, born in Berlin in 1749 as the son of a Huguenot refugee, had learned his trade first at the Königliche Porzellan-Manufaktur (KPM, Royal Porcelain Manufactory) and then acquired further experience in St. Petersburg, Marseille, Stralsund (Germany), and Marieberg (Sweden). At that time, a command of foreign—and historical—models was a prerequisite for any porcelain manufacturer. As the Rococo-style tableware produced by Schooren went out of fashion despite its more sober Zurich touch, Dortu relied instead on contemporary, elegant neoclassicism. The lack of aristocratic or royal patrons dictated not only his austere budget but also the artistic program. The sleek design of the Nyon porcelain thus seems like a bellwether announcing the ascendancy of the middle class, even while its elegant proportions, consummate forms, and expressive embellishments attest to the exquisite taste of the Ancien Régime. The Nyon cup strewn with hand-painted flowers still looks as fresh as if it had been launched on the market just yesterday. AR

Molin 1904; Helvetica 2011.

54

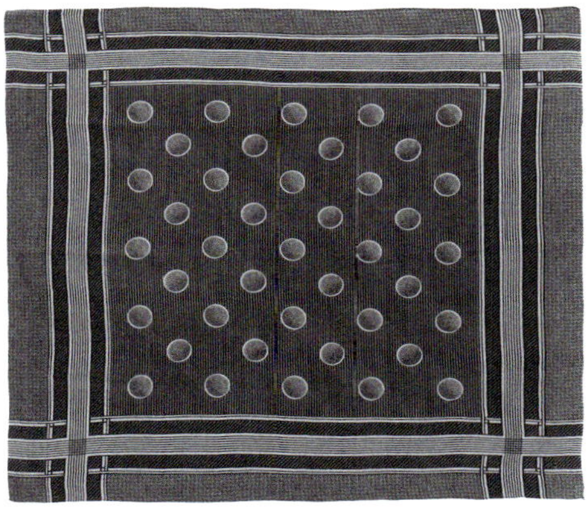

55

56

54 Anonymous, Scarf, 1900–25
Printed cotton with discharge-printed flower pattern, 63 × 64 cm, F. Blumer & Cie., Schwanden
Swiss National Museum (LM-58864)

55 Cotton scarves with geometric and floral patterns,
56 1900–50
Collection of Margrit Rüegg

"Glarus scarves" can be found today in every souvenir shop. But these square, typically red cotton Swiss scarves are only a faint echo of all that was once produced in Glarus or Schwanden in the nineteenth century, when the textile industry there rose to a leading position in the global market. Cotton printing, the first modern reproduction technique, enabled the industrial production of replicas of handcrafted originals from all over the world. Far Eastern fabrics in particular were often imitated using sophisticated means and successfully resold in the same countries. If on today's "Glarus scarves" the simplified India patterns seem to us emblematic of the Swiss identity, this is the result of this early cultural crossover. The old head or neck scarf (a "Fazenettli") pictured was made in the Blumer textile printing factory in Schwanden. The floral ornament is etched into a background in Turkey red. Other patterns were executed with rolls, plates, or stencils, preprinted with black aniline dye, overprinted with zinc white, overdyed with yellow, marbled or illuminated, or embellished with a subsequent manual imprint. Another success story was the plaid and striped fabrics manufactured by Fröhlich Brunnschweiler & Cie. in Ennenda, destined for export to Africa, Turkey, and Java. The geometric patterns of these "handkerchiefs" anticipate trends in modern art, but they were never sold in Switzerland itself. AR

von Moos 1992, pp. 59–62; Barten 1997.

École d'Art
Student works

Four Watchcases in the Style Sapin
ca. 1906

57 Anonymous, Four Watchcases in the Style Sapin, ca. 1906
Metal, in part enameled or gilded, various dimensions, student works from the École d'Art, some of them exhibited in Milan in 1906
École d'arts appliqués, La Chaux-de-Fonds

58 Milan 1906 Diplome d'honneuer avec médaille d'or. Envelope of Charles-Edouard Jeanneret's diploma, awarded to the École d'Art.
Collection of Arthur Rüegg

59 Charles-Edouard Jeanneret, Pocket watch, 1906
Silver, copper, steel, gold, diamonds, student work from the École d'Art, exhibited in Milan in 1906
Fondation Le Corbusier, Paris

57

58

59

In the early twentieth century, La Chaux-de-Fonds was at a high point in its development. The booming city—with 40,000 inhabitants, situated in a high Jurassic valley at an altitude of 1,000 meters—was one of Switzerland's foremost industrial locations. Shortly before World War I, more than half of the world's clocks were made, or at least assembled, in La Chaux-de-Fonds. The enameled dials and engraved cases played a decisive role in the successful marketing of these products. This is no wonder, because an art school for the specialized training of the required skilled personnel was already founded here in 1870. However, the emergence of industrially manufactured wrist watches threatened the existence of the many at-home workers who were involved in the production of the richly decorated pocket watches. The charismatic painter and sculptor Charles L'Eplattenier (1874–1946), director of the École d'Art, tried to counteract this trend by bringing the most progressive art movement of the day to the Swiss Jura Mountains—Art Nouveau. Together with his students—among them Charles-Edouard Jeanneret, who would later achieve world renown as Le Corbusier—he sought to establish a genuinely Jurassic brand of Art Nouveau, the Style sapin. Geometrically stylized evocations of the limestone formations and the flora and fauna of the Jura mountains decorated the watchcases created under his aegis, soon attracting international recognition—in 1906, for example, in the form of an honorary diploma with a gold medal at the Esposizione Internazionale di Milano. AR

Brooks 1997; *Der junge Le Corbusier* 2002; Thomson 2006.

60

60 61 Edouard Wanner (attributed), Napoleon III Armchair, ca. 1870
Glued poplar and walnut veneers, upholstered seat with fabric cover (new), 80 × 61.5 × 59 cm, Guichard et Wanner, Fabrique de meubles et scierie de placages, Yverdon
Museum für Gestaltung

62 Rue de la Plaine, Yverdon, on the left the J. P. Guichard furniture store, ca. 1880

61

62

This armchair is typical of the successful late phase of the Yverdon-based manufactory of Jean Pierre Moïse Guichard (1792–1860) and Edouard Wanner (1833–1904). Its appeal lies in the expansive proportions, the elegant sweep of the flat armrests, and the discreet ornamentation. Ever since the 1855 Exposition Universelle in Paris, the objects produced in series by the cabinetmaker Guichard—who had founded the company in 1829—and his nephew Wanner had been featured frequently at national and international exhibitions. The manufactory's goal was to offer the bourgeoisie an affordable yet distinguished range of furniture by employing ingenious modes of production. The individual components are made of thin veneers executed using special saws—with soft poplar wood inside and typically walnut on the exterior—which with presses were molded into a sculptural form and then glued. The pieces were connected using dowels. Clients could select their own decorative elements: rather than carvings laboriously worked out of solid wood, these consisted of stock ornaments produced with presses and then simply glued on. The division of labor was just as forward-looking as the technique used, considering that the plywood industry wouldn't emerge until around 1900. Guichard and Wanner produced not only armchairs but also sofas, consoles, buffets, and even praying chairs. A few magnificent specimens produced for exhibitions can be regarded as built "sample books." AR

Monnier/Vuille 1979; Köhler 2002b.

63

64

65

63 Anonymous, Bouillon Cube (presentation model), 1920s
Wooden cube with printed paper and cardboard, 1.4 × 1.4 × 1.4 cm, Maggi AG, Kemptthal
Museum für Gestaltung/Donation of Kurt Fässli

64 Anonymous, Display stand for Maggi products, 1920s, Maggi AG, Kemptthal
Museum für Gestaltung

65 Maggi pavilion at the Swiss National Exhibition in Bern, 1914. Postcard

Julius Michael Johannes Maggi (1846–1912) was the son of an Italian immigrant who had risen to prosperity in Switzerland as a mill owner and businessman. In 1869, he took over his father's hammer mill in Kemptthal as well as acquiring a vegetable-growing business and founded the company J. Maggi & Cie. in 1872. Under the aegis of the Schweizerische Gemeinnützige Gesellschaft (SGG, Swiss Society for Public Utility) and the physician Fridolin Schuler, a believer in scientifically based nutrition, the Maggi company developed affordable, protein-rich legume meal to improve workers' diets. This led to the production of ready-made soups, followed in 1886 by the Maggi seasoning sauce, plant-based but with a meaty taste. By 1890, Maggi had factories and branches in Switzerland, Europe, and the United States. Thanks to early and extensive advertising (including ads by Frank Wedekind), the brand became well known worldwide. The progressive company also introduced a number of social measures. Julius Maggi did a great deal of business in France and in 1897 moved his office to Paris, where he set up controlled distribution of pasteurized milk to combat infantile cholera. In 1907, Julius Maggi founded the Société du Bouillon Kub to introduce the new product. Emblazoned with a "cross-star" logo designed by Maggi himself and bolstered by plenty of advertising and a tasting office, six million units of this early form of convenience food were being sold in France by 1912. RM

Vinçon 1992.

66

67

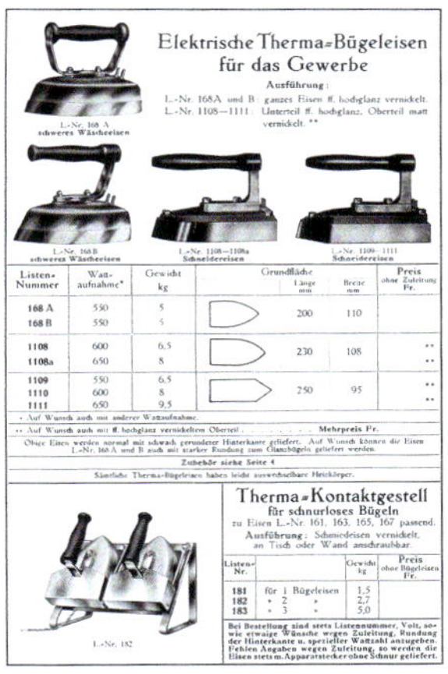

Elektrische Therma-Bügeleisen
für das Gewerbe

Ausführung:
L.-Nr. 168A und B: ganzes Eisen ff. hochglanz vernickelt.
L.-Nr. 1108–1111: Unterteil ff. hochglanz, Oberteil matt vernickelt. **

L.-Nr. 168 A schweres Wäscheeisen
L.-Nr. 168 B schweres Wäscheeisen
L.-Nr. 1108–1108a Schneidereisen
L.-Nr. 1109–1111 Schneidereisen

1

Listen-Nummer	Wattaufnahme*	Gewicht kg	Grundfläche Länge mm	Grundfläche Breite mm	Preis ohne Zuleitung Fr.
168 A	550	5	200	110	
168 B	550	5			
1108	600	6,5	230	108	**
1108a	650	8			**
1109	550	6,5	250	95	**
1110	600	8			**
1111	650	9,5			

* Auf Wunsch auch mit anderer Wattaufnahme.
** Auf Wunsch auch mit ff. hochglanz vernickeltem Oberteil . . . . Mehrpreis Fr.
Obige Eisen werden normal mit schwach gerundeter Hinterkante geliefert. Auf Wunsch können die Eisen L.-Nr. 168 A und B auch mit starker Rundung zum Glanzbügeln geliefert werden.
Zubehör siehe Seite 4
Sämtliche Therma-Bügeleisen haben leicht auswechselbare Heizkörper.

Therma-Kontaktgestell
für schnurloses Bügeln
zu Eisen L.-Nr. 161, 163, 165, 167 passend.
Ausführung: Schmiedeisen vernickelt, an Tisch oder Wand anschraubbar.

L.-Nr. 182

Listen-Nr.		Gewicht kg	Preis ohne Bügeleisen Fr.
181	für 1 Bügeleisen	1,5	
182	" 2 "	2,7	
183	" 3 "	5,0	

Bei Bestellung sind stets Listennummer, Volt, sowie etwaige Wünsche wegen Zuleitung, Rundung der Hinterkante u. spezieller Wattzahl anzugeben. Fehlen Angaben wegen Zuleitung, so werden die Eisen stets m. Apparatstecker ohne Schnur geliefert.

68

66 Samuel Blumer (attributed), Therma Electric Iron,
67 1904
High-gloss nickel-plated iron body, replaceable heating element, enameled cover, wooden handle, 14 × 19 × 10 cm, Therma AG, Schwanden. Models shown date from ca. 1927.
Museum für Gestaltung/
Donation of Arthur Rüegg

68 Therma Iron, models for household and commercial use, catalogue, 1927, Therma AG, Schwanden

In 1904, the young inventor Samuel Blumer (1881–1959), a self-taught electrical engineer, opened a modest factory "for electric heating and cooking appliances" in Schwanden. The first chafing dishes and irons sold well, such that the small enterprise was able to move into a new building in 1907 as a joint stock company. The name "Therma" was coined by the local pastor. Blumer came up with two groundbreaking inventions: the tape-wound heating element of the electric iron and the grooved cooking plate filled with hardened chamotte powder. Both of these functional innovations brought a major boost to Therma—and to the entire household appliances industry. It is to Blumer's credit that the electric iron found its way to the wider public as the first commonly used electrical appliance. In 1914, there were only 1,600 electric stoves in use in Switzerland—but 85,000 electric irons! Therma dominated the Swiss market and also exported goods to France, Italy, Denmark, and Norway. Soon, specialized appliances were being developed, such as the egg-shaped iron for laundries, the shoe iron, and the hatmaker's iron. In 1911, a chrome and nickel alloy came into use for the heating resistors, but it was not until 1931 that Therma came out with its first iron with adjustable controls. For a long time, the development in irons was visible in the design of the handle only—from elaborate wooden clamp to lathed rod and then onward to today's ergonomic plastic handle—while the elegant form of the highly polished nickel-plated and, in some cases, color-enameled iron body remained a constant. AR

Therma 1925–27; *Therma* 1957.

Charles-Edouard Jeanneret (Le Corbusier)

Chair for Anatole Schwob 1914–15

69 Charles-Edouard Jeanneret, Chair for Anatole
70 Schwob, 1914–15
Solid walnut and plywood, fabric seat cover (not original), 89.5 × 47 × 47 cm, Ebénisterie Jean Egger, La Chaux-de-Fonds (?)
Collection of Arthur Rüegg

71 Immeuble Léopold-Robert 73, La Chaux-de-Fonds, 1906. Home of Camille and Anatole Schwob from 1913 to 1917.

69

70

71

As the young Charles Edouard Jeanneret (1887–1965) began to explore his talents on the way toward his bright future, furnishings and interior decoration were evidently a major area of interest. The son of a watch dial enameller, he trained first as a watchcase engraver and then tried his hand as an architect while attending L'Eplattenier's Cours supérieur at the École d'Art in 1906–07. Five years later, after an extended stay in Berlin, the decorative artist working in the Art Nouveau style had turned into a neoclassicist concentrating on standard production. Unlike his contemporaries, however, he was not interested in reinventing objects of daily use but instead based his design procedure on the careful selection of available models. Between 1912 and 1916, he focused intensively on cherry picking and then refining and playing variations on a palette of neoclassical furniture types of predominantly French provenance whose elegant forms graphically embody a specific function. The business elite of La Chaux-de-Fonds soon came to admire the talents of the young architect. In 1913, the clock manufacturer Anatole Schwob hired him to furnish his luxurious home near the railway station. For the smoking room, Le Corbusier developed a refined version of the *siège en gondole,* whose back features, in place of the conventional carved diamond pattern, a bent plywood sheet with triangular and rhomboid-shaped cutouts. AR

Brooks 1997; Rüegg 2012a.

72

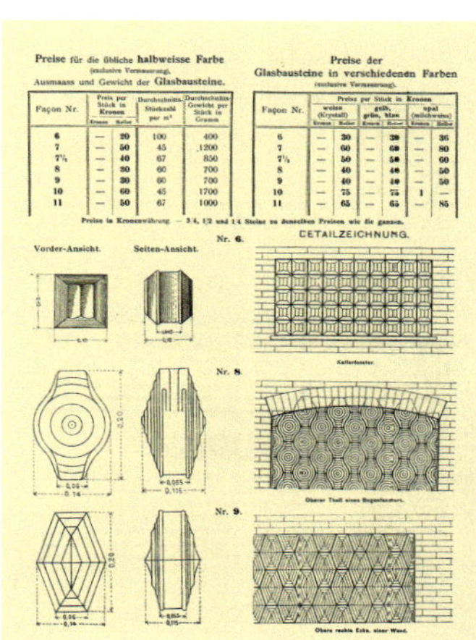

Preise für die übliche halbweisse Farbe
(exclusive Vermauerung).
Ausmaass und Gewicht der Glasbausteine.

Façon Nr.	Preis per Stück in Kronen (Kronen)	Preis per Stück in Kronen (Heller)	Durchschnitts-Stückzahl per m²	Durchschnitts-Gewicht per Stück in Gramm
6	—	20	100	400
7	—	50	45	1200
7½	—	40	67	850
8	—	30	60	700
9	—	30	60	700
10	—	60	45	1700
11	—	50	67	1000

Preise der
Glasbausteine in verschiedenen Farben
(exclusive Vermauerung).

Façon Nr.	weiss (Krystall) Kronen	weiss (Krystall) Heller	gelb, grün, blau Kronen	gelb, grün, blau Heller	opal (milchweiss) Kronen	opal (milchweiss) Heller
6	—	30	—	[illegible]	—	36
7	—	60	—	[illegible]	—	80
7½	—	50	—	[illegible]	—	60
8	—	40	—	[illegible]	—	50
9	—	40	—	[illegible]	—	50
10	—	75	—	75	1	—
11	—	65	—	65	—	85

Preise in Kronenwährung. — 3/4, 1/2 und 1/4 Steine zu denselben Preisen wie die ganzen.

Nr. 6.

Vorder-Ansicht. Seiten-Ansicht.

DETAILZEICHNUNG.

Kellerfenster.

Nr. 8.

Oberer Theil eines Bogenfensters.

Nr. 9.

Obere rechte Ecke. einer Wand.

73

74

72 Gustave Falconnier, Glass Brick Models 6, 8, and 9, patented in 1886
Green-tinted glass, Model 6: 10 × 10 × 9 cm, Models 8 and 9: 20 × 14 × 11.5 cm, various manufacturers including Gaston Blanpain Massonet, Brussels, and Hayward Brothers & Eckstein Ltd., London
Collection of Arthur Rüegg

73 Catalogue page, n.d.

74 Gustav Gull, Stadthaus Zürich, 1898–1900. Atrium

In 1914, the poet Paul Scheerbart dreamt of the "glass architecture" to come, which would let in "the light of the sun, the moon, and the stars, not merely through a few windows, but through every possible wall, which will be made entirely of glass." Nearly thirty years earlier, the Nyon-based architect, inventor, and politician Gustave Falconnier (1845–1913) had applied for a revolutionary patent for glass bricks from which just such translucent walls could be made. Like bottles, their hollow glass volumes were blown in iron molds and then closed with a glass plug. Four different models, labeled with the numbers 6 to 9, were sold—usually made of the greenish glass fashionable at the time, but also, for better light quality, of white glass or, for decorative purposes, in red, blue, yellow, or green. The inventor applied for patents throughout Europe and even in the United States, where he was awarded an honorary diploma at the 1893 Chicago World's Fair. The Art Nouveau architect Hector Guimard created a monument to Falconnier's products in 1898 with his Castel Béranger in Paris and the Perret brothers used model 9 in 1904 in the stairway of their just as famous, but neoclassical-style, house at 25 bis rue Franklin. The perfectly preserved glass vault over the atrium of Zurich's historicist city hall, completed in 1900 by Gustav Gull, is still an insider's tip. AR

Scheerbart 1914, p. 41; Vellay 1985; Bourban 2009.

75

76

77

75 Sophie Taeuber, Guard Marionette, 1918
Turning work by Carl Fischer
Turned wood, painted, double metal eyelets, 40.5 × 18 cm (w/o arms), Schweizerisches Marionettentheater
Photo: Heinrich Helfenstein
Museum für Gestaltung

76 Scene from the marionette play *König Hirsch* (*The Stag King*), Swiss Werkbund Exhibition, Zurich, 1918
Photo: Ernst Linck

77 Sophie Taeuber, Reconstruction of the stage set for *König Hirsch* (1918), central cabinet, 1992
Museum für Gestaltung

If there was one sight at the Zurich Werkbund exhibition of 1918 that could be said to have burned itself into collective memory, then it was the marionette play produced by Sophie Taeuber-Arp (1889–1943, married Hans Arp in 1922). Alfred Altherr, director of the Kunstgewerbeschule Zürich and the first chairman of the SWB, had established a marionette theater in the lead-up to the show, making it the centerpiece of the exhibition grounds. He enlisted Sophie Taeuber, who had taught textile design at the Kunstgewerbeschule since 1916, to produce the play *The Stag King* by Carlo Gozzi (1762). She chose an updated version of the tragicomedy—a mixture of Oriental fairy tale and commedia dell'arte—by dramaturgists René Morax and Werner Wolff that parodied psychoanalysis, at the time a controversial practice (the beetle-like figures hunt in the Burghölzli Forest—"Burghölzli" is a nickname for the psychiatric hospital of the University of Zurich—and the magician is called Freudanalyticus). On the border between high culture and folk art, an opportunity arose for an avant-garde experiment such as this. Sophie Taeuber made her seventeen marionettes out of rotating elements painted in different colors, crafted for her on the lathe by Carl Fischer, a teaching colleague. She counteracts the figures' high degree of abstraction with feathers, pieces of cloth, and decorative elements, though without veiling their naked forms or obstructing their grotesquely unrealistic movements. Her five-legged spider- or robot-like guard figure is a masterful Dadaist parody of the military machinery of the German/Austrian Imperial era while also anticipating a future design aesthetic. AR

Minges 1996. Afuhs/Reble 2007.

Wooden Chair for the Kunsthaus Zürich
ca. 1910

78 Karl Moser (attributed), Wooden Chair for the Kunsthaus Zürich, ca. 1910
Stained and varnished solid wood and plywood, 85 × 60 × 60 cm
Museum für Gestaltung

79 Karl Moser, Wooden Chair for the Kunsthaus Zürich
Photo: Philipp and Ernst Linck, 1910

80 Karl Moser, Kunsthaus Zürich, exhibition wing, central first-floor gallery
Photo: Philipp and Ernst Linck, 1910

78

79

80

Karl Moser (1860–1936), first with his Kunsthaus (1910) and then with the Universität Zürich (1914), managed to execute in swift succession two seminal cultural edifices that still impress us today. The compact exteriors conceal on the inside richly differentiated sequences of spaces; in the Kunsthaus the rooms are "suffused with the whiff of Secessionist incense and orchestrated with reddish Nassau marble and ornament that alternately murmurs and sumptuously proclaims its splendor" (Stanislaus von Moos). The integration of art was part of the architectural program and furniture a natural component of the Gesamtkunstwerk the building was meant to embody. While the stately ensembles of upholstered furniture—which are today still used in the central hall of the Kunsthaus—can be readily identified on the basis of old photographs, there is unfortunately no visual documentation of the use of the wooden chair model that was produced in a small series for the project. Was it destined for the first library, for the conference rooms, or, as today's management surmises, "for mobile deployment"? Moser already furnished the second Kunsthaus library, which opened in 1928, with conventional wooden "type" chairs. Had he tried in 1910 to design his own simple general-purpose chair, one with a casual elegance that could stand up to his interiors? With its straight, sharp-edged front legs and slightly curving back, which ends in a flat profile and is filled with straight rungs, this minimalist seat has a captivating simplicity that takes it far beyond any "whiff of Secessionist incest"—even today. AR

Hildebrand 2010; von Moos 2010.

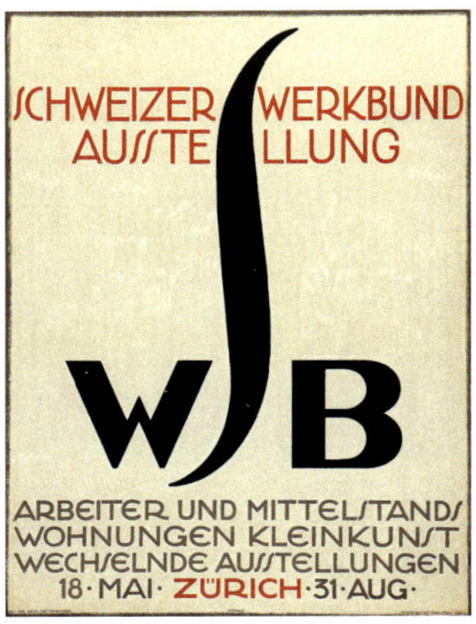

81

82

83

84

86

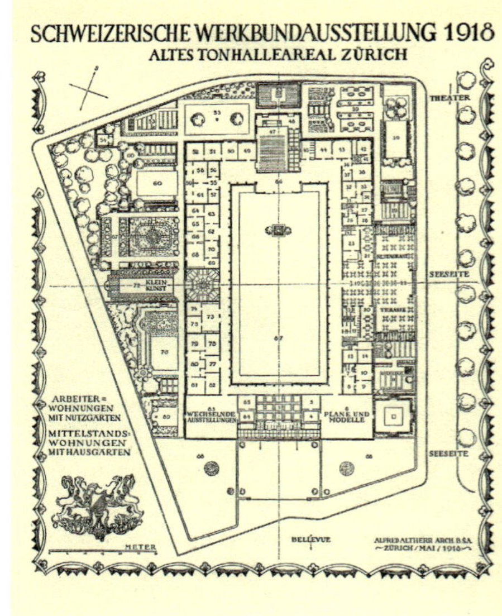

85

Swiss Werkbund Exhibition, Zurich, 1918
81 Otto Morach, *Schweizerische Werkbundausstellung,* Zurich, 1918. Poster
82 Rendering of the exhibition building with the design still showing pitched roofs, 1917
83 Alfred Altherr, Entrance hall, statue by Paul Osswald, wall paintings by Carl Roesch
84 Alfred Altherr, Entrance hall, view into the garden courtyard, sculpture by J. Gisler, in the rear the Marionettentheater
85 Old Tonhalle complex, Zurich. Site plan
86 Paul Ami Bonifas, Pitcher, before 1918

Zurich during World War I: whether Lenin and Mussolini really played chess together at Café Odeon and whether Robert Walser and Lenin actually met in the city—this is the stuff of literary speculation.[1] It is however a fact that, thanks to the course of world history, neutral Switzerland and especially Zurich suddenly rose to unusual prominence. In terms of intellectual history, it was the spontaneous emergence of Dada in particular that bestowed on the city a role it never would have wanted, and for which it had no use. Why Zurich? Shortly after the war began, in 1914, the poet Albert Ehrenstein suggested to his friends Hans Richter—who had just received his marching orders—and Ferdinand Hardekopf: "If the three of us are still alive, let us meet at the Café de la Terrasse in Zurich in exactly two years from now, on 15th September 1916, at three in the afternoon."[2] Zurich as a dream destination for peacetime. Richter, who would become one of the founders of Dada, wrote in retrospect, "I knew neither Zurich nor the Café de la Terrasse, and my hopes of being there seemed more than slender."[3] Even feeble jokes can make history. Four years later, in 1918, the Dada movement had already left a notorious trail of scandal in its wake, in the process sending out offshoots from Zurich to Berlin, Paris, Hannover, Cologne, and New York.[4]

One hundred meters from the Terrasse, likewise on Bellevueplatz, there rose up in the summer of 1918 the stately quadrangle that was to house the Swiss Werkbund exhibition. Only five years after its founding, and while the war was still raging—a situation to which the Swiss had simply become accustomed in their inimitable manner—the Swiss Werkbund (SWB) found the strength to present itself confidently in the most prominent of urban locations. With hindsight, we might even be tempted to see a genuine connection between this exhibition and Dada, even more so since Sophie Taeuber (1889–1943), an important artist, had a hand in both. But such a link does not in fact exist.

Unlike Dada, the Werkbund had no ambition to be considered avant-garde. The desperate sarcasm of a figure like Hugo Ball and the "constructive" work of the SWB, which set out to gain broad acceptance, were worlds apart. As anti-art, Dada changed our concept of what constitutes art through, for example, interdisciplinary performance. The new approach aspired to by the Werkbund was likewise interdisciplinary, but in this case affirmative and even mainstream. How new were its goals then?

A statement such as the following provides an idea of how the Werkbund's founding generation saw themselves: "Where formerly lone figures pursued a solitary creative path, now a free community of fellow-artists has come together who all ... pursue a common objective—improved professional work. New, more rational attitudes to food, clothing, and housing will give the creative forces in applied arts fields new problems to solve ... which we can very well try to do in a unique way ... in the Swiss manner pervaded by the same Gothic German and classical Roman essence and spirit that we can already admire today in countless well-preserved buildings, interiors, furniture, and appliances from the heydays of bygone cultural periods in our homeland."[5] No iconoclast would speak in such solemn tones. The emphasis on Swissness was intended to check the diverging sympathies of the various parts of the country with the wartime enemies Germany and France—German-speaking Switzerland sided with Germany and the French-speaking region with France. In the cause it took up, the SWB plainly modeled itself on the Deutscher Werkbund (DWB) and on Hermann Muthesius's notion of *unadorned, rational form.* But the famous rhetorical controversy provoked by Muthesius's question of what was *typical,* which nearly brought an end to the DWB in 1914 in Cologne, served as a warning to the SWB not to try to shine through words but through deeds. This is why its early years lacked any emphatic proclamations of a new theory of design.[6]

The 1918 exhibition was one such deed.[7] The exhibition building was designed by Alfred Altherr (1875–1945), the founder and chairman of the SWB and from 1912 director of the Kunstgewerbeschule and Gewerbemuseum Zürich. Altherr was also director of the exhibition. The entrance to the single-story quadrangle was in the north, on Bellevueplatz, with galleries for temporary exhibitions at either side.[8] In the middle of the lake side, on Utoquai, there was the Café de la Terrasse and two working-class houses complete with domestic gardens, designed by Hans Bernoulli and Hector Egger, respectively. The wooden building had smooth façades and plastered interior walls; the arcade looking onto the planted courtyard was flanked by a pathway of yellowish gravel. With hipped roofs rising on the inner side and towering walls over the arcade, the building with its neoclassical grammar looked monumental in the contemporary manner. The Marionettentheater (Puppet Theater) opposite the central axis of the foyer formed the prime focus of the inner courtyard.

Most of the exhibition consisted of sample rooms designed by the members and arranged according to local groups. The interiors were the work of architects or interior designers and featured furniture made by cabinetmakers or apprentice workshops supplemented with woven carpets, curtains, and lighting fixtures with mostly textile lampshades.[9] A grand space in the transverse central axis was dedicated to the *Kleinkunst* ("minor arts"), a term used to refer to ceramics by artists such as Berta Tappolet and Paul Ami Bonifas, as well as handcrafted glassware.[10] Garden design played a major role in the show—evoking the vision of the Garden City with its low-density development and the back-to-nature lifestyle it embodied. The SWB viewed the rampant real estate speculation as a challenge that could be surmounted, and public housing projects in various Swiss cities in the years preceding the world financial crisis temporarily bore out this assumption.

The meager photographic documentation of this elaborate exhibition astounds us present-day picture addicts.[11] Taking photographs was a complex proposition at the time, and with printing costs extremely high, many were probably deemed unworthy of publication. In the June 1918 issue of *Das Werk,* there are pictures of the entrance front, the foyer, and the room displaying "small sculptures," as well as two axial shots of the leafy courtyard with the exterior of the Marionettentheater, along with photographs of small gardens and marionettes, including some by Carl Fischer and Otto Morach. The August issue featured an illustrated report on various garden designs, as well as two western (French-speaking) Swiss dining rooms and a beautiful tribute to the Marionettentheater, illustrated with Sophie Taeuber's stage design for *König Hirsch* (*The Stag King*). Finally, the October issue presented without commentary the room for female students designed by the young Lux Guyer,[12] as well as an article about the two working-class houses, particularly the one designed by Hans Bernoulli.

Some of the interiors were presumably quite unremarkable. The wallpaper probably exhibited more restrained patterns than the ones that were popular at the time, the furniture was likely simpler and more tasteful, the decorations not lavishly naturalistic but rather stylized. If it had been any different, we would have heard about it. The Werkbund was not inventive but rather reformist.

A Marionette Avant-Garde

The only part of the show to take a different tone was the Schweizerisches Marionettentheater (Swiss Puppet Theater). It was founded expressly for the exhibition and formed the compositional vertex directly opposite the main façade of the Stadttheater (today the Opernhaus).[13] With its 160 seats, it was probably the section of the exhibition where the SWB and the French-speaking Swiss sister organization, L'Œuvre, converged most closely. And if there really was an intersection between the Werkbund and the avant-garde, here is where it could be found. The puppet play was a portable medium, an inexpensive way to bring art to the working population and the lower middle class and to delight both young and old. It was autonomous in its creative substance and, as a "minor art," could be considered a full-fledged art genre. Here, the Werkbund was totally in its element, considering how Heinrich von Kleist had postulated in 1810 in his epoch-making text *Über das Marionettentheater* (*On the Marionette Theater*) that only through artistic awareness was it possible for people to find their way (back) to the authenticity of expression they sought (which was inevitably lost in the process of growing up). The review of Sophie Taeuber's *König Hirsch* presents the puppet theater as a modern alternative to conventional dramatic theater with its alleged naturalism, claiming that, due to its very abstraction, the puppet theater was able to express deeper feelings. "With artistic boldness, the world itself is perceived as a fabric of undreamt-of fairy-tale rhythms; a law of simple body mass holds sway over it all.... Every form seems guided by an intellectual center, merely a paraphrase of one and the same

88

87

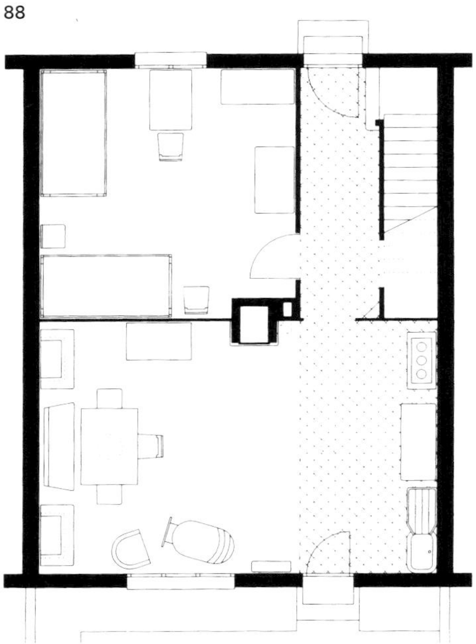

89

90

91

Swiss Werkbund Exhibition, Zurich, 1918
Hans Bernoulli, Worker's House
87 Eat-in kitchen
88 Façade on Utoquai, with domestic garden
89 Ground-floor plan

Interior Design Class/Gewerbeschule Zürich, teacher Wilhelm Kienzle, Worker's Apartment, executed by the Städtische Lehrwerkstätten
90 Eat-in kitchen
91 Bedroom

The presentation also featured stoves, dishware and ceramics, kitchen utensils, and bathroom fixtures, all designed according to Werkbund principles and made in Switzerland.

92

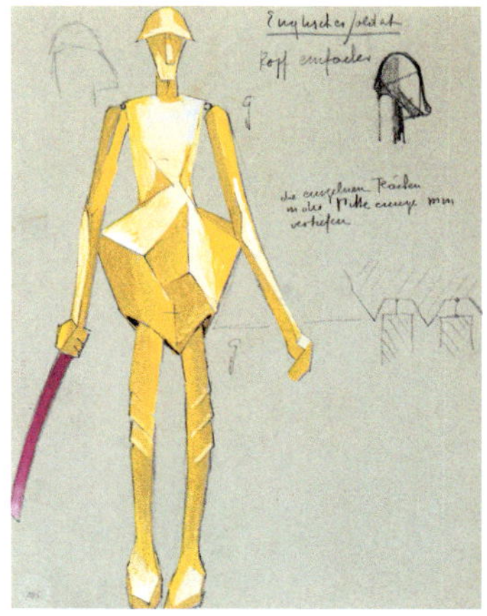

93

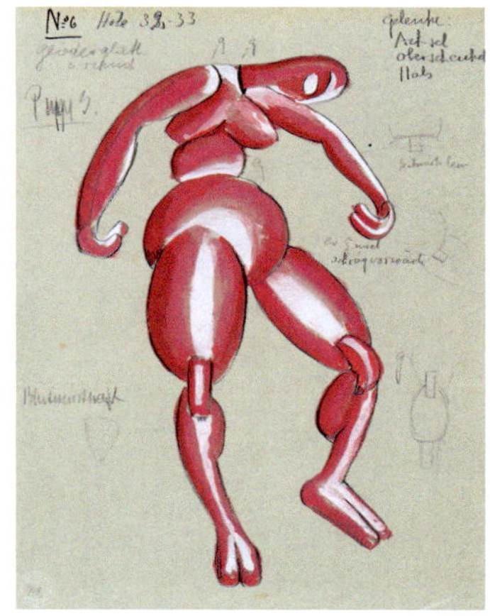

94

95

92 Sophie Taeuber and Hans Arp with the marionettes from *König Hirsch,* 1918

Otto Morach, staging of Claude Debussy's *La boite à joujoux,* 1918
93 Design for the English Soldier marionette Hugo Stüdeli, Solothurn
94 Design for the La poupée rosée marionette Hugo Stüdeli, Solothurn
95 Stage set with Eiffel Tower Museum für Gestaltung

subject.... More is thus achieved in these puppets, which—narrow and pointed or broad and round—pick up and play variations on the rhythm of the simplest forms, than in the mere floating drama of the human creature."[14]

Reading such sentences was like experiencing the epiphany of a different, wider world than that of, for example, the tombstone art or architectural sculpture that otherwise oddly figures so strongly in *Das Werk.* We must remember that during the same period the De Stijl group in the Netherlands was proclaiming a never-before-seen universalism for its design ethos! Such a grand agenda was alien to the Werkbund. Nevertheless, commissioning Sophie Taeuber with the set design for *König Hirsch* may indeed indicate a cautious opening up toward the avant-garde (which had by then become a force to be reckoned with, going beyond mere slapstick). Just a short time before, Altherr had sharply censured "Fräulein Täuber's" collaboration in Dada group events.[15] Hans Richter's later comment, by contrast, constitutes a tribute, although it still falls short of really doing her justice: "Sophie was studying dancing under Laban, in whose hallowed halls I was a frequent visitor.... She usually obviated the necessity of speech with one of her shy or thoughtful smiles.... During the Dada period she had a job teaching at the school of arts and crafts, as well as being a painter and a dancer.... There were abstract drawings, extraordinary Dada heads of painted wood, and tapestries, all of which could hold their own alongside the work of her male colleagues.... She had lectured at the Zurich museum of arts and crafts for years, and had by necessity acquired the skill of reducing the world of lines, surfaces, forms and colours to its simplest and most exact form, and formulating ideas in the simplest way."[16]

This last sentence, written at a distance of forty years from the others, turns the inventor that Taeuber was into a mere representative of a style. In fact, it was the other way around: the possibilities for such stereometric, "simplest formulations" had up to that point not been grasped by anyone at the Kunstgewerbeschule or the SWB except for Taeuber. Could Altherr ever have imagined that this woman would become one of the most important figures in the institution?[17]

If the puppet theater can be seen as an opportunity for artistic experimentation, almost all other areas of the exhibition aimed to strike a mediatory tone in order to gain widespread acceptance. They all share a touch of neoclassicism, which also characterized the work of the DWB, indicating a renewed interest in the Biedermeier style. The age of Goethe was regarded at the time as the last period with a solid and serene middle class, before industrialization and the associated acceleration of goods production drove the emerging entrepreneurial class to embrace eclecticism and the quest for prestige.[18] This notion was the economic justification for mounting such an elaborate exhibition in the most urban part of town, which also explains why policymakers supported the project so strongly.[19] The public was to be alerted to the problem of the ever-increasing number of imported goods and encouraged to buy local products. The importance of high-quality commodities was joined by the issue of maintaining jobs.[20] And in addition, only when workers and employees had suitable housing would they be able to deliver the quality work that Switzerland had to produce in order to prevail on a market that was increasingly being infiltrated by the world abroad. Under these circumstances, refined interior decoration for the elite was now of secondary importance.[21]

The Theme of Working-Class Furniture

The sections of the exhibition devoted to the working-class home gave the Werkbund itself the best orientation for the future. Hans Bernoulli described in retrospect the public popularity of the exhibition—achieved despite adverse circumstances—as a double-edged sword.[22] Its "inner success" for the SWB itself was after all questionable: "The task of procuring cheap, solid working-class furniture has not or has only been partially solved due to a lack of time for practical testing."[23]

That said, "working-class furniture" is something that never came about afterward either. The "type furniture" that Bernouilli envisioned and that in fact began to be developed around 1930 was, as we know, adopted not by workers but by the small portion of the middle class that was culturally aware. Social developments resulted anyway in the whole idea of a "working class" being abolished, and the actual alternative that arose to the tenement was not the Garden City but the apartment house.

96

97

Swiss Werkbund Exhibition, Zurich, 1918
96 Rudolf S. Rütschi, Lady's study
97 Lux Guyer, Living room/study for a female student, executed by the Städtische Lehrwerkstätten.
The blue-painted firwood furniture, bed with curtains and open canopy, and window seat with a desk set at a right angle to it are all nonconformist, even in terms of the Swiss Werkbund guidelines.

How Do We View the Exhibition Today?

The style of living presented by the Werkbund at the exhibition was basically *à l'ancienne.* There were tables, chairs, beds, nightstands, sewing tables, benches, dressers, open shelves for kitchen utensils, and highboys with glass doors, accompanied by home textiles such as curtains, woven rugs, tablecloths, and lampshades. Together with the master-builder-designed architecture, figural sculptures, handmade ceramics, and silverwork, all the items on display embodied the preindustrial world of the Werkbund. This was not a case of nostalgia, though. The SWB did look to the future, but from a bygone perspective, which is why it didn't see what was coming, in other words, what was really new. A short time later, L'Esprit Nouveau would lend civilization new momentum, and Hannes Meyer would celebrate in his manifesto "Die neue Welt" ("The New World") all the new standardized products, such as paper formats, lightbulbs, or sleeping cars, as demonstrations of the internationalization of the economy.[24]

One look at Bernoulli's kitchen/living room for working-class families shows to what extent the standards we take for granted today first had to be developed over time. The kitchen sink and the drain board are still two different things, and the work surface for cooking is a table with an additional shelf just above the ground. Another shelf above head height provides storage space for pots and pans and the coffee mill. With the electric lamp, a transparent glass element with enameled shade and visible lightbulb, it would have been impossible to be in the room without being blinded by the glare or casting a shadow on the rest of the space. These are faults that were not perceived as such and for which therefore no products had yet been developed. And for that matter: How many things were still unheard of back then! The refrigerator still lay far in the future. Only the well-to-do could afford a telephone or gramophone, only the rich an automobile, and even for them there were no radio broadcasts yet and thus no radios either.[25] Gas water heater or household boiler? Only for the better half; the others used a wash basin set atop the bedroom dresser. The furniture was to a great extent typologically fixed, but only occasionally were the custom-made pieces actually designed as models for furniture types that would be suitable for serial production.[26]

The SWB at that time was basically an association with conservative values. It saw its mission as being the "refinement" of familiar models, in keeping with the wording of its 1913 statutes: "The goal of the SWB is the artistic refinement of professional and industrial work."[27] That which already existed was to be improved on, but anything really innovative had not yet been envisioned.[28] This attitude is revealed, for example, by the maxim of *Materialgerechtigkeit,* or "doing justice to the material," that was so important to the Werkbund as a supposed recipe against the knee-jerk application of ornament. The fact of the matter, however, is that the major steps were not taken through *justice to the material* at all, but rather on behalf of and by way of *justice to fabrication techniques:* with Michael Thonet and Henry Ford as prime examples. The Werkbund practiced forms of cooperation "of all artisans" just as Walter Gropius proposed in his 1919 Bauhaus manifesto, but this idealized leitmotif of handcraftsmanship was already being replaced circa 1925 by serial production (toward which the compass needle would thereafter point—even during the economic crisis). The criterion of faithfulness to materials may be compatible with the division of labor in (interdisciplinary) cooperation between different trades, but it loses its cogency in (transdisciplinary) industrial processes with their much more profound transformation of the materials at hand.

Cosmologically, the last hundred years are nothing; geologically, they are but a grain of dust; in terms of the history of civilization, though, they are half an eternity. And for us today, the said period is twice removed: much of what we know today did not yet exist back then, and much of what was new or upcoming back then we have long since forgotten.[29]

We still today have no richly nuanced name for the time period between Jugendstil/Art Nouveau and modernism, from 1900 to 1920: it is a world in twilight—for the modernists the hour before dawn, for their antipodes that before dusk. For the former SWB, it was simply the new and improved real world.

98

99

100

98 Wilhelm Kienzle, Secretary desk in cherrywood, ca. 1918
99 Alfred Altherr, Two painted nightstands, before 1923
100 Wilhelm Kienzle, Painted wardrobe, ca. 1923

1 Max Frisch noted a dream he had in Berzona in 1968: "Someone reports of a verified encounter between Robert Walser and Lenin on Spiegelgasse during which Robert Walser asked Lenin only one question: Do you love the pear bread from Glarus as much as I do? In my dream, I do not doubt the authenticity of this report and I defend Robert Walser until I awaken—I am still defending Robert Walser while shaving." Frisch 1979, p. 165.
2 Richter (1964) 1997, p. 27.
3 Ibid.
4 The Futurists in Marinetti's circle were ahead of the Dadaists by several years in their courage to break with taboos in Italy, taking an unpalatable contradictory (war-glorifying) approach.
5 Linder 1918, p. 14.
6 This lively debate became increasingly pedantic and was fueled largely by resentments against Muthesius. See the transcript in Posener 1964, pp. 204–5.
7 See note 27.
8 One of these exhibitions documented, by way of plans and photographs, exemplary buildings by architects from the circle of the SWB (no pictures available).
9 As had been hoped, a large proportion of these items found buyers by the close of the exhibition. See SWB Annual Report 1918–19, SWB archives.
10 See *Das Werk* 5, no. 8 (1918), pp. 126–27.
11 The exhibition budget was 600,000 francs, of which construction costs accounted for 328,000 francs, sizable sums at the time. Source: "Schweizerische Werkbundausstellung Zürich 1918. 1. Flugblatt," undated (presumably late 1917), SWB archives.
12 The designer was still named Louise Guyer (1894–1955) at the time and was a student at the Kunstgewerbeschule Zürich. A short time later, she took courses at the ETH and was one of the first female architects to run her own office in Switzerland. See Huber/Rucki 1998.
13 The Marionettentheater continued to exist afterward as an institution at the Kunstgewerbemuseum Zürich, up until 1935.
14 Jollos 1918, p. 131. Waldemar Jollos (b. 1886 in Moscow, d. 1953 in Melide), an art historian, writer, poet, translator, playwright, and political commentator, was another political refugee in Switzerland; see Jollos 1955.
15 Altherr, director of the Kunstgewerbeschule, had even threatened to fire her from her job (she taught composition as well as embroidery and weaving techniques), whereupon she thereafter took part in Dada evenings under an assumed name and wearing a mask. See Arp 1955, cited in Staber 1970, p. 16.
16 Richter (1964) 1997, pp. 45–46.
17 Sophie Taeuber (Täuber) remained at the Kunstgewerbeschule Zürich until emigrating to France in 1929. After her marriage in 1922 she called herself Taeuber Arp.
18 See Mebes 1908.
19 Sitting on the "honorary board" of the exhibition were federal councilor Calonder (Department of the Interior), councilor Mousson (Canton of Zurich), and Zurich mayor Nägeli. The sixteen-member "organizing committee" included four additional Zurich city councilors. Source: Wegleitung zur Ausstellung, SWB archives, n.p.
20 The newspaper *Neue Zürcher Zeitung* wrote, "When the arms are laid down, an economic battle just as fierce will take hold in commerce and industry. And this war knows no neutral zones; we, too, will be drawn in. Since we are poor in mineral resources and, as a land-locked country, must reckon with high wage costs for the import of raw materials and coal, it is impossible for us to compete ... with cheap prices.... Only with top-quality pieces that exemplify our better insights, our tough character, and our good taste will we be able to prevail." *Neue Zürcher Zeitung,* May 12, 1918, SWB archive.
21 The choice of such a prominent location ensured that what was originally intended as an "interior decoration" show became an exhibition of dwellings for the middle and working classes.
22 One of the main reasons for this, in addition to the hasty preparations, the war, and the restrictions in rail transport, was the Spanish Flu, which claimed millions of victims worldwide, including 25,000 fatalities in Switzerland (and 740,000 infected, in a population of just 3.8 million!), making it impossible to hold performances at the puppet theater for weeks on end. Source: SWB Annual Report 1918–19, p. 12, SWB archive.
23 Bernoulli's account is summarized in the SWB Annual Report 1918–19.
24 See Meyer (1926) 1994, pp. 445–47.
25 First radio broadcasts in Switzerland: 1923.
26 According to *Das Werk,* the Globus department store was interested in serial production of some of the models from the exhibition, with cabinetmaking shops (members of the SWB) as suppliers. It is not known whether anything came of this interest.
27 "The SWB tries to achieve this goal through the cooperation of artists, craftsmen, and industrialists. Through general education about its efforts, in particular by organizing competitions and exhibitions, and by making statements on economic and artistic issues insofar as they touch on the objectives of the SWB." The SWB thus adopted almost verbatim the wording of the German Werkbund, founded in 1907, Article 2 of whose statutes reads: "The goal of the association is the refinement of professional work through the cooperation of art, industry, and the crafts, through education, propaganda, and united attitudes to pertinent questions."
28 Examples: radio, amateur photography and compact cameras, sound film, jazz, electric locomotives, airports, oil heating, gas water heaters, bathing beaches, escalators, buses, jars.
29 For us, the following belong to the past: the mechanical calculator, the small offset printer, single-story heating, the "instant camera" as a special case, the wall phone, the blueprint, the pneumatic letter shoot.

* I would like to thank the secretarial office of the SWB (Managing Director Dr. Monika Imboden) for permitting me to do research in the files.

101

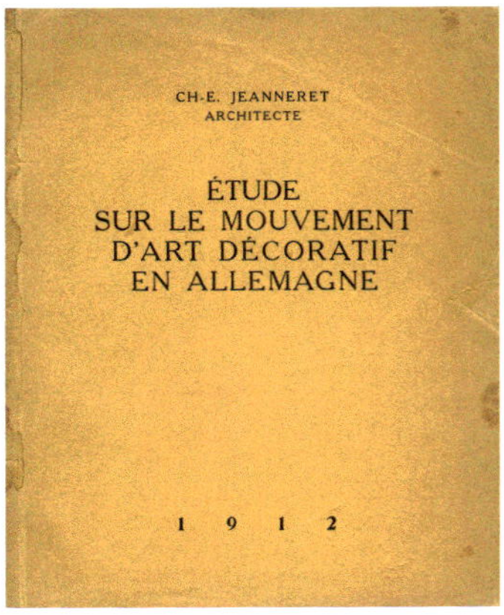

102

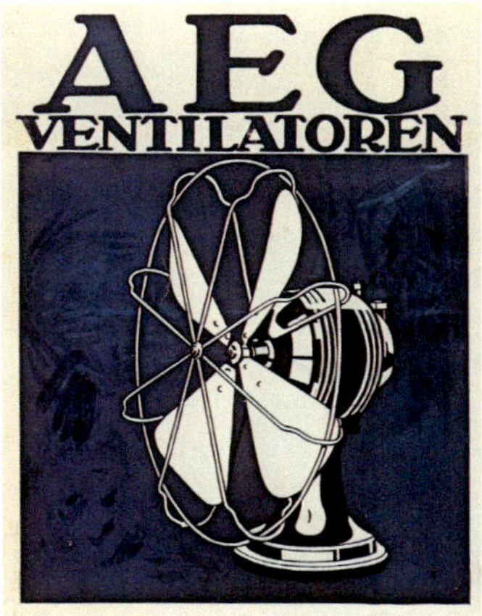

103

104

105

106

101 Charles-Edouard Jeanneret (Le Corbusier), Design for a decorative hair comb, ca. 1903–04, gouache on paper
Private collection

102 Charles-Edouard Jeanneret (Le Corbusier), *Étude sur le mouvement d'art décoratif en Allemagne,* La Chaux-de-Fonds, 1912. Cover
Archive of Arthur Rüegg. Copy from the collection of William Ritter

103 Peter Behrens, Brochure for AEG direct-current fan, ca. 1908. Cover design

104 Charles-Edouard Jeanneret (Le Corbusier), Desk chair for Hermann Ditisheim, 1915–16, mahogany with fabric-covered seat (original)
Photo: Eveline Perroud
Musée des beaux-arts, La Chaux-de-Fonds

105 Bergère à paille, 19th century. From the Villa Jeanneret-Perret, La Chaux-de-Fonds, 1912
Fondation Le Corbusier, Paris

106 Desk Chair Model 1224 from the company Robert Mey Paris, with brandmark GLARIS from the AG Möbelfabrik Horgen-Glarus
From the Villa Albert Jeanneret-Rääf, Paris, 1925
Collection of Arthur Rüegg

"The left half of a wood carving will never be fully identical to the right," asserted the influential architect Hermann Muthesius (1861–1927) in a polemical statement made shortly after the founding of the German Werkbund in 1907. "Each side will reveal small nuances.... It's quite a different story with the machine-made product. Instead of the life that radiates from the handmade product, the machine presents of this life only a cold death mask."[1] Circa 1900, popular enthusiasm for the Neo-Renaissance style faced the dilemma of the cheap machine copy. "The machine should not deign to mimic the techniques of handcraftsmanship. It should practice machine techniques and produce only such forms as correspond to its nature. And these are smooth, precise, so to speak artless forms, as we can see from a bicycle—a genuine machine product."[2]

"Genuine machine products" instead of the lifeless fabrications of industrial art: with this maxim in mind, Muthesius wanted to shake the foundations of the prevailing historicism and set a goal for the future work of the Werkbund. Machine work, however, inevitably leads to a standardization of the goods produced. In 1914, at a conference that took place at the Cologne Werkbund exhibition, Muthesius encountered bitter resistance. In a famous controversy, Henry van de Velde (1863–1957) defended the artist as creative individualist: "By his innermost essence the artist is a burning idealist, a free spontaneous creator. Of his own free will he will never subordinate himself to a discipline that imposes upon him a type, a canon. Instinctively he distrusts everything that might sterilize his actions, and everyone who preaches a rule that might prevent him from thinking his thoughts through to their own free end."[3] The dispute waged within the German Werkbund also made a mark on developments in Switzerland—albeit in a more moderate Helvetian form.

From Precious Unicum to Serial Production

As early as 1903, when the charismatic Charles L'Eplattenier (1874–1946) from Neuchâtel took over as director of the École d'Art in La Chaux-de-Fonds, historicism was already being eclipsed there as well. As an artist and craftsman, L'Eplattenier took van de Velde's position. And as a teacher, he tried not only to modernize the look of the world's premier clocks with a regional Jura version of Art Nouveau but also to breach the boundaries of a curriculum limited to the decoration of pocket watch cases. With great dedication, he succeeded at turning La Chaux-de-Fonds into the only center of Art Nouveau in Switzerland—on a more modest scale, but by all means comparable to the centers of design reform abroad with which he was familiar in most cases from firsthand experience: Nancy, Brussels, Paris, Glasgow, Vienna. On his quest for a genuinely "indigenous and autochthonous"[4] style for the Jura region, he oriented his teaching on the detailed study of the tectonic limestone formations of the Jura Mountains and the native flora and fauna. Drawing lessons were always based on a vocabulary of forms suitable for the decorative arts, flowing lines à la van de Velde increasingly giving way over time to the geometric stylization of natural elements. The first interiors produced by the "Cours supérieur" he founded in 1905 drew on the English Arts and Crafts movement, whose raison d'être was "home decor designed by an artist as deluxe one-of-a-kind creation, featuring the handcrafted and thus precious unicum."[5]

Charles-Edouard Jeanneret (1887–1965), who would later rise to world fame as an architect under the name Le Corbusier, attended L'Eplattenier's classes from 1902 to 1907.[6] After undertaking his initial projects as craftsman and architect as part of the almost medieval-seeming artisan community of the Cours supérieur (Mathey-Doret music room, 1906; Villa Fallet, 1906–07), his journeyman years took him to Florence and Vienna and then onward to Paris and finally Berlin. It speaks for L'Eplattenier's broad-mindedness that he sent his former student on his way with a scholarship in hand and the politically charged order to write a report for the École d'Art on progressive German applied arts.[7] In 1910–11, Jeanneret worked with Peter Behrens (1868–1940), who was responsible for industrial design and corporate identity at the Allgemeine Elektricitäts-Gesellschaft (AEG), a company that at the time had 60,000 employees. Jeanneret admired the "modest, sober, almost impersonal appearance" of the heating and lighting devices produced by AEG, which were designed by way of an "authoritarian intervention of art."[8] Like many others of his generation, he would subsequently focus his attention not on the unique handcrafted piece but on more economically and socially relevant serial production.

But how does one arrive at a standardized series product? In contrast to most of his peers, Jeanneret was hardly interested in inventing and artistically formulating objects of use for industrial production. Instead of designing "genuine machine products" himself, he therefore began compiling a "sélection" of items that had demonstrated their usefulness by being constantly revised and which had thus already achieved a perfect form. His goal was to determine what was typical about them—the iconic prototype—as a specific category of man-made object. After breaking away from L'Eplattenier and publishing his report in La Chaux-de-Fonds,[9] Jeanneret worked with considerable success as an interior decorator and furniture designer, selecting and redesigning formally distinctive French furniture types from the late eighteenth century. This eclectic design process soon gave birth to a range of furnishings that were suitable for various domestic functions while at the same time vividly embodying these uses. Around 1915, when Jeanneret began to replace the luxurious period furniture with manufacturally produced models from the nineteenth century, he had finally taken the last step from artisanal unicum to standardized serial product while simultaneously discovering the precursors of the anonymous industrial culture of the present.[10]

Homogeneous versus Heterogeneous Furnishings

Furnishings and interiors stand *pars pro toto* for the respective architecture, or a specific architectural style. This is why the step from unicum to series product is also significant in a more general sense. During the eighteenth and nineteenth centuries, France provided the binding formulations and also the style names for the development of interior decor as something that merged architecture and furnishings into a uniform, inseparable whole.[11] By the 1920s, the days of the traditional "ensemble" were over. Within the European avant-garde only the Dutch De Stijl movement still insisted on the fusion of interior and furniture—as demonstrated by Gerrit Rietveld (1888–1964) with his Schröder House from 1924, where the design of a chair is subject to the same rules as the house itself. By contrast, the forward-looking interiors of the Neues Bauen (New Building) movement were characterized by the programmatic separation of the interior and its furnishings, by the elimination of anything superfluous, and by the use of *Typenmöbel,* or "type furniture": early standardized pieces that could very well possess an individual character. This development stakes out the whole breadth of the field: on the one hand, the homogeneous interior with a single design theme; and on the other, heterogeneous furnishing using diverse objects for which it seems unnecessary to question their common denominator.

At the Werkbund exhibition in Zurich in 1918, more moderate reformist furnishings were still on view that were hard to localize between the two poles. In the *Wohn- und Arbeitszimmer einer Studentin* (Study and Bedroom for a Female Student) created by the young Zurich architect Lux Guyer (1894–1955), wavy motifs and a coat of blue paint provide optical cohesion between a rustic desk and a four-poster bed. Airy curtains on the dressing table and the windows lent a lively feminine touch to the otherwise severe room with its strong Biedermeier flavor.[12] More resolute is the impression made by Basel designer Hans Bernoulli (1876–1959)—who had ten years of experience as an architect and lecturer in Berlin under his belt—in his spartan working-class house, its furnishings oriented consistently on functionalism and faithfulness to materials. Similar to Jeanneret, he also played with variations on the anonymous, standardized furniture types from the Biedermeier era, which anticipated the ideals of Neues Bauen. Bernouilli's furniture thus, according to Leza Dosch, "from today's aesthetic viewpoint, exhibits a rough charm."[13]

Meanwhile, Le Corbusier had begun in Paris to conjoin the old idea of the formal refinement of utilitarian objects with their commercial and industrial production in series. As the avant-garde standard-bearer par excellence, he used almost exclusively anonymous brands, but he had no intention of ceding control over their formal cohesion. He defined the aesthetic criteria for the precise

107

108

107 Gerrit Rietveld, Schröder House, Utrecht, 1924. Kitchen. Example of a homogeneous interior conceived by a single designer.

108 Hannes Meyer, *Co-op. Interieur,* 1926. Example of a heterogeneous interior with industrially produced furnishings.

selection, further development, and arrangement of these so-called *objets-types* primarily according to artistic considerations. While on the one hand crusading against the decorative arts, which reached their high point with the Exposition des arts décoratifs et industriels modernes in 1925, he was at the same time testing out in the Pavillon de L'Esprit Nouveau the formal harmony between his readymades—selected based on their emblematic character—and their architectural framework, applying compositional principles that he had developed in Purist painting starting in 1918. His theme was the arrangement or assembly of basically autonomous elements—furniture, pictures, sculptures, objects—into walkable three-dimensional "still lifes."[14]

Almost concurrently, there arose—in Switzerland of all places—the most radical manifesto of *Neues Wohnen* (New Living), which described par excellence the opposing position to the homogeneous Gesamtkunstwerk: the *Co-op. Interieur.* In 1926, Hannes Meyer (1889–1954), future director of the Bauhaus Dessau, lined a corner of his office in Basel's Luftgässlein with paper in order to strikingly showcase a few products suggesting industrial production in large series. The springboard for this deliberately improvised-looking art installation was—as in the "Environments" of the fifties and sixties—a conceptual approach. In the spirit of Russian agitprop, Meyer arranged together a few basic furnishing pieces for the "semi-nomad of contemporary economic life,"[15] which with their symbolic form as good as trumpeted the message of "mechanization in our daily life."[16] These included the expressive Chantal-brand funnel gramophone, which could never be replaced by the new suitcase model. If there is any cohesion to be found among these objects of such disparate form and design, it must lie in the ideology they represented.

Swiss "Type Furniture"

Hannes Meyer's small garden table, his mattress resting on truncated cones, and his improvised glass shelf were proxies for modern consumer goods that did not yet exist at the time. Standardization of objects of everyday use, eradication of luxury goods, assembly-line production: Meyer's postulates and those of his like-minded peers took some time to make inroads in Switzerland. Although a large number of painstakingly designed interiors were created, the debate on modern standardized furniture was limited to a small circle of mostly very young designers. The first "appetizers" as a prelude of things to come could be seen at the end of 1926 in the exhibition *Das Neue Heim I* (The New Home I) at the Kunstgewerbemuseum Zürich. The magazine *Das Werk* spoke in its review first of the usual tasteful "furniture and interiors"—"leftovers from the ornamental arts and crafts trend that is slowly fading." These were now joined, however, "for example, in the works of Wilhelm Kienzle and Ernst F. Burckhardt, by furniture strongly concentrating on the functional aspect, which will undoubtedly become the main distinguishing feature of this 'world style.'... The living room furniture, chair and table especially, have achieved a refined form that emphasizes primarily mobility and comfort, without imitating the specific machine-like nakedness of 'Bauhaus' furniture."[17] With prophetic foresight, *Das Werk*'s editor Josef Gantner neatly encapsulated the Swiss position.

The year 1926 then saw the first Swiss standardized furniture. The Zurich architect Max Ernst Haefeli (1901–1976), who had recently returned home from Berlin, developed in collaboration with the director of the AG Möbelfabrik Horgen-Glarus, Ernst Kadler-Vögeli (1900–1972), several chairs and stools made out of partly bent beech or walnut, with seats and backs of plywood or woven cane. Serially produced in Glarus, their factory character did not proclaim itself in "cheap manifesto pomposity" (Peter Meyer), and they were in fact still largely made by hand.[18] Since the models were not made of rounded wood profiles like the Viennese Thonet chairs popular since the mid-nineteenth century, Haefeli was able to draw directly on conventional domestic chair forms. Shortly thereafter, he also ventured to use metal,[19] but unlike Le Corbusier (or Hannes Meyer) he had no plans to transfer ready-made models from the office, hospital, or travel field to the domestic realm. It's no wonder that the new models seemed so familiar and sold quite well, often with small modifications by the manufacturer. Although Haefeli saw through the transition from handcraftsmanship to industrial design in direct cooperation with industry, his goal was not to apply technical advances to, for example, reinventing seating furniture—which was however definitely the ambition of designers in the Bauhaus orbit, such as Marcel Breuer and Mart Stam.[20] The first Swiss

109

109 Le Corbusier and Pierre Jeanneret, Villa Albert Jeanneret-Rääf, Paris, 1925. Living room. Example of a composition with "type furniture," Berber rugs, and Purist art.

TI 206
gustav hassenpflug
klappsessel, stoffbespannung

auf der suche nach dem standard

die stühle auf seite 12—15 gehören zu einer neuen versuchsreihe der bauhaus-tischlerei. sie sind auf der ausstellung „der stuhl" in stuttgart ausgestellt.

ein beliebter vorwurf gegen das bauhaus

und seine zeitschrift: wir treiben zu viel weltanschauung, wir vermischen gestaltung mit fragen der gesellschaft und philosophie usw. hoffentlich ist das vorliegende heft geeignet, diesen vorwurf recht nachdrücklich zu bestätigen. wir wollen und werden uns auch weiterhin gedanken machen über das soziale wieso und wofür der bauhaus-arbeit. das bauhaus kann sich keineswegs mit der vermittlung von baustofflichen, bautechnischen und sonstigen bauwerklichen kenntnissen zufrieden geben, so selbstverständlich dieser werkunterricht (auch für die übrigen werkstätten) im vordergrund der bauhaus-arbeit zu stehen hat. die bauhaus-arbeit muß einen geistigen orientierungspunkt haben, sonst erzieht sie lediglich fachsimpel, die für alles andere außer ihrem engsten spezialkreis ignoranten bleiben. ein mangel an weitgespannter geistiger orientierung bedeutet soziale und menschliche desorientierung. das ist der trübe zustand, in dem unsere zeit sich befindet, zur größten zufriedenheit der gewalten, die in solcher trübe ungestört gedeihen können.

die vorwürfe wegen der weltanschaulichen gepflogenheiten des bauhauses kommen von einer seite, wo man alle gesichtskreise möglichst einschränken möchte,

besonders wenn diese gesichtskreise zu menschen gehören, die mit werkstätten und betrieben zu tun haben. man will eben aus bekannten gründen nur spezialisierte arbeitsautomaten haben, keineswegs menschen mit offenem und kritischem geist. für den geist sind andere fachkreise da: die staatlich approbierten philosophieprofessoren, die kirchen und die familienblätter.

das bauhaus aber muß eine pflegestätte sein,

nicht nur der werkarbeit, sondern auch des geistes, der sich weder auf das wohnen, noch auf sonstige, tausendmal gemeinnützige dinge beschränken läßt. daß wir unter geist nicht das kauwerk philiströser stubenhocker und auch nicht den nervenreiz übersättigter allerweltsnobs verstehen, dürfte aus allen lebensäußerungen des bauhauses genügend hervorgehen. aber auch das kann unmöglich unser letztes ziel sein, einer unternehmungslustigen industrie folgsame handlanger zu liefern. wir müssen uns immer darüber klar sein, was von den materiellen notwendigkeiten unserer zeit lediglich als notwendiges übel zu betrachten ist, sonst wird unsere praktische betriebsamkeit zum selbstzweck. es wäre ein übel, nicht minder verwerflich wie der faulste kunstakademismus.

e. k.

---

junge menschen!

was sucht ihr an den kunstakademien? kommt ans bauhaus!

das bauhaus, hochschule für gestaltung in dessau, leitung hannes meyer, beginnt sein wintersemester am 30. oktober.

lehrgänge:
gestaltungslehre, werkstattlehre, architektur, reklame und druckerei, bühne, tischlerei, weberei, wandmalerei, metallwerkstatt, freie malerische und plastische gestaltung.

lehrkräfte:
l. feininger, w. kandinsky, p. klee, hannes meyer, o. schlemmer, j. albers, h. scheper, j. schmidt, gunta stölzl, h. wittwer, m. stam, zwei ingenieure, drei dozenten im nebenamt, zwei dozenten für sport.

zugelassen sind:
auch ausgebildete handwerker, techniker und architekten.

aufnahmegebühr 10.— rm.
1. und 2. semester je 60.— rm.

nähere bedingungen
durch das bauhaus-sekretariat, dessau.

15

110

111

112

110 Gustav Hassenpflug, TI 206 Folding Chair, 1928, wood with fabric. "Type" chair from a trial series designed at the Bauhaus in Dessau under the direction of Hannes Meyer.

111 Display window in Paul Matzinger's linoleum and carpet store, Basel, ca. 1925. Series of anonymous wood and canvas folding chairs.

112 Anonymous wood and canvas folding chair, ca. 1925. Model used by Hannes Meyer in his *Co-op. Interieur.*
Museum für Gestaltung/
Donation of Ruggero Tropeano

113

114

115

116

117

Max Ernst Haefeli, Adjustable type furniture, 1926
113 Bookshelf for Friedrich Traugott Gubler, wood
Museum für Gestaltung
114 Floor lamp for Friedrich Traugott Gubler, brass and raw silk
Museum für Gestaltung/
Donation of Ruggero Tropeano
115 Reading desk for the Bücherstube Girsberger, Zurich, wood, brass, and iron
Private collection

116 Bentwood fabrication at the Thonet factory in Bistritz, ca. 1920
117 Flora Steiger-Crawford and Rudolf Steiger, Carl Steiger House, Kilchberg, 1927–28. Living room with dining alcove. Type chairs by Max Ernst Haefeli, lamp over dining table by Flora Steiger-Crawford and Rudolf Steiger.

118

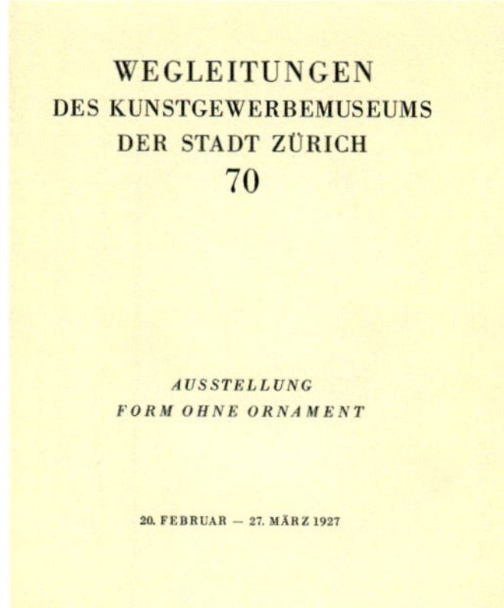

WEGLEITUNGEN
DES KUNSTGEWERBEMUSEUMS
DER STADT ZÜRICH
70

AUSSTELLUNG
FORM OHNE ORNAMENT

20. FEBRUAR – 27. MÄRZ 1927

119

120

118 Max Ernst Haefeli, Type chairs in bentwood with cane seats and lights made of cast iron, brass, and Jena glass. *Form ohne Ornament* exhibition at the Kunstgewerbemuseum Zürich, 1927

119 *Form ohne Ornament* exhibition, Wegleitung 70, Kunstgewerbemuseum Zürich, 1927

120 Max Ernst Haefeli, Type chair ("Haefeli Chair"), 1926, Wohnbedarf Model 5, Horgen-Glarus Model 293
Museum für Gestaltung

standardized furniture is thus the result of a middle path between traditional and avant-garde ideas—and thus veritably exemplary of a specifically Swiss position within modernism, one that would also shape the overall character of interiors.

The design of inexpensive serially produced furniture types that can be combined at will and that, thanks to their straightforward lines, make a congenial addition to any domestic interior is an issue that still preoccupies designers today. At the same time, the urge continually arises to invent spectacular and exclusive unique pieces with which consumers can express their individuality and social status. One thing seems certain, however: home decor will remain a key theme of everyday culture and also thus the question of the significance and coherence of interior furnishings.

1 Muthesius 1908, p. 9.
2 Ibid., p. 10.
3 van de Velde (1914) 1975, p. 29.
4 "Indigène et autochtone," see L'Eplattenier 1910, p. 1.
5 von Moos 1992, p. 228; von Moos discusses the Swiss applied arts reforms in detail in this volume.
6 On Jeanneret as craftsman, see Rüegg 2012a, pp. 14–75. See also the fundamental study by Brooks 1997.
7 "L'organisation des métiers d'art, la création, la fabrication et la vente of the produits artistiques," *Procès verbal de la commission de l'École d'Art,* May 16, 1910, Bibliothèque de la Ville, La Chaux-de-Fonds.
8 Le Corbusier (1912) 2008, p. 177.
9 Jeanneret 1912; English: Le Corbusier (1912) 2008.
10 These were the so-called *bergères â paille;* see Rüegg 2012a, pp. 56–57.
11 See Rüegg/Steinmann 1983, p. 34. On the developments in France between 1900 and 1925, see Troy 1991.
12 See Rüegg 2009, pp. 107–19.
13 Dosch 2002, p. 304.
14 See Rüegg 2012a, pp. 76–105; see also Rüegg 1987 and von Moos 1987.
15 Meyer (1926) 1994, p. 448.
16 Ibid.
17 Gtr. 1926, p. 366.
18 For more on this, see Rüegg 2002b.
19 In 1927 he exhibited furniture made from an aluminum alloy called Elektron at the German Werkbund exhibition *Die Wohnung* (The Dwelling) in Stuttgart, and in 1928 individual tubular steel pieces were produced for the model homes on Wasserwerkstrasse in Zurich.
20 For example, Marcel Breuer's *Wassily Chair* (1925–26), the first chair made entirely of bent and bolted steel tubes, as well as Mart Stam's *Chair without Back Legs* (1927), the translation of the cantilevering phenomenon in architecture into furniture.

121

Max Ernst Haefeli, Model homes on Wasserwerkstrasse (Rotach Houses), Zurich, 1927–28. Workplace with a wall-arm lamp for both direct and indirect light, produced by the company Eberth & Co., Zurich, 1928. Set on the banks of the Limmat River, the model homes were built by the Rotach cooperative for the exhibition *Das Neue Heim II* (The New Home II) at the Kunstgewerbemuseum Zürich. The houses had interiors and furnishings by Haefeli himself, as well as by Richard Haechler, Hans Buser, and Wilhelm Kienzle and his interior design class.

Photo: Heinrich Wolf-Bender

1920 – Mission

In the mid-twenties, the internationally networked, Basel-based ABC group calls for a "dictatorship of the machine" as an alternative to an applied arts reform movement that has become increasingly mired in staid conservatism. The goal is for the standardization and serial production that have proven so successful in the machine industry to now make inroads in the consumer goods industry as well. The *Co-op. Interieur* represents this utopian, uncompromising everyday aesthetic. The German-Swiss Werkbund takes on the mission, in contrast to its sister organization in French-speaking Switzerland, L'Œuvre, which remains true to handcraftsmanship. Instead of investing in "manifesto pomposity," however, the Werkbund aims at actual feasibility. In the second half of the decade, the domestic industry brings the first no-frills standard products onto the market: ceramic goods, light fixtures, and *Typenmöbel* ("type furniture"). With their carefully calibrated combination of traditional and avant-garde ideas, they embody a specifically Swiss position within classical modernism.

122

Ernst Mumenthaler and Otto Meier, 3m furniture, 1927/28–1935. The add-on cabinet furniture brought the two Basel architects first prize in the competition held for the exhibition *Das Neue Heim II* (The New Home II). The picture shows cabinetmaker Elisabeth Bernasconi (who later married Otto Meier) before a Drei-Kantstabil Type 1 cupboard, an advanced version in lightweight construction from ca. 1935.

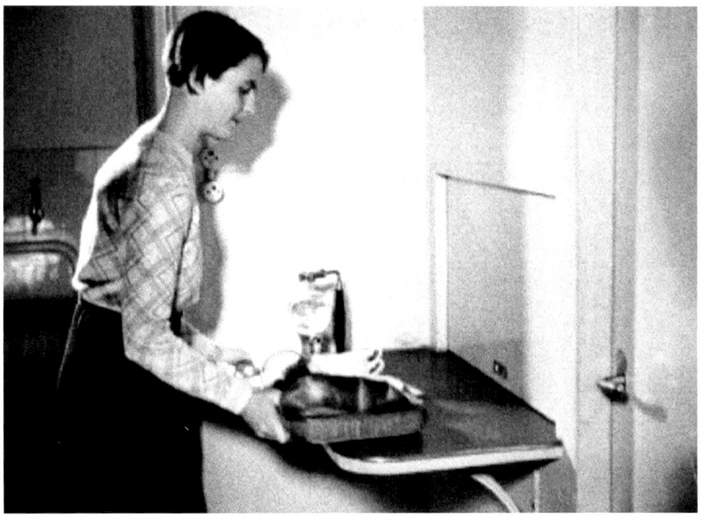

123

Hans Richter, Werkbund film *Die neue Wohnung* (New Living), 1930. The SWB planned to use the film to promote modern home furnishings at the Woba, the first Swiss housing exhibition, in Basel. Richter shot various scenes in Haefeli's model homes on Wasserwerkstrasse. In the kitchen of House No. 29, the perfectly organized kitchen workflow was demonstrated.

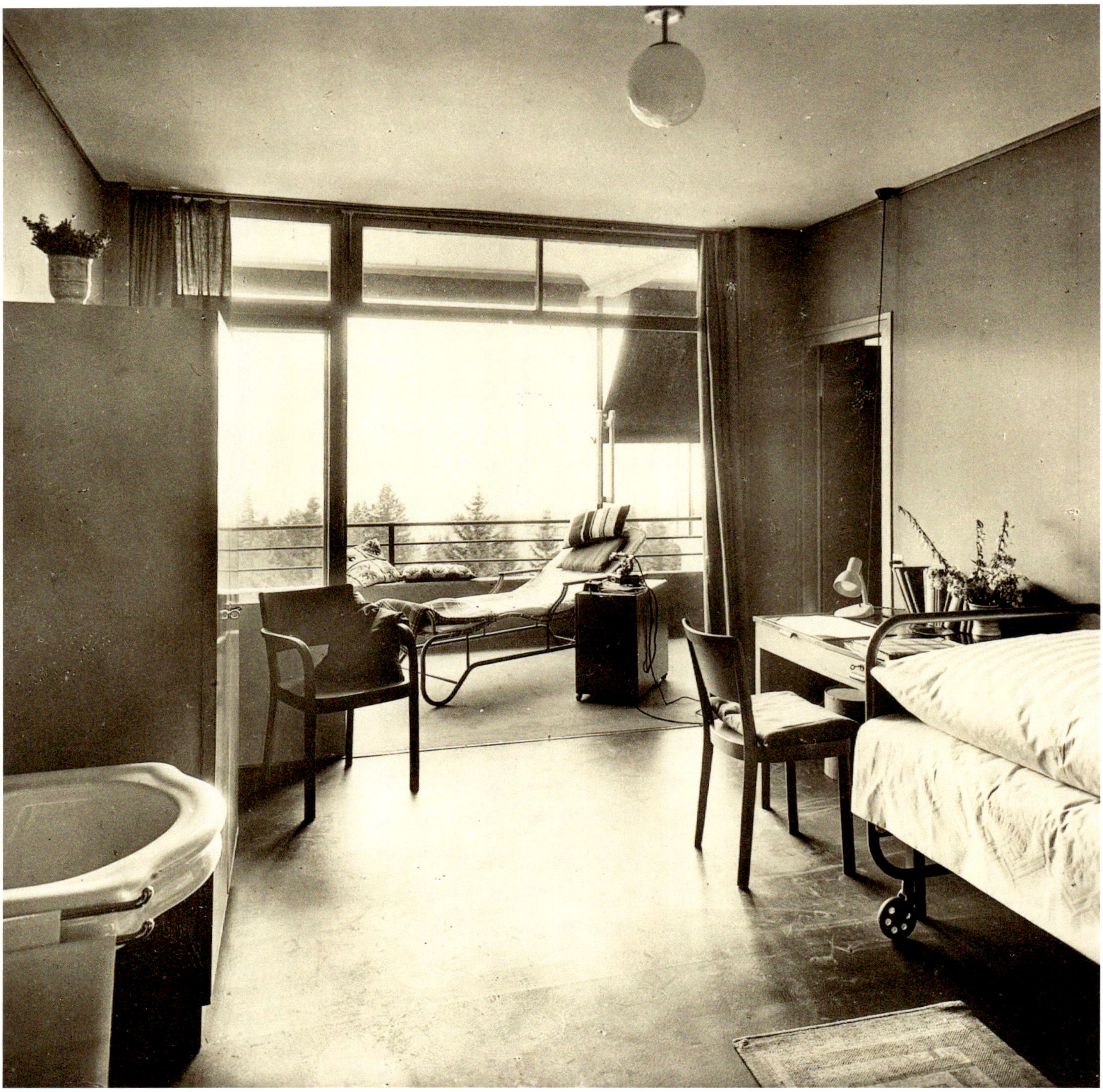

124

Rudolf Steiger and Flora Steiger-Crawford, Arnold Itten, Bella Lui Sanatorium, Crans-Montana, 1928–30. Patient's room furnished by Flora Steiger-Crawford, chairs by Max Ernst Haefeli. Particularly captivating among the furniture pieces designed especially for Montana, executed in gray-green and black painted wood, is the table that functions as vanity, reading table, or desk and is equipped with a flip-up mirror, a drawer, and a precisely fitted rolltop—the very prototype of a multifunctional accessory for temporary living spaces.

125

The Jaray Ley Car (left) and the Jaray Chrysler Car at Lake Lucerne, ca. 1928. The Austrian engineer and aerodynamics expert Paul Jaray, who moved to Switzerland in 1923, is one of the fathers of the streamlined form. His aerodynamically optimized car bodies anticipated the shapes of today's cars while integrating the complete mechanical systems of the conventional automobiles of their day.

126

Vegetable repacking area at Maggi & Co., Kemptthal, ca. 1905. Industrially processed nature: starting in 1885, Julius Maggi developed the first instant soups with financial support from the Schweizerische Gemeinnützige Gesellschaft (Swiss Charitable Society). “Good, thrifty cooking” was the company slogan in the 1920s. The logo was omnipresent in the production plant.

127

Patent Ochsner garbage collection system, ca. 1926. The counterpart to industrial food production. Introduced in Zurich in 1908, the system consisted, on the one hand, of a garbage wagon with a wooden collecting box, a drop bottom or fold-down tailgate, and a patented cambered top with sheet metal lids arranged along rails, and, on the other, the matching garbage cans known as "Ochsner buckets." Starting in 1926, use of the hot-dip galvanized Ochsner II garbage can with a round lid was mandatory in Zurich.

128 Gebrüder Vadàsz, Mikiphone Portable Record
129 Player, patented in 1924, produced 1925–27
130 Nickel-plated brass, felt, bakelite, paper,
4.5 × 11.5 cm (can), Paillard S.A., Sainte-Croix
Museum für Gestaltung

131 Detail of the operating instructions

128

129

130

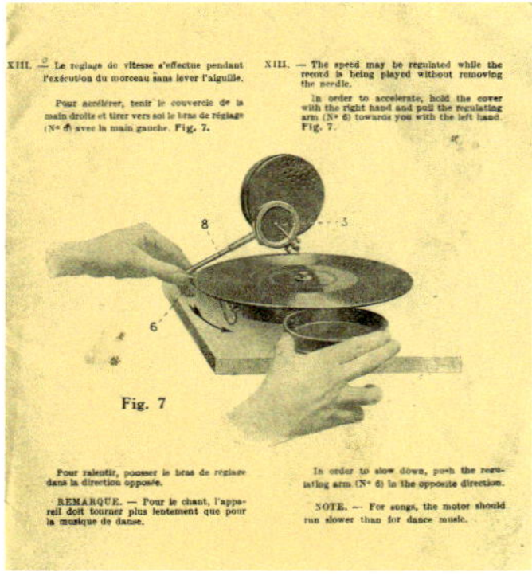

XIII. — Le reglage de vitesse s'effectue pendant l'exécution du morceau sans lever l'aiguille.

Pour accélérer, tenir le couvercle de la main droite et tirer vers soi le bras de réglage (N° 6) avec la main gauche. Fig. 7.

XIII. — The speed may be regulated while the record is being played without removing the needle.

In order to accelerate, hold the cover with the right hand and pull the regulating arm (N° 6) towards you with the left hand. Fig. 7.

8

5

6

Fig. 7

Pour ralentir, pousser le bras de réglage dans la direction opposée.

REMARQUE. — Pour le chant, l'appareil doit tourner plus lentement que pour la musique de danse.

In order to slow down, push the regulating arm (N° 6) in the opposite direction.

NOTE. — For songs, the motor should run slower than for dance music.

131

After the revolutionary invention of the gramophone and the records to go with it, manufacturers focused on the miniaturization of music players. By the mid-twenties, there was already a portable gramophone so small that it could be taken along on a picnic. Still not satisfied, the successful precision industry in the Jura Mountains, in the canton of Vaud—which exported fully assembled gramophones and components in series of hundreds of thousands—strove to develop devices that could be transported in a briefcase or even a trouser pocket. Thorens in Sainte-Croix produced a case with a crackle-painted finish, measuring 5 by 26.6 by 11.5 centimeters, which held the tone arm, pickup cartridge, speakers, and crank. The Paillard company, located in the same town and run by relatives, came out with a rival to this miniature gramophone in the form of the much more compact *Mikiphone,* which was housed in a hand-size nickel-plated box. The "first pocket gramophone in the world," fabricated in a series of 180,000 after a patent belonging to the Hungarian Vadàsz brothers, a cumbersome assembly before use. The recording head and a two-piece bakelite resonator had to be connected to the foldout tone arm before the shellac disc could be placed on the turntable's central pin. This precision engineering feat was awarded first prize at an international music exhibition in Geneva in 1927. Le Corbusier went so far as to tout the *Mikiphone* in 1926 as a "prime demonstration of the spirit of a machine age."
AR

Schilder Bär 1995; Christian R. Fonck, Portable Gramophones, Geneva.

Armchair
1924–25

132 Henri-Robert Von der Mühll, Armchair, 1924–25
133 Solid natural walnut, 80 × 60 × 70 cm, Ébeniste Frédéric Ballenegger, Lausanne
Museum für Gestaltung

134 Reception room of Henri-Robert Von der Mühll's architecture studio, Lausanne, photo ca. 1925

132

133

134

The two armchairs constructed of walnut boards once stood in the reception area of the Lausanne architecture firm of Henri-Robert Von der Mühll (1898–1980), together with an oblong coffee table, a plan cabinet, and a geometrically patterned carpet of woven coconut fiber. The severe elegance of the seating arrangement could hold its own with those in the poshest Parisian studios. Von der Mühll did not take his cue from French Art Deco, however, despite the fact that the style was de rigueur in French-speaking Switzerland in his day. The young architect, a graduate of the ETH Zürich, had instead developed his own personal cosmopolitan approach—first in Berlin (with Bruno Paul), and then in Paris, Brussels, and Vienna (where he had contact with Frederick Kiesler)—on which he was able to build further in 1928 as cofounder of the Congrès internationaux d'architecture moderne (CIAM). For his young architecture office's "calling card" he relied in 1924 on a stylistic distillation of all his contradictory experiences abroad—a distillation that is likely to have embodied for him the very essence of modernism. The design of his ensemble is reminiscent of the elementary furniture produced by the Dutch De Stijl group, but he chose to renounce the vibrant colorfulness they introduced the previous year. Although the simplicity of the concept allows for a few small extravagances, such as extending the armrests beyond the straight back, the focus is on the commanding presence of the fine wood and its precision craftsmanship. AR

Von der Mühll 1973; Baudin 2002.

Elektron Chair
1927

135 Max Ernst Haefeli, Elektron Chair, 1927
Frame made of the aluminum alloy Elektron, stained plywood seat and back, 80 × 38 × 43 cm, AG Möbelfabrik Horgen-Glarus; Giesserei Oederlin & Co. AG, Baden; Suter-Strehler Söhne & Co. AG, Zurich
Museum für Gestaltung/
Donation of Max Ernst Haefeli

136 Detail of Elektron Chair. The sculpturally formed cast Elektron elements are screwed together.

137 Kollektivgruppe SWB, Living and dining room in the apartment block designed by Mies van der Rohe for the exhibition *Die Wohnung* (The Dwelling), Stuttgart-Weissenhof, 1927

135

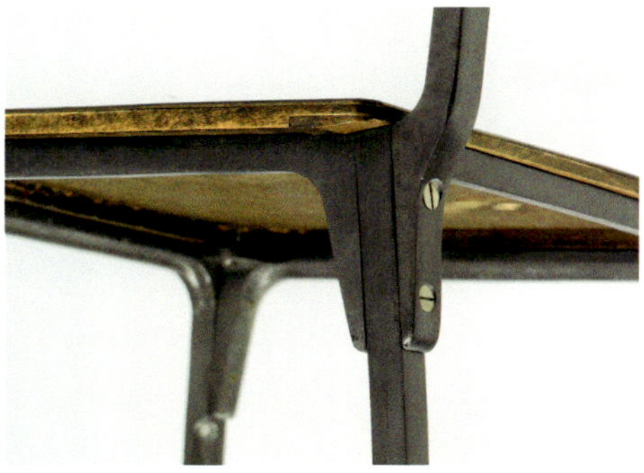

136

137

The development of modern Swiss "type furniture" began in the pivotal year of 1926. And as early as 1927 the Kunstgewerbemuseum Zürich was already presenting the first industrial designs by the architect Max Ernst Haefeli (1901–1976) in the exhibition *Form ohne Ornament* (Form without Ornament). That same year, the SWB commissioned designs from a dozen promising young architects with an international standing for the, later legendary, exhibition *Die Wohnung* (The Dwelling), organized by the German Werkbund in Stuttgart-Weissenhof. The big hit in the six collectively planned apartments was Haefeli's metal type furniture. Its elaborate production process called for a combination of cast parts and plywood elements. Haefeli applied the cast-iron technique commonly used at the time for city and hospital furniture but used instead an aluminum alloy called "Elektron." In addition to chair frames, the promising silver metal was also used for cast table legs, arms for wall lamps, and even supports for a grand piano. Although the German company Esch and Anke expressed interest, Haefeli's Elektron line never went into series production—in contrast to the cantilever chairs that were likewise introduced for the first time in Stuttgart by Mart Stam and Ludwig Mies van der Rohe. In Switzerland, Marcel Breuer developed a successful line of lightweight metal furniture in 1933 made of slotted aluminum profiles with cast aluminum connecting elements. AR

Mehlau-Wiebking et al. 1989, pp. 56–57; Tropeano 2002a.

Sliding Pendant Lamp
1927

138 Max Ernst Haefeli, Sliding Pendant Lamp, 1927
139 Nickel-plated brass, opal glass, diameter of sphere: 17 cm, shade: 29.6 cm, Jenaer Glaswerke Schott & Gen., Jena (DE); Paul Eberth & Co., Zurich
Museum für Gestaltung/
Donation of Arthur Rüegg

140 Max Ernst Haefeli, Swivel wall lamp, 1927. First shown at the exhibition *Form ohne Ornament,* Kunstgewerbemuseum Zürich, 1927

138

139

140

This lamp by Max Ernst Haefeli, which can be slid along a curved rail on casters, was a custom design made for the architect Karl Egender and his wife Trudy Egender-Wintsch, a painter. Until 1985, it hung over a dining table, likewise designed by Haefeli, in a stately period apartment in Meilen that the couple had filled with a wonderful mix of furnishings by designer friends as well as their own creations. The oldest pieces—a small Elektron table as well as the sliding ceiling lamp—probably date from Egender's membership in the Swiss collective that was established by the SWB for the Stuttgart Werkbund exhibition *Die Wohnung* (The Dwelling) in 1927. Haefeli first presented this lamp type with its opal glass sphere topped by an opal glass shade in February–March 1927, in the exhibition *Form ohne Ornament* (Form without Ornament) at the Kunstgewerbemuseum Zürich, accompanying his first "type furniture." Conceived for series production, the Series II lamp group included a pendant, a wall, and a floor lamp. The glass shades came from Schott in Jena, as did those for two further series, designed for the *Die Wohnung* exhibition and *Das Neue Heim II* (The New Home II) exhibition in Zurich in 1928. As is typical for Haefeli, despite some markedly technoid elements and the unadorned glass shades, neither the standard models nor the one-off edition announce their factory-made quality with a cheap-seeming "manifesto pomposity" (Peter Meyer). AR

"Das gute Industrieprodukt," special issue of *Werk* 14, no. 8 (1927), p. 237; Bürkle/Tropeano 1995.

Drei-Kantstabil Type 1 Cupboard
1928–35

141

142

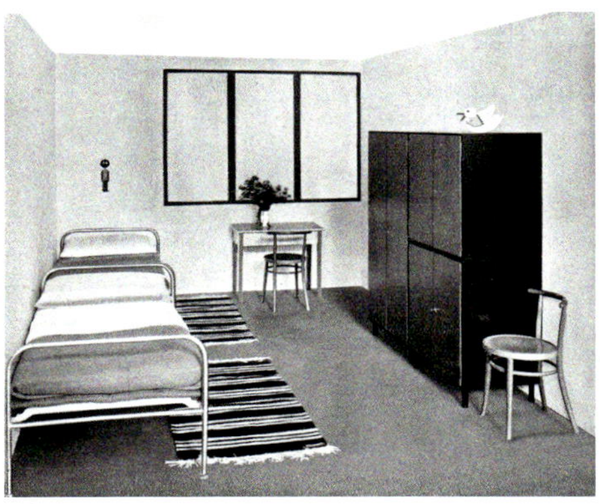
143

144

141 Ernst Mumenthaler and Otto Meier,
142 Drei-Kantstabil Type 1 Cupboard, 1928–35
Fir, lightweight plywood, chrome-plated tubular steel, wood tenons, 168 × 84 × 51 cm,
E. Mumenthaler & O. Meier (Ernst Mumenthaler, Otto Meier, Elisabeth Meier-Bernasconi), Basel, ca. 1935
Museum für Gestaltung

143 Interior with 3m furniture, *Typenmöbel* exhibition, Gewerbemuseum Basel, 1929

144 Ernst Mumenthaler, Poster for the exhibition *Typenmöbel*, Gewerbemuseum Basel, 1929

In 1928, the Kunstgewerbemuseum Zürich hosted a competition for simple contemporary furniture in the context of the exhibition *Das neue Heim II* (The New Home II). The young Basel architects Ernst Mumenthaler (1901–1978) and Otto Meier (1901–1982)—technically and artistically gifted graduates of the Gewerbeschule Basel—impressed the jury with the combinability of their standardized cupboards. The use of twelve-centimeter-high metal feet also made it easier to clean the floor. At first glance quite austere, these furnishings came to symbolize the Neues Bauen, obsessed as it was with functionality and hygiene. After 1933, lightweight, affordable variants of the plywood *3m* furniture (*3m* for "Mumenthaler, Meier, and *Möbel,*" or furniture) were created along with the closely related Wohnbedarf *Inkombi* cupboards developed by Max Ernst Haefeli. For their *Drei-Kantstabil* (Stable Triangular) version, Mumenthaler and Meier used a structural frame of triangular wooden boards to which thin plywood panels were attached. Even the shelves were made of lightweight plywood in order to reduce weight, material consumption, and price, while not however sacrificing stability. The height of the basic element shown here—the *Type 1* cupboard—was double its width. The combinable and stackable elements were also available in half-width and half-height. AR

Suter 1995; Rüegg/Mumenthaler-Grisard 2000.

J. Ochsner AG
Factory design

Ochsner II Garbage Can with Round Lid
Introduced 1926

145 Anonymous, Ochsner II Garbage Can with Round Lid, introduced 1926
Galvanized sheet metal, 48 × 42 × 46 cm,
J. Ochsner AG, Zurich
Museum für Gestaltung

146 The Ochsner bucket is hung into the sliding lid on the garbage wagon and emptied, London, 1930s, J. Ochsner AG, Zurich

147 Five Ochsner rear-loading garbage trucks before the triumphal arch on the Piazza della Vittoria, Genoa. Photo from the 1930s, J. Ochsner AG, Zurich

145

146

147

The Patent Ochsner is regarded as an early example of Swiss system design. Rather than just a single product, this was a concept for the systematic disposal of household waste not found anywhere else in Europe. Originally wainwrights in eastern Switzerland, the Ochsner company developed for the municipal administration in Zurich (incidentally one of the first cities on the continent with a waste incineration plant, since 1904) vehicles with custom containers for garbage collection. Jakob Ochsner jun. succeeded in efficiently coordinating the various workflows involved, and in the process improved the hygienic and ergonomic working conditions of the garbage collectors. In 1926, the Ochsner garbage can became mandatory, an absolute novelty in Europe. The characteristic perforated tab on the hinged lid of the hot-dip galvanized iron can served as a direct link to the public waste disposal system. The *Ochsner II* garbage can with round lid (which until 1972 came in three sizes: 25, 35, and 55 liters) would be lined with old newspapers and placed at the edge of the street, where the garbage collector could retrieve it and hang it directly on the sliding cover of the garbage truck, emptying it automatically using a lever. The patented system spread beyond Swiss cities and was followed by further inventions in the thirties, such as the rear loader with hydraulic compacting device, which brought the company international success.
RM

Lichtenstein 1987; www.ochsner.ch.

Paul Ami Bonifas

Terre lustrée noire Vessel
ca. 1930

148

149

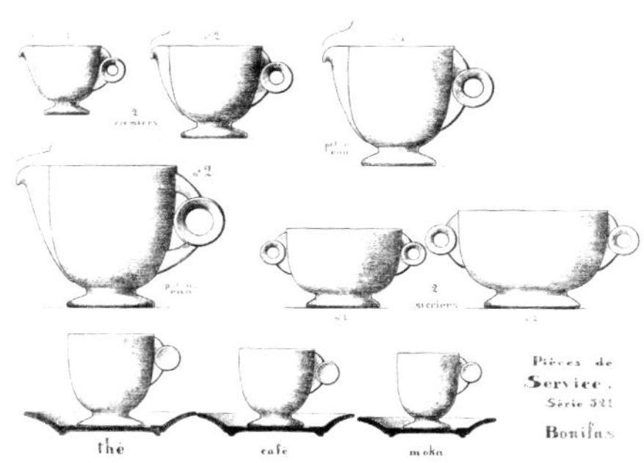

150

148 Paul Ami Bonifas, Terre lustrée noire Vessel, ca. 1930
Earthenware, black enamel glaze, 14 × 21.1 cm, Paul Ami Bonifas, Ferney-Voltaire (FR)
Collection of Arthur Rüegg

149 Paul Ami Bonifas, Terre lustrée noire Vessel, ca. 1930, Paul Ami Bonifas, Ferney-Voltaire (FR)
Museum für Gestaltung

150 Household ceramic goods, lithographic plate from a sales catalogue, ca. 1931–32

The Genevan ceramicist Paul Ami Bonifas (1893–1967) was actively involved in the western (French-speaking) association L'Œuvre before moving to Paris in 1921, where he was secretary of the editorial board of the avant-garde magazine *L'Esprit nouveau.* The radical ideas of the circle surrounding Le Corbusier and Amédée Ozenfant influenced and inspired the experiments with forms, techniques, and glazes he undertook between 1922 and 1930 in the French border town of Fernay-Voltaire. This gleaming black table vessel is exemplary of earthenware production at the time. Thrown on the wheel in a negative form, it attests to a desire for low-cost serial production. The enamel glaze, reminiscent of the lustrous surfaces of archaic bronze vessels, gave the high-end product series its name: *Terre lustrée noire.* The classically balanced vessel, inspired by Oriental models, is not a utilitarian ceramic object in the strict sense; its function consists wholly in the aesthetic pleasure conveyed by its sculptural presence. Bonifas also produced tableware, as evidenced by a few surviving sales catalogues (for example, the page shown, with the note "Pièces de Service/Série 521"). When in 1931 he embarked on a collaboration—albeit short-lived—with the Verrerie de Saint-Prex for the design of a new line of glassware, he became an early product designer in the modern sense of the term—long before moving to America in 1945 to work as a teacher and journalist. AR

Beaujon 1961; *Paul Bonifas* 1997; Mauderli 2002.

151 Anonymous, Suva Safety Goggles, 1929
Cellulose derivate, glass, textile band,
6 × 14 × 3 cm, Schweizerische
Unfallversicherungsanstalt (SUVA), Lucerne
Museum für Gestaltung

152 Construction diagram showing the individual components of the goggles, 1929 model
Museum für Gestaltung

153 Suva Safety Goggles, 1947. Also used as sunglasses.
Museum für Gestaltung

154 Suva Safety Goggles, 1947. The design permits the easy replacement of individual parts.
Museum für Gestaltung

151

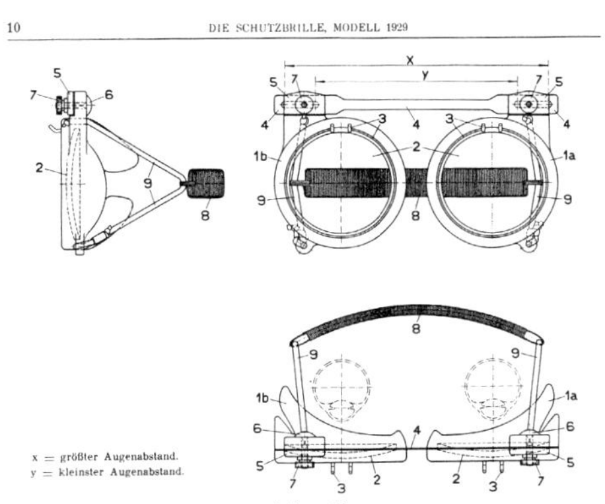

152

153

154

Only with the introduction of compulsory accident insurance in 1918 by the Schweizerische Unfallversicherungsanstalt (Swiss National Accident Insurance Fund), or Suva for short, was greater attention placed on the physical protection of workers. With one in seven industrial accidents in Switzerland affecting the eyes, Suva distributed safety goggles as one of its first campaigns. In 1928, it then developed various models (made by Swiss companies) geared to user needs and had them patented. Over three million pairs of goggles based on the same design principle were sold. All models can be flexibly adapted to different face shapes and are easy to use. A clip of fine steel wire encircles the round lenses and connects them via a resilient bridge. All components—lenses, headband, and lens frames that fit close to the skin and are made out of molded fiber material (a cellulose derivative)—can be replaced individually.

The 1929 model was available in a sturdy variation with an elastic strap or with eyeglass earpieces. This first model, as well as the other variants from 1932 and 1947, was worn not only as safety goggles but also as sunglasses by young men enamored of the machine aesthetic—an early example of the symbolic use of functional design. RM

Unbekannt – Vertraut 1987, pp. 110–11.

Therma AG
Factory design

Siedlungsherd
1930

155

156

157

155 Anonymous, Siedlungsherd
(Housing Development Stove), 1930
Enameled sheet iron, cast iron, plastic,
81 × 103 × 55 cm, Therma AG, Schwanden
Museum für Gestaltung/
Donation of Arthur Rüegg

156 Small kitchen in the Eglisee housing development,
Basel, 1930.
Ida stove and kitchen sink made of nickeline
(Designed by August Künzel)

157 Siedlungsherd handle with bakelite knob
(detail of fig. 155)

The rational organization of the modern household was a topic of burning interest not only for the Swiss Werkbund and CIAM circles but also for the general public. Housewives became the subject of intense study. The length and geometry of the "walking lines" in the kitchen were meticulously recorded and analyzed. The 1926 book *Der neue Haushalt* (The New Household) by the German expert Erna Meyer was also a bestseller in Switzerland. In 1930, discussions on the topic in Basel culminated in the exhibition *Die Küche* (The Kitchen) at the Gewerbemuseum and the Eglisee model housing development built for the Schweizerische Wohnungsausstellung (Woba, Swiss Housing Exhibition). While on the advertising pages of the museum's catalogue conventional narrow electric stoves were used to promote the advanced "electric kitchen," in the minimalist homes in the housing development a novel kind of stove that integrated several functions caused an international sensation. "Electric stove from Therma AG, Schwanden (Glarus). Oven adjacent to the burners, so that the top offers an extra work surface. The space under the burners offers a very convenient further storage shelf," wrote Hans Eckstein of Munich in his 1934 book *Die schöne Wohnung,* extolling the innovative appliance. A particularly nice touch is the oven vent with a round knob as handle. The *Siedlungsherd* set a new standard. In the Werkbund's Neubühl housing development in Zurich, it was mounted in 1931–32 on wall brackets for easier floor cleaning—thus losing its Biedermeier-like legs. AR

Die praktische Küche 1930; *Woba* 1930; Eckstein 1934.

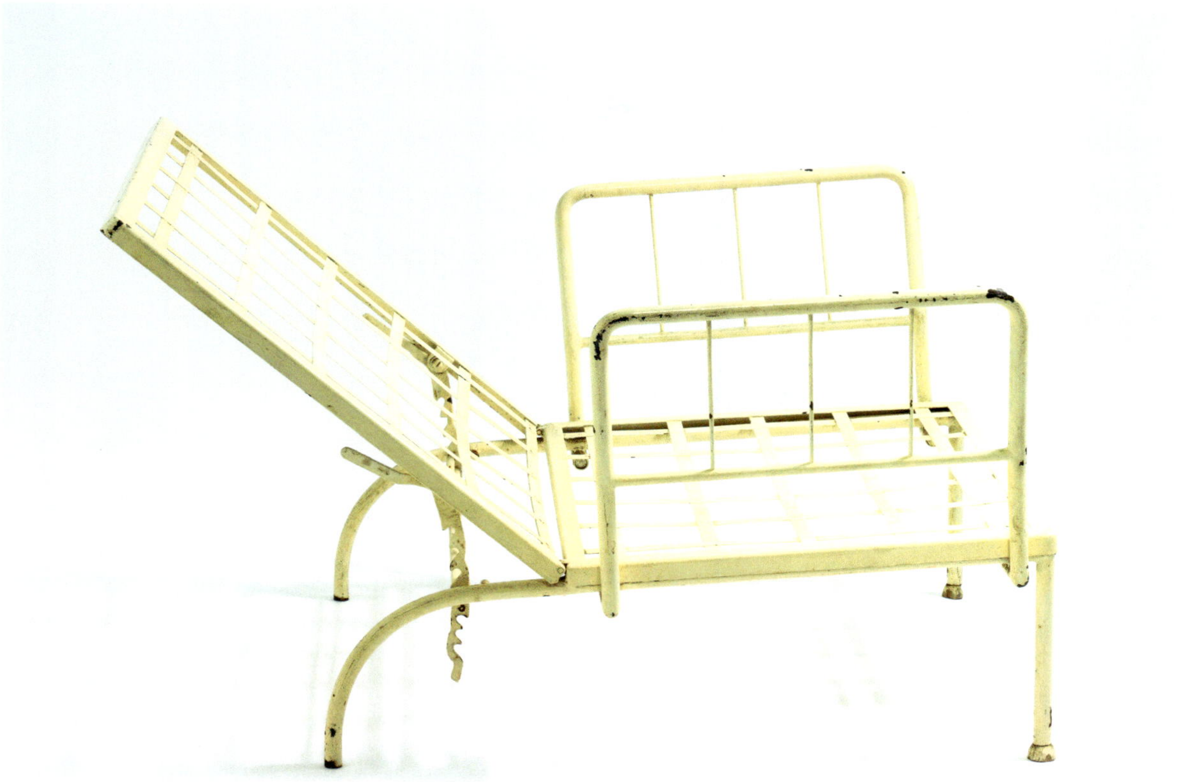

158

159

160

158 Max Ernst Haefeli, Garden Chair, 1928
159 Painted tubular and band steel, upholstered cushions with fabric cover (original), 68 × 89 × 95 cm, Embru-Werke AG, Rüti
Museum für Gestaltung/
Donation of Ruggero Tropeano

160 Max Ernst Haefeli relaxing in his Garden Chair talking with Alice Hugelshofer
Photo: Sigfried Giedion, 1928

1927–28 is considered the *annus mirabilis* of tubular steel furniture. Within the period of a single year all the new furniture types came out in mature form, from cantilever chairs by Mart Stam, Ludwig Mies van der Rohe, and Marcel Breuer to the distinctive pieces created by Le Corbusier, Pierre Jeanneret, and Charlotte Perriand. Max Ernst Haefeli (1901–1976)—at the time the motor behind Swiss furniture design—erected three model homes in 1928 in Zurich's Wasserwerkstrasse for the exhibition *Das neue Heim II* (The New Home II), impressively demonstrating the interplay between the latest architecture and the new furniture styles. He too was experimenting at the time with tubular steel, most enduringly perhaps in the design of a broad sofa that later went into series production (*Embru Model 20*). His unwieldy garden chair makes a more improvised impression: a white-painted assemblage of tubes of varying thickness, brackets, and iron bands, with an adjustable seat back and thick cushions covered in coarse fabric—undoubtedly a one-off piece. A contemporary photograph shows Haefeli, totally relaxed, nestled deep in the cushions of his garden chair and surrounded by broad-leaved plants in the garden room of the house at Wasserwerkstrasse 29. With this atmospheric photo, suggestive at once of coziness and openness, the art historian Sigfried Giedion pursued primarily a didactic aim: namely, a vivid depiction of what he advocated as "Befreites Wohnen," or liberated living. AR

Mehlau-Wiebking et al. 1989; Bürkle/Tropeano 1994.

161

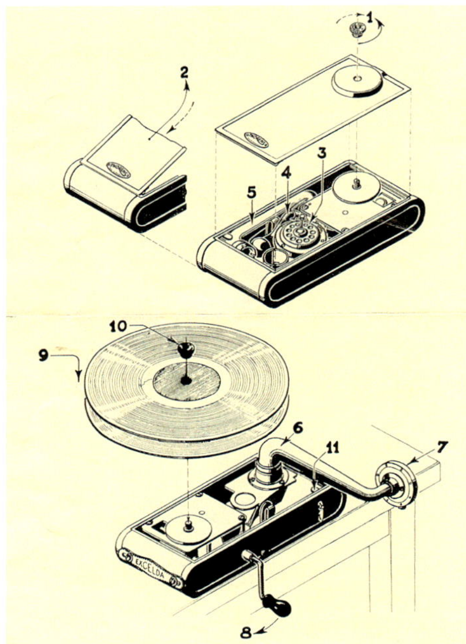

162

161 Instructions for using the Mikiphone portable record player (designed by Gebrüder Vadàsz), patented in 1924, produced 1925–27, diameter of can: 11.5 cm, Paillard & Cie., Sainte-Croix

162 Instructions for using the Thorens Excelda portable record player, n.d. (produced from 1935), Thorens S.A., Sainte-Croix. The case has room for the tone arm, pickup, speaker, and crank handle.

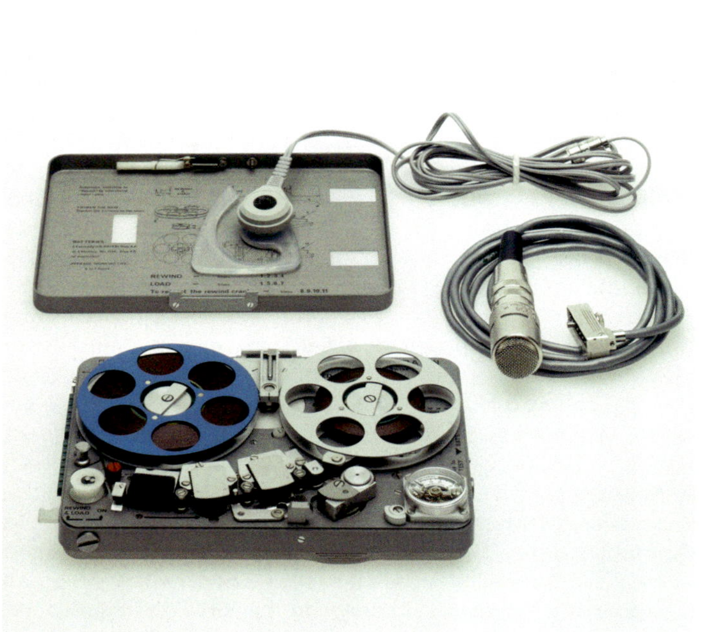

163

164

163 Stefan Kudelski, Nagra SN reel-to-reel tape recorder, 1960, 2.5 × 14.5 × 10 cm, Kudelski S.A., Cheseaux-sur-Lausanne
Museum für Gestaltung*

164 Factory design, Thorens Excelda portable record player, 1934, Thorens S.A., Sainte-Croix.
With crackle-painted case, 26.5 × 11.5 × 5 cm, in the colors red, green, blue, brown, or black.
Museum für Gestaltung

When speaking in terms of design, to call something *small* can mean a number of things. Hearing aids that disappear inside the inner ear, for example, or a nanocoating that adds an extra function to a given tool. In his famous adage, Charles Eames insists that these small details are not details at all—because they are what makes the design.[1] Small is what we would consider a coffee machine with a more compact brewing unit than that of its predecessors. This smallness is the outcome of a wave of miniaturization made possible through advances in technology, and it is welcomed by designers. Since the 1990s, this trend has become particularly apparent in electronic devices. In any case, when it comes to consumer goods, the human scale is always the measure of user-friendliness. A touchscreen that can no longer be operated by our fingertips may be small but it ceases to be a touchscreen.

Toward the end of the nineteenth and in the early years of the twentieth century, electric sewing machines, irons, and household appliances began to appear on the market. Not only industrial but private users as well made the switch from mechanical to electric and electromechanical devices, which since the fifties no average household can do without.[2] While in the early days their functions were still comprehensible to most users, today's devices, especially in the area of communication technology, lack this sort of transparency: their functions are understood solely by IT specialists. Users who attempt to open their mobile phones using a screwdriver are confronted with a virtually indecipherable circuit board.[3] The user interface alone reveals how the device is to be used. This development has led to a differentiation in the areas of responsibility of developers versus designers: whereas technological innovation is driven by tinkerers and engineers, the dematerialization of functions poses a special challenge to designers. Their role is to work out the usability and semantic expression of a device that to us is at first nothing but a black box.

Before electronics came onto the scene, however, small-scale precision engineering had its day in the sun. Fine mechanical devices made in Switzerland were, and still are, meticulously fabricated, inventively constructed, and luxuriously equipped. The watches and machines that have brought Swiss inventors and producers worldwide fame since the seventeenth century, later joined by audio and visual playback devices, perfectly exemplify this fascination with a type of design that works on a small scale.[4] Nowadays, even scientific measuring devices, precision scales, and hearing aids are part of this panoply of products. Their conception is motivated equally by a desire for miniaturization and a drive toward extended functionality and simplified usage.

The Watch as Paradigm

In terms of timepieces, mechanical detail and formal workmanship have always been two sides of the same coin. It was not without reason that watches were named "montres" in French: up until the early nineteenth century, the opulently decorated cases of pendant and pocket watches, which could often be opened up, made them into highly distinctive showpieces.[5] The lack of accuracy was compensated for by designing the watches as pieces of jewelry. However, their future lay not solely in their decorative form. Technological inventions in the history of watchmaking that solved the problem of keeping the watch running accurately made them what they are today: daily companion, dependable time-telling device, and status symbol. Their development began with various types of sun, sand, and water clocks and then moved toward the use of different mechanical movements—at first using weights, then tension springs, and finally electronic motors. The necessary inventions, which contributed to increasing miniaturization and ease of use, were obviously not only Swiss made. In 1675, the Dutch astronomer and physicist Christiaan Huygens patented the wearable balance spring watch, which capitalized on the finding that a balance spring, much like a pendulum, is a regulator that is generally not very sensitive to movement.[6] It winds and unwinds in a periodic rhythm. Attached to the balance, a small flywheel, the spring's winding and unwinding is translated into a swinging motion. This allowed the pendulum to be used as a pulse generator in the same way it was used in floor clocks, which in turn enabled the invention of small-scale clocks in general. Thanks to the invention of the balance, these now wearable clocks no longer required daily winding and only lost around ten minutes a day—a hitherto unknown level of precision that had been sought after by seamen, astronomers, and scientists alike. Their demands were most often met by British watchmakers.

Swiss timepieces were being purchased by royal courts as early as the sixteenth century. It was then also a queen who opened up the watch market for the company Patek Philippe, founded in Geneva in 1851 by a Polish exile and a French watchmaker. After Queen Victoria purchased three of the company's pocket watches, which she had caught sight of at the Great Exhibition of 1851 in London, orders from further European courts followed.[7] This was a boon for the partnership between watchmaker Jean Adrien Philippe (1815–1894) and merchant Antoine Norbert de Patek (1812–1877), which had been ongoing since 1845. However, this development also began with a technological innovation. In 1840, Philippe invented the "keyless winding system," a pocket watch with a winding wheel hidden in its stem.[8] Aside from this solution's mechanical elegance, it was also a way to reduce costs and increase production quantities. This development, as well as many other later inventions, heralded the watch's move away from its status as luxury item toward serving as a timepiece for the broader masses. The industrialization of production was facilitated not only by institutionalized training for watchmakers[9] but also through manufacturers' efforts to rationalize production—by simplifying the movements and standardizing the calibers. In this sense, Omega was a pioneer, introducing mechanical production methods in 1880, and, beginning in 1913, the company was also in a position to make its pocket watches in a uniform caliber or standardized size.[10]

In parallel with this development in production, the use of watches began to change. The fin de siècle witnessed a move away from the highly popular pocket watch to wristwatches, for practical reasons. This new fashion was started by upper-class ladies who took to wearing their pocket watches on a ribbon around their wrists when they went out. Soon, men followed suit and began to opt for the practical wristwatch on the battlefields of the Boer War and World War I, as well as in the increasingly popular automobiles and later in the cockpit. With the rise of the wristwatch, however, the possibility of decorating the obverse and the lid with precious artworks, as in pocket watches, was lost.

Nowadays, it is not a diamond-encrusted case alone that makes a watch a luxury product and sought-after value object. Prestige comes from accuracy, and especially from all sorts of "complications," additional functions such as chronographs, eternal calendars, multiple time zones, striking mechanisms, and power reserve displays. They drive up the price of a model, turning watches produced in very small series into luxury items. Models with complex mechanical movements, marketed as "grand complications" or ultra-complicated watches, are therefore extremely rare. Anyone wishing to own such a watch has to apply for the privilege, including a letter of motivation and proof of serious collecting activity, much like the procedure common on the art market. The number of these mechanical watches produced each year is accordingly small.[11] The current auction record in all categories was achieved fifteen years ago by the supercomplication *Henry Graves* pocket watch (1933) by Patek Philippe, which brought 11 million dollars at a Sotheby's auction in New York in December 1999.[12] Aside from telling time, it boasts no less than twenty-four complications.

These grand complications ushered in a new relationship between exterior and movement, between appearance and features. The designer's ambition is no longer restricted to the case, dial, hands, crown, glass, and band. Even the ticking clock movement with its mysteriously steady pulse can be aesthetically transformed. Such watches are produced by manufactories that fashion all the brass, nickel silver, steel, and gold components in-house, down to the tiniest screws. The parts are polished, sanded, fractured, and smoothed by hand—even when they measure no more than tenths of millimeters. Watchmakers have always prided themselves not only on the perfect construction of these parts but also their workmanlike production. Even today, the prime components in an expensive watch are decorated by hand, with circular graining and guilloche engraving.[13]

A watch's exquisite aura is today carefully orchestrated: the glass-backed watch, for example, showcases the movement's excellent craftsmanship.[14] Similar models began to appear toward

165

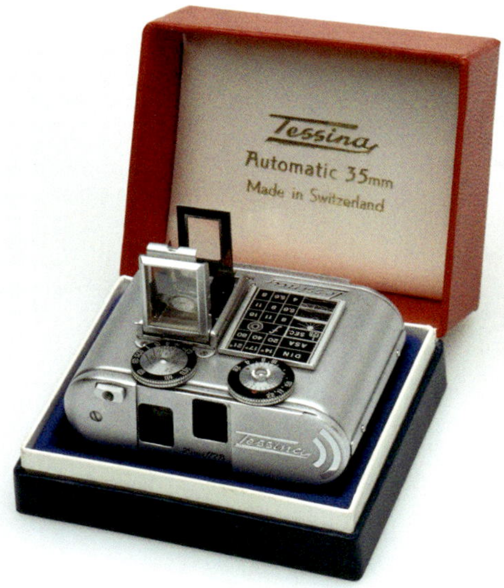

166

167

165 Prospectus for the Tessina 35mm subminiature viewfinder camera, n.d. (market launch in 1960). Attached to a wristband, the camera was also used by secret services.

166 Factory design, Tessina Automatic 35 mm subminiature viewfinder camera, 1954–75, Siegrist & Co., Grenchen; Concava S.A., Lugano. Developed with precision engineering expertise from the watchmaking industry, with pop-up viewfinder and decorative box.
Museum für Gestaltung

167 Noel Pemberton-Billing, Compass subminiature viewfinder camera, produced 1936–37, Manufacture Jaeger-Le Coultre S.A., Le Sentier. Equipped with all of the technical refinements of the day and with a form recalling a chronometer.
Museum für Gestaltung/Donation of M. Käser

168

169

170

168 Instructions for the M8 film projector, n.d. (produced 1950–60), Paillard-Bolex S.A., Sainte-Croix

169 Instructions for the Swissa Piccola typewriter, ca. 1955 (designed by Max Bill, 1944), Schreibmaschinenfabrik Aug. Birchmeier's Söhne, Murgenthal

170 Instructions for the Curta Type 1 calculating machine, n.d. (produced 1948–72), Contina Bureaux- & Rechenmaschinenfabrik AG, Mauren (LI)

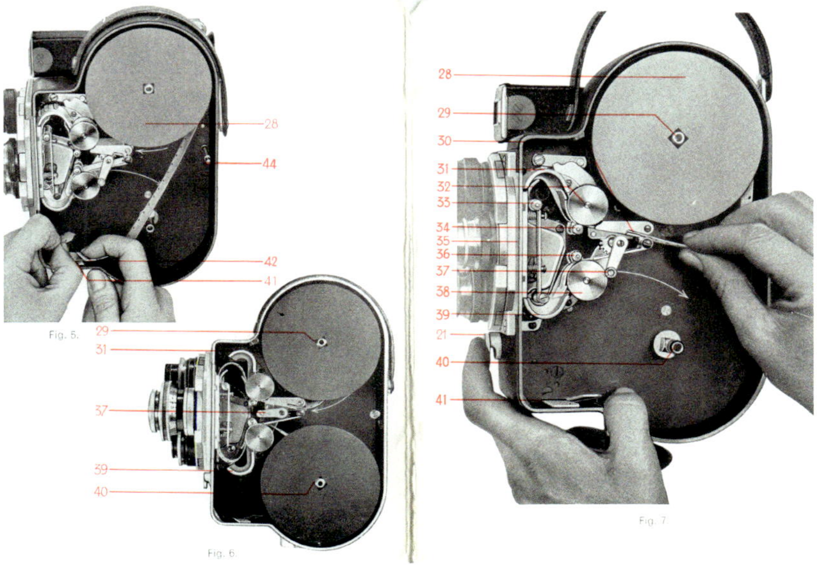

171

172

171 Double-page spread from the instructions for the Paillard H16 16mm film camera, n.d. (produced 1946–58), Paillard S.A., Sainte-Croix

172 Page from the instructions for the Bolex 155 Super 8 film camera, slide holder accessory for filming stills, n.d. (produced 1967–70), Paillard S.A., Sainte-Croix

the end of the eighties—at a time when the structural crisis of the Swiss watch industry, triggered by the development of quartz watches, was slowly being overcome.[15] At one end of the price spectrum, the path out of the crisis led to a renaissance of mechanical watches. Through the glass backing, the eye can appreciate the artfully interlocking gears, the rubies and carefully polished bridges. The time it takes to create such a complex caliber far exceeds what is customary in the cycles of industrial production. Swatch managed to perfect the latter in 1983 with its analog quartz watch—showing that there was an alternative way to escape the crisis at the other end of the price spectrum.[16] The Swatch's plastic body, which so cleverly incorporated all the various parts, was differentiated and distinguished through its formal design. The transparent model *Jelly Fish* (1985), which put the watch's mechanical movement on display, illustrates how the trend of making inner workings visible could, as a concept, be emulated in order to escape from the crisis in low-cost production.[17]

The serial rethinking of the way in which movement and exterior were integrated was not the only remarkable feature of the Swatch. It was also groundbreaking in creating the first uninterrupted chain of design. This chain covers everything from the idea of a watch that is inexpensive to manufacture to its technical execution, its production in large quantities, the models' graphic designs and collection themes, as well as the versatile ways in which sales are orchestrated and collecting is encouraged—further increasing demand. All elements coalesced perfectly. And graphic design played a central role. It made the watch into an accessory that could be adapted to the latest fashions thanks to its ever-changing form.[18] The Swatch received its highest honor yet when Max Bill declared shortly before his death that he had always wanted to make a standard watch like the Swatch.[19]

Watchmakers Who Make Other Things

The fine mechanics that had been perfected step by step in mechanical watches also benefited other industries. In Geneva and Vallée de Joux, for example, the watchmakers and jewelry creators turned their attention to the production of music boxes. Their golden age was during the years 1875 to 1896, when around thirty companies were active in the field, creating some highly popular exports among their music boxes. Boxes by Lassueur, Reuge, Thorens, Mermod, Paillard, and Vidoudez from Sainte-Croix, and by Cuendet from L'Auberson, can still be found on collectors' markets today.[20] Nowadays, Reuge, a family business from Sainte-Croix, leads the market in the production of high-end music boxes. The company has been manufacturing music boxes, singing bird boxes, and musical pocket watches since 1865. Musical mechanisms made by Reuge are used in a wide variety of objects—even in powder compacts and lighters, and often in jewelry boxes and snuff boxes.

The miniaturization of the musical mechanism has been an important factor in the production of music boxes. However, while a watch's defining feature was keeping accurate time, music boxes were intended to induce a sort of synesthesia. These ornately decorated boxes, inlaid with precious woods, set the scene for the melody that sounds upon opening. The regularity of this sound succeeds in transmuting our fascination with mechanics into something that we generally consider antithetical to technology and its rational nature—music. Herein lies the allure that attracts a market and many active collectors to this day.

Similarly to the development of clock movements, technological advances in music boxes did not transpire independently of the design of their outer form. Just as in the watch industry, those who failed to react to technological progress would be punished by history. When music boxes with exchangeable cylinders began to appear around 1890, sales at first increased, but a perforated plate invented and manufactured in Leipzig would shortly thereafter transform the market. Producers who did not realize its potential soon went bankrupt. Anyone able to perfect this new technology, however, was able to recapture their share of the market.[21] Much like what had occurred in the watch industry, the next new threat was posed by the rationalized production of music boxes, particularly in Japan, and soon in China. Due to these developments, by 1995 there were only three manufacturers left in the former center of the music box industry, the Jura town of Sainte-Croix. One of these was Reuge.[22]

Nowadays, the music box industry has developed into a specialist niche. On a larger scale, the purpose that music boxes once

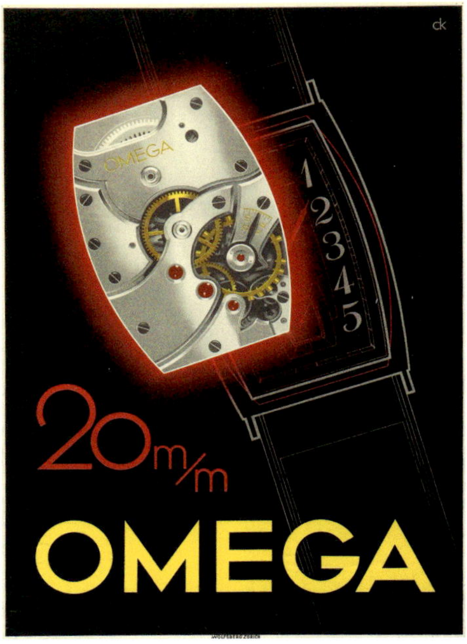

173

174

175

176

177

178

179

173 Charles Kuhn, Omega 20 m/m, 1936, Omega S.A., Biel. Poster

174 Factory design, Marine Chronometer watch, 1974, Omega S.A., Biel. The smallest marine chronometer watch ever produced, from the beginning of the age of the electronic watch. Museum für Gestaltung

175 Poster for Digi Quartz watch, ca. 1972, designed by Knaus & Knaus Volketswil for Mondaine Watch Ltd., Zurich. Erwin Bernheim developed and produced with his company Mondaine Watch Ltd. the first industrially made quartz watch with polarizing LCD display. The new technology (from Hoffman La Roche) enabled the watch to run much more precisely than its mechanical counterparts.

176 Factory design, Tissot Plastik, transparent plastic watch, 1971, Tissot S.A., Le Locle
Museum für Gestaltung/
Donation of Gottfried Kopf

177 Factory design, Jelly Fish, transparent watch, 1985, Swatch AG, Biel
Museum für Gestaltung

178 Patek Philippe & Co., Exploded assembly drawing of the Nautilus Travel Time Chronograph, ref. 5990/1A, 2014. Launched on the market in 1976, the watch gets its form from the portholes of old ocean liners.

179 Tissot 2270 Astrolon, transparent plastic movement, integrated into the Tissot Plastik watch, 1971.

served has long been superseded by the possibilities offered to us by audiovisual media. Smartphone playlists have now replaced record collections, which for their part once supplanted music boxes. Not only mobile devices can help us achieve the same synesthetic delight today; so can the music videos we play on them.

Invisible

Measuring time and listening to music: these two purposes do not cover by far all of the applications of precision engineering. The optics industry was also able to benefit from the insights gained by the watch industry. It was a crucial factor in setting off the development of other products, such as the compact *Compass Camera.* Designed by Noel Pemberton-Billing, it was produced by Jaeger-Le Coultre in 1936–37. From 1956 on, the *Tessina* subminiature camera was manufactured by the company Siegrist & Co., which was originally geared solely toward supplying the Swiss watch industry. The world's smallest 35mm camera was manufactured and sold until 1990.[23] Today, the company makes precision turned parts and gears. The *Tessina*'s second life as a spy camera prototype goes on—even though the cameras integrated into today's smartphones are able to provide high-definition digital images just as inconspicuously as a mini camera. The smallest Super 8 film camera ever created, by the company Bolex in Yverdon-les-Bains, suffered a similar fate. The *Bolex 233 Compact S* did not survive, despite its expert miniaturization—but once again, its fine mechanical capabilities, which made the camera a popular choice among amateurs, were not to blame. It was instead the media transformation that has replaced analog recording devices with their digital counterparts that doomed this perfectly designed camera to disappear from the market. Regretfully, it fell victim to what was admittedly a higher power.

The *Tessina* mini camera shares not only its dimensions with a hearing aid. As a spy camera, it also aims to be as invisible as possible. And the same is true for a hearing aid, which is worn inside the ear canal. However, an opposing design attitude could easily be imagined. Why hide the device, when it can simply be reinterpreted? This new approach takes the hearing aid out of the realm of prosthetics and puts it into the company of jewelry, allowing it to be appreciated as a fashionable accessory, just like eyeglasses. New ground has been broken through the ambitious design of such hearing aids. Milani Design, for example, convinced the manufacturer Phonak that the lack of acceptance for hearing aids in society is due to their unattractive design, which signalizes disability. With the prize-winning *Savia* (2005), *Smart* (2006), and *Audéo* (2007) models,[24] Milani developed hearing aids to be worn behind the ear, with color choices and soft forms reminiscent of a piece of jewelry.

Watches, music boxes, cameras, and hearing aids present different challenges to precision-engineered construction and its relation to outer form. Invariably, the main issue is how to integrate technology and design in a way that ensures devices are up to speed with the technological state of the art as well as being fit for purposeful and high-demand use. Design has taken on a strategic role in the development of many of these small devices: it has given us devices that today we could hardly live without. It is for this reason that what were once considered prostheses—if they are conceptually positioned and designed correctly—can now become part of an individual's personality. They are an accessory, a fashionable accoutrement that not only serves a useful function but can also play a central role in helping us to express and continually redefine our sense of identity.

1 "In the narration he wrote and recorded himself for a film he and Ray made to explain a storage system they designed, Charles Eames says, 'The details are not the details. They make the product. The connections, the connections, the connections.'" Caplan 1976, p. 15.
2 On the role played by electrification in the Swiss household, see Joris 1990, pp. 99–116.
3 Furthermore, manufacturers discourage this kind of tampering by voiding the guarantee should an unauthorized person open the device, or by using special screws in the housing that can only be opened with expensive screwdrivers. See "Your Hardware Warranty," Apple, http://www.apple.com/legal/warranty/products/iphone-english.html (accessed April 4, 2014); also see Kremp 2011.
4 See the summary in Schilder Bär 1995.
5 The Calvinist ban on jewelry forced goldsmiths in Geneva and non-native Huguenots to transfer their skills to the design of watches. Donzé 2009, pp. 13–14.
6 Trueb 1999, p. 25. "With this invention miniaturisation begins which can be compared with today's developments in electronics." Delfs 2007, p. 10.
7 Trueb 1999, pp. 335–36.
8 Ibid., p. 65.
9 In 1824, the Comité de la Société pour l'avancement des arts opened the first watchmaking school in Geneva. In 1865, a school in La Chaux-de-Fonds followed. Schilder Bär/Wild (2001) 2003, p. 149.
10 "The individuality of the watches lay solely in the size and design of their cases." Schilder Bär/Wild (2001) 2003, p. 150.
11 Patek Philippe produces around 20,000 watches per year. Trueb 1999, p. 336.
12 For the auction record, see http://www.sothebys.com/de/departments/watches.html (accessed April 11, 2014).
13 Trueb 1999, p. 337.
14 There were in fact already pocket watches that were transparent on both sides, in order to showcase their elaborately engraved or skeletonized movements. This includes the *Marie Antoinette* by Abraham-Louis Breguet, which he originally created for the queen who would later be beheaded. The watch had a back and dial made of rock crystal. Suggestion made by Timm Delfs, Basel, www.breguet.com (accessed April 11, 2014).
15 "The introduction of electronics to the watch industry caused a structural crisis with grave consequences for the Jura region. The number of employees shrank from around 70,000 in the mid-sixties to just 30,000–35,000 in the eighties. Although the quartz watch was well-known in Switzerland, nobody here believed that it would be successful. The opposite was true in Japan, which soon overtook Switzerland in the production of electronic watches. This forced the Swiss watch industry to turn to microelectronics. The first prototypes for a quartz wristwatch came from the research labs of the Centre électronique horloger SA in Neuchâtel in 1968. From 1979 to 1980, the production of mechanical and electronic watches was about equal." Fallet/Veyrassat 2014.
16 On the causes of the crisis in the Swiss watch industry from 1975 until the end of the eighties, see Donzé 2009, p. 159, as well as Trueb 2013.
17 Suggestion by Timm Delfs, Basel. With its *Sistem 51* collection, launched in autumn 2013, Swatch brought a self-winding mechanical movement to the market consisting of only fifty-one parts in five modules. With the slogan "The Front Tells the Time, the Back the Story," the manufacturer got to the heart of what watches are all about.
18 "For a surprisingly long time, the Swatch's graphic design guaranteed its financial success. Jean Robert and Käti Robert-Durrer developed the collection from 1983 to 1989, after which the creative department—which had now moved to Milan—was led by Alessandro Mendini. Only after the first hundred million watches had been produced was the body of the watch altered, in 1992. Today, the Swatch has become trivialized within the narrowing market for lifestyle products. As one of many, it remains what it always was: a cheap watch." Ernst 2006, pp. 340–41.
19 Bill/Rothenbühler 1994, pp. 34–36.
20 Information sheet on the special exhibition *The Art of Sound—Early Days, Golden Age and Decline of Swiss Music Boxes,* Museum of Music Automatons, Seewen, Solothurn, 2009.
21 Troquet 2009.
22 At the end of the twentieth century, Japan produced 55 to 60 million music box mechanisms each year, while China produced 30 million and the businesses in Sainte-Croix and Yverdon that specialized in making large, luxury music boxes produced around 3 million. Troquet 2009.
23 Data from W. Siegrist AG, http://www.siegrist-ag.ch/english/company.htm (accessed December 4, 2013).
24 Scharf 2007.

180

181

182

180 Hannes Meyer, *Co-op* display case, 1924–25. Second display case, cooperative house in the Freidorf development in Muttenz. Standard items in a rigid arrangement

181 Hannes Meyer's and Hans Wittwer's "laboratory," Luftgässlein 3, Basel. This is where the text "Die neue Welt" ("The New World") was written.

182 Hannes Meyer, *Co-op. Interieur,* 1926. Installed in the Luftgässlein studio and commented on in the text "Die neue Welt."

Whether or not it would be pleasant to actually *live* in the *Co-op. Interieur* pictured on page 219 of the June 1926 issue of *Das Werk – Schweizer Monatsschrift für Architektur, Kunstgewerbe und Freie Kunst* was almost certainly not a question the thirty-seven-year-old Hannes Meyer (1889–1954) asked himself as he went about "procuring" the "collection of material" for that issue.[1] Judging by the tacks holding up the plain paper covering the walls, we can assume that the items on display were hastily set up for a programmatic photograph in the Basel office that Meyer and his architecture partner Hans Wittwer (1894–1952) had just moved into.

This photo, together with thirty-three further images, illustrates an essay by the editor titled "Die neue Welt" ("The New World"), which takes up over half the issue. The interior, captioned "The home," was joined by photos of a "Diesel engine factory," "Housing for munitions workers," a "Model design from the 1925 Hamburg export fair," high-voltage pylons in "the St. Jakob battlefield near Basel in 1926," and "Von Meyenburg's rotary tiller"—all meant to illustrate a central tenet of Meyer's text (and the world he lived in). This was the concept of the "standard" whose dimensions would provide a social setting for the orphaned individual: "The truest mark of community is the gratification of such needs by equivalent means. The result of such collective demands is the *standard product.* Typical standard wares of international origin and uniformity are: the folding chair, the rolltop desk, the light bulb, the bathtub, the portable gramophone. They are the instruments of mechanization in our daily life. Their standardized form is impersonal. Their manufacture proceeds serially. As serial item, serial equipment, serial component, serial house."

Contrary to the initial impression of a strict systematic way of thinking made by the language Meyer uses in his essay, the pictures he chooses do not really illustrate his concepts—of the five examples of "standard products" cited in the text, only the folding chair (two of them) and gramophone are shown in the sample interior. Meyer was particularly fond of folding chairs: after being appointed head of the architecture school at the Bauhaus the following year, he commissioned the development of a basic folding chair model. The fact that the funnel gramophone has little to do with the latest technical developments available in 1926 and is certainly not "portable" must be taken as a concession to the metaphorical impact of the staging. Everything else in the scene, especially the folding garden table on which the gramophone stands, the metal shelf with fourteen glasses holding implements found in an architecture studio, and the metal mattress frame with its peculiar cone-shaped feet, is meant to suggest mechanization and serial production: the essence of "standard products." This fits in well, as is already indicated by the abbreviation "Co-op" in the image's title, with the contemporary project of social housing, in which Meyer had already been involved between 1919 and 1921 while working on the Freidorf development in Muttenz (in the canton of Basel-Landschaft). On the other hand, the scenario fails to counter the impression of coldness and "impersonality," as "Architecture is not an individual act performed by an artist-architect and charged with his emotions."[2]

Hannes Meyer's essay and his illustrations from the mid-twenties are part of a culture of spirited manifestos of various tenors that were only rarely followed by the concrete realization of the demands framed. In the wake of World War I, the faith in progress and rapid change that had burgeoned in the course of the nineteenth century was accompanied by a sense of dramatic loss. All of the values and principles that had once reigned in the traditional world as the underpinnings for collective and individual existence seemed to have evaporated, leaving people longing for new foundations on which to build their lives. Discourses promising guaranteed progress and revolution—usually in socialist form—fell on open ears, as did the conservative, and also soon enough fascist, call for the creation of "new myths." Martin Heidegger's work *Sein und Zeit* (*Being and Time*), written in manuscript form in 1926, converges with this second context in its effort to restore to human existence one of its primordial dimensions through a renewed concentration on "being."

Especially in German-speaking Switzerland, intellectuals and artists engaged with and actively responded to this historical dynamic. Literally every page of the June 1926 issue of *Das Werk* betrays traces of the experience of a world that is undergoing rapid change, and at the same time the symptoms of an obsession with the "contemporary." At the end of the issue there is an allusion to "a few contemporary gramophone records" (mainly fox trots and tangos), as well as lists of "a few contemporary books" (by Le Corbusier, Lissitzky, Schwitters, and Tzara, among others), and "a few contemporary magazines," covering a truly intercontinental panorama (along with *La Révolution surréaliste* and *Il Futurismo,* they also include *Ariel* from Santiago de Chile, *Klaxon* from São Paulo, and *The Little Review* from New York). This breadth of interest corresponds with the motif unfurled in Hannes Meyer's manifesto on "The New World" of an ongoing conquering of space made possible by cars ("Ford" and "Rolls-Royce") and airplanes ("Fokker" and "Farman"), which will ostensibly ultimately lead to the "simultaneity of events." The text had already started out along these lines: "The North Pole voyage of the *Norge,* the Zeiss Planetarium in Jena, and [Anton] Flettner's propeller ship are the most recent heralds of the step-by-step mechanization of our planet."

Yet, while such projections on the geographic level stage progress and its phenomena as a stable order of things that recalls the static juxtaposition of furniture in the photograph of the *Co-op. Interieur,* Meyer's description of the individual and above all the social life of his day is pervaded by the fervid rhythm of an obsession with the future—evidently the same hectic momentum behind the signs of careless haste in the photo of the *Co-op. Interieur.* Meyer sees the "straight lines of mechanical and scientific provenance," the straight lines of the rationality leading into the future, as being threatened in his present day by the "tangled web of its social and economic force fields." This is why his restless counting-off of diverse signs of progress seems like a kind of flight before any reality that would interfere with that forward motion—an effect that thwarts all of his gestures of logic and objectivity, transforming them into an entropic muddle.

Meyer expresses hope that accelerated movement will lead to a new convergence among nations—and thus undermine the concept of the homeland: "We learn Esperanto. *We become citizens of the world.*" But he already fails to achieve conceptual transparency with his next attempt at outlining the symptoms of the present, the construction of this present in a form all its own "by modern means." This is because "the burden of our knowledge of the past weighs upon us, and our institutions of higher education betray the tragedy of obstacles strewn along our path." In other words, for Meyer, an awareness of the past stands in the way of a clear vision of progress. The formula "function times economy" (which he illustrates with compound terms such as "trade fair, grain silo, music hall, and airport") should therefore replace the preeminence of aesthetic experience and art as "composition." It remains unclear, however, for what reasons "the purposeful function of a building always contradicts artistic composition."

Mechanization, finally, and the standardization that follows from it lend themselves to association with the values of equality, community, and even a cosmopolitan outlook. But how then on the other hand can it be claimed that "psychoanalysis explodes the all-too-narrow edifice of the soul" and "graphology exposes the essence of individual being"? And that both are promoted, based on the theory of relativity, through the "outer masculinization of woman" and therefore the "inner equality of the sexes"? The line of reasoning becomes more and more obscure with each successive paragraph. In the end, the text breaks off all attempts to keep its explosive outpouring of observations, values, and catchwords under control or to superimpose some order and instead proceeds to distance itself from everything in the past through long sequences of negations. Meyer's rationality loses patience with the past and cuts itself off from it: "Yesterday is dead: dead, the bohemian; dead, the mood and the value, the glaze and the brushstrokes of the accidental. Dead, the novel: we lack belief and time to read. Dead, painting and sculpture as a likeness of the empirical world." There then follows for one last time Meyer's programmatic instructions for the future, which through constant repetition have become a kind of ritual chant: "Today is ousting yesterday in material, form, and tools. Instead of the random blow with an axe, we have the chain mortiser. Instead of the scumbled line of the charcoal pencil, we have the clean-cut line produced with the T-square. Instead of easel-work, we have the drafting machine."[3]

The more passionately Hannes Meyer invoked his "straight lines of scientific and mechanical provenance," the less they seemed

DIE INTERNATIONALE

J. J. P. OUD
»Café de Unie«, Rotterdam, 1925 / Halbdauernde Baukonstruktion

A. Melnikoff, Moskau
Sovjetpavillon der Pariser Kunstgewerbeausstellung 1925

LE CORBUSIER, PARIS
Haus Laroche in Auteuil

206

DAS HOCHHAUS

MIES VAN DER ROHE, BERLIN
Hochhaus aus Eisen und Glas, 1921

207

DAS MATERIAL

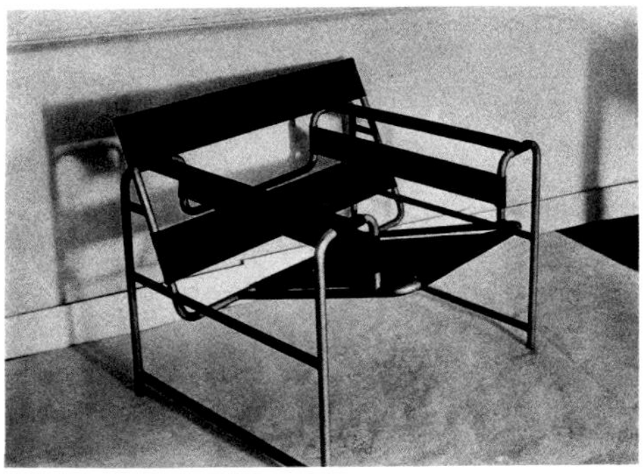

MARCEL BREUER, BAUHAUS DESSAU
Stahlstuhl mit Stoffgurten, 1925 / Phot. Lucia Moholy, Dessau

STADTBAURAT A. BERG, BRESLAU
Der Messehof zu Breslau, 1925

210

DIE BÜHNE

THEATER MEIERHOLD, MOSKAU
»Der Wald« von Ostrowsky / Aus der Zeitschrift »Das neue Russland« (Berlin-Pankow)

THEATER MEIERHOLD, MOSKAU
»Tarelkin's Tod« / Aus der Zeitschrift »Das neue Russland« (Berlin-Pankow)

211

183

183 Hannes Meyer, Double-page spreads from "Die neue Welt," *Das Werk* 13, no. 7 (1926)

DIE PHOTOGRAPHIE

FLIEGERBILD WATERLOO-STATION LONDON
The Central Aerophoto Co. Ltd.

212

DAS PHOTOGRAPHISCHE BILD

Positiv

Negativ

L. MOHOLY-NAGY, BAUHAUS DESSAU PHOTOGRAMM

213

DIE PROPAGANDA

HANNES MEYER, BASEL
Die Vitrine Co-op, 192[illegible] · Phot. Th. Hoffmann, Basel

M. BURCHARTZ & J. CANIS, BOCHUM
Zeitungsprospekt, 192[illegible]

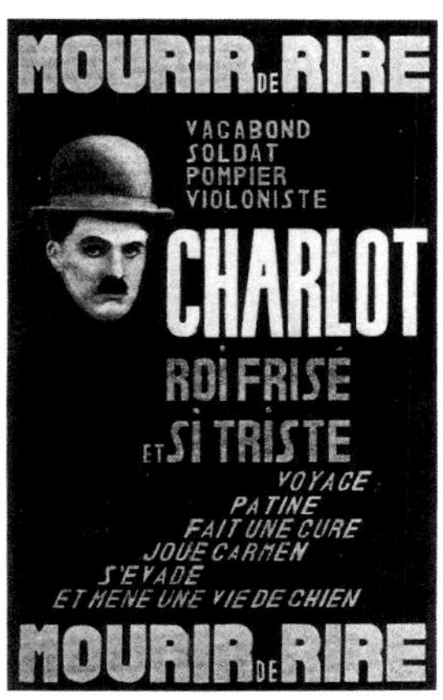

KINOPLAKAT

216

DER TANZ

GRET PALUCCA, DRESDEN
Sprungmoment · Phot. Frau Fiedler, Dresden

217

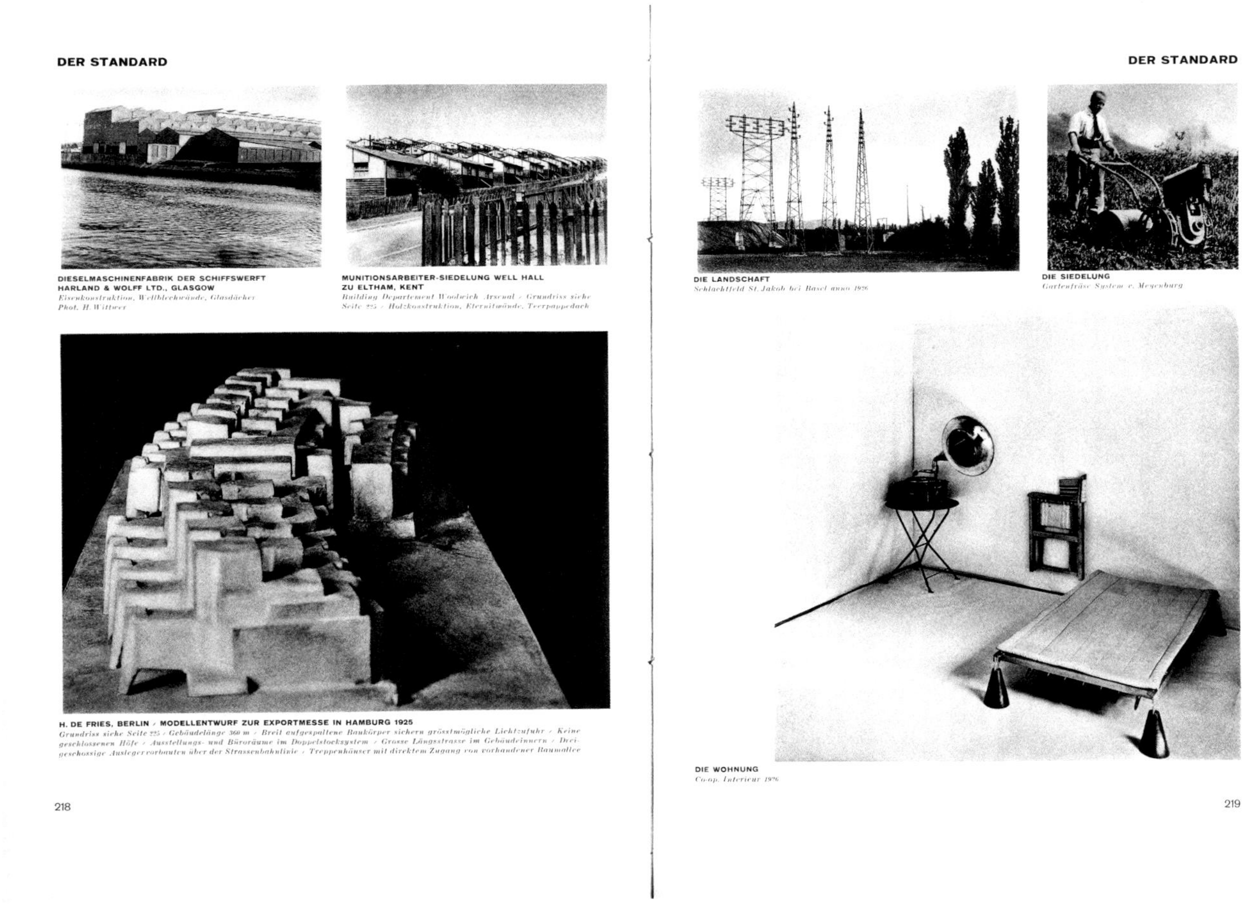

183

to figure in his future life. The architectural practice he established with Hans Wittwer was known chiefly for two projects whose claim to fame is their uncompromising innovation: the designs for the Petersschule in Basel and for the League of Nations building in Geneva. Neither was ever built. As architect for the Leipzig-Halle airport company, Hans Wittwer completed a restaurant building in 1931 that attracted a great deal of attention (later destroyed in World War II), but the construction of a new reinforced-concrete reception and administration wing had to be cancelled due to the world economic crisis of the early thirties. Meyer's career seemed at first to be under a lucky star. After Walter Gropius had summoned him to Dessau in 1927 as head of the new Bauhaus department of architecture, Meyer carried out a housing project for the city of Dessau, and built headquarters for the Allgemeiner Deutscher Gewerkschaftsbund (Confederation of German Trade Unions) in Bernau near Berlin, before replacing Gropius as director of the Bauhaus in 1928. He soon came into conflict with the municipal administration, however, and was fired in 1930, because—true to his political convictions—he was unwilling to take a stand against the growing agitation by the communist students' group.

Together with some students he emigrated in 1930 to the Soviet Union, where he hoped to discover the horizon of his own future. He worked there for several years as a lecturer, as a consultant for the planning of complex urban development projects, and even as an international representative of the new Soviet culture. In 1937, however, the year that saw the great Stalinist show trials, the political "purges," and the definitive departure from an avant-garde state aesthetic, Meyer was compelled to return to Switzerland. In 1939 he received an appointment as director of city planning at the Instituto Politécnico Nacional in Mexico City, and he was once again ready to believe in the future promise of a totalitarian state. Meyer was dismissed in 1941 but remained in Central America until 1949. He died in 1954 in Switzerland, where he had not received any further commissions.

In an era whose acceleration he had embraced with so much enthusiasm and tried so impatiently to promote, Hannes Meyer was destined to remain behind the times. His tragedy became not "the burden of the past" but rather, as he had written in 1926, the pull of the future. Today, in a new millennium, the traces left behind by his projects are no longer regarded as instructive for our own reality. They illustrate only the future that was once envisioned by a past that is slipping further and further away. They have become, of all things—in a constellation of historical irony—an object of the very aesthetic experience Meyer believed he had overcome through functionalism.

1 Meyer (1926) 1994, editorial note on p. 236; all citations from this text are from Meyer (1926) 1994, except where noted.
2 Meyer (1938) 1965, p. 55.
3 Meyer (1926) 1965, p. 95.

DIE ZEITSCHRIFT

EINIGE ZEITGEMÄSSE ZEITSCHRIFTEN

Phot. Th. Hoffmann, Basel

235

184

184 Hannes Meyer, "Die neue Welt," *Das Werk* 13, no. 7 (1926). Under the heading "The Magazine," page 235 features a typographic montage by Hannes Meyer of "a few contemporary magazines."

185

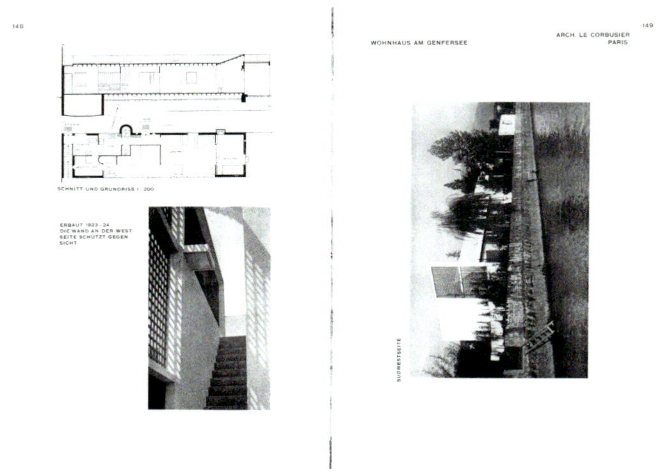

186

187

185 Peter Meyer, *Moderne Schweizer Wohnhäuser,* Girsberger, Zurich, 1928. Cover

186 Double-page spread from Peter Meyer's *Moderne Schweizer Wohnhäuser* showing the Villa Le Lac by Le Corbusier and Pierre Jeanneret, Corseaux, 1924. Example of a "living, organic relation of spaces."

187 Lux Guyer, Sunnebüel House, Küsnacht-Itschnach, 1929–30. View through the living room to the hallway coat closet. Example of modern domestic culture combined with traditional elements.

The title of this text refers to an alleged contradiction, and to attempts to overcome it. A contradiction, because the smoothly functioning but radically purged and "airy"[1] houses of the avant-garde met with the enthusiasm of only a few insiders in the 1920s, usually encountering instead baffled incomprehension and even being deemed inhumane and uninhabitable by a tradition-loyal public. Approaches geared toward bridging this gap could in this respect be found everywhere. Clearing away the old-fashioned bombast of historicism and instead making "daily convenience, keeping house with a minimum of time, money, and personnel, in short the internal organization of private life ... the main thing"[2]: such were the guidelines that were to point the way out of the predicament.

There is no doubt that (German-speaking) Switzerland offered fertile ground for attempts to mediate between avant-garde ideas and more conventional tastes. The Swiss tended to keep their distance out of principle "from the stylistic extravagances that had run wild since the war."[3] Besides, "the suddenly emerging, hothouse-cultivated nouveau riche, whose desire for splendor in disregard of tradition has bred in Germany a special variety of decorative architecture" was utterly lacking in Switzerland.[4] The architect and writer Peter Meyer (1894–1984), a militant advocate of a new domestic culture firmly rooted in tradition, framed a possible alternative as follows: "Impartiality toward all historical, existing forms is an essential feature, impartiality toward all new and perhaps merely faddish forms a necessary complement for true living architecture."[5] However, caution was advised: "Like every movement, the modern one, too, has its dangers: the danger of whimsical exaggeration, of a machine-like quality in the applied arts of fashionable modernity, of the vain avant-garde gesture."[6] Meyer's compendium *Moderne Schweizer Wohnhäuser,* that "gentle manifesto"[7] of a modernity held in strict check by Germanic deliberation, was published by Hans Girsberger in 1928, coincidentally the *annus mirabilis* that saw the advent of tubular steel furniture and the first Congrès internationaux d'architecture moderne (CIAM). We will return to Meyer's arguments later.

The Home as Housing

The interior furnishing concept was, proclaimed Le Corbusier, "the Gordian knot ... of the renewal of the plan of the modern house."[8] He was referring to the Gordian knot that always has to be severed when cherished domestic habits and the associated "innumerable varieties of superannuated furniture that were known by half-a-hundred different names"[9] stood in the way of his visionary architectural concept. In concrete terms, he aimed his thunderbolts at the historicist interior, in which an "impression of its occupant"[10] (Walter Benjamin) was generated by an accumulation of useless forms and things. Individualism in furnishing the home had in fact been elevated to the status of an aesthetic program in the nineteenth century, one that would culminate in the legendary ornamental orgies of the Viennese painter-prince Hans Makart. At the same time, however, a generally comprehensible, symbolically and sensually charged specification of the various domestic functions—the hall, the living room, the salon, the dining room, the breakfast room, the library, the study, and the billiard room—was achieved through nuanced distinctions in floor plans and styles, as well as different moods created through the use of color and materials.[11] The catchword here is differentiation.

Salvaging the specific characteristics of the rooms for contemporary domestic culture was the mission of some of the devotees of an "alternative" modernity. Exemplary here is the work of the Bernese architect Otto Rudolf Salvisberg (1882–1940). During his time in Berlin (1908–30), he achieved a contemporary formulation of characteristic room situations constituting the essence of upper-middle-class villas and country houses.[12] His floor plans seem to have been made up of individual modular elements that one could also imagine in other constellations. The prestigious character of sequential rooms was enhanced by a circulation system that usually placed the doors in the corners, helping to stage each room as a self-contained rectangle. The gaze fell either on furniture placed symmetrically in the room (for example a heavy desk) or on a big central opening leading to the next room (from the living to the dining room with its dining table set in the center). The home he built for himself in 1930 after being appointed as professor at the ETH Zürich displays the final stage in this series of developments. The dining room that has been developed out of the winter garden—an important element in the bourgeois home—features a massive table designed by Salvisberg himself and bentwood chairs painted salmon red; the living room has a fireplace and opposite a wall of windows with an upholstered window seat; the hall is dominated by an elegant spiral staircase (a "stairway room"); the spacious study has paneled walls, glass book cabinets, and tables, all made of the same tropical wood. Bookshelves, sofas, and benches, the fireplace with armchairs, and the sliding doors with opulent sapwood veneer: all are Salvisberg's own painstakingly elaborated designs. Walnut was used along with macassar for furniture, and the walls were covered with light beige Salubra wallpaper. Despite the inclusion of bentwood chairs by Adolf Schneck in the dining room and their tubular steel counterparts by Erich Dieckmann in the garden room, Salvisberg did not conceive here neutral envelopes for the heterogeneous furniture types of the modern inhabitant but instead distinctive ensembles of a "spirited conservative" nature[13] that create different ambiences through carefully selected materials and references to traditional domestic iconography.

The Home as Furniture

Zurich architect Lux Guyer (1894–1955) also focused on the middle-class household with its built-in furniture, cozy fireplace, built-in closet in the bedroom, and period furniture coordinated with the respective interior design style. She was adept at formulating her homes as a continuous sequence of spaces with numerous usage options, but in her case, in contrast to Salvisberg, their standardized components were enhanced by her own "inventions."[14]

In 1926 she proved in the exhibition *Das neue Heim I* (The New Home I) at the Kunstgewerbemuseum Zürich "how masterfully she was able to adapt rooms and furnishings to modern needs as well as to industrial advances."[15] She showed a four-room apartment made up of two prototypical room components—two bedrooms with a bathroom in between plus a living area with a kitchen, dining alcove, and adjoining room arranged to one side—all of which were linked by several different visual axes and circulation options.[16] Glazed doors regulated the relationship between the main living area and the adjoining rooms. Double doors allowed the bathroom between the parent's and children's rooms to be partitioned in a variety of ways for different uses—a Taylorist idea that would become Lux Guyer's trademark. This intelligently designed home was indeed like a "machine for living" that challenged its inhabitants to make full use of its possibilities but without pandering in formal terms to a modernist "machine aesthetic."[17] In the same exhibition, Wilhelm Kienzle (1886–1958) showed a widely acclaimed minimalist apartment with a foldout table between the kitchenette and bedroom/ living room recalling something from an old charterhouse; the kitchenette itself could be closed off by a sliding door. Both of these designs are based on making spaces adaptable through movable fixtures; the apartment thus itself becomes, so to speak, a piece of furniture. It was not until the following year that Le Corbusier and Pierre Jeanneret would try to provoke the public with their "transformable two-family home" at the Werkbund exhibition *Die Wohnung* (The Dwelling) in Stuttgart.

Lux Guyer tested her home designs by living there herself, and in the process found herself enriching them more and more with older furniture models, anonymous finds, and handcrafted pieces. The mixture of old and new furniture she advocated increasingly took on the character of an autobiographical arrangement that embodied and commemorated all of the ideas that meant something to her personally—in fact, not unlike the juxtaposition of iconographic references that evidently appealed to the bourgeois taste of the nineteenth century. In Lux Guyer's magnum opus, the Sunnebüel House in Itschnach,[18] where she lived from 1930 to 1939, there can still be found today a wonderful collection of ceramics by her friends Berta Tappolet and Luise Meyer-Strasser, paintings by Cornelia Forster, and embroidery by Lux's sister Claire, and also—in addition to a great number of highly original furniture pieces she designed herself—a small collection of historical chair types. This amalgamation of personal references undoubtedly enhanced the domesticity as well as the color differentiation of the interiors, whose porosity increased from bottom to top. Upstairs, a circuit

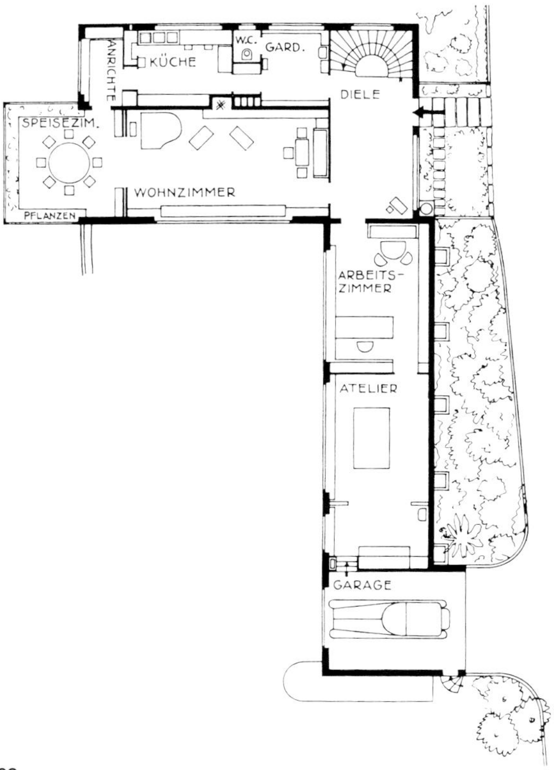

188

189

190

191

with diverse views through the rooms and out the windows leads through various room clusters that, with their alternate niches and open halls, can be combined in a variety of ways. Although the English influence can still be felt, Sunnebüel House betrays a striking personal vocabulary of forms indebted neither to "clamorous modernity"[19] nor to a false historicism.

The Home as Organism

Otto Rudolf Salvisberg, based as he was in Berlin, does not appear in Meyer's picture book, but Lux Guyer does, as well as—of all people—"Le Corbusier Paris (Charles E. Jeanneret from La Chaux-de-Fonds)."[20] The minimalist house he built in 1924 on Lake Geneva seems to have found favor with Meyer. Its longitudinal rectangular floor plan arranged freely along an interior circuit evidently corresponded to Meyer's idea of a house in which "the whole inner organization of private life" had become the main concern. In 1929, on the occasion of the second CIAM congress, whose theme was "Die Wohnung für das Existenzminimum" (The Minimum Subsistence Dwelling), Le Corbusier himself would define the apartment as a "phénomène biologique":[21] the goal was to work through the limits imposed by structural engineering and connect only the areas that were strictly necessary—just like in a natural organism—by way of an efficient circulation system ("circulation exact, économe, rapide").[22]

The common denominator in Meyer's modern Swiss homes is in fact—along with the sleek outward appearance of the buildings—the "living, organic relation of spaces."[23] Already before World War I, the horizontally organized "country home" of the English variety had supplanted the "villa," and with it the functionality of the floor plans. The relevant Berlin versions of the English models were highly differentiated spatial aggregates, grouped relatively freely in different wings in a "picturesque" fashion.[24] The residences Karl Moser (1860–1936) designed for the directors of the company Brown, Boveri & Cie. in Baden are good illustrations of the development from villa to country house. For his neo-Gothic Villa Boveri (1895–97), the architect still employed a rusticated base. Arranged around a central hall were the reception rooms, which were typical examples of historicism oriented on medieval and Renaissance models. With his home for Sydney and Jenny Brown (Langmatt, 1899–1901), Moser then replaced the "chivalric romanticism" of his earlier houses with a "decidedly 'modern' solution"[25]—a "country home instead of a castle" that was organized according to the functions it should fulfill.[26] English models (Mackay Hugh Baillie Scott) were now the obvious inspiration for furnishings in the style of the Arts and Crafts movement and rooms with white plaster walls, articulated through pine fittings and accented with ceramic tiles by Max Läuger.

By 1928, Moser's zeitgeist-perfused masterpiece was already history. Meyer now preferred the home Moser's contemporary Max Haefeli senior (1869–1941) had built for himself six years before, which joined traditionally self-contained rooms as well as a terrace and veranda by way of doors and differentiated stairways to create a circuit of varied spaces. Its markedly conventional interior decor, with Persian rugs and period furniture showcased as unique pieces, was similar to what could be found in the simple gabled-roof buildings by Albert Bodmer (1893–1990), Alfred Gradmann (1893–1962), and even in the homes designed by their younger colleagues Ernst F. Burckhardt (1900–1958) and Hans Hofmann (1897–1957). Paul Artaria (1892–1959) in Basel and Hans Schmidt (1893–1972) formed a rationalist counterpoint with their early use of timber- and steel-frame construction. The rooms in their homes were often painted sunny yellow and framed with black moldings, and they were furnished with Artaria's black- or gray-painted "box furniture" with its red or black handles, as well as with commonly available bentwood chairs (or with *Stabellen,* rustic board chairs!). Like their more conventional counterparts, such interiors—unexpectedly—radiate a Biedermeier-like domesticity.

Whether antique furniture types or anonymous industrial products: the use of standardized furniture is without doubt a key feature of modern interior design culture. Around the dining table in Max Ernst Haefeli's (1901–1976) Ritter House (1924–25) in Erlenbach were, in 1925–26, arranged woven cane Haefeli Chairs, the first Swiss *Typenmöbel,* or "type furniture," produced by the Horgen-Glarus furniture factory. The raw-silk lampshade is likewise

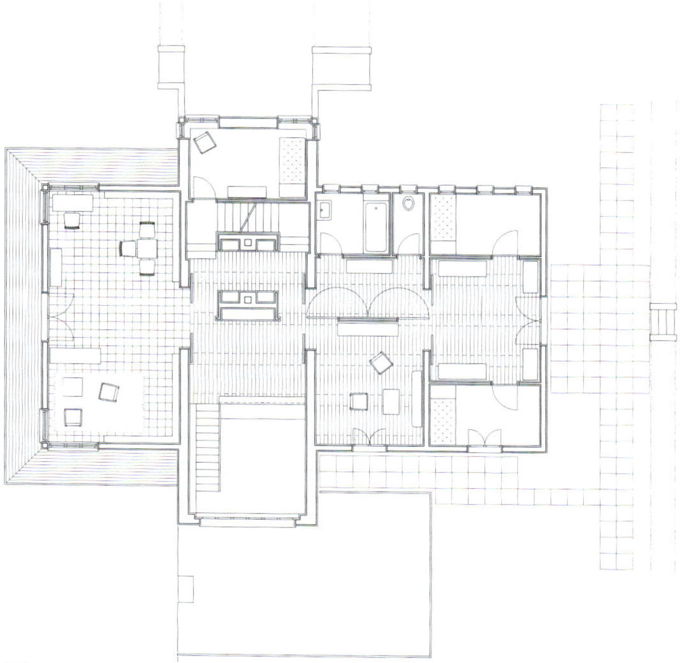

192

193

194

195

Otto Rudolf Salvisberg, Salvisberg House, Zurich, 1930
Photos: Heinrich Wolf-Bender

188 Ground-floor plan. Standard rooms are developed here from the spatial program of the upper-class country house.

189 Living room with serpentine stone fireplace, view toward dining room

190 Dining room as winter garden. Salmon-red bentwood chairs by Adolf Schneck

191 Sliding door between living and dining room in walnut burl

Lux Guyer, Sunnebüel House, Küsnacht-Itschnach, 1929–30

192 Main-floor plan, original state with living room, hallway, guestrooms, and staff rooms. Variably combinable room groups provide diverse sight lines of the interior and exterior.

193 Southeast elevation. Red Eternit shingles with white window frames and shutters

194 View from the upper into the lower hall. On the right, an "Antimott" armchair designed by Guyer, ca. 1930

195 Lux Guyer, Reclining Armchair, 1917–30 Photographed in the lower hall of the Sunnebüel House. Photo: Heinrich Helfenstein

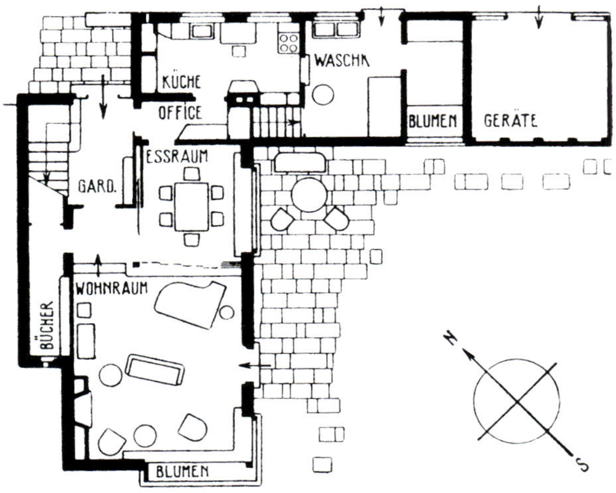

198

196

197

one of the first Swiss lighting fixtures for the modern household. Unlike Hannes Meyer with his nearly concurrent *Co-op. Interieur,* Haefeli was by no means trying to break with tradition in his widely acclaimed—even by Sigfried Giedion[27]—debut. All of the details and furnishings, right down to the lamps, had a familiar look, even though they were almost without exception new designs. With a finely attuned range of materials—fir stained in a uniform tone, with oak ceiling moldings; walls and ceilings in beige, yellow, and red fine plaster; oak parquet; and curtains in olive green Shantung silk—Haefeli accomplished a seamless transition from the interior decor of his parents' generation. The lightweight sliding walls covered with Japanese mats between the dining area and the sunken living room two steps down, and, next to the fireplace, a picture window that rounds a corner and features slim cross-bars and a low sill—a "flower window"—still convey even today a feeling of extreme spaciousness thanks to the skillful linking of the economically sized living spaces. The furnishings demonstrate—as J. Christoph Bürkle and Ruggero Tropeano put it—"Haefeli's obsession with detail and the implementation of the Werkbund ideals he had learned in his days with Otto Bartning in Berlin."[28]

Peter Meyer completed his picture book with a ten-page documentation of the appearance of the Schweizer Kollektivgruppe (Swiss Collective Group) at the Stuttgart Werkbund exhibition *Die Wohnung* (1927). A good dozen young designers from Swiss Werkbund circles were responsible for the well-received layout and interiors of House 4, a building type with two apartments per floor (*Zweispänner*) in the Mies van der Rohe block. They presented themselves as a cohesive group without insisting that the individual authors of the design be identified. However, the furniture and lighting designs were, as can be seen from surviving sketches, almost exclusively by Max Ernst Haefeli.[29] All of them were space-saving models with a natural appearance that were suitable for series production and for being combined at will in any setting. Despite their everyday practicality, Haefeli's innovative sample furnishings, oriented on making a concrete contribution rather than embodying an ideological manifesto, no doubt exercised an enormous influence—albeit under different auspices—on Swiss domestic culture in the coming decade, which would find itself caught up in a continual push and pull between tradition and innovation.

Max Ernst Haefeli, Model homes on Wasserwerkstrasse (Rotach Houses), Zurich, 1927–28
196 View from dining to living room, House 31. Sliding partitions, aluminum window frames with wide, low-set sills. Photo: Gasser und Eckert, 1994
197 Living room in House 29. Interiors by Max Ernst Haefeli and Hans Neisse, 1928
Photo: Heinrich Wolf-Bender

199

200

Max Ernst Haefeli, Ritter House, Erlenbach, 1924–25
Photos: Heinrich Wolf-Bender
198 Ground-floor plan. Example of "living, organic relation of spaces."
199 View from living to dining room. Dining table with Haefeli's first "type furniture," produced by the AG Möbelfabrik Horgen-Glarus. Sliding partitions covered in Japanese matting
200 View into living room. Change in level from dining to living room

1 Sigfried Giedion writes in *Building in France, Building in Iron, Building in Ferroconcrete,* "Corbusier's houses are neither spatial nor plastic: air flows through them! Air becomes a constituent factor! Neither space nor plastic form counts, only RELATION and INTERPENETRATION! There is only a single, indivisible space. The shells fall away between interior and exterior." Giedion (1900) 1995, p. 85.
2 Meyer 1928, p. 13.
3 Ibid., pp. 12–13.
4 Ibid., p. 12.
5 Ibid., p. 14. The most authoritative work on Peter Meyer is still Medici-Mall 1998.
6 Ibid.
7 See Robert Venturi, "Nonstraightforward Architecture: A Gentle Manifesto," in Venturi 1966, p. 22.
8 Le Corbusier (1930) 1991, p. 105: "The renewal of the plan of the modern house cannot be undertaken efficiently without laying bare the question of furniture. This is the Gordian knot."
9 Boesiger/Stonorov (1929) 1995, p. 96.
10 Benjamin (1982) 1999, p. 220.
11 See the detailed description of each room type in the middle-class house in Muthesius 1915; as well as, for the period 1870–1900, Köhler 2002a.
12 See Rüegg 1995b and Lichtenstein 1995.
13 Westheim 1925, p. 8: "He opts for certainty over uncertainty.... It is this attitude that gives the spirited conservative his resolve."
14 See Rüegg 2009a.
15 Baer 1927.
16 See Claus et al. 2009, pp. 186–87.
17 Peter Meyer speaks of the "machine-lovers' catchword 'machine for living,'" in Meyer 1928, p. 13.
18 See Claus et al. 2009, pp. 233–36; and Rüegg/Felder 2007, pp. 44–45.
19 Peter Meyer remarks on the staid attitude (of the Swiss citizens), on which "clamorous modernity" did not make much of an impression, in Meyer 1928, p. 9.
20 Ibid., p. 147; the entry on Le Corbusier is on pages 147–53.
21 Le Corbusier/Jeanneret 1930, p. 20.
22 Ibid., p. 21.
23 Meyer 1928, p. 11; Meyer refers to the country houses of Hermann Muthesius and—above all—Theodor Fischer.
24 See, for example, Muthesius 1922.
25 von Moos (2001) 2005, p. 41.
26 Oechslin/Hildebrand 2010, p. 75 (catalogue entry by Isabel Haupt).
27 Sigfried Giedion called it "pretty much the only house ... one shows when someone in Zurich asks to see an example of Neues Bauen," *Neue Zürcher Zeitung,* no. 1508, September 9, 1927. Despite protests, the Ritter House was demolished in 2014.
28 Bürkle/Tropeano 1995, p. 44.
29 Mehlau-Wiebking et al. 1989, pp. 16–17 and 34–35.

201

Teatime in a garden in the Werkbund's Neubühl housing development, Zürich-Wollishofen, ca. 1935. From left to right: Lita Finsler, lawyer Elsa Girsberger, graphic and stage designer Fritz Butz (Cabaret Cornichon, Corso-Theater) and his wife Gret Butz, below the art historian and dramaturgist Hans Curjel and his wife Jella Curjel.

1930 — Pragmatics

“Away from fashionable seasonal items and on toward enduring standard products!” This Werkbund maxim heralded the furnishing needs of “liberated living.” Exemplary of this change in priorities is the furniture program designed for the Werkbund’s Neubühl housing development in Zurich and thereafter marketed commercially. For most household functions, a single piece of furniture solidly built of tubular steel is offered. Foldable, collapsible, lightweight, and multifunctional: each piece is tailored precisely to the modestly sized living spaces. Thanks to an extended product range, freedom of choice soon returns, however, until finally the world financial crisis and popular taste bring the cultural revolution to an abrupt end. The order of the day is to now carefully consider each individual case without compromising on quality. The pragmatic maneuvering involved in attempting to revive traditional, regional forms becomes the general theme of the 1939 Swiss National Exhibition.

202

Advertising photo for the Indi lamp series (designed by Sigfried Giedion and Hin Bredendieck, produced by the BAG Bronzewarenfabrik, Turgi, 1932). At the center of the atmospheric evening scene is the clamping lamp with its flexible arm. Photographed at the Werkbund's Neubühl housing development, 1932.

Photo: Hans Finsler

203

Indi Lamp, patented pull-switch made of Galalith (designed by Sigfried Giedion and Hin Bredendieck, produced by the BAG Bronzewarenfabrik, Turgi, 1932). With his striking staging of the technological innovations offered by the Indi lamp series, Finsler succeeded at creating some iconic advertising images.

Photo: Hans Finsler

204

Max Bill, entrance to the Swiss section of the 6th Milan Triennale, 1936. Bill's presentation, which won the grand prize at the Triennale, relied on spare—mostly graphic and sculptural—elements and a dynamic spatial structure. Just a few products from the various areas were shown. A common denominator was provided by a large-format photograph of a Swiss landscape. It was a high point in Switzerland's self-promotion abroad that still impresses today.

Photo: Binia Bill

205

Max Bill, curator of the *Weihnachts-Verkaufs-Ausstellung. 1000 Geschenke ausgestellt von SWB* (Christmas Sale Exhibition: 1,000 Gifts Displayed by the SWB) at the Kunstgewerbemuseum Zürich, 1940, examines an Alsiko teapot from the ceramics studio of Fritz and Helene Haussmann, Uster. Bill also designed the exhibition poster with its star motifs.

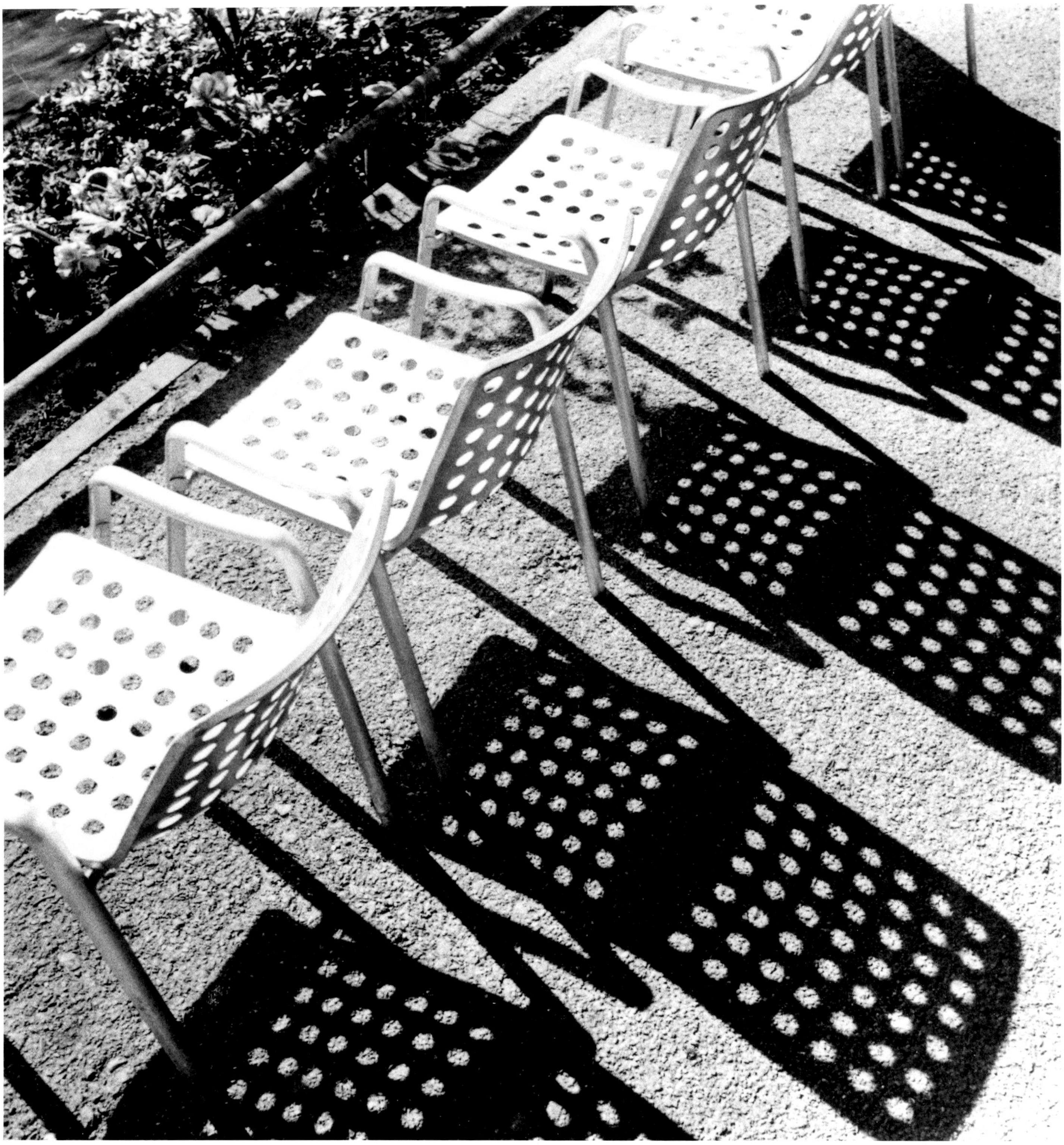

206

Hans Coray, Landi Chair, 1939, tempered aluminum, MEWA Blattmann, Wädenswil. Coray's shell chair anticipated the principle of the plastic seat shell—later taken up, for example, by Willy Guhl and Charles and Ray Eames—by several years. Made of the "Swiss" material of aluminum, the lightweight garden chair was popular among the most diverse visitor groups at the 1939 Swiss National Exhibition (Schweizerische Landesausstellung, or "Landi").

Photo: Hans Baumgartner

207

Jacob Müller (Bauwerkstatt, Zurich), roofed terrace of a rural home at the Swiss National Exhibition, Zurich, 1939. The clean-lined geometry of the spare building structure contrasts with the rural romanticism of the lovingly arranged rustic furniture and frames an idyllic view of the slightly upward-sloping garden.

Adjustable Garden Chair
1931

208

209

210

208 Werner Max Moser, Adjustable Garden Chair,
209 Wohnbedarf Model 25, 1931
Tubular steel, ash slats, solid wood armrests (all parts repainted), 68–75 × 58 × 89 cm, Wohnbedarf AG, Zurich/Basel; Embru-Werke AG, Rüti. Shown in the housing exhibition at the Leimenegg housing development, Winterthur, 1932.
Museum für Gestaltung/ Donation of Arthur Rüegg

210 Pierre Gauchat, Prospectus for Embru garden furniture, 1942, Embru-Werke AG, Rüti

As the Neubühl housing development (1928–31/32) started to take shape, the Zurich Werkbund took up the question of its interior design and furnishings. The architects were tired of "planning houses for furniture that does not exist on the market or is still too expensive." Under their influence, interior design became "a quick-change artist, a picture puzzle that through resourceful tricks could serve any number of purposes," wrote the newspaper *Neue Zürcher Zeitung.* Werner Max Moser (1896–1970) designed most of the models classifiable as lightweight, movable, collapsible, stackable, or adjustable—in other words, that were made to fit the relatively compact rooms of the Neubühl apartments. His garden chair consists of two individual parts: a load-bearing frame made of tubular steel, and a lightweight, anthropomorphically shaped seat made of iron profiles and ash slats. The nearly transparent seat seems to be hung in the frame. It can be adjusted using pins to different positions, allowing the same chair to be used for reading as well as relaxing or sunbathing. Moser's garden chair is thus another multifunctional piece of furniture, unlike the upholstered version designed for the indoors, the so-called *Volkssessel* (People's Armchair) with its fixed seat. The price of the armchair was its most convincing argument, the cheapest version costing only 75 Swiss francs, not much more than half the price of comparable German products in terms of comfort, such as the *KS 46* by Anton Lorenz (at a price of 125 reichsmarks). AR

Mehlau-Wiebking et al. 1989; Tropeano 2002b.

Giuseppe Prezioso

Hermes Baby Typewriter
1935

211 Giuseppe Prezioso, Hermes Baby Typewriter, 1935
Sheet iron body, case of sheet aluminum with wrinkle-painted finish, plastic keyboard, leather handle, 7 × 28 × 28 cm, Ernest Paillard & Cie. S.A., Yverdon
Museum für Gestaltung

212 Hermes Baby components, 1935, Ernest Paillard & Cie. S.A., Yverdon. The typewriters were produced using interchangeable parts.

213 Anonymous, Poster for Hermes, 1958, Ernest Paillard & Cie. S.A., Yverdon

211

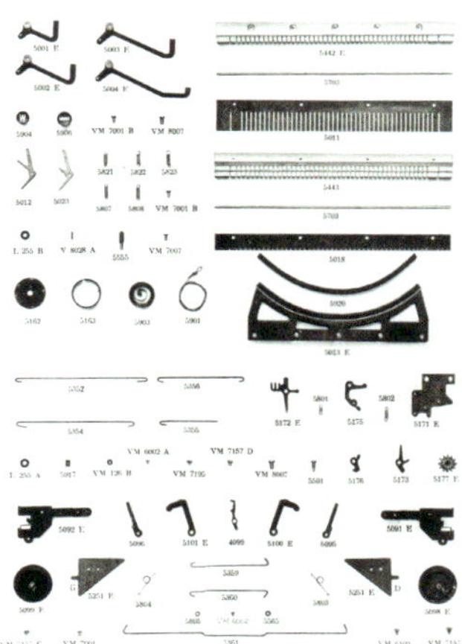

212

213

Introduced in 1924, the Hermes typewriter, named after the messenger god and protector of travelers in Greek mythology, allowed Ernest Paillard & Cie. S.A. to build up an entirely new industry, making Switzerland the world's third largest exporter of typewriters. In 1939, the business specializing in mechanical precision parts employed 1,100 workers from the defunct watchmaking industry in the canton of Vaud. The patented portable typewriter model was the primary reason for this economic success. The *Hermes Baby* also achieved great popularity with writers, including Ernest Hemingway, John Steinbeck, and Max Frisch. At a height of only six centimeters, it was the world's slimmest mini typewriter; it weighed in at only 3.6 kilos, was comparatively inexpensive, and fit inside a briefcase. 42,000 units of this mechanical precursor to the laptop had been exported by 1938. The *Hermes Baby,* often referred to as a "people's typewriter," is an excellent example of the judicious application of the advantages of industrial production (*Werkgerechtigkeit*), with a reduced number of parts (1,700 in the 1931 *Hermes*) and individual elements performing multiple functions in order to keep the weight and cost to a minimum. The chassis frame, for example, which is made of three punched iron sheets, is screwed directly onto the bottom of the case, giving it extra stability; the four screws used for this purpose serve the additional function of locking the lid. The economic example set by Fordist mass production was not the only seminal aspect of the *Hermes Baby;* its looks also self-confidently proclaim the beauty of precision engineering. RM

Bignens 1987.

214

215

216

214 Traugott Stauss, Entryway Unit, 1933
215 Color-painted solid wood, 110.5 × 107.5 × 41 cm, designed for the Blöchlinger House, Uznach Museum für Gestaltung*

216 Dining room in the Blöchlinger House, Uznach, view ca. 1996. Photo: Hanspeter Schiess

The teachers Alfred Altherr and Sophie Taeuber-Arp no doubt exerted a lasting influence on the work of the graphic, stage, textile, interior, and furniture designer Traugott Stauss (1898–1952) from Toggenburg. His involvement with the artistic avant-garde, which began during his time at the Kunstgewerbeschule Zürich and was intensified in Germany, is reflected not only in this maverick's graphic design work and his few architectural designs (Eos Cinema, Wattwil, 1929–30), but also in his cubist furniture often painted in color. In 1931, at around the same time that the Zurich branch of Wohnbedarf started business, he opened a shop in Lichtensteig called Raumkunst ("Art in Space"), in which he sold his own furniture pieces. It survived for only a year, although Stauss was given the opportunity to design the interiors and color scheme of the Blöchlinger House in Uznach in 1933. For the orange, yellow, and blue-gray entry hall of this Gesamtkunstwerk, he created in his father's wainwright shop an orange-red piece of furniture offering both shelf space and a seat. Shoes could be placed either in the open shelves or in the chest, umbrellas in the vertical section, and gloves inside the narrow drawer. Stauss's color schemes gave every room in the house its own unique identity. He placed minimalist dark blue furniture in a bright yellow dining room, while he immersed one of the bedrooms in a pale periwinkle tone. In 1934, after giving up hope of profitably breaking into the furniture business, Stauss shifted his focus to textile design. AR

Kirchgraber et al. 1996; Hanhart 1997; Selden 1998.

Model 2722 Bookshelf
1931

217 Wilhelm Kienzle, Model 2722 Bookshelf, 1931
218 Stained maple, painted metal, 101 × 93 × 22 cm, Embru-Werke AG, Rüti
Museum für Gestaltung

219 Pierre Gauchat, Drawing for an advertisement for the Kienzle bookshelf, ca. 1942. Proof

217

218

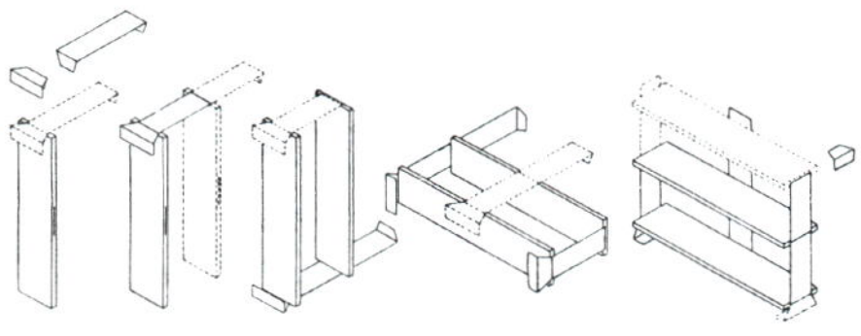

219

The ingenious designer and teacher Wilhelm Kienzle (1886–1958) described the "first step toward sensible furniture design" as "normalization." What he meant by this, in 1927, was primarily the move toward standardization, similar to what was occurring at the same time in the clothing industry. He used this idea to differentiate between the principles behind the design and construction of serially produced pieces of furniture and whole interiors created by craftsmen. According to Kienzle, the furnishings in rental housing should be lightweight and versatile, allowing them to be arranged in a variety of different ways. Set up freely in a room, they enter into a direct relationship with the inhabitant and fulfill their practical function wherever they are needed. In 1931, Kienzle designed the easy-to-disassemble bookshelf in accordance with the ethos and in the context of the Neues Bauen, perhaps for the housing exhibition at the Neubühl housing development in Zurich. Manufactured by the Embru-Werke in Rüti, it was one of their best-selling pieces of furniture for years and was produced continuously, with very few interruptions. The rectangular sheet metal shelves with their folded edges are fitted into grooves in the upright sides and in wooden panels at the back for greater stability. The bookshelf can be expanded upward as desired, with a range of shelves and sheet metal elements in different sizes allowing for a multitude of different combinations. The simple construction, economical use of materials, and flexibility of use all make the Kienzle bookshelf a prime example of unpretentious Swiss design. RM

Wilhelm Kienzle 1991; www.embru.ch.

Sigfried Giedion (attributed)

Adjustable Indirect Lamp 1932

220

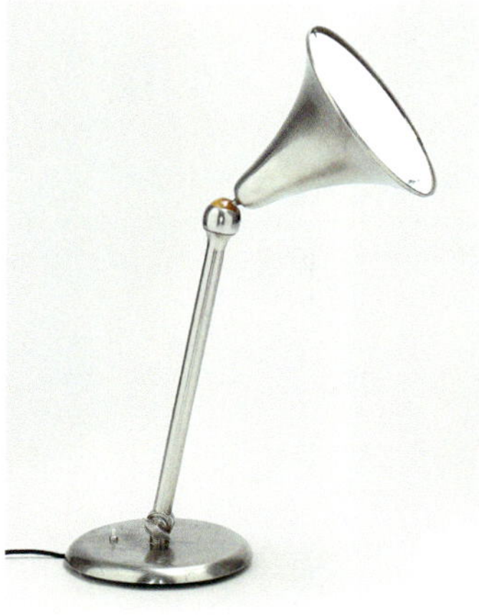

221

222

220 Galalith ball joint

221 Sigfried Giedion (attributed), Adjustable Indirect Lamp, Model 4.344, 1932
Nickel-plated brass, cast iron, Galalith ball joint, enameled inner reflector, H: 80 cm, D: 29 cm, BAG Bronzewarenfabrik AG, Turgi. Shade labeled "BAG-wohnbedarf typ."
Museum für Gestaltung

222 Sigfried Giedion and Hin Bredendieck, Indi Lamp, 1932
Photo: Hans Finsler, 1932

Le Corbusier and Pierre Jeanneret missed the opportunity to inaugurate their Genevan apartment building Immeuble Clarté with their own interior design exhibition. As a result, this demonstration of French elegance was left up to the local decorators, while Wohnbedarf AG in Zurich promoted its own simple "type furniture" with two complete interiors. Between May 25 and June 15, 1932, the first indirect lighting made by the company BAG Turgi was also presented to the public. These dates are of importance because the trained mechanical engineer Sigfried Giedion—with Werner Max Moser one of the two main stockholders of Wohnbedarf AG from 1931 to 1935—offered his services to BAG in July 1932 as official "advisor" on further developments. For this reason, we might suspect the lamps shown in Geneva, ennobled with the label "BAG-wohnbedarf typ," to be merely in-house designs by BAG if we weren't already aware of Giedion's well-known enthusiasm for indirect light and for the socket and friction joints found in French industrial models invented by Bernard-Albin Gras that would be determinant on the development of Giedion's *Indi* lamp series that began in July. The adjustable indirect lamp *BAG 4.344* not only offers a charming, if slightly cumbersome, interpretation of both of Gras's joints, but also boasts a light reflector hidden inside a thick brass funnel-shaped shade that Giedion would use later on as a shade for his legendary *Indi* lamps. Giedion was presumably behind the first indirect lamps produced by BAG in the spring of 1932, before he was able to take full responsibility for the design. AR

Le Corbusier 1933; *BAG* 1933; Rüegg 1999.

Glashütte Bülach AG
Factory design

Bülach Preserving Jar
1924–72

223

224

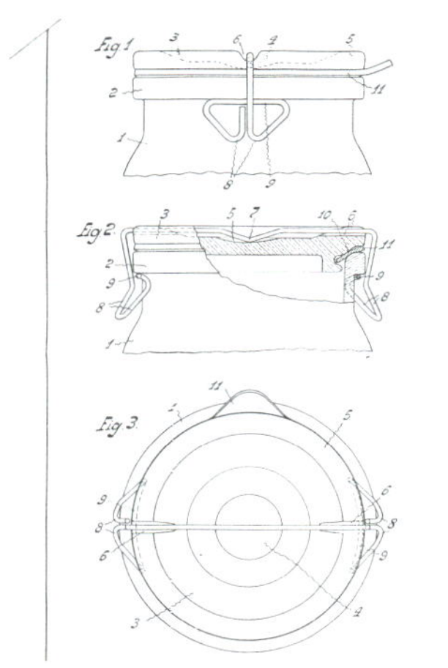

225

223 Anonymous, Bülach Universal Preserving Jar, 1948, clamp patent 1952
Colored pressed glass, lid of cast glass, wire clamp, rubber ring, holds 1.5 liters; opening 8 cm, Glashütte Bülach AG
Museum für Gestaltung/
Donation of Glashütte Bülach

224 Peter Birkhäuser, Bülach, 1934, Glashütte Bülach AG. Poster

225 Drawing of clamp, patented in 1952
Museum für Gestaltung/
Donation of Glashütte Bülach

During World War I, preserving jars of the German brand Weck were in such short supply that the Swiss national authorities initiated the launch of a Bülach *Kochflasche* (cooking bottle). It came onto the market in 1920, making use of a rubber ring and porcelain lid to create a seal (similar to the beer bottle closures that had come into use shortly before). In order to make the parts easier to clean when making preserves, the Bülach glassworks invented a cast glass lid complete with matching rubber seal and removable wire clamp. The wire rested in the lid's groove, so it could be clamped to the bottleneck. This preserving bottle, which in 1924 was made of the green glass typical for Bülach and had a four-centimeter-wide opening, became a nationally renowned brand-name item thanks to numerous posters, recipe books, and public presentations. Unlike in neighboring countries, where bottling preserves called for sterilization of the jar, in Switzerland one merely needed to fill the bottle with hot contents, which required less effort yet guaranteed scrupulous hygiene. In 1939, a glass with an opening measuring six centimeters was introduced, in order to allow whole fruits to be preserved. During World War II, Bülach steadily turned out jars at full capacity, and in 1948 the Universal-Konservenglas Bülach (Bülach Universal preserving jar) with an eight-centimeter opening was offered, which, beginning in 1952, featured a patented clamp closure. The jar was produced until 1972, when it eventually began to disappear from pantries, replaced by the freezer. RM

Unbekannt – Vertraut 1987, pp. 124–26.

226 Alfred Roth, Roth Aluminum Chair, 1933
227 Aluminum, 77 × 38 × 45 cm, Aluminium-Schweisswerk AG, Schlieren (?)
Museum für Gestaltung

228 Aluminum series by Alfred Roth in the official documentation of competition entries

226

227

228

In 1933, the Parisian office of Alliance Aluminium Cie. held an international competition to explore the ways in which the brittle metal could be put to use in the furniture sector. Four of the thirty-one participants that can be identified today were Swiss, and the Embru company in Rüti was responsible for the full development of the revolutionary winning project by "Monsieur Breuer (Hongrie)." The architect Alfred Roth (1903–1998) from Zurich, a disciple of Le Corbusier, did not garner an award but he did receive publicity for the "finesse de la ligne" of the three models he submitted—one chair with and one without armrests and a matching footstool. Meanwhile, nobody seemed to have noticed that these rustproof chairs could be stacked, making them perfect for use in the garden. The frame, bent from a single round profile, was—like that of the French iron *fauteuil*—furnished with aluminum slats that ceded somewhat on pressure, providing comfortable seating. However, the few models that were produced lacked stability and the aluminum slats tended to break, which prompted Roth to undertake further attempts using perforated metal sheets instead. In 1933, Roth was busy working with Marcel Breuer and his cousin Emil Roth on the design of the apartment buildings in Zurich's Doldertal neighborhood, commissioned by Sigfried Giedion. Giedion, secretary general of the Congrès internationaux d'architecture moderne (CIAM), sat on the aluminum competition's expert jury, along with Walter Gropius and Le Corbusier. AR

Rüegg 1996; Lepel 2009.

229

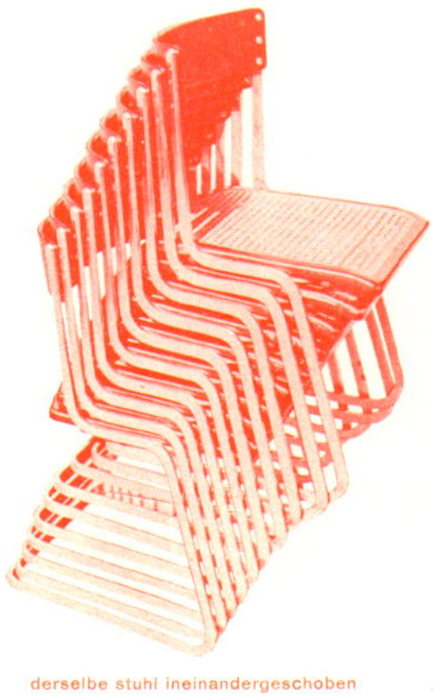

230

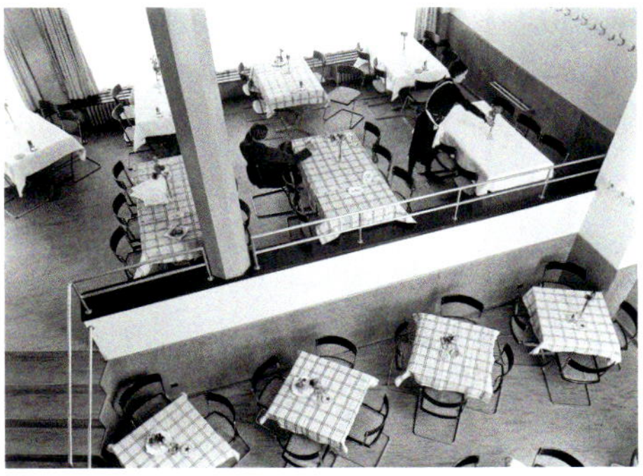

231

229 Flora Steiger-Crawford, Steiger Stacking Chair, 1931–32
Galvanized flat steel, (newly) painted plywood, 77.5 × 48.5 × 48.5 cm, Embru-Werke AG, Rüti
Early model with broad backrest from the Zett House, Zurich, 1932
Museum für Gestaltung/
Donation of Arthur Rüegg

230 Herbert Bayer, Page from the Wohnbedarf catalogue, 1932–33, with Flora Steiger-Crawford's Model 7 Stacking Chair with cane seat

231 Flora Steiger-Crawford, Restaurant in Zett House, Zurich, 1932. Tables with cast-iron feet and Steiger Stacking Chairs
Photo: Hans Finsler

From 1930 to 1932, the architects Hubacher & Steiger built Zurich's first modern commercial building at the Stauffacher junction. Flora Steiger-Crawford (1899–1991), Rudolf Steiger's wife, commented of her project for a living and dining tract wedged in between a generously sprawling, curved office block and a cinema with a novel sliding roof: "It was a hellish mess." She was, however, pleased with her role as independent planner of the Zett Restaurant along with its garden and furnishings. The relatively sober rooms were enlivened by being split into different levels; the menu was emblazoned on a multistory column; Max Bill designed the exterior signage. The most fascinating design subject, though, was the compact stacking chair, which would determine the entire character of the space. This sophisticated yet simple model with its offset steel-band frame bent to form a back and its seat made of woven cane (for the restaurant) or plywood (for the garden) is the compact counterpart to Alvar Aalto's "hybrid" stacking chair, likewise released by Wohnbedarf AG in 1932. The Embru-Werke in Rüti produced Steiger-Crawford's design with a slightly narrower wooden back as well as an additional model with armrests, to the designer's chagrin. Although her magnum opus did not elicit the same eulogies from the more internationally oriented Sigfried Giedion as Aalto's "Columbus egg for restaurants, bars, cafés," it outsold it nearly three times over in 1932. AR

Rüegg 1998; Rüegg 2003.

Alsiko Tableware
ca. 1933

232

233

234

232 Fritz and Helene Haussmann, Alsiko Tableware, ca. 1933
Red and black unglazed stoneware and yellow-glazed stoneware, various dimensions (teapot: 12.5 × 25 × 17 cm), Keramische Werkstätte Alsiko, Uster. Various sales outlets.
Collection of Arthur Rüegg

233 Haussmann plate painted by Berta Tappolet for the Cornelius shop, after 1937
Collection of Arthur Rüegg

234 Alsiko pottery workshop in Uster, built in 1933, ca. 1935

The German-trained ceramicists Fritz (1900–1968) and Helene (1905–1989) Haussmann derived the name of their pottery workshop Alsiko, which they founded in 1928 in Niederuster, from a primary component in the manufacture of ceramics: aluminum silicon oxide. In 1933, they moved to a modern new building in Uster designed by Max Ernst Haefeli, which housed not only a spacious manufactory but also a shop and their own apartment. After initially creating mainly studio ceramics that were rather severe in form, they were soon producing simpler vessels intended for everyday use. Their Alsiko tea service is classic design par excellence. The patented teapot—which is round, much like Wilhelm Wagenfeld's model made from Jena glass—has a downward-pointing, non-drip spout. The various pieces of the service, made from extremely fine-grained clay, were available either unglazed in red or black or with a variety of different color glazes. Thanks to numerous sales outlets in Zurich (including the sales collective Zur Spindel and, from 1937, Cornelius), Basel, and Klosters, the Haussmanns' ceramics were distributed well beyond Werkbund circles, allowing the factory to hire up to fifteen employees. Famous decorative artists such as Berta Tappolet, Luise Strasser, Cornelia Forster, and Amata Good worked there as painters. Helene Haussmann also created figurative works. She was in charge of the painting workshop and experimented with a multitude of colors, glazes, and firing cycles. AR

"Keramische Werkstätte" 1935; Information provided by Steffen Kuhn, Zurich.

Wohnbedarf Model 42 Folding Table
1931

235 Werner Max Moser, Wohnbedarf Model 42
236 Folding Table, 1931
Hardened tubular steel, crystal glass, Marbrite glass, H: 60 cm, D: 73.5 cm, Wohnbedarf AG, Zurich/Basel; Embru-Werke AG, Rüti
Museum für Gestaltung/
Donation of Arthur Rüegg

237 Herbert Bayer, Wohnbedarf catalogue, 1932–33. Rudolf Graber demonstrates how to fold the three-legged model of the Moser table.
Photo: Hans Finsler

235

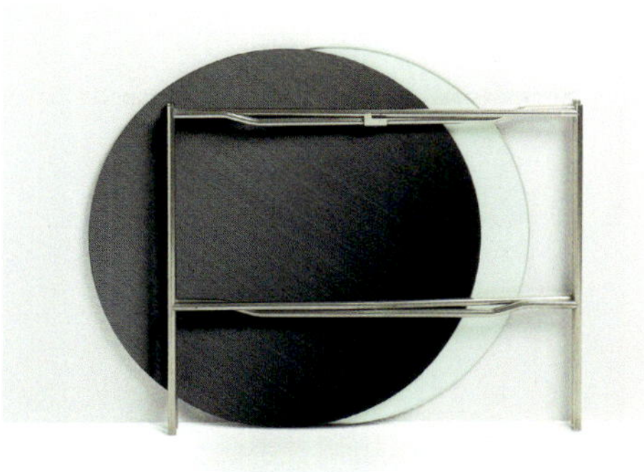

236

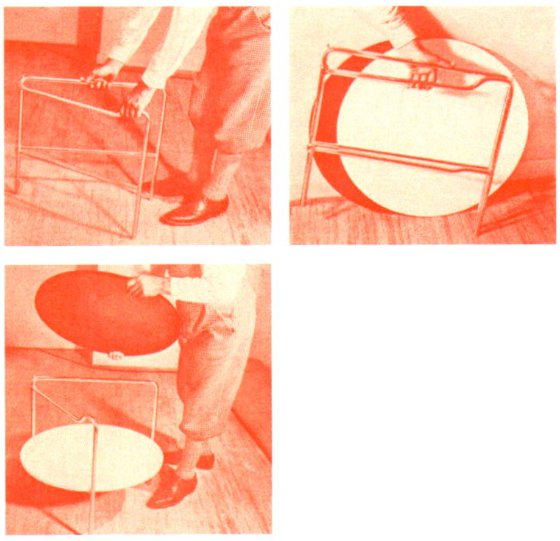

237

A key scene in Hans Richter's 1930 Werkbund film *Die neue Wohnung* (New Living) features avant-garde Zurich designers dancing around a model apartment. The cupboards were built in to save space, the lightweight wooden furniture could be quickly pushed aside, and a low windowsill was even available as a seat. The living room could thus be effortlessly transformed into a party venue. When the designers of the Werkbund's Neubühl housing development in Zurich set about planning their own range of metal furniture soon thereafter, transformability and multifunctionality became crucial focal points of the program. Werner Max Moser (1896–1970) presented his three-legged round folding table for the first time at the opening of the development in September 1931. Together with his wife Silva, he illustrated a new style of living with his "work and living space for a journalist couple." As the knickerbocker-clad man in Hans Finsler's advertising photograph vividly demonstrates, the tubular steel frame of the small round table can be folded up with a simple movement and quickly set aside with its two panels in order to make room for festivities. The upper panel can also be used as a serving tray. By the end of 1930, Moser had already drafted a four-legged version of the table with a slightly larger diameter. Both models were a core part of the Wohnbedarf range from then on. The color-lacquered, "duralized," or chrome-plated frame could be ordered with either linoleum-covered wood panels or—in its more elegant versions—with crystal or opaque glass. AR

Mehlau-Wiebking et al. 1989; Rüegg/Tropeano 1992.

238

239

240

241

238 Mille-fleurs service, 1781–1813, Manufacture de porcelaine, Nyon. Neoclassical porcelain tableware exuding a bourgeois charm

239 Factory design, Carona serving pieces, ca. 1955, Rössler Porzellanfabrik AG, Ersigen. Produced until 1976, this tableware could be found in many a Swiss kitchen.
Museum für Gestaltung

240 Frau Bachmann-Kienast, "Rebhüsli" table, 1913, *Der gedeckte Tisch* (The Table Setting) exhibition at the Kunstgewerbemuseum Zürich. "Table decoration should not dominate the occasion but rather serve it."

241 Hans Finsler, Porcelain exhibition with Langenthal products at the 1939 Swiss National Exhibition. The appeal of understatement in a geometric configuration.

Ceramics are inextricably linked to the evolution of the arts of the table throughout all strata of society. From rustic earthenware plates to fine china, tableware offers revealing insights into the lifestyle, habits, and tastes of the people who choose this crockery and use it in their everyday lives. Dining habits in the course of the twentieth century cannot readily be examined in the light of Swiss production alone. Switzerland is not an island: tableware from abroad (imported from neighboring countries or from the Far East) has of course influenced the market and had an impact on consumer taste with respect to both industrially manufactured and designer ceramics. While not wishing to make overly reductionist generalizations, we would argue that there are nevertheless good reasons to identify a "clay wall" between German-speaking Switzerland, which is influenced by ceramic production from northern Europe, in particular from Germany, and the French-speaking cantons, where France is the most pronounced influence.

Since the early twentieth century, the boundaries demarcating the realm of ceramics have been extended in response to ceramic artists' (legitimate) ambitions to move beyond the world of craft and utility production and make their mark as full-fledged artists. This is linked to a wish to overcome the compartmentalization and hierarchical structures defining relationships between the fine and the applied arts (though one might ask whether this goal has been attained). Parallel to this, developments in the arts of the table (nowadays often reduced to a simple meal on a tray in front of the television or at the computer) have not only changed how we eat but have also transformed rituals associated with meals and the type of dishes in which food is served. Rather than drawing up an exhaustive inventory of the ceramic artists, studios, and manufacturers that have made a contribution to Swiss ceramics design, this essay seeks to present a subjective account of the challenges facing ceramics design in Switzerland and the milestones in its history.

1913–1960: *Gute Form* and Well-Set Tables

The arts of the table in the early twentieth century were dominated by a respectable folk *Heimatstil* rooted in Swiss rural identity. Family meals were served on simple, tasteful crockery. Model housewives and mothers were expected to take great care with their house while avoiding any ostentation. A distinction was drawn between everyday crockery—simple, practical, hygienic, and functional—and the "good china," which was more luxurious and sophisticated, intended for special occasions. The 1913 exhibition *Der gedeckte Tisch* (The Laid Table) at the Kunstgewerbemuseum Zürich gives a good sense of these well-set tables. Growing out of this fertile ground, and with a view to offering tableware that would reflect local taste and needs, the Langenthal porcelain manufactory came into being in 1906, after a caesura of more than a century in the history of Swiss china production. The first dinner services and tea or coffee sets were inspired by historic forms and paid homage to their precursors: the historic manufactories of Zurich (Kilchberg-Schooren) and Nyon.[1] The names given to these services—Rhône, Bellinzona—and the meadow-flower patterns decorating them appeal to patriotic feelings and underscore the company's desire to maintain strong roots in Switzerland.

The year 1913 saw the birth of the Schweizerische Werkbund (SWB) in Zurich and its French-speaking Swiss counterpart L'Œuvre (OEV) in Yverdon. These associations, inspired respectively by German and French models, adopted sometimes diverging approaches. Both sought to promote links between art and industry, developing the concept of *Gute Form* ("good form," or good design), which aims to strike an optimum balance between an object's function and its aesthetics, in the case of both one-off works and mass-produced pieces. Numerous artists, ceramicists, and manufactories joined these movements, taking part in exhibitions and competitions. The Langenthal manufactory was no laggard in this respect and the services created by Fernand Renfer (1898–1979), the firm's artistic director, and his son Pierre (1925–1995) won a series of awards. The numerous high-quality services from the manufactory include, to name just the most striking examples, *Landi,* created by Fernand Renfer in cooperation with the SWB in 1939 on the occasion of the Swiss National Exhibition (Schweizerische Landesausstellung, or "Landi" for short), and above all *City,* designed by Pierre Renfer in 1965. With its stackable modular system and its timeless rationality, *City* is a perfect illustration of the *Gute Form* concept—and is still part of Langenthal's range today.

The Renfers played a major role in the company's artistic development. In addition to its regular collaborations with artists, designers, sculptors, and ceramicists,[2] the manufactory set up an artistic studio as early as 1956, for it always aimed to position itself firmly within the modernist movement, in terms of both the shapes and the ornamentation of its output. Concerns about commercial viability meant Langenthal was constantly engaged in a delicate balancing act, aiming to reconcile its bourgeois clients' conservative tastes with the audacious experimentation demanded by the imperatives of international competition.

The focus on formal rationalism was naturally not the prerogative of the manufactories alone. When ceramicist Paul Ami Bonifas (1893–1967) returned from Paris in 1922 to set up his studio in Ferney-Voltaire (France), he had assimilated the Purist theories developed by his friends Amédée Ozenfant (1886–1966) and Charles-Edouard Jeanneret[3] (1887–1965). Purity of form and rationalization lie at the center of Bonifas's explorations, manifested in particular in his famous black luster-glaze pieces and in his work on mechanizing production processes (sizing and molding).

Ceramics studios active in Switzerland from 1920 to 1930, such as Ziegler in Schaffhausen, Bodmer in Zurich, or Noverraz in Geneva, generally focused their attention on "decorative ceramics." Schools of ceramics were set up to train skilled workers for the industry (Bern in 1906, Chavannes-Renens in 1912). In contrast, Fritz Haussmann (1900–1968), who opened a studio in 1928 in Uster near Zurich, was a fervent champion of pure form, even if he did collaborate on decorative vases and pots with, for example, Luise Strasser (1894–1974) and Berta Tappolet (1897–1947). Haussmann's rigorous and ingenious *Alsiko* no-drip teapot is a good example of his avant-garde work, which was sold in the shops Cornelius[4] and Spindel.[5]

Figures such as Margrit Linck (1897–1983) or Mario Mascarin (1901–1966) appear to be outsiders in this context. Linck's traditional pots increasingly developed toward an eminently personal plastic language, and evolve into anthropomorphic sculptures. Mascarin introduced high-temperature firing techniques into Switzerland, applying subtle monochromatic enamel glazes to the extremely pure lines of his utilitarian stoneware.

1970–2000: The Impact of Expressive Ceramics on Functional Ceramics

In the seventies, there was quite a craze for functional ceramics. Particular mention should be given to Ursula Luginbühl (b. 1936), who developed a formal language in her work that might be said to evoke a Bauhaus aesthetic. Her monochrome receptacles with clean, pure lines and thoroughly sculptural forms invariably underline the object's functional purpose.

At the same time, the Langenthal manufactory, today relocated entirely to the Czech Republic, entered a period of slow decline, undermined by cutthroat competition in the sector and ongoing developments in the arts of the table. The bold decision to launch *Bopla!* in 1993 with a view to gaining and retaining younger customers proved less successful than hoped.

In the late eighties, the parameters defining the world of ceramics shifted: the Swiss ceramics industry had to adapt rapidly to a new economic context. Today there are a handful of larger studios, which make few formal innovations. However, potters find it more and more difficult to make a living from small series production. Increasingly, the market niche for functional ceramics has shrunk to craft markets. Ceramicists therefore tend to concentrate on figurative or abstract sculptural work, paying scant attention to the utilitarian function of their work.

At the same time, consumers' relationship with product pricing has shifted: rather than making a long-term investment, people prefer to change their dinner service regularly, even every season. Tableware is certainly still supposed to be practical, although at times the forms assumed by contemporary crockery are influenced by international culinary traditions that enrich European gastronomy. Consumers also pay less and less attention to where ceramics are produced. Furthermore, more affluent consumers tend to opt

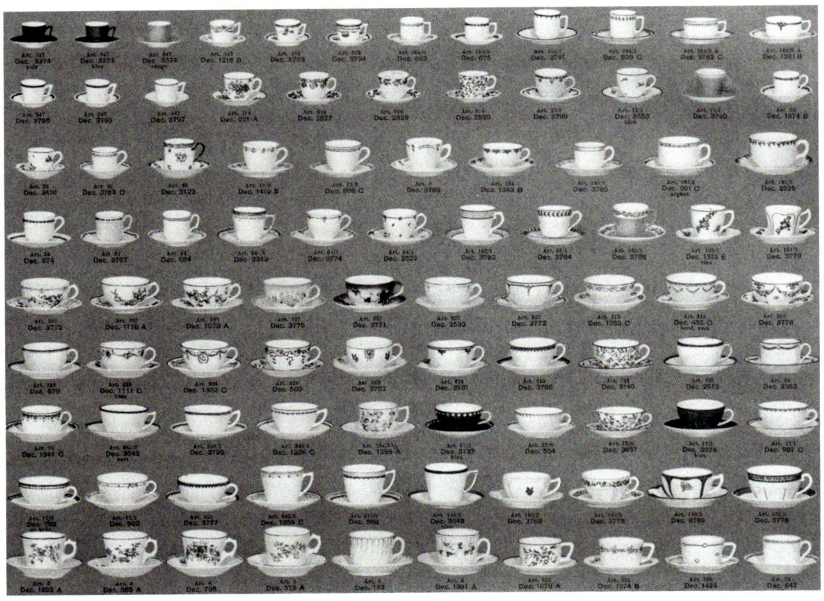

242

243

244

245

246

242 Sample board no. 6, ca. 1920, Porzellanfabrik Langenthal. No differences are discernible between household and hotel tableware.

243 Faience table service, painted by Berta Tappolet, n.d., Porzellanfabrik Langenthal. Photo: Hans Finsler. From the SWB picture collection of well-designed consumer products.

244 Arbeitsgemeinschaft Schweizer Werkbund, “Landi service,” 1939, Porzellanfabrik Langenthal. Photo: Hans Finsler. A Swiss classic. From the SWB picture collection of well-designed consumer products.

245 Factory design, Resista ovenproof cookware, 1918, brown-glazed porcelain, Porzellanfabrik Langenthal. Produced until 1997 with minimal formal changes.
Museum für Gestaltung

246 Factory design, Platter No. 30 and Plate No. 35, ca. 1913, Porzellanfabrik Langenthal. Produced until 1997, these were the Porzellanfabrik Langenthal’s most enduring items.
Museum für Gestaltung

247

248

247 Porzellanfabrik Langenthal, Prospectus, ca. 1935
Photo: Hans Finsler

248 Hans Finsler, Porcelain production at the Porzellanfabrik Langenthal, ca. mid-1930s. Finsler documented the entire fabrication process of the tableware from raw material to decorated finished piece.

for lines from major international brands, with aspirational lifestyle associations, rather than choosing design work by Swiss ceramicists.

In the nineties, two ceramicists were emblematic of artist-driven designed work. Marco Mumenthaler (b. 1948) remains faithful to the potter's traditional tools and role, working with the wheel to create various vessels. Nevertheless, the way he works gives his pieces a touch of modernity that chimes entirely with twenty-first-century tastes. His boxes, vases, and bowls convey a subtle reflection on the forms per se. Ceramicist Setsuko Nagasawa (b. 1941), born in Japan but now based in Paris after many years in Geneva, has incorporated a design facet into her practice, and since 1993 has produced a collection of design objects in addition to her ceramic sculptures. Blending various media, she works with ceramicists, lacquer specialists, and goldsmiths to produce the pieces.[6] Keen to maintain the inimitable quality of artisanal production, she has never considered having her objects produced using industrial methods.

2013: Industrial Design or Artist-Driven Design?

Ceramics training in Switzerland reflects the creative and industrial context of the early twenty-first century: career routes intersect and blur, with growing exchanges between disciplines and an intermingling of different fields. Although schools of ceramics cannot guarantee graduates a secure professional future, they equip students with a sound artistic grounding, as well as offering postgraduate diplomas in ceramics research for artists who have no previous experience with these materials. This open-minded approach is encouraging, yet it is also worth considering the issue of the increasing erosion of technical expertise.

On the whole, after completing their studies ceramicists either specialize in expressive work or focus on artist-driven design with a view to shaking off outdated clichés about potters.[7] This involves working on limited-edition series and building up a private or corporate customer base, as exemplified by Simone Stocker (b. 1974), whose *Torlo* series won the Swiss Federal Design Award in 2006.[8]

Since the early 2000s, Margareta Daepp (b. 1959) has been creating receptacles inspired by motifs from the cultures she has encountered during study trips (in Europe, Japan, China, and Turkey), also as limited-edition series; they are pieces that invite the viewer to consider urban and industrial images and textures.

Ceramicists nonetheless also play a vital role in the production of design objects: their technical expertise is much in demand with designers who want to create ceramic products. The list of designers with a keen interest in ceramics is remarkably long. Alfredo Häberli (b. 1964) designed *Origo* (Iittala, Finland) in 1999, crockery with functional contemporary lines and playful colors that dovetails perfectly with the changing contours of modern everyday life.

Familiarity with the materials deployed enables artists such as Peter Fink (b. 1962) to satisfy designers' technical expectations.[9] The École cantonale d'art de Lausanne (ECAL), in particular, draws on his services to produce ceramic objects following its design students' specifications. Artists and product designers from Atelier Pfister[10] have also commissioned him to produce their prototypes. Like Peter Fink, Felix Vogler (b. 1956) trains apprentices and provides technical assistance to designers' projects. Both also create work for series production and aim to recoup the manufacturing costs by formal and technical research work.

In 2012, the Swiss umbrella association for the applied arts, Form Forum, which has over six hundred members, launched its second design competition, with the title "Error—qui se trompe trouve"; Estelle Gassmann (b. 1981), in particular, stood out among the award winners. The young ceramicist has developed two lines of porcelain objets trouvés, giving the found pieces a second life by coating them with tragacanth[11] or combining them with run-of-the-mill plastic baskets. This produces strange structures whose design blends nostalgic and sculptural elements, semi-functional objects quite at home in a multidisciplinary era.

We should be careful not to draw too sharp a distinction between design and ceramic art. Nowadays, sculptors and designers dream up forms that are produced in ceramic material in an industrial or artisanal context; sculptors working with ceramics are developing design production; firms seeking to enhance their image hire well-known artists for specific projects... There are numerous examples of fruitful interdisciplinary experiments and research, moving ceramics design beyond the realm of simple tableware. Over and above the fascination that ceramics exert on contemporary

249

250

249 Ursula Luginbühl, Black-glazed household ceramics, 1960s, Ursula Luginbühl, Mötschwil. Photo: Leonardo Bezzola. From the SWB picture collection of well-designed consumer products.
250 Margrit Linck, Vases and bowls, from left to right: S243, V78, V87, S285, V4, V61, 1960s, white-glazed ceramics, produced by Linck Keramik, Zollikofen, 2012. Photo: David Willen

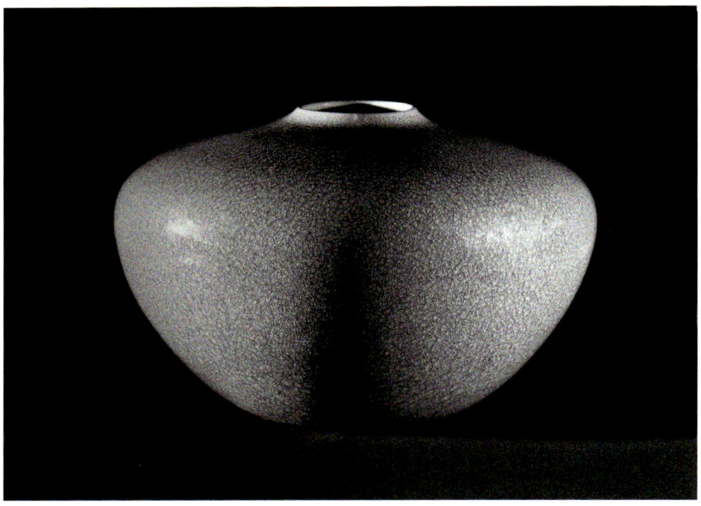

251

252

251 Helene Haussmann, Ceramic vessel, ca. 1950, Keramische Werkstätten Alsiko, Uster. Max Bill included this in his 1952 compendium *Form.*
252 Paul Ami Bonifas, Terre lustrée noire vessel, ca. 1930, earthenware pottery, Ferney-Voltaire (FR)
Private collection
253 Simone Stocker, Torlo tableware, 2003–04, Bottoni Porzellan, Bern
Museum für Gestaltung*
254 Trix and Robert Haussmann, Broken, Stripes, and Stars underplates, ca. 1984, Swid Powell Collection (USA). Trompe l'oeil with strong color contrasts
Museum für Gestaltung
255 Saucer, 1781–1813, Manufacture de porcelaine, Nyon. Faux-bois decoration
Musée historique et des porcelaines, Château de Nyon
256 Berta Tappolet, Teacup with saucer, ca. 1917–27, hand-painted. Delicate artisanal decoration
Museum für Gestaltung
257 Olaf Breuning for Atelier Pfister, Hallau ceramic bowls, 2011, Keramik-Manufaktur Potsfink, Ependes-Fribourg. Each bowl is unique. Breuning decorated another series for Pfister with a wild profusion of black fingerprints.

253

254

255

256

257

artists and designers, the medium testifies to the permeable boundaries between artistic genres, giving rise to a fertile creative dynamism and hinting at still greater openness to innovative developments in the future. It is exactly this kind of cross-fertilization that enriches creative work and gives vital support to creative artists.[12]

1 The Zürich-Schooren manufactory operated from 1763 to 1790; the one in Nyon from 1781 to 1813.
2 These included Paul Boesch (1889–1969), Leo Steck (1883–1960), Aldo Patocchi (1907–1986), Édouard-Marcel Sandoz (1881–1971), and Hans Erni (b. 1909).
3 Better known as Le Corbusier.
4 Company established by the ceramic artists of the Zurich Werkbund in 1937 (Cornelia Forster, Amata Good, Fritz Haussmann, Luise Strasser, and Berta Tappolet) in order to sell their work.
5 Spindel, originally based in the "Haus zur Spindel" in Zurich, was a sales outlet for Swiss crafts set up in 1916 by the Schweizerische Frauenzentrale. It merged with the Schweizer Heimatwerk in 1947.
6 Wolfgang Vegas and Fabienne Gioria for the ceramics, Hikaru Tsuji for the lacquerwork, Fabrice Schaeffer and Heidi Röthlin for the metalworking.
7 It is also interesting to note that the federal qualification certificate in pottery has vanished and been replaced by a federal qualification certificate in ceramics, and that "ceramics decorators" are now described as "3-D polydesigners."
8 Others succeed in launching some of their work abroad. For example, *Camera Obscura* (2010) by Magdalena Gerber (b. 1966), released by Bernardaud; *Brocca Trottola* (2011), pitcher by Claudia Napoleone (b. 1968), released by Esprit Porcelaine.
9 We would like to thank Claire Favre Maxwell, curator at mudac, Lausanne, and Maurizio Ferrari, ceramicist and teacher at the Schule für Gestaltung Bern.
10 Atelier Pfister is a brand of Pfister Möbel with collections designed by young Swiss designers.
11 Gum extracted from the stalks of plants of the genus *Astragulus,* which is used in the pharmaceutical and textile sectors, as well as for culinary purposes.
12 Here we could cite, in particular, Sèvres – Cité de la céramique, Maison Bernardaud, and, in Switzerland, Atelier Pfister.

258

259

260

261

258 Max Bill, Entrance to the exhibition *Das Licht in Heim, Büro und Werkstatt* (Lighting in the Home, Office, and Workshop), Kunstgewerbemuseum Zürich (housed at the time in the Swiss National Museum), 1932. Poster by Alfred Willimann, photo: Hans Finsler

259 Hans Finsler, *Mein Weg zur Fotografie. 30 Aufnahmen aus den zwanziger Jahren,* Pendo Verlag, Zurich, 1971. Cover. Luminograms: Siegfried Zingg, 1969

260 Hans Finsler, Indi Lamp, 1932 (product design by Sigfried Giedion and Hin Bredendieck, produced by BAG Bronzewarenfabrik AG, Turgi, 1932)

261 Sigfried Giedion (design), Hans Finsler (photography), Indi Lamps, prospectus produced by BAG Bronzewarenfabrik AG, Turgi and Wohnbedarf AG, Zurich, 1932. Front cover

Adorning the cover of one of the most beautiful lighting prospectuses from the thirties, in white on black, is a magical luminogram—an "abstract" image created when a beam of light collides with a photographic film. Hans Finsler (1891–1971), a master of subtle brightness modulation, made light itself the theme here by photographing directly into the funnel of a floor lamp. Only the brand logo seeming to float inside the funnel gives the picture a touch of realism. The message is clear: only the light itself is important and not the visible form of the lamp. Today, with light-emitting electronic components replacing conventional lightbulbs, the old dream of lampless artificial light appears to be coming true. "Light is no longer equated with a light-emitting source such as the light bulb and its characteristic form, but rather radiates from bodies of light and can be shaped and designed and even transformed according to needs and times of day"—such is the current state of the art.[1] Nevertheless, consistently downplaying the design aspect in favor of maximal light output was of course never the only goal of lighting design—and certainly not in keeping with popular taste. The "Apollonian ideal of brightness"—says Silvia Berselli[2]—has always been opposed by "the Dionysian principle of evocative chiaroscuro," just as the light fixture as a minimalist apparatus contrasts with the lamp as a concrete furnishing object.

Art Industry and Applied Arts

In 1916, L'Œuvre, the French-speaking branch of the Swiss Werkbund, showed in its first traveling exhibition[3] not only glass, ceramics, enamel work, glass painting, and mosaics, but also the current state of artisanal lighting design, for example the precious stained-glass table lamps produced by Société "Le Phare" in Lausanne. The goal was to also cleverly showcase the gentler qualities of electric light, robbing it of its harshness. Charles-Edouard Jeanneret (later known as Le Corbusier, 1887–1965) contributed an imaginative lamp object that combined a three-legged table frame with a goldfish bowl made of jade glass and a rose-filled lighting fixture.[4] The company he had founded with some friends, Lumière – Société pour la fabrication de lustrerie d'art, produced a range of idiosyncratic lamps between 1914 and 1917 in La Chaux-de-Fonds, including carved wall fixtures, floor lamps with pleated fabric shades rising out of small neoclassical wooden tables, and table lamps made of painted ceramics combined with drinking cups.

Apart from unique pieces such as these—and perhaps also the elegant work lamp used by clockmakers in the Jura region[5]—it is difficult to discern any specifically Swiss achievements in the early days of lighting design. Still untouched by the reformist zeal of the Werkbund designers, the products of the lighting industry fit very well into the traditional "genteel bourgeois home," the standards for which were set in the German Reich. As early as 1908, the Zurich company Wilhelm Egloff & Co. was selling electric light fixtures for every conceivable purpose: chandeliers with electric candles, voluminous train lamps with lead-weighted fabric shades, electrified Dutch chandeliers, glass-shaded directional lamps, bureau lamps, sliding lamps, workshop lamps, watchmakers' lamps, bakers' lamps, weavers' and passementerie lamps, piano lamps, draftsmans' lamps; only shop window lighting was still offered in combination with gas lamps. New were fabric shades, made possible by the fireproof electric light. In formal terms, these models differ little from those subsequently produced by the successor company founded in 1909, the Swiss bronze ware manufacturer BAG Turgi. The fact that electrification had scarcely altered the shape of the kerosene lamp or the Dutch or crystal chandelier no doubt contributed to the continuity of both forms and lighting concepts.

Indirect and Directional Light

It was not until the mid-twenties that designers began to recognize the potential offered by a lightbulb that could be attached anywhere in a room and in any position. In 1923, the Swiss Carl Jucker (1902–1997) was still tinkering with an electric version of the glass oil lamp at the Bauhaus in Weimar. He made the visible wiring running through the glass stem a constituent element of his design (the spherical shade was passed on to the—later legendary—"Bauhaus lamp" by Jucker's successor Wilhelm Wagenfeld).[6] In Switzerland, the architect Max Ernst Haefeli (1901–1976) likewise experimented in 1926, 1927, and 1928 with the glare-free milk glass produced by Schott in Jena. He conceived three separate series in those years, which, depending on the type and purpose, could be hung, stood, or screwed on and directed. The 1927 series, with bases made of the ultralight metal Elektron, were used in the Weissenhof Settlement in Stuttgart. Flora Steiger-Crawford (1899–1991) and Rudolf Steiger (1900–1982), for their part, domesticated a multidirectional doctor's lamp using a conical parchment shade. These carefully detailed models, which exude an engineering vibe and yet still make congenial accessories for the home, were manufactured only in very small series, while in Germany, the Netherlands, and Denmark industrial production of modern lighting fixtures had already begun.[7]

It was then art historian Sigfried Giedion (1888–1968), cofounder of Wohnbedarf AG, who in 1931 sounded the charge for an all-out attack against the "chandelier hung in the middle of the room": "To fit our modern needs, we must have the freedom to deal with our rooms flexibly and not to always be compelled to place the dining table in the middle.... The centering of the room around the dining table lamp, or any kind of hanging lamp for that matter, is thus perceived as a constraint. Instead of the dining table lamp, we need easily movable wall arms ... that can if necessary also be directed toward the sofa or a comfortable chair. In many cases, indirect lighting can also be used to immerse the room in a general, shadow-free brightness."[8] To realize his vision, Giedion was at first dependent on foreign products, pictures of which he compiled on index cards according to the categories "standing floor lamps," "miscellaneous wall arms," and also "clamping lamps" and "table lamps."[9] The lamps made by the French company Didier des Gachons & Ravel out of molded sheet metal with cast metal parts and aluminum shades were represented in all of Giedion's categories; they could be rotated thanks to ball joints at the base and their arms were also adjustable via disk joints.[10] Industrial products like these were employed in particular by Le Corbusier, representing designs that, as Roger Ginsburger wrote in 1931 on the occasion of the *Französische Ausstellung* (French Exhibition) at the Kunstgewerbemuseum Zürich, obviously did not arise from a formal ideal of the inventor but were instead dictated by the cheapest and most practical method of production. This contrasted with similar products made in Germany in which one could often sense in the details "a formal intention, not openly avowed but operating on a subconscious level, to make something 'beautiful.'"[11]

Giedion, who approached BAG Turgi in 1932 in search of sensible lamps and a production facility for new models, suggested without further ado a reorganization of the company and a three-month trial phase, which he himself would make use of as a guest designer for the development of a modern sales range. Hin Bredendieck (1904–1995), a Bauhaus designer recommended by László Moholy-Nagy, worked at BAG from July to October on this signal Swiss contribution to modern lighting. He exhibited an initial prototype series at the show *Das Licht im Heim, Büro und Werkstatt* (Lighting in the Home, Office, and Workshop), which opened on October 1 at the Kunstgewerbemuseum Zürich. The time was ripe: the whole city celebrated the "Lighting Week" conceived as a folk festival by the Zentrale für Lichtwirtschaft (Center for the Lighting Industry).[12] The markedly minimalist lamp series produced by BAG betrays with its many ball joints Giedion's ambition to build directly on the French industrial products. The most important of the designs was undoubtedly the *Indi Lamp,* which could be used either as a floor or table lamp for indirect lighting. An early model had both an inner reflector and an elegantly shaped outer funnel. Giedion and Bredendieck then retained only what was required by the law of maximum light production—the reflector—and separated it from the shaft, so that "mosquitoes, flies, dirt, etc. could not collect at the bottom of the funnel." What's more, when the lamp was switched on, a small ring of light was formed "that prevents the lamp from becoming a relatively dark object in the room."[13] A ring-pull switch made of Galalith emphasized the transition from shaft to stand. A rubber ring was wedged into the rim of the plate-like foot, which—along with the light ring and the pull switch—was immediately registered for a patent. Together with the photographer Hans Finsler, Giedion then also designed the advertising prospectus that, as a supplement in the magazine *Das Werk,* was ultimately intended to reach all members of the Werkbund and the Bundes Schweizer Architekten (Federation of Swiss Architects).

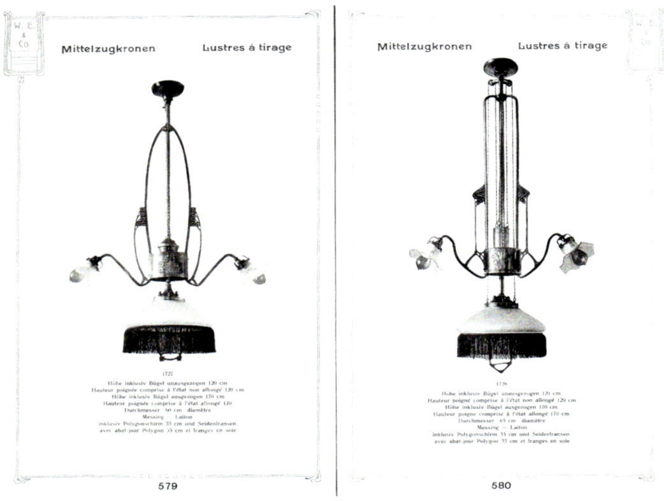

262

263

264

265

266

262 Center-pull chandeliers, 1729 and 1730, from the catalogue of Wilhelm Egloff & Co., Zurich, 1908

263 Ceiling lamp in the dining room of Villa Mooser-Nef, Zurich. Architecture by Niedermann & Staehli, 1912–13

264 Charles-Edouard Jeanneret (Le Corbusier), Table lamp, 1916, painted ceramic, brass, and fabric, location unknown. Photo: H. Allen Brooks, before 1978

265 Watchmaker's lamp 685 from the catalogue of Wilhelm Egloff & Co., Zurich, 1908

266 Max Ernst Haefeli, Standing lamp from *Das neue Heim II,* 1928, iron, brass, and Jena glass, Eberth & Co., Zurich. Photographed at the model home at Wasserwerkstrasse 29, Zurich, 1928
Photo: Heinrich Wolf-Bender

267

268

267 Hans Finsler, BAG 4.836 Indi Lamp with reading arm (product design by Sigfried Giedion and Hin Bredendieck, 1932), iron, brass, and Galalith. Photographed at the Neubühl housing development, Zürich-Wollishofen, 1932
268 Max Bill, Indirect pendant lamp, 1944–46, painted aluminum, BAG Bronzewarenfabrik AG, Turgi

Industrial Design

In Geneva, the metalworker Edmond Wanner (1898–1965)—the client for Le Corbusier and Pierre Jeanneret's project Immeuble Clarté—also ventured in 1932 to take the step from hand-forged floor lamp to a version for indirect lighting plated in mirror-finish chrome, although not yet able to fully cast off the trappings of Art Deco design.[14] The Zurich graphic designer, architect, and artist Max Bill (1908–1994) examined programmatic variations on the theme of the *Indi Lamp,* but by no means sought—like Giedion—to conceive the sparest formal representation of pure function. He gave the indirect pendant lamp he designed in 1944 an outer housing that sheds light through slits at the upper and lower edges, thus staging its own robust physical presence. His goal was a form that, although derived from functional design, was still vividly sculptural. In accordance with his now famous maxim for future industrial design, the beauty of an object by no means resulted automatically from perfect function but was itself a function requiring deliberate design.[15]

Most of the lamps designed in the spirit of *Gute Form* follow Bill's dictum. The light source was still the lightbulb and the respective lamp type already existed; the only question to be solved was the design.[16] The scissors-arm lamp *Lux* (1949), an in-house design by the Basel company Amba, had a timeless, unpretentious air, while Gian Franco Legler's (b. 1922) *Movalux* floor or ceiling lamp (1950), a variation on Edouard-Wilfrid Buquet's jointed lamps from the twenties, anticipated the fifties fashion for lamps that looked as lightweight as mobiles. A decisive role in the further development of lighting design was played by the interior designer Rosmarie Baltensweiler-Schwarz (b. 1927) and the electrical engineer Rico Baltensweiler (1920–1987), who worked together as designers and producers first in Lucerne from 1951, and then in Ebikon.[17] Their very first product, the *Type 600* floor lamp, went on to become an icon of nomadic living that is still sold today. Thanks to its patented spring joints, it can be swiveled in all directions. With an aluminum reflector that came to be known as the "Chinese hat," the lamp can be positioned to create either indirect lighting or an atmospheric accent. The high-quality components developed by the Baltensweilers formed the basis for a homogeneous-looking family of flexible models, such as the *60 S* table lamp (1956). The series has only a few rarities, such as the *Guggerli* (1966) with its glass spheres cradled in an openwork plastic tube.

From Edi Franz (b. 1924) and Reni Trüdinger (1927–2000) to Alfred Hablützel (b. 1931) and Andreas Christen (1936–2006), key players in the fifties and sixties devoted their energies to designing metal lamps. Hans Eichenberger's (b. 1926) height-adjustable floor lamp made of square-section tubes with a white fiberglass shade (1954) has a cross-shaped foot that perhaps alludes to its Swiss origins, or else pays homage to his great role model Mies van der Rohe. In 1965, Robert Haussmann (b. 1931) emulated the trend toward replacing standardized industrial furniture types with modular furnishing systems in his *Light Structure,* which can be expanded in all directions, while the *Cloud Lamp* created by Susi (b. 1938) and Ueli (1937–2008) Berger in 1970 and the colored floor lamp *18149* launched on the market by Swisslamps International in 1971—both made of polystyrene—signaled the advent of plastics in the world of interior decoration.

New Kinds of Lightbulbs

Around 1970, what had once been regarded as the secure basis for any lamp design changed abruptly, because the classic lightbulb suddenly encountered some serious competition. A period of experimentation with new technologies began that still continues today. While the fluorescent tubes invented shortly before World War II had long since made inroads in the workplace, in the early sixties the halogen lamps developed by General Electric took homes, shops, and museums by storm with their intense white light. The delicate fixtures fitted with low-voltage halogen lamps share formal similarities in terms of the minimal size of the lamp head and the heavy transformer required. In addition to numerous popular Italian models—such as Richard Sapper's low-voltage *Tizio* halogen lamp from 1972—a few Swiss designs also achieved international renown. Of note here is the world's first low-voltage halogen system using lights clamped onto tensile contact cables—a maximum of ten meters long—with which the young designer Hannes Wettstein (1958–2008) sparked a major new lighting trend in the eighties.[18] Wettstein's *Metro-45* series was produced by the Belux company, founded in 1970 by Thomas Egloff in Gebenstorf, which also brought a number of innovative low-voltage halogen table lamps onto the market in 1982–83. Along with Wettstein's minimalist *Ball,* also available as a floor lamp, these include the multi-award-winning *Lifto* by Benjamin Thut (b. 1959), in which a gas-pressure spring patented by Belux moves two joints,[19] and Jo Niemeyer's (b. 1946) *Tubo,* the product of an international competition, in which a fixed strut can be shifted to adjust the lamp's height.

The ever-increasing demand for lighting has been accompanied by rapidly rising electricity consumption. No wonder that the call for greater energy efficiency ushered in the next burst of innovation. On the office scene, metal halide lamps seemed at first to promise a breakthrough, but their long switch-on time and the large ballast required had a negative impact on their popularity. The lighting program *Rectangolo* designed in 1999 by Silvio Schmed (b. 1952) for Prolux thus pursued a different strategy—and attained the status of a formally rigorous milestone in the race for a new type of lamp. It is intriguing not only for the sharp precision-welded edges of its sheet-metal body, but also for how it casts light not only up toward the ceiling, but also downward through a light-refracting linear prism. In contrast, the gas-discharge bulbs with integrated ballast launched on the market by Philips in 1980 as "energy-saving lamps" managed to prevail in the end, at least once their initial flicker was eliminated and the color reproduction was improved—and since the Zurich inventor Jürg Nigg patented his first lamp with a corresponding bulb socket in 1984.[20] For their Hebelstrasse House, the Basel architects Jacques Herzog and Pierre de Meuron (both born 1950) created in 1988 a variation on the theme of the candlestick that captivates with its metaphorical flair, made simply of a ballast concealed in a cube of copper or aluminum sheet metal and an energy-saving bulb. (Their later designs are likewise emblematic: the 2001 *Jingzi* for Belux, for example, clad in a teardrop-shaped silicone shell, blatantly recalls a sperm cell.[21]) Energy-saving bulbs can be used to replace incandescent bulbs in the imaginative sheet-metal pieces by the Neue Werkstatt in Winterthur (*Sissi,* 1988; *Penguin,* 1989) and for the technically updated versions of forties aluminum lamps that have been sold since 1984 by the company Licht + Raum in Bern.

At the moment, light-emitting diodes (LEDs) are replacing both lightbulbs and fluorescent lights. In the seventies, glowing green LEDs lit up the faces of digital watches for a few seconds at the touch of a button.[22] Today, an enormous outpouring of innovative energy lasting some ten years has given birth to a light source that is superior to all others when it comes to durability, energy efficiency, and now also color reproduction. The spectrum of applications ranges from lightbulb substitute to luminous construction component. Young designers are floating gleaming matte discs with bright edges over our dining tables. Baltensweiler already developed a versatile folding LED light in 2003, *Zett USB,* made up simply of an upper and a lower plate connected by a rod. "Playful ingenuity" (*Spielwitz*)[23] was also demonstrated by Michel Charlot (b. 1984) in his family of LED spots designed for Belux in 2012, called *U-Turn.* Using a magnetic ball joint, the spot can be swiveled with a maximum radius or even turned right around; along with its variable light angle, this feature makes *U-Turn* a "light tool with a big fun factor."[24]

1 Kugler 2013, p. 7.
2 Berselli 2013, pp. 20–27.
3 *Les Arts du Feu,* with stops in Geneva, La Chaux-de-Fonds, Zurich (under the title *Die schweizerische Glasindustrie*), Winterthur, and Lausanne. See also Wegleitung 17 published in 1917 by the Kunstgewerbemuseum Zürich for the exhibition *Beleuchtungskörper.*
4 The design survives only as a sketch (*Carnet Al-26*), but Jeanneret described it in detail in a letter to Fritz-Ernst Jeker dated September 11, 1919 (Fondation Le Corbusier, Paris); see Rüegg 2012a, pp. 230 and 67–70.
5 The "watchmaker's lamp" was already included in the 1908 catalogue of Wilhelm Egloff & Cie., Zurich (models 683, 685; catalogue: Collection of Arthur Rüegg). The design was adopted in 1909 by the company BAG Turgi (catalogue: Collection of Arthur Rüegg).
6 See Rüegg 2010, pp. 16–21. The text was reviewed by Dr. Meike Noll-Wagenfeld.
7 To cite just one example: the *PH* lamp developed by Poul Henningsen in Denmark in 1925 was immediately put into production by Louis Poulsen, whose initial workforce of thirty nearly exploded in 1926, with 372 smaller and larger resellers registered thereafter from Bergen to Cape Town; see Jørstian/Munk Nielsen 1994.
8 Sigfried Giedion, "Die nächste Etappe der Wohnbeleuchtung," typescript, Estate of Sigfried Giedion, gta Archiv, ETH Zürich.
9 See Rüegg 1989.
10 The models were designed by Albin Gras starting in 1921 and were sold by Wohnbedarf AG Zurich from 1931 on; see Teissonnière 2012.
11 Ginsburger 1931.
12 See Bignens 2013.
13 Both quotes: Sigfried Giedion, *BAG Turgi – wohnbedarf typ – indi-leuchten,* prospectus, photography by Hans Finsler, 1932.
14 See Le Corbusier 1933. The illustrations juxtapose the furnishing ensembles put together by the Geneva set dresser with furniture from Wohnbedarf AG in Zurich. On the photos taken by Boissonnas in October 1932, Wanner's lamps can already be seen, while on Finsler's June photos of the Wohnbedarf ensemble, only the previous model of the *Indi* lamp is visible.
15 Bill (1948) 2010.
16 Details on the lamps cited can be found in Rüegg 2002, "Catalogue of Objects," pp. 336–99.
17 Rico Baltensweiler worked for the Swiss Federal Railways as an electrical engineer and Rosmarie Schwarz was an assistant to Max Bill in Zurich. They designed their first two lamps in 1951 for their own use. The *Type 600* floor lamp made its way into Le Corbusier's Villa Shodan in the Indian town of Ahmedabad as well as Jacques Tati's film *Mon oncle.* Later, the Baltensweiler company was able to incorporate the changing forms taken by the lightbulb into groundbreaking new models.
18 See Meier 2002b.
19 See Meier 2002a.
20 Jürg Nigg now runs the company Arcotronic AG in Zurich. Along with his adapter for energy-saving bulbs, he has seventy additional patents.
21 The *Jingzi,* together with the *Lucifer* by Hannes Wettstein (1992), are the only Swiss lamps that made it into the "Masterpieces" selection of one hundred lamps in the 2013 Vitra catalogue *Lightopia* (Kries/Kugler 2013).
22 A good summary of the current state of the art is provided in Wiseman 2013 and Kries/Kugler 2013.
23 *Spielwitz* is a term coupling "play" and "wit" used by Claude Lichtenstein that eludes simple translation; see Lichtenstein 2007.
24 According to Belux's online brochure, October 20, 2012.

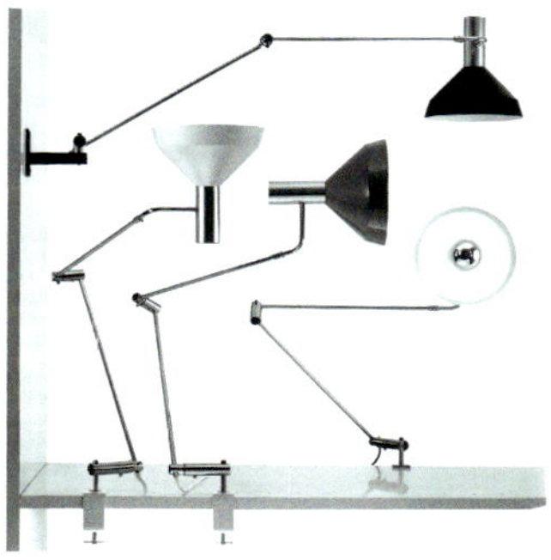
269

270

271

272

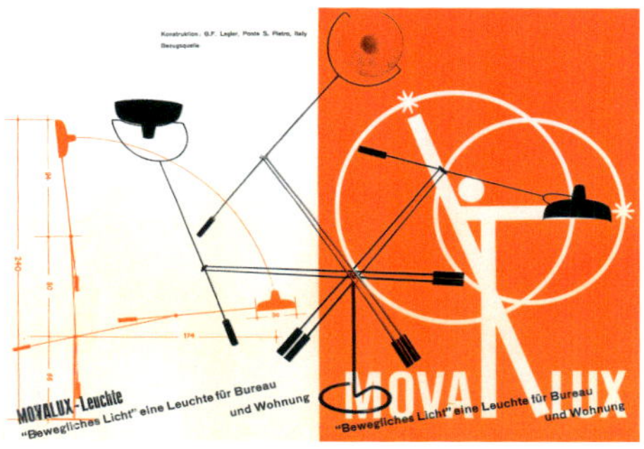

273

274

275

276

277

269 Rosmarie and Rico Baltensweiler, Type 60 S desk and wall lamps, 1954, anodized/painted aluminum, chrome-plated metal rods, Baltensweiler AG, Ebikon. Photo: Siegfried Zingg

270 Edi Franz, Floor lamp 18090, 1956, chrome-plated pole and side arm; foot, reflector shaft, and aperture ring in black wrinkle-finish paint, Waldvogel, Zurich; Swisslamps International, Zurich

271 Alfred Hablützel, La Ligne table lamp, 1957, chrome-plated brass, white satin glass globe, Swisslamps International, Zurich
Photo: Michael Wolgensinger

272 Andreas Christen, Floor lamp, 1958, chrome-plated steel pole, cast-iron foot, white acrylic glass shade, Lehni AG, Dübendorf
Museum für Gestaltung

273 Max Huber, Prospectus for the Movalux floor lamp (product design by Franco Legler), 1950. Front cover

274 Robert Haussmann, Lichtstruktur (Light Structure), 1965, mirror-finish aluminum, black plastic nodes, Swisslamps International, Zurich

275 Jacques Herzog and Pierre de Meuron, Wall fixture for the Hebelstrasse House, Basel, 1987–88, aluminum with energy-saving bulb
Collection of Arthur Rüegg

276 Jacques Herzog and Pierre de Meuron, Jingzi ceiling lamp, 2001, silicon, Belux, Birsfelden. Photographed at the SBB Restaurant Elsässertor, architecture by Herzog & de Meuron, 2006

277 Baltensweiler Niederberger, Zett USB color LED reading lamp, colored carbon fiber, Baltensweiler AG, Ebikon

278

279

280

281

278 The Aluminium-Industrie AG (AIAG) founded the world's first aluminum works in 1888 at the Rhine Falls near Neuhausen.
279 Charles Christofle, Centerpiece made of aluminum, 1858. The first aluminum objects were luxury items.
280 View of the aluminum pavilion at the Swiss National Exhibition, Zurich, 1939 (architect Josef Schütz). The exhibition displayed the use of the material in vehicle, aircraft, and container construction.
281 Josef Schütz, Aluminum pavilion at the Swiss National Exhibition, Zurich, 1939. View of entrance with aluminum awning.

"With the power of its underground water, Switzerland is producing its own metal," proclaimed the sign in the aluminum pavilion at the Swiss National Exhibition in 1939.[1] The ability to generate cheap electricity using hydropower was a major advantage for Switzerland in the race toward economical aluminum production. This is because fused-salt electrolysis—a process developed in 1886 that is still used today to produce aluminum—requires huge amounts of electricity. Back then, Switzerland had the pioneering spirit and the technical know-how to transform French bauxite into aluminum on a grand scale. Founded in 1888, the Aluminium-Industrie AG (AIAG) operated the world's first aluminum smelter at the Rhine Falls near Neuhausen and was for a few decades the leading aluminum producer, followed by the United States and France.[2] Switzerland transformed aluminum from a luxury metal into an affordable universal material for everyday objects. Apart from wood, aluminum was the only raw material of Swiss origin—even though the natural resources that went into making it had to be imported. At the 1939 National Exhibition, two hundred Swiss manufacturers of aluminum products did their best to cultivate the myths surrounding their national metal.

Between Euphoria and Reality

The discovery of aluminum in 1807 sparked a genuine euphoria. For the first time, there was a material that was at once hard, lightweight, and extremely malleable. Scientists were convinced that just about anything could be made from it. Until that could happen, however, it took some major technological developments and educational work to dispel people's skepticism toward the unknown metal. Aluminum's heyday coincided with the advent of modernism, helped along by the way that products made of aluminum perfectly answered the concerns of the functionalists: they were sober, true to their material, and capable of being serially produced. Aluminum was equated like hardly any other material with aspects such as modernity, dynamism, and purity. Thanks to its material properties, it would leave a lasting mark on twentieth-century design.[3]

Lightness Is Everything

The great catalyst for the boom in aluminum was the wartime economy. In 1892, four out of five orders came from the defense industry, and during World War I, some 90 percent of the aluminum produced was likely used for military purposes.[4] In the nineteenth century, the AIAG produced flasks and boilers to be used in the field for the war ministries of Prussia and Russia. In the Swiss Army, the kidney-shaped aluminum mess tin replaced the tin-plated steel models from about 1898. A huge test field had opened up for the new material. By 1902, the Belgian, Danish, German, Norwegian, Russian, Swedish, and Swiss armies were using eating and drinking utensils made of aluminum, while steel pots and pans were still the rule in France, England, and the US. However, aluminum did not pass the test for every use. The field flasks were too thin-walled and liable to burst, and the aluminum contaminated the water. First doubts surfaced as to the metal's unrestricted suitability.

The field of aviation was the area to benefit the most from the new metal. There was nothing like it in terms of hardness, lightness, and low combustibility. The Zeppelin and Junkers factories expanded their—defense-related—production several times over, and the AIAG delivered 80 percent of its production to Germany during World War I. After the end of the war came a search for new markets, leading to an increase in the domestic processing of aluminum. The knowledge gained in the aircraft industry in terms of lightweight construction and aerodynamic design was later reflected in diverse civilian industrial applications, whether in the Swiss Railways' CLe 2/4 light railcar, the postal service's *Car Alpin,* or the bicycles produced under the Stella brand. Streamlined auto bodies could be built with aluminum, and aluminum components also made the vehicles lighter and faster. Modern curvaceous forms as well as increased capacity and fuel savings were further motivating factors for the use of aluminum.[5] Even today, the transport and aviation industries would be hard put to do without this metal. Although the use of plastics has been steadily increasing since the eighties, aircraft still consist of around 70 percent aluminum.

Fabrication and Form

Aluminum's properties allowed for new manufacturing techniques. It was so ductile that it lent itself not only to the formation of open, conical containers from a single piece but also to the rational production of bottles in just a few working steps. The impact extrusion process was one of the technical prerequisites for the mass spread of pharmaceutical packaging and kitchenware. Pots and pans made of aluminum were lightweight and thin-walled. They had no welds vulnerable to corrosion and displayed better heat conduction than cast-iron pans. Manufacturers of electric stoves promoted the use of aluminum cooking utensils, teaming up with their makers in order to make sure that unsatisfactory results from cooking with poorly conductive iron pans would not discredit the new cooking method from the very start. With the inception of electric cooking, the sale of aluminum pans thus rose steeply.[6]

In contrast to other metals, aluminum could be injected. The company Injecta AG began in 1919 to develop and perfect a method for high-pressure injection of very exact forms. This so-called injection molding was a world first. The compact cast parts could be used to build machines and housing. The inner mechanics and the body were fused in the cast aluminum part into a single formal unit. A major advantage was that injection-molded parts were so precise that they did not require any post-processing. They were ready to use right out of the mold. This direct connection between fabrication and form distinguished many aluminum objects. They were prime examples of functional design while also expressing Swiss product ethics.

Supplanting and Being Supplanted

Aluminum crowded out several other materials from the market. In the kitchen, these were copper and iron. But despite its widespread use at times—two-thirds of Swiss households were still using aluminum cookware in the fifties—it was in fact a transitional material that managed to prevail for a long time for economic reasons but was ultimately replaced for reasons of hygiene. Today, aluminum is used only for the undersides of pots and pans, where its good thermal conductivity is still appreciated. As was already the case with the field flasks used in the army, aluminum can be problematic when it comes into prolonged direct contact with food and drink or is not cared for properly. Producers tried to prevent unwanted contact through surface coatings, but the economic boom and the associated price pressure in the sixties finally made it possible to switch to stainless steel and plastic.[7]

The Sigg bottle and the Rex peeler are products that have survived to this day. They represent the uncompromising functionalism typical of aluminum. Not much has changed since the beginnings in terms of their fabrication and form. The manufacturers did however have to respond to increased demands with respect to use and cleaning. The aluminum bottle is today coated inside with a varnish to prevent any metallic taste or oxidation. The original Rex peeler in aluminum was joined in the seventies and eighties by additional models in stainless steel and plastic—with more to follow. However, regular users still swear by the original aluminum peeler due to its low weight, nonslip edges, and warm grip.

Old from New

Aluminum has been recycled ever since the beginning of its commercial production. Waste generated during aluminum production and processing is separated according to alloy, melted down, and returned to the production cycle. In addition to this new scrap, old scrap from used products has been piling up substantially since the eighties. Most of this used aluminum, 90 percent, comes from the construction and automotive industries. As far as packaging is concerned, in Switzerland nine out of ten beverage cans are recycled but only half of used coffee capsules. Aluminum scrap can be recycled without loss of quality, requiring an energy consumption of approximately 5 percent of that of the production of primary aluminum—also taking into account the mining of the raw material bauxite and the transport of the finished products.[8]

282

283

284

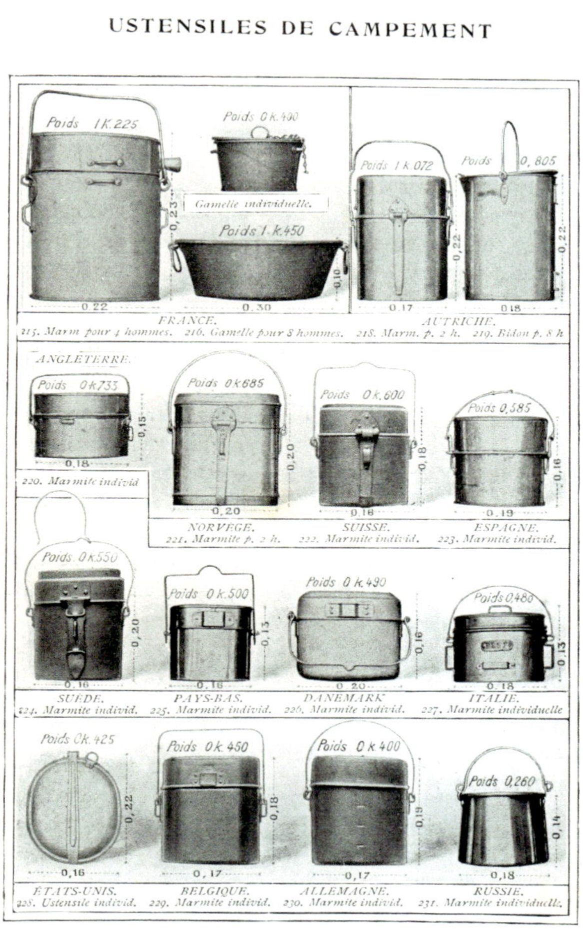

285

286

282 Aluminum casting at the 1939 Swiss National Exhibition. In the injection molding machine the aluminum alloys are injected into molds made of a special type of steel.

283 Aluminum casting at the 1939 Swiss National Exhibition. In addition to sheet metal processing, molding was the second most common procedure used to make household items.

284 Manufacture of pans in the deep drawing press, 1943, Sigg AG Aluminiumwarenfabrik, Frauenfeld

285 Ustensiles de Campement, catalogue for aluminum military tableware, 1902

286 Hermann Eidenbenz, Pour cuire à l'électricité, HK-Duro-Aluminium. Poster for Heinrich Kuhn Metallwaren AG, Rikon, 1952

287

288

289

290

291

292

287 Claude Dornier, Do X flying passenger boat, 1929, Dornier-Metallbauten, Friedrichshafen (DE). The Do X was at the time the world's biggest passenger aircraft, made entirely of aluminum.
288 Zeppelin-Werke, Friedrichshafen, Interior view of hall with support framework for an airship, 1931
289 Hans Erni, Picture of a passenger airplane for the classroom wall, 1941, Ernst Ingold & Co.
Swiss National Museum (LM-105963.1)

290 Alfred Neweczerzal, Rex economy peeler, 1936, Zena AG, Affoltern am Albis
Museum für Gestaltung/Donation of Zena AG
291 Borde cookware, ca. 1952. Three-piece set, for use with a gas cooker developed by the Borde company in Zurich. Used by the Swiss military (mountain infantry) during World War II.
Museum für Gestaltung/
Donation of Chris Niemeyer

292 Factory design, Model ALA 1823 canteen, 1940s, Sigg AG Aluminiumwarenfabrik, Frauenfeld
Museum für Gestaltung/
Donation of Marcel Herbst

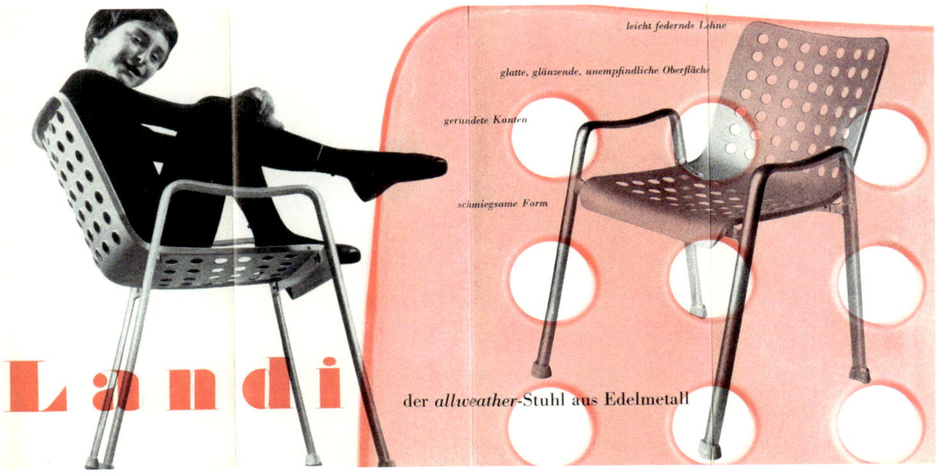

293

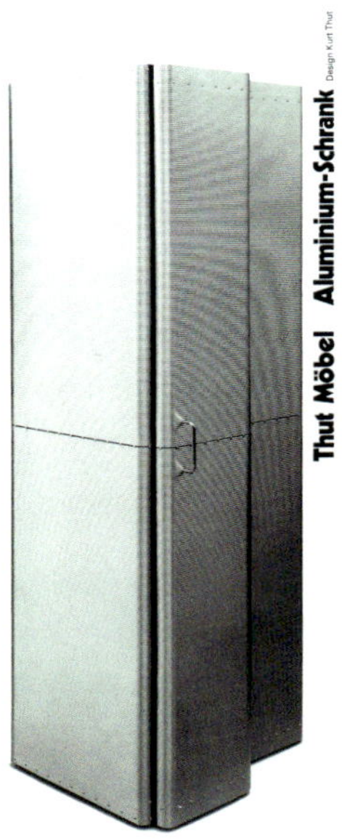

294

295

296

297

298

293 Prospectus for Hans Coray's Landi Chair, 1939, MEWA Blattmann, Wädenswil

294 Prospectus for Kurt Thut's Aluminum Cupboard, 1986, Thut Möbel AG, Möriken. The construction method is derived from metal aircraft like the Do X.

295 Prospectus for Andreas Christen's Regalsystem (shelf system), 1964, Lehni AG, Dübendorf (prospectus designed by Hans Peter Dubacher, ca. 1979). The shelf made of bent and folded aluminum sheets was developed for Expo 64.

296 Factory design, Teca water kettle, 1949, anodized aluminum, MEWA Blattmann, Wädenswil Museum für Gestaltung/Donation of MEWA

297 Valérie Jomini and Stanislas Zimmermann, itbox, 1998, anodized aluminum, Tara Möbel, Burgdorf Museum für Gestaltung*

298 Alfredo Häberli, Gordon armchair, 2001, anodized aluminum, Remo Trunz AG, Arbon Museum für Gestaltung*

Furniture Made of Sheet Aluminum

Interest in aluminum furniture has undergone major fluctuations. The metal could be found in material experiments in the modernist era and was valued in the eighties and nineties for its restrained air and high-tech appeal. Products from the pioneer era of aluminum furniture include the Elektron furniture (1927) designed by Max Ernst Haefeli, with frames made of the ultralight aircraft metal Elektron, the cantilevered chairs Marcel Breuer constructed out of aluminum profiles (see Lepel, "The Aluminum Industry Competition of 1933," pp. 130–31), as well as Hans Coray's garden chair for the 1939 National Exhibition. Coray went a step further than his predecessors and fused seat and backrest into one ergonomic shell made of a single sheet of aluminum, setting it atop a frame of bent aluminum profiles. The flanged perforations lend the sheet metal added stability. Shell chairs were a novelty at the time and helped aluminum to live up to its reputation as a state-of-the-art material.

Folded sheets and profiles are the preferred aluminum elements used by Swiss designers. International designers such as Jorge Pensi, Philippe Starck, Ron Arad, and Marc Newson tend to use instead embossed forms or cast parts for their furniture. Andreas Christen (1936–2006) revisited the subject of aluminum in an all-metal bookshelf he conceived for the Expo 64 Swiss National Exhibition. The idea of metal furniture for the home was at the time something completely new. Christen's product, known today as the *Lehni* shelf after its manufacturer, still consists of aluminum sheets that are able retain their shape and stability through folding and bending. This bookshelf led to a long line of aluminum furniture made from bent aluminum sheets. In the late eighties and during the nineties, high-tech aluminum furniture formed a counterpoint to colorful and eclectic postmodernist pieces. With their unity of form and design and modular construction, they can be regarded as modernist invocations. Kurt Thut (1931–2011) borrowed engineering principles from metal aircraft construction for the design of his 1986 *Aluminum Cupboard.* A skeleton of plywood and aluminum profiles is sheathed in finely ribbed aluminum sheeting. The decoratively placed rivets are not just another reference to aircraft technology but in fact the preferred method for affixing aluminum sheets.

Every time a piece of furniture is to be produced in a small series with minimal technical effort, aluminum is the material of choice. A furniture item made of aluminum sheet is less expensive to produce in small quantities than one made of wood. The *Pepe* rolling cabinet (2002) designed by Hanspeter Weidmann is based on the principle of the dressmaker's pattern. The form is cut out of a single aluminum sheet in such a way that the sheet can then be folded into a three-dimensional piece of furniture with minimal waste. Spatial visualization and the ability to minimize and to optimally turn structural engineering details into eye-catching aesthetic features are indispensable requirements for this type of design process. And aluminum is the ideal material for the job: because it doesn't have a front or back side, it can be bent into tight radii, and its edges are less sharp than those of sheet steel.

Indispensable and Invisible

In 2001, the Algroup, formerly known as Alusuisse and originally AIAG, was sold to the Canadian firm Alcan. The last aluminum smelters in the Valais region closed down a few years later.[9] Today, raw aluminum is imported into Switzerland and processed into high-quality semifinished products. The comparatively few manufacturers still using it specialize in products meeting the most demanding quality standards. Modernist designers prized aluminum for its lightness in both the constructive and aesthetic sense. Aluminum lent itself to minimizing material and form, revealing the construction, and creating pared-down, sober shapes. However, in those applications where steel or tubular steel already fulfilled modernist requirements—such as in the furniture industry—aluminum had a difficult time supplanting the heavier metal. Nonetheless, this lightweight material is still important today wherever reduced weight saves costs. In automotive and aircraft construction, in engines and machines, aluminum components, though increasingly invisible, have become more essential than ever.

1 *Die Schweiz im Spiegel* 1940, pp. 743–70.
2 On the history of the aluminum industry, see AIAG 1890, AIAG 1942–43, Falk/Schwarz 1991, Bauer et al. 1989, Knöpfli 2010.
3 On the international design products, see Nichols 2001.
4 Bauer et al. 1989, pp. 29 and 89.
5 For more on this subject, see Jost 1981.
6 Large parts of the present text are based on Sonderegger 1997 (with detailed bibliography).
7 Statements by former employees of the companies Kuhn-Rikon and Sigg AG, Frauenfeld, as well as remarks in the sales catalogues (Kuhn, Sigg, Stöckli).
8 Information from IGORA-Genossenschaft Zürich and Aluminium-Verband Schweiz.
9 For more on this subject, see Bärtschi 2011.

299

300

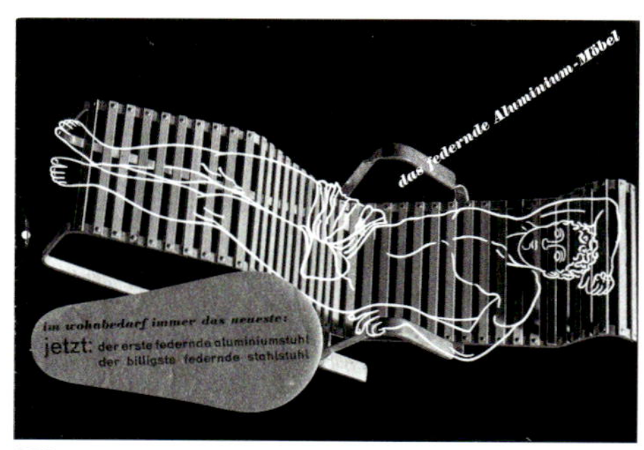

301

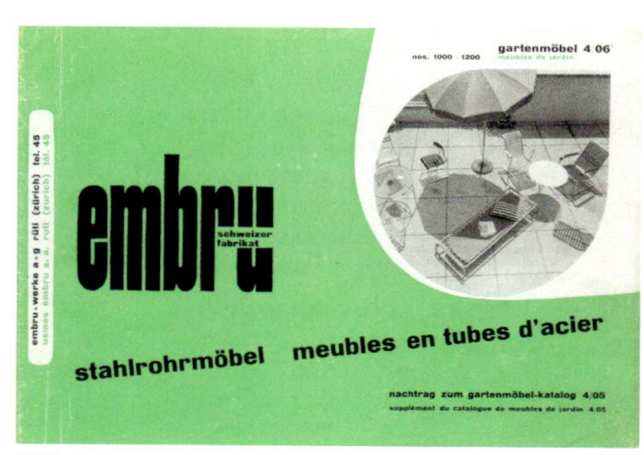

302

299 Four of five prototypes for the Breuer aluminum series, Embru-Werke AG, Rüti, 1933. The series was awarded first prize by both juries in the international competition hosted by the Alliance Aluminium Cie., Paris. Photo from the official documentation of the competition entries.

300 Marcel Breuer, Aluminum Armchair, 1933, Embru-Werke AG, Rüti
Museum für Gestaltung*

301 Herbert Bayer, Resilient aluminum furniture, 1934, Wohnbedarf AG, Zurich/Basel/Bern. Catalogue

302 Pierre Gauchat, Embru tubular steel furniture, Garden furniture 4/06, 1936, Embru-Werke AG, Rüti. Catalogue

303

303 Swiss pavilion at the Brussels International Exposition, 1935 (architecture by Hans Hofmann). Lightweight garden furniture made of aluminum and steel positioned in the passageway between the clock and machine halls; in the background a mountain panorama with nighttime illumination (designed by Max Bill).

When world aluminum sales dropped dramatically in the years 1930 to 1933 as a consequence of the 1929 world economic crisis, the Basel-based aluminum producers' cartel Alliance Aluminium Compagnie (AAC) decided to hold a special competition for seating furniture made of aluminum. The Concours international du meilleur siège en aluminium was announced in January 1933, with entries to be accepted as of October 1 of the same year.

In May 1933, Sigfried Giedion, in his capacity as secretary-general of the Congrès internationaux d'architecture moderne (CIAM), contacted the manager of the Aluminium-Industrie AG in Neuhausen, Dr. von Schulthess, to say that he had heard about the competition and wanted to suggest that CIAM should have one or two delegates in the jury. Giedion's letter came quite late, because the conditions for the AAC competition had already been set at the end of 1932. Von Schulthess thus expressed his regrets to Giedion that the jury had already been constituted before the competition opened and that a change in its composition was no longer possible. The jury comprised, in addition to AAC representatives, several furniture industry professionals.

Giedion did not approve of the jury composition, so he vehemently campaigned for some progressive modernist architects to be added to the mix. After some intense discussions with the AIAG Neuhausen, he managed to persuade the AAC to agree to his idea and to allow the establishment of a "special jury" recommended by Giedion, composed of himself, Walter Gropius, and Le Corbusier. The AAC also provided the special jury with a budget of 1,000 Swiss francs for an extra award as well as pledging to cover all the expenses of its three members.

In accordance with the conditions set down by the AAC, entrants could choose one of two ways to participate in the competition. For the first category, "finished" designs were demanded, in other words, full-size models along with cost calculations; while in the second category only drawings or sketches were requested. The considerable sum of 6,000 Swiss francs was made available as prize money, with the addition of the aforementioned extra award of 1,000 Swiss francs to be bestowed by the CIAM jury.

Giedion thereupon encouraged Marcel Breuer, who in those years spent much of his time in Switzerland, to participate in the competition. Indeed, Breuer possessed an international patent that made it possible to circumvent the copyrights held by Mart Stam and Mies van der Rohe to the cantilevered chair. In order to obtain his patent, Breuer had given the chair an additional support element, which also allowed him to use materials with reduced load-bearing capacity, such as wood, or for that matter aluminum, for the construction. The company Embru AG in Rüti now developed prototypes and samples for Paris in close collaboration with Marcel Breuer, and Embru also covered the costs for the development, tools, and all expenses.

The jurors met in Paris on November 24, 1933, to appraise the approximately two hundred entries from fifteen countries. The five models submitted by Breuer and Embru fulfilled in an exemplary manner the essential requirements set by Gropius. The two juries unanimously agreed, and Marcel Breuer was awarded both first prizes. Breuer, unlike other entrants, had indeed not just converted existing tubular steel furniture into aluminum versions but had succeeded in optimally exploiting the singular properties of the prescribed material.

In addition to Breuer's designs, the Swiss perspective demands that we also recognize the contribution made by the architect Alfred Roth. Since very few entries dealt with outdoor furniture, special attention was devoted to the three models submitted by Roth in the competition overview published by the trade journal *Aluminium Revue,* which especially praised their "finesse de la ligne," the delicacy of the lines.

Alfred Roth's pieces did in fact make very good use of aluminum's material properties; they were light in weight and rustproof, making them ideal as garden furniture. However, as Roth himself admitted, they were not sufficiently stable and therefore did not go into serial production after the competition.

Marcel Breuer by contrast was able to award the licenses for producing his designs in Switzerland to Embru. Unfortunately, their sales did not live up to the expectations of the chronically empty-pocketed architect.

304

305

Stuhl aus gebogenem Holz
Schleiflack wasserfest

mit Armlehne

Bürosessel aus gebogenem Holz und
Rohrgeflecht

Fauteuil-Sessel mit rundem Tischchen / Entwurf aller Möbel: M. E. Haefeli, Arch., Zürich und E. Kadler, Glarus

Fahrbarer Servier-Tisch, Schleiflack, mit Ausziehtablaren

Allseitig glatt fournierter Kinostuhl, staubfrei

239

306

307

308

304 Wilhelm Kienzle, Net shopping bag made of rope, storable in the wooden handle, ca. 1950
Illustration by Warja Lavater, ca. 1950
Museum für Gestaltung/
Donation of Edith Kleinberger

305 Max Ernst Haefeli, Rolling reading desk for the Bücherstube Girsberger, Zurich, 1926. Stained larchwood, adjustable matte-varnished desktop and display shelf
Private collection

306 Max Ernst Haefeli and Ernst Kadler-Vögeli, Chairs and occasional pieces for the AG Möbelfabrik Horgen-Glarus, ca. 1927. In the middle, the "Bücherstube chair"

307 Max Ernst Haefeli, Bookshop furnishings

308 Girsberger & Co., Zurich, 1926. Reading niche and "Bücherbar" (book bar) with adjustable display shelves. On the left, Hans Girsberger; on the right, Willy Zahn

What do a sewing machine and a mesh shopping bag, or a pencil and a tea table, have in common? Although this question might at first summon thoughts of Lautréamont's neat summation of Surrealism ("as beautiful as the chance encounter of an umbrella and a sewing machine on a dissecting table"), it leads in the present context in exactly the opposite direction: to a pronounced realism. This is a specifically Swiss attitude. Yes, there is a common denominator between the Elna sewing machine in its foldout metal case and Wilhelm Kienzle's mesh shopping bag with a hollow wooden handle, as well as between the *Fixpencil* "architect's pencil" from Caran d'Ache (with a lead sharpener concealed in the button) and the folding tea table by Werner M. Moser.[1] It consists in a certain propensity for solving a problem before others have even become aware of it. Is this part of the Swiss DNA? The diversity, density, and quality of such solutions are certainly striking. This theme is likewise typically Swiss: an emotional penchant for little things that seem to loom so large when regarded close up and remain just as important when appraised from the outside. In a 1950 report on the telephone booths of the Swiss Post-, Telefon- und Telegrafenbetriebe (PTT), the London-based magazine *Design* referred to them as "efficient and unassuming" and noted in envious admiration "the unfailing attention to detail which was once said to be the mark of genius."[2] There may be other definitions of genius, but this one is particularly apt for Switzerland.

This essay will focus on ingeniously designed furniture as versatile all-around pieces, based on the early years of Wohnbedarf. The theme was of course already pertinent before 1931, when the furniture store was founded, and is just as topical today, as evidenced by the great interest shown everywhere in furniture design.

The new architecture and the new interior decorating style came onto the scene later in Switzerland than, for example, in the Netherlands or Germany, but what resulted was more than just the product of a delay or phase shift. Swiss modernism took a different tone from the outset. The villa as a laboratory and a manifesto for "flowing" space behind display-window-sized glass fronts was nonexistent in (German) Swiss modernism—in contrast to the imposing examples on the outskirts of Paris, Berlin, or Brno. The frame of reference for developments in this country was instead the rental apartment, and its reconsideration was driven primarily by like-minded spirits in Basel and Zurich. The most articulate among them, Hans Schmidt, noted in 1929, "Consumers (buyers and users) will learn to regard furniture as the simplest utilitarian object rather than as the showpiece it is taken for today.... Modern furniture ... will be based on ordinary everyday usefulness for sitting, reclining, clearing away, and storing, and will seek to respond to these needs with the least possible technical effort and expense."[3] This prediction obviously turned out to be wrong, but mistakes sometimes have very positive consequences: the furniture that was developed as a result is a good example, with its unobtrusive elegance, economical gestures, and discerning love of fine structural details.

As one milestone in this development, the young Max Ernst Haefeli (1901–1976) redecorated Hans Girsberger's Bücherstube bookstore on Kirchgasse in Zurich in 1926.[4] His goal was to make the most of the space at hand while catering to customer needs. Portable lecterns on castors had two opposite tops that could be inclined individually as required. The adjustment mechanism was similar to that of a deck chair, with supporting poles that could be notched in at varying heights by way of an iron rod with brass knobs (a functional detail with an almost expressionist touch, a quality that is further emphasized by the reddish brown of the varnished wood surfaces). Brass trim protected the varnished feet. A "Bücherbar" (book bar) had additional foldout surfaces on all sides and also a floor lamp attached at one corner, with a fabric shade that widened toward the top. Haefeli also designed for the shop wall-mounted lamps on swivel arms, equipped with a counterweight and a looped cable so they could be adjusted in height. Haefeli's signature throughout—both in the whole and in the material details—is that of a forward-looking interior designer gifted with evident technical inventiveness. His material repertoire consisted of woodwork, painted iron rods, and brass handles. In contrast with the approaches taken by other designers that year (Hannes Meyer, De Stijl, and so on), Haefeli operated on the basis of cabinetmaking and with an apparent love for traditional craftsmanship. And yet he expands on artisanal skills to develop unconventional solutions from vividly conceived usage scenarios. But not only Haefeli mastered this ability. The same idea was pursued in the six apartments that the Schweizer Kollektivgruppe of the Schweizerische Werkbund (SWB) designed and furnished one year later (1927) in the apartment block Mies van der Rohe designed for Stuttgart's Weissenhof Settlement. These interiors were widely acclaimed for their clear-sightedness and conceptual independence.[5] Critics discerned here a quality that was arguably the key resource of the Swiss avant-garde: analytical intelligence and sensitivity rather than showiness; *a culture of function rather than one of status symbols.* This behavior can most likely be traced in part to the Swiss farmhouses with their ingenious space-saving devices, such as drawers concealed in the steps of staircases, flaps in the floor for heating, chests that did double duty as benches, and pull-out beds. These are all things that were determined by practical considerations in the face of material limitations under (non-feudal) conditions.[6]

From Furniture as Prototype...

When Haefeli's "model homes" on Wasserwerkstrasse went on display in 1928 as the first example of Neues Bauen (New Building) in Zurich, visitors encountered here, again, rooms with furniture that for the Swiss public embodied the objectives of *liberated living*[7] but which predominantly remained unavailable to them. Haefeli's dining table was a one-off, as were the bookshelves, and the exceptionally low sofa was probably the first piece of Swiss tubular steel furniture of this type.[8]

The year 1928 was also when the Basel architects Mumenthaler & Meier designed their first examples of *Typenmöbel,* or "type furniture": space-saving buffets and cupboards with modules of different dimensions that could be combined in varying ways. These pieces won first prize in a "type furniture" competition, after which they went into production at a cabinetmaking shop in Basel and were marketed under the name *3m* (Mumenthaler-Meier-Möbel).[9]

Since the new architecture came onto the scene in Switzerland later than in other countries and with more cautious (or prudent?) buildings, it lacked the instances of polemical exaggeration found elsewhere—a lack that was offset by a solid foundation. (Avant-garde as "patricide" seems not to have been an issue here, perhaps because knowledgeable and highly revered figures such as Karl Moser, Max Haefeli, Carl Steiger, and also the Basel professor of geology Carl Schmidt encouraged the next generation rather than trying to curtail their heirs' creativity.)

Furniture took the same path as architecture. The Bella Lui Sanatorium in Crans-Montana is a prime example.[10] The contract to design furnishings for the patients' rooms went to the architect Flora Steiger-Crawford (1899–1991)—the first woman to earn an architecture diploma at the ETH. (Curiously, she never mentioned the Bella Lui project in the memoirs she wrote late in life).[11] The tables, cabinets, shoe cupboards, and occasional furniture pieces she designed for the sanatorium, made of pine and plywood painted an elegant gray-green, with black linoleum tops and black-painted handles, were in part the precursors of the type furniture that would a short time later be sold by Wohnbedarf. Steiger-Crawford's furniture was rich in ingenious and cleverly implemented ideas. A table with a flap that folds out to reveal a vanity mirror on the inside is just as remarkable as a bedside table with extra shelves revealed by pulling up a retractable panel,[12] or wardrobes with open handles for easy transport.

One recurring theme in furniture developed circa 1930 is the extensive replacement of wood with metal. Kienzle's iron stool with webbing seat from 1926 was a forerunner, and in 1927 Haefeli designed a group of furniture (chair, stool, table) made of the lightweight alloy Elektron in which the bolting together of precast parts basically transferred the additive construction familiar from wood to the material of metal. All pieces of the Elektron furniture were employed in Stuttgart in 1927, impressing visitors. Haefeli's three-seater sofa for the Wasserwerk houses has a tubular steel frame that is still handled like traditional metalwork with axial joints. It was Haefeli's friend and colleague Werner M. Moser who in 1930 would take the next step, toward the "industrial-like" method of tube bending, which he cultivated in incomparable fashion in his occasional furniture.[13]

309

310

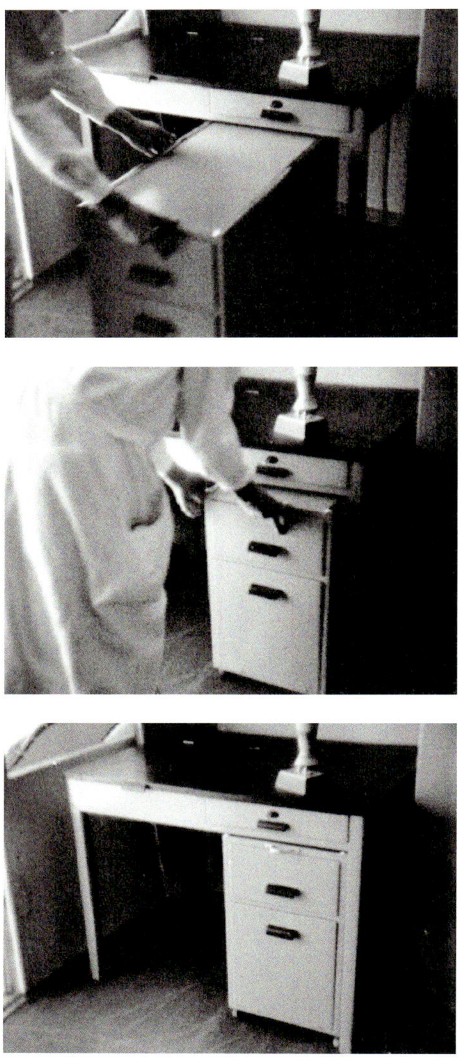

311

312

309 Marcel Breuer, Vanity with movable mirror for the Haus am Horn, Weimar, 1923

Flora Steiger-Crawford, Furnishings for patient rooms at the Bella Lui Sanatorium, 1929–30 (architecture by Rudolf and Flora Steiger-Crawford, Arnold Itten, 1928–30)

310 Flora Steiger-Crawford, Nightstand Museum für Gestaltung

311 Hans Richter, Werkbund film *Die neue Wohnung* (New Living), 1930, stills. The multifunctional rolling container fits perfectly underneath the desk.

312 Flora Steiger-Crawford, Desk with flip-up mirror Museum für Gestaltung

Alfred Altherr jun., Lounge chair Model 2064, 1931, Embru-Werke AG, Rüti

313 Advertising photos, ca. 1932. The chair can be adjusted into several different sitting and reclining positions and can be folded up for use as a bench.

314 Details of tube and joint geometry; flip-up armrests. Photo: Michael Lio

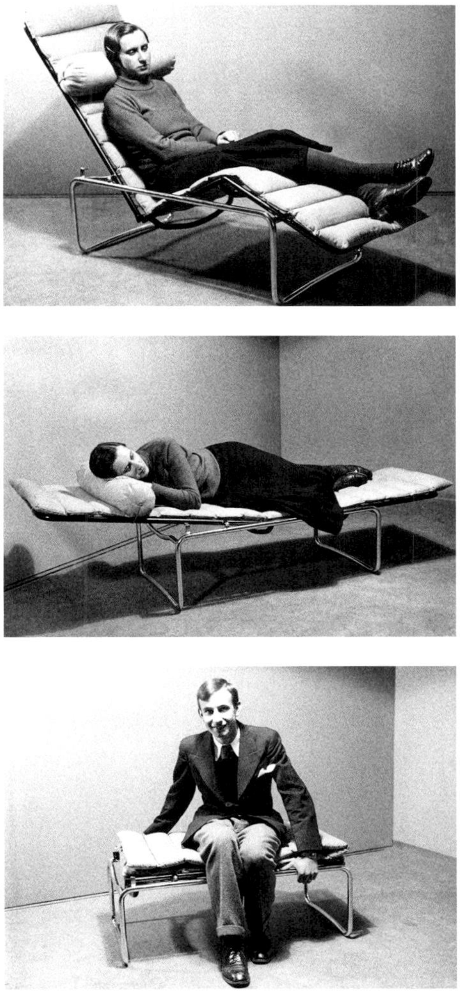

313

314

... to Furniture "To Go"

The situation had now changed. With the construction of whole housing developments, such as the Schorenmatten in Basel and especially the Neubühl development in Zurich, dozens of modern apartments suddenly had to be furnished in a fitting style. Furniture was expected not only to supply inspiration but to actually be available in large quantities. This rise in demand led to the founding of the Wohnbedarf AG shortly before the first stage of construction at Neubühl was concluded in May 1931. The partners in the enterprise were Sigfried Giedion (1888–1968), Werner M. Moser (1896–1970), and the merchandiser Rudolf Graber (1902–1971). "Wohnbedarf," as it was soon being referred to, would be the first furniture store that produced its own furniture types in all categories: seating and beds, tables, storage units and lamps, plus home textiles.[14] The concept behind the product range was, however, the antithesis to the bourgeois ideal of the "complete ensemble." Instead of stylistic uniformity among solitary pieces, the aim here was to achieve an *inner coherence* between the various furniture types. This coherence was thought to result from the functionality, restrained elegance, independence from fashion, and last but not least the space-saving design of the furniture; in short: its *rightness* for the purpose at hand. While up to this point in history the efforts of interior designers had focused on the single-family home or villa—that is, a well-known spatial situation—now "type furniture" came into its own: serially produced, practical, inexpensive, and commercially available models that were designed to harmonize with any home furnishing scheme as well as with each other. This was comfortable prêt à-porter rather than the earlier, heavy haute couture. In accordance with these intentions, the furniture had to be modest in appearance. The term and the logo "Wohnbedarf Typ" expressed the fact that a development division was at work behind the retail shop. The Wohnbedarf AG was a pioneer in conjoining the functions of design, development, manufacturing, sales, and customer service, and with the signature "Wohnbedarf Type" for its products also recognized early on the value of professional branding.

Moser had begun to delve intensively into the topic of type furniture in 1930, and was the first to design a cantilevered armchair along these lines, followed by folding tables and occasional furniture (nesting tables and side tables). These pieces anticipated Wohnbedarf's future product line: compact, sensitively designed furniture exuding refined elegance.

In Hans Eckstein's informative book *Die schöne Wohnung* (The Beautiful Home), published in Munich in 1934—it was ready to go to press before Nazi cultural policy began exercising its stranglehold—a living room in Neubühl (from an interior design exhibition in 1931) with furniture by Moser was described as follows: "A bright, light-flooded room with easily movable, comfortable, but extremely simple furnishings that aspire to be nothing more than objects of use. There are no conspicuous, bold colors in this room. Each bouquet, the color of each article of clothing (the human being himself, that is) as well as the green of nature thus stands out even more.... For all their capacious comfort, the armchairs nevertheless take up a minimum of space, making them suitable especially for the relatively small rooms of the homes in the new development. The room should after all belong to the people, not the furniture. Thanks to simple disassembly, these armchairs are easy to clean (the cushions are removable). The small round table can be folded up and effortlessly taken out onto the terrace."[15]

It is thus worthy of note how, even in this book with its diverse international examples and the obviously open-minded attitude of its editor, the Wohnbedarf models stand out with their "different character." It would be an amusing task to search for a connection between the bending radii and the provenance of a piece of tubular steel furniture.[16] What is obvious here is that the Swiss type furniture managed to achieve refined elegance without the peristaltic tube convolutions devised in other European countries.[17] This is no coincidence.

When Sigfried Giedion wrote one of his first texts, in 1923, about the "Bauhaus Week" in Weimar,[18] his report expressed a certain reservation: "It is good that at this point people are trying for once to take inspiration from the machine.... Floor lamps made of iron and glass tubes, without a dimly lit silk skirt, are reminiscent of scientific equipment, seating pieces resemble looms, furniture recalls printing presses, teapots look like glass water gauges. Only time will tell how far this radical clean sweep of all our romantic residues will prove to be creatively fruitful; for now, these are but

315 Herbert Bayer, First Wohnbedarf catalogue, 1932–33. Catalogue cover and double-page spread showing pieces by Werner M. Moser and Immeuble Clarté exhibition furnishings, Geneva, 1932

315

experiments, and we would indeed even be grateful were they merely to fulfill the function of freeing us from unnecessary ballast."[19] These are the words not of an opponent but of a comrade-in-arms and future friend of Gropius and Breuer—fascinated, but also detached. Giedion draws attention here to a Helvetic idiosyncrasy. By the time he cofounded Wohnbedarf in 1931, its cultural profile would be diametrically opposed to the experience that in 1923 had revealed itself to him as a kind of covert exorcism.

Modernism for Everyone

The compactness of the furniture was an important part of it.[20] With this goal in mind, the Swiss architect/designers took a fundamentally different approach than, for example, Marcel Breuer, whose epochal *B3* (*Wassily*) *Chair* embraces within its frame with its slung textile surfaces the expansive width of the traditional leather armchair as extravagant void. Steiger-Crawford's stacking steel chair, originally designed for the restaurant in the Zett House, was space-saving in use and even more so when stacked and put away. If you were to sit down at the table in this chair there was space to "park," unlike, for example, with Mies van der Rohe's dining room chair with its semicircular base extending at the front.[21] Compact dimensions were also the main feature of the dressers and cabinets, whose low height of only 155 centimeters meant that they took up a minimum of space. Storage units were originally not a high priority for designers, probably because the new housing developments were equipped with built-in cupboards and closets. In 1933, Wohnbedarf cabinets and buffets were revised and made "combinable from within": prefabricated attachment points allowed the internal compartments to be reorganized and expanded through plug-in tracks, removable shelves, inner drawers, and a metal shoe rack. The term "Inkombi" was supposedly coined for this system by László Moholy-Nagy.[22]

The *Moser Chair* is perhaps the most typical piece in the Swiss line of furniture for Wohnbedarf. Unlike many German, French, Dutch, Belgian, or Italian models, it does not rely on the dialectic of tubular frame and slung textile surfaces.[23] And unlike many other designers, Moser did not aim to make the seat, backrest, and armrests from a single line of tubing. His chair instead displays the graduated interplay of six actors: tube support, angled frame, chain mesh seat, cushions, upholstery fabric, and wooden armrests.[24] Aiming at the targeted interaction of various components, Moser's design experience must have been much different than that of Breuer in his model *B3,* whose logic led to the necessary conclusion that the designer need only, as an "operator," listen to reality in order to elicit the fantasy already residing therein.[25]

Such "fantasy" found no air to breathe in Switzerland. Instead, ideas arose here based on a brand of pragmatism unmatched elsewhere, because it never strays from the purely rational. Willy Boesiger's small folding tables for Café Select (1935) or Kienzle's ingenious Embru bookshelf with its wood shelves inserted into the upright metal sides (1931) could be cited here, as could from later years Kienzle's popular *Shoe Rack* (Wohnhilfe, 1951), Max Bill's versatile *Square-Round Table* (1952), or Jürg Bally's height-adjustable *S.T. Table* (1956), all the way to the Lehni aluminum shelf system by Andreas Christen (1964) and even the furniture systems by Stefan Irion (since 2001) or the first lamp models by the Baltensweiler company (since 1951). I would like, however, to close this list of tributes with the Embru lounger by Alfred Altherr jun. (1911–1972), an astounding invention by a twenty-year-old who had just returned from working in the studio of Le Corbusier and Pierre Jeanneret in Paris. While in the lounge chairs designed by the latter designers and by Charlotte Perriand the adjustment options were limited to changing the positions of the two rigid elements of the frame and reclining surface, in Altherr's model (1931) the reclining surface itself can be adjusted in several ways thanks to its flexibility. Using cleverly applied rotation axes and lever effects, an ingenious folding and shifting mechanism allows for numerous different positions between the extremes of horizontal lounger and being folded up for storage.[26]

Like the other examples, this piece of furniture attests to the designer's unshakeable faith in the beauty of function.

316

317

316 Max Bill, Wohnbedarf Typ, full-page ad, *schweizerischer Baukatalog,* 1932. Rudolf Graber demonstrates how the Wohnbedarf furniture works.

317 Werner Max Moser, Garden and work table, 1932, Wohnbedarf Model 81, with folding and sliding mechanisms. Collapsible for storage or transport.
Photo: Michael Lio
Museum für Gestaltung/
Donation of Arthur Rüegg

1 The metal carrying case for the first compact portable sewing machine (Elna, 1940) could be turned into a sewing table if the free arm wasn't used. See *Unbekannt – Vertraut* 1987. *Fixpencil* lead holder: Ibid. The circular tabletop of Moser's tea table could also be used as a serving tray.
2 H. L. 1951.
3 *Typenmöbel,* exh. cat. Gewerbemuseum Basel 1929, cited in Flierl 1965, pp. 46–47.
4 Factory drawings for the furniture, dated July 22, 1926. See Mehlau-Wiebking et al. 1989. Dr. Hans Girsberger was the founder of the Verlag für Architektur.
5 Members of the working group: Paul Artaria, E. F. Burckhardt, Karl Egender, Alfred Gradmann, M. E. Haefeli, Hans Hofmann, W. Kienzle, W. M. Moser, Hans Neisse, R. S. Rütschi, Franz Scheibler, Hans Schmidt, R. Steiger. See Schmidt 1927.
6 In 1948, Giedion associated American patent furniture with a young post-feudal democracy/civilization, limited space and financial resources, and technical resourcefulness, which he subsequently and self-critically connected with his own contribution to Wohnbedarf AG—one and a half decades before. American patent furniture of the eighteenth and early nineteenth centuries is related in its ideological distance from any culture of prestige to the Swiss avant-garde one hundred years later. See Giedion 1948, pp. 419–38, esp. 428, 430–32.
7 Giedion 1929.
8 The sofa was subsequently featured in the Embru company's catalogue, but soon replaced by Haefeli's sofa bed, as "Wohnbedarf Typ," which in turn was supplanted shortly thereafter by Aalto's sofa bed.
9 See Rüegg/Mumenthaler-Grisard 2000.
10 Bella Lui: sixty beds, architect: Rudolf Steiger, collaboration: Robert Winkler, furniture: Flora Steiger-Crawford, construction period: 1928–30.
11 Glanzmann et al. 2003.
12 This refers to both foldout and horizontally retractable doors.
13 The folding tables probably served as methodological and technical inspiration here. The first four-legged version of the tea table still had eight welded joints, while the second three-legged version had only three (and three bending points).
14 The company was founded on short notice, so that it was unable to offer a comprehensive product range from the very start. See Graber 1932.
15 Eckstein 1934, p. 19.
16 See Máčel/van Geest 1980.
17 Hans Schmidt took an ambivalent view of the vogue for tubular steel furniture, referring to it as tubular steel macaroni.
18 The first Bauhaus exhibition, in September 1923, was accompanied by a weeklong program of events.
19 See Giedion 1923.
20 In Neubühl the biggest living room measured twenty square meters, in other residence types about seventeen square meters, and a master bedroom around thirteen square meters. Many other apartments at the time had similarly sized rooms.
21 Rudolf Graber wrote of the Zett House chair: "It arose from the need to keep a large number of chairs in reserve in as compact a space as possible and also from the intention to finally create a small steel chair that didn't take up more space than a wooden one." See Graber 1932, p. 336.
22 Verbal remark made by Walter Custer, June 1986.
23 See Máčel/Geest 1980.
24 See von Moos 1980, p. 19.
25 Breuer: "We no longer have a need to create or venerate fantastic … flourishes of taste or of style (even the 'modern' style) at the expense of reality. We know that reality itself is inexhaustibly fantastic." See Breuer 1928, p. 133.
26 See Rüegg 2013.

318

Classroom at the Pestalozzi Children's Village, Trogen, 1945–49, architect: Hans Fischli, furnishings: Jacob Müller (Werkgenossenschaft Wohnhilfe, Zurich). The Trogen school desk with adjustable tilting desktop had a frame constructed according to the principles of the first Wohnhilfe program. The separation of desk and chair anticipated the school furniture that would become widespread a few years later. The ingeniously simple construction of the early Wohnhilfe furnishings—of which the chair was one—allowed for the division of labor between several trades (turner, wood bender, cabinetmaker).

Photo: Anita Niesz

1940 — Réduit

The army withdraws into its Alpine redoubt (*Réduit*), the family to the comforts of the cozy home. There, the modern postulate of an intimate connection between nature and architecture is balanced with the needs of traditional bourgeois domesticity. Meanwhile, the economy is plagued by wartime shortages. The prospect of the armaments industry being freed up after the war prompts preparations for exporting to areas affected by the conflict. The assumption of reconstruction aid captures the imagination of designers and encourages a symbiosis of industrial and craft techniques. Collapsible wooden furniture that comes in kits points the way toward escaping the folk *Heimatstil,* becoming in the early postwar years the leitmotif of Swiss product design. In the Swiss Werkbund (SWB), the artist Max Bill comes to the fore as a pioneer of what will come to be known as “industrial design.” At the end of the decade, he is able for the first time to present his ideas of *Gute Form* to a large audience.

319

Max Ernst Haefeli, Heberlein House, Wattwil, 1940–41. Living room with fireplace. The female client's distaste for the "hospital atmosphere of the Neues Bauen" correlated here with an architectural trend to respond to the personal needs of residents by formulating a customized interior ambience. Haefeli succeeded in translating the unpretentious appeal of his early interiors into a contemporary, organic form without succumbing to the temptations of the traditional, rustic *Heimatstil.*

320

Aldo van Eyck, Room remodeling, Zurich, 1946. For the remodeling of the tower room in the Villa Löffler, built in 1926, the young Dutch ETH graduate invented a light and airy interior that created a new room within the old—by shifting everything a few degrees from the geometry of the walls. The floating, "mobile" elements of this interior provide a foretaste of the repertoire of an entire era.

321

Wilhelm Kienzle, Strub-Gestell (Strub Frame), 1944–45, Robert Strub, Zurich. Advertising photo. Frugal use of materials, light weight, variability: the "collapsible frame for merchandise" found its way into many an office or home because it perfectly fit the lifestyle of the immediate postwar years. It was among the few exhibits featured in the special exhibition *Die gute Form* at the Basel Trade Fair (Muba) in 1949.

322

Architektengemeinschaft Kantonsspital Zürich (AKZ, Architects' Association of the Canton Hospital, Zurich), Hospital room sink, 1940–42. The goal was to achieve "dimensions and detailing familiar to people from their private daily environment." (Max Ernst Haefeli). "A rational arrangement of separate objects that have all been equally well designed for their different purposes results in a significant as well as a practical combination." (Max Bill)

323

Otto and Heinrich Pfister, Dining car for the Swiss Federal Railways (SBB), 1943–44, Schweizerische Wagons- und Aufzügefabrik AG, Schlieren-Zürich. The Pfister brothers were dedicated as architects to rational design down to the last detail. They also remodeled in the same unpretentious fashion the dining cars for the Rhaetian Railway, as well as the Zurich city buses (1927) and the "Elephant" motor trams for the Zurich tram line (1930).

324

Hans Hilfiker, Loading crane, ca. 1950, Swiss Federal Railways (SBB) and Konstruktionswerkstätten Wartmann AG, Brugg. Solid welded tower, Differdange steel beams. “Where will an era whose intellectual strength can be found in technical design lead us if it is not able to express the simple, honest, and conscientious fulfillment of duty?” (Hilfiker)

325

326

327

325 Antonio Vitali, Toy Figure Pairs, 1946
Wooden toy figures, 13/12.5 × 4 × 2 cm, sampling models in cast brass and aluminum (from 1953, for machine production), 19/12.5 × 4.5 × 2 cm
Museum für Gestaltung/
Donation of Antonio Vitali

326 Antonio Vitali, Fox toy figure, 1944.
Goose toy figure, 1960
Museum für Gestaltung/
Donation of Antonio Vitali

327 Documentation photo from the exhibition *Das Spielzeug. Spielsachen aus alter und neuer Zeit* (The Toy: Playthings from Ancient and Modern Times) with Deer by Antonio Vitali, Kunstgewerbemuseum Zürich, 1951

After training as a sculptor in Zurich and Paris, Antonio Vitali (1909–2008) became a father at the young age of twenty-three and then, shortly after the birth of his second child, a widower. During these years of crisis, he eked out a living as a photojournalist, but then war broke out. Vitali went into active duty, married again, and by 1942 had four children. Because money was always tight, he made his own furniture for the family and carved simple wooden animals as toys for his children. Watching them at play, he realized that rather than needing naturalistic depictions they were capable of seeing animals or other objects in randomly shaped pieces of scrap wood. This experience, as well as the influence of Rudolf Steiner's anthroposophy, encouraged him to pursue a path toward formal abstraction. He opened his own workshop in 1944. Johannes Itten exhibited Vitali's toys at the Kunstgewerbemuseum Zürich in 1951. Max Bill described them in his book *Form:* "This toy roe-deer is so admirably delineated that it is really a perfect little piece of sculpture. The graining of the wood emphasizes its formal expression, which is not only aesthetically right but also adapted to the size of a child's hand." Vitali toys also made a name for themselves in the US. To meet the growing demand, Vitali tried using carving machines. However, as simple as the sculptural forms may appear, they could not be produced in the same high quality by machine and had to be individually hand-finished by someone with the requisite skill.
RM

Bill 1952, pp. 102–3; Vitali 1994.

328

329

328 Ramon Casas Robert, Elna Sewing Machine, 1940
Cast and painted aluminum alloy, Tavaro S.A., Geneva
Museum für Gestaltung/
Donation of Reta Sonderegger-Dudli

329 Elna: The first and only All New sewing machine in 50 years, 1940s, Elna International Sewing Machine Company Inc., New York. Prospectus cover and inside page

The Spanish precision engineer Ramon Casas Robert (b. 1905) set out to combine the benefits of his mother's pedal-operated sewing machine with those of the free-arm electric sewing machines used in industry. Fleeing the chaos of the Spanish Civil War, he emigrated to Switzerland in the late thirties and worked in his Geneva hotel room on the prototype for a portable electric sewing machine with a free arm. The Geneva firm Tavaro S.A., a supplier to the armaments industry, was seeking to expand its product range to the civilian sector and acquired Casas's patents. It was able to produce the sewing machine on the assembly line using the principle of interchangeable parts common in weapons manufacturing. Tavaro brought out the *Elna* as a world first in 1940, advertising it specifically as a product for the modern housewife. The first portable free-arm sewing machine, it could be stored in its space-saving carrying case and quickly set up when needed. The lid of the case became a sewing table and a built-in, glare-free light completed the temporary workspace. The sewing machine's unusual sharp-edged body and color reflect a machine aesthetic. The different functions can be read in the design: a vent shows where the motor housing is located and a plastic belt indicates the position of the drivetrain. The sliding lever for regulating stitch length and the foldout knee control are clearly distinguished from the main body of the machine. The moving parts are chrome-plated and reflect the light, while the body is painted in an industrial green that is easy on the eyes and exudes calm and stability. RM

Schilder Bär 1987; www.elna.ch.

330

331

332

330 Willy Guhl, Seat shell for the Scobalit Chair, 1948
Fiberglass-reinforced polyester, gel-coated, 34 × 49 × 60 cm, Scott Bader & Co. AG, Zurich
Museum für Gestaltung*

331 Willy Guhl, Scobalit Chair, 1948, produced from 1951, Scott Bader & Co. AG, Zurich
Museum für Gestaltung

332 Willy Guhl, Scobalit Chair, first model of the seat shell, jute-reinforced clay, 1947
Museum für Gestaltung*

In 1946–47, Willy Guhl, assisted by his brother Emil, developed three ergonomic seat forms, making models out of clay. They were the basis for a lounger, an armchair, and a garden chair. Because the shape of their shells was derived from an anatomically correct mold adapted to individual body shapes and sitting positions, the chairs were so comfortable that they didn't need cushions. With a series of design studies made from jute-reinforced plaster and plastic, the Guhl brothers studied the technical possibilities for producing a comfortable garden chair while giving the seat a natural and aesthetically pleasing shape. Taking his cue from the postwar design approach practiced in America and Scandinavia, Willy Guhl became a Swiss pioneer of the new "soft objectivity," passing on his knowledge as a teacher and the founder of the first product design course at the Kunstgewerbeschule Zürich. His garden chair, intended as a weatherproof chair for outdoor use, can be disassembled by loosening two screws for stacking away in tight spaces. The Zurich firm Scott Bader produced Europe's first plastic shell chair in a small edition made of glass fiber-reinforced polyester, marketing it under the name *Scobalit*—derived from the names of the company and the manufacturer of the material used. Despite widespread skepticism among designers regarding the use of the infinitely malleable synthetic material, in 1952 the *Scobalit* was the first plastic product to be awarded the "Die gute Form" seal by the SWB. RM

Erni 1985.

Tripod Chair
1949

333 Max Bill, Tripod Chair, 1949
Frame of varnished solid wood, seat and backrest of varnished molded birch plywood,
75 × 43.5 × 52 cm, Wohnbedarf AG, Zurich/Basel; AG Möbelfabrik Horgen-Glarus
Museum für Gestaltung

334 Tripod Chair and Three-Circles Table (dining table model), 1949
Photo: Hans Finsler

335 In his four-legged Cross-Frame Chair as well as in the Tripod Chair, Bill experimented with angular support profiles.
Photo: Ernst Scheidegger, ca. 1951

333

334

335

The artist Max Bill (1908–1994)—who in 1949 had just launched the campaign "Die gute Form"—believed that design tasks could not be solved through functional and structural means alone. His geometrically conceived pictures and sculptures provide a key to understanding the "aesthetics of puritanical frugality" (von Moos) that would influence furniture design as well as utilitarian objects and architecture. The outline of his *Three-Circles Table* is thus determined by a geometric shape constructed from three equal circles, which in turn relates to the triangular wood frame. The seductive, freely drawn forms in the matching *Tripod Chair* are derived from a similar substructure, but the outlines of the seat and backrest are—by necessity—constructed with two different radii each. The free forms of table and chair complement one another because they originate from the same process. The same can be said of the quite common conical tapering legs, which display a unique variation in the seatback—likewise due to the dictates of the specific situation—and thus also become characteristic features of both designs. With his "concrete design" method, Bill managed to blend the different elements into a self-evident furniture ensemble with a very distinctive presence and which looks as through "cast from a single mold." AR

von Moos 1995; Rüegg 1995g.

Kandahar Schuhmanufaktur
Factory design

Après-Ski Boots with Riri Zipper
1948

336

337

338

336 Anonymous, Après-Ski Boot with Riri Zipper, 1948
Suede, wool lining, 17 × 10 × 28 cm, Kandahar Schuhmanufaktur AG, Thun-Gwatt
Museum für Gestaltung/
Donation of Lotte Schilder Bär

337 Kandahar Schuhmanufaktur AG, Alpina boots, 1955, 2003 reedition
Museum für Gestaltung

338 Drawings of shoe models as artwork for the catalogue, ca. 1955, Kandahar Schuhmanufaktur AG, Thun-Gwatt
Museum für Gestaltung/
Donation of Kandahar Schuhmanufaktur

In 1928, the ski instructor, ski racer, and shoemaker Fritz von Allmen designed ski boots for the British Kandahar Ski Club in Mürren. The members were so impressed with the solid footwear that they allowed Allmen to use the name of the exquisite club as a brand. He launched serial production in 1932 and thereafter continued to expand his product range. Well aware of the stresses and strains of skiing, he invented comfortable après-ski boots made of sheepskin and crêpe rubber with an insulating cork midsole. Those who frequented the fashionable winter sports resorts found them very chic, and they soon came to represent the best in comfort and Swiss quality. Particularly successful were the models from the forties and fifties in cowhide, horsehide, or sealskin with a high, reinforced sole edge. Their form is derived from functional demands and their largely handcrafted production. With the exception of the sewing thread, all materials are organic. Leather is breathable and, when tanned, waterproof; the lining of fine calfskin or sheepskin is gentle to the skin. The midsoles retain moisture and insulate against the cold, while the round toe caps allow freedom of movement and keep the feet warm. In 2003, the family-owned company gained a higher profile in exclusive European shoe stores with a vintage collection based on two proven models: the boot with the typical Riri zipper running vertically up the instep, and another boot with laces. RM

Hummel 2004; www.kandahar.ch.

339

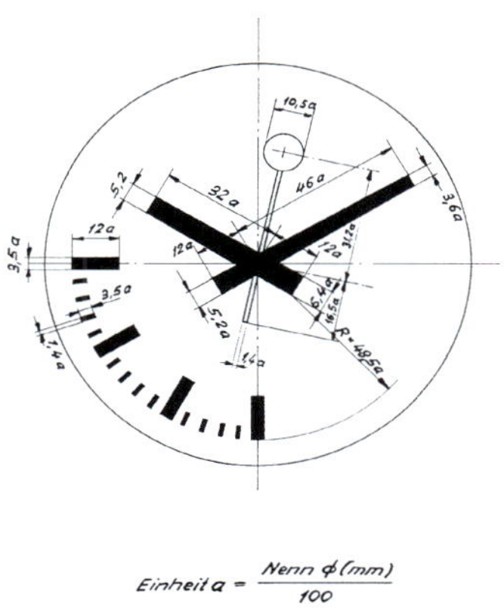

340

341

339 Hans Hilfiker, Swiss Railway Clock, 1944, redesign 1952
Light-metal housing with hammered paint finish, Depth: 28 cm, Diam.: 87 cm, Schweizerische Bundesbahnen (SBB), Bern; Uhrenfabrik W. Moser-Baer AG, Sumiswald
Museum für Gestaltung/Donation of the Schweizerische Bundesbahnen (SBB)

340 Standard model for clock dials, Swiss Railway Clock, 1952, Uhrenfabrik W. Moser-Baer AG, Sumiswald
Museum für Gestaltung/Donation of the Schweizerische Bundesbahnen (SBB)

341 Martin Drechsel, Stop2Go, 2013, Mondaine Watch Ltd., Zurich, electronic watch with pausing second hand
Museum für Gestaltung/
Donation of Mondaine Watch Ltd. Zurich

Hans Hilfiker (1901–1993) worked at the SBB as an engineer and deputy head of the construction department. At the beginning of his tenure there, in 1932, he developed a public clock for the Bahnhofplatz outside Zurich's main railway station, his debut as a creative engineer with a feeling for corporate design. In 1944, the SBB standardized their clock faces and synchronized their timekeeping. As the timetable uses only whole minutes, station clocks could be driven by a shared system, receiving each minute an electric impulse from the central master clock via the telephone network. In 1943, Hilfiker began testing for the SBB an electrically operated station clock with a red second hand. In 1952 he designed the clock face launched in 1955 and that to this day remains unchanged. Recalling the red signaling disk used by train crews, the second hand indicates to rushing passengers the time left for boarding as a circle segment. When the hand reaches twelve o'clock, it pauses for a second—the minute hand then jumps forward and sets the trains in motion. The SBB uses the station clock with its red second hand in thousands of locations, and it is legally protected as a registered trademark. In 1986, the SBB granted watchmakers Mondaine in Zurich the license to transfer the official Swiss Railway Clock to the wrist as the "Official Swiss Railway Watch," which since 2013 also features electronically controlled second-hand movement. Apple was so taken with the clock as a global design icon that it integrated its circular second hand into the world clock on its iPads in 2012. Meanwhile, the familiar clock at every Swiss station continues to faithfully attest to the punctuality of the SBB. RM

Hans Hilfiker 1984; Gantenbein 2013.

Spannstützengestell 1946

342

343

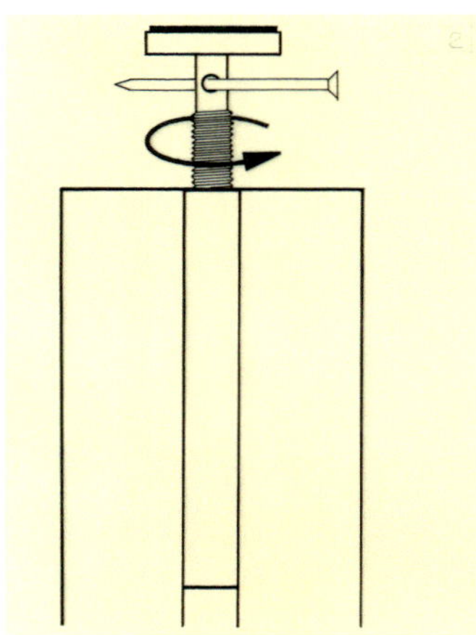

344

345

342 Hans Bellmann, Spannstützengestell
343 (Screw-Tension Shelves), 1946
Poles and shelves in varnished natural maple,
240 × 85 × 23 cm, various manufacturers for
Wohnbedarf AG, Zurich/Basel
Museum für Gestaltung

344 Hans Bellmann, Detail of the construction plan for
the Spannstützengestell, 1946
Museum für Gestaltung/
Donation of Rosmarie Bellmann
345 Early version with profiled lower pole as well as
table and cabinet elements
Photo: Hans Finsler

The Bauhaus graduate Hans George Bellmann (1911–1990) was employed by Wohnbedarf AG in Zurich and also worked for such renowned architects as Ludwig Mies van der Rohe and Lilly Reich, Leopold Boedecker, Rino Tami, Hans Brechbühler, and Alfred Roth before rising like a phoenix from the ashes of war in 1946 to become one of the foremost "industrial designers" on the scene. His first furniture range, realized in a single year, laid the groundwork for the postwar revival of Wohnbedarf AG in Zurich and Basel. Made mostly of wood, with their economic construction, easy dismantling, and visually comprehensible design ideas the pieces embody the new way of life that began to take hold in the postwar period, expressed by the concept of "nomadic living." In the freestanding *Spannstützengestell* (Screw-Tension Shelves) with the counter-sliding maple poles, the metaphor is evident. The two poles are fixed in place using wire brackets that at the same time support the shelves. Slender concealed tension rods hold the whole structure together. It is the first wall unit, to our knowledge, to be tension-fixed between floor and ceiling, and the shelving can either be placed up against the wall or used as a transparent room divider. With its integrated cabinet and table elements, it is a precursor to the wall units that would later become so popular. Formally, it anticipates the architectural language to come, in which the depth of the "brise-soleil" would become a distinguishing feature of a resurgent modernism. AR

Rüegg/Wieser 1991; Rüegg 1995a.

Carl Hans Koch

Sinar View Camera
Patent 1947

346

347

348

346 Werner Bührer, Sinar p2, 1982, anodized aluminum, 50 × 38 × 63 cm, Sinar AG, Feuerthalen Museum für Gestaltung*

347 Gerstner + Kutter, Alexander von Steiger (photo), Sinar professional camera, 1961. Advertisement proof

348 Instructions for the Sinar Norma/Sinar Fach, ca. 1952

It was with some reluctance that Carl Koch (1916–2005) took over his grandparents' photography shop and studio in 1947—the profession of photographer did not enjoy much prestige at the time, not least due to the clunky professional cameras photographers were forced to hide behind, concealed by a cloth. Koch soon set about developing a new system that would eliminate the mechanical limitations on photographic reproduction. His *Sinar* introduced the principle of modular construction to the professional camera. Koch divided the camera box into an axial "optical bench" with adjustable "standards" (coupling elements) and divisible bellows. He standardized the parts and gathered them together to create a combinable, expandable system. The first cameras of this type were made by watchmakers. In order to ensure the interchangeability of parts and subsequent possibility of upgrading the systems, Koch built his own production facilities in Feuerthalen in 1968. New accessories were continually developed and adapted to the latest technical advances, right up to digital photography. The *Sinar*'s formal details—visible screws and no ornamental elements or colors—convincingly convey its inner qualities to the specialists. In 1970, the designer Werner Bührer, who received his diploma from the Hochschule für Gestaltung Ulm in 1962, was recruited for product development and thereafter played a major role in shaping the appearance of the *Sinar p* and *Sinar p2.* The uniform design of the handles, scales, and labels as well as the relocation of the structural elements to the inside attest to Bührer's expertise in product design.
RM

Unbekannt – Vertraut 1987, pp. 136–39; www.sinar.ch.

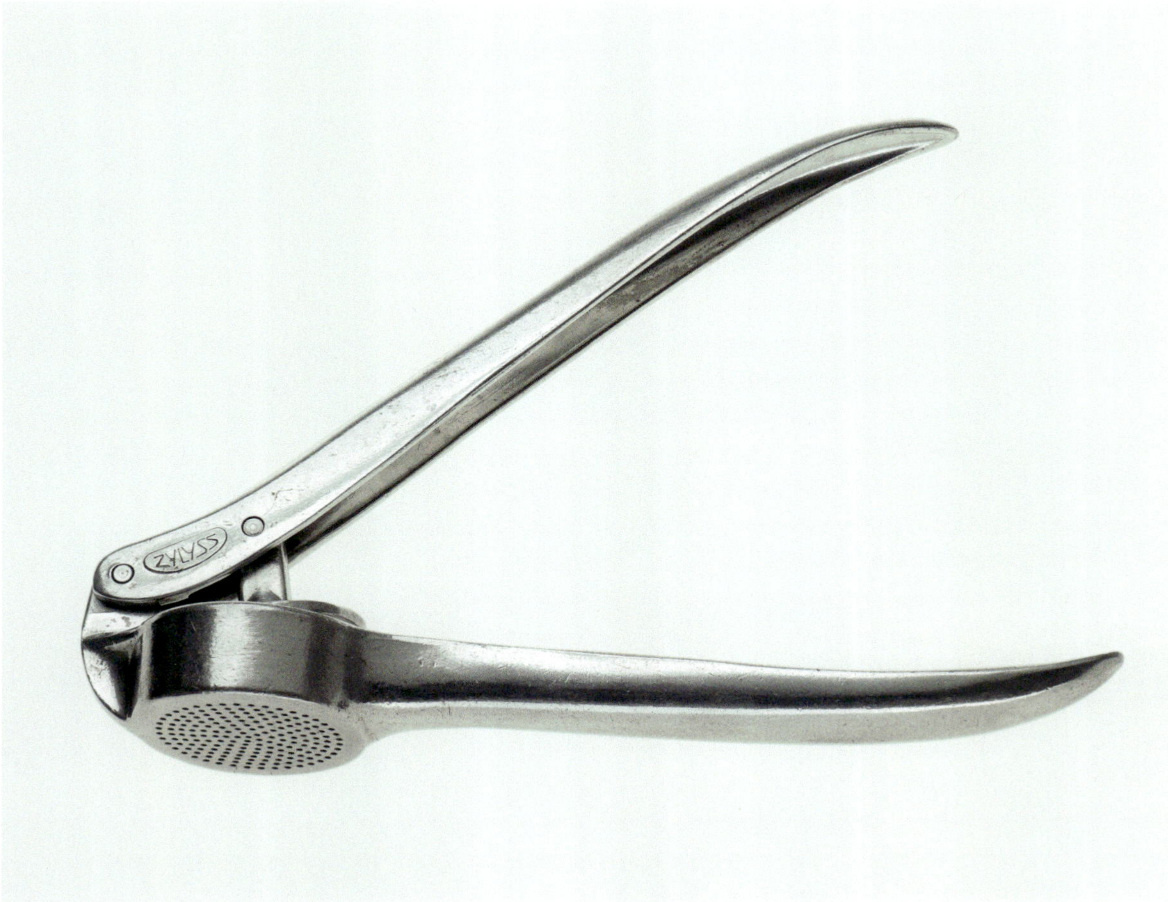

349

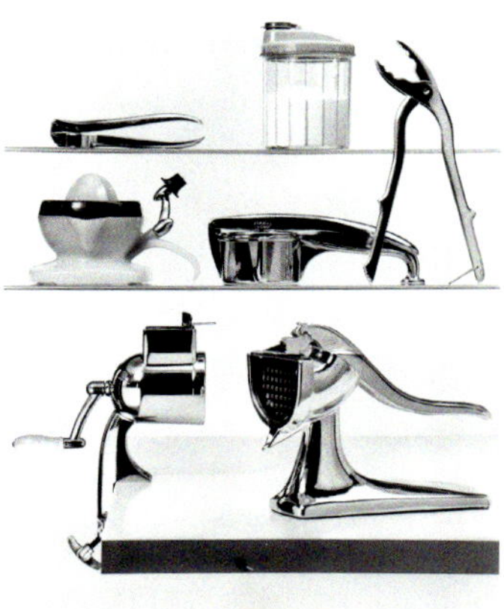

350

351

349 Karl Zysset, Zylyss Garlic Press, ca. 1950
Aluminum, 3.5 × 5 × 19 cm, Zysset & Co., Lyss
Photo: Max Buchmann
Museum für Gestaltung/
Donation of Knecht Arredamenti

350 Appliance product group, ca. 1950, Zysset & Co., Lyss. From the SWB picture collection of well-designed consumer products.

351 Reinhart Morscher, Jürg Bernhardt (photo), Hacken Zyliss, 1981, Zyliss AG, Lyss. Poster

Karl Zysset (1907–1988) came up with his first inventions while a bicycle mechanic and proprietor of the Velohaus Zysset in Lyss: a foldout, side-mounted kickstand and a luggage rack. Nothing is known today about Zysset's garlic consumption, but around 1950 he developed a simple device for pressing garlic cloves into a fine puree, similar to a juice press. This first garlic press featured slightly curved aluminum handles that lay comfortably in the hand just like a bicycle brake. It consists of three parts: a lever with a piston and a cylinder with a perforated bottom. The garlic clove is placed in the cylinder and pressed through the small holes in the bottom by the force of the lever. The simple practicality and convincing ergonomics of the utensil impressed the jury of the Swiss Werkbund, which awarded it the "Die gute Form" seal in 1952. Zysset gave up the bicycle shop and founded Zylyss (a combination of his family name and the company's location, lated changed to "Zyliss"), subsequently devoting himself to the development, manufacture, and sale of kitchen utensils. In addition to citrus juicers with rotating and lever mechanisms, along with bread and sausage slicing machines, Zysset was also responsible for two additional world-renowned technical innovations: the salad spinner and the onion chopper (1953). The latter, with the slogan "Zick-Zick-Zyliss," made the company name synonymous with Swiss kitchen utensils in the 1960s.
RM

Monnerat 2003; www.zyliss.ch.

352

353

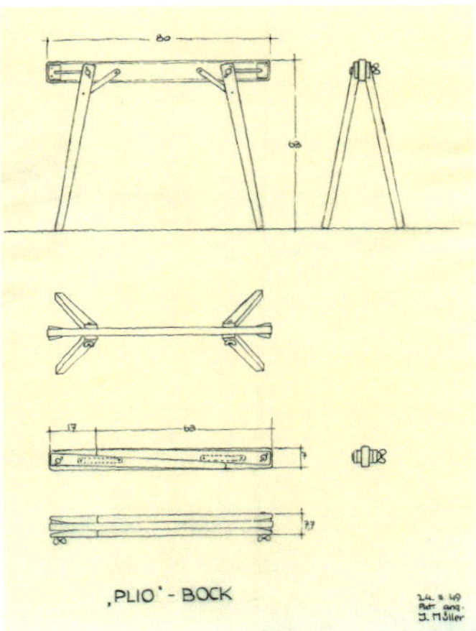

354

352 Jacob Müller, Plio Folding Furniture, 1948,
353 produced until ca. 1980
Seat in beech plywood, frame of varnished solid beechwood, table: 65 × 90 × 62 cm, stool: 36 × 45 × 33 cm, Aermo-Möbel GmbH, Zurich
Museum für Gestaltung

354 Jacob Müller, Plio trestle construction plan, 1949
Museum für Gestaltung

Just after the end of World War II, a number of self-employed master cabinetmakers from German-speaking Switzerland joined forces to establish the Werkgenossenschaft Wohnhilfe. Although the orders from the refugee assistance authorities they had initially hoped for did not materialize, this work cooperative was hired to furnish all the bedrooms, common rooms, and classrooms of the Pestalozzi Children's Village in Trogen. The initiative for the project came from the well-traveled cabinetmaker and craft teacher Jacob Müller (1905–1998), who also supplied all the designs for the solid natural wood furniture. Around 1947, Müller devoted himself to analyzing the "altered domestic needs of modern man" and began propagating ultra-lightweight furniture "that is truly mobile"—robust and lightweight enough for outdoor use and yet still sufficiently aesthetic "so as not to appear as an alien body, provisional and primitive, within the home." In an effort to use materials as economically as possible while ensuring the utmost ease of production, he arrived at his iconic *Plio* program. The basic set, weighing in at just 7.8 kilograms, consists of a folding table and two chairs that can be folded up and inserted underneath the table so the whole can be taken along "to the beach, camping, or to a vacation cottage." The small *Plio* folding stool even fits into a briefcase. In all models—including the chair and bench—the pairs of legs form two-arm levers whose ends latch into holes in the thin airplane plywood tops, forming a stable triangular connection. AR

Müller 1949; *Jacob Müller* 1988.

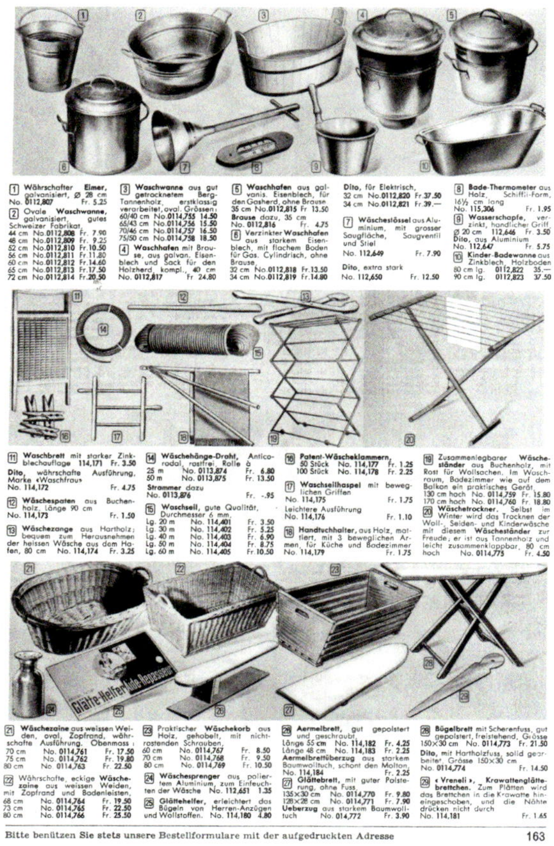

355

356

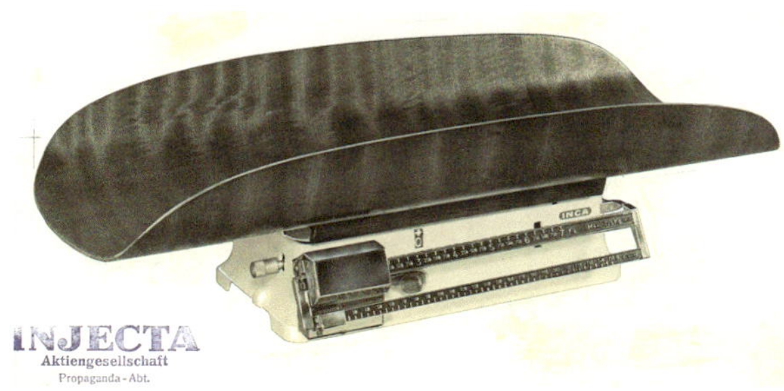

357

358

355 Grands Magasins Jelmoli, Page from the 1949 summer catalogue. Simple household items made of wood, wicker, sheet zinc, and aluminum.
356 Factory design, Iron with duroplast handle, 1940s, Therma AG, Schwanden
Museum für Gestaltung/
Donation of Margrit Rageth
357 Factory design, Inca de Luxe baby scale with molded plywood tray, 1940s, Injecta AG Spritzgusswerke und Apparatefabrik, Teufenthal. Artwork
358 Hermes Baby portable typewriter, advertisement in *Time* magazine, June 1946

The 1940s, Switzerland, and design—this constellation does not immediately send out sparks. We may assume that people had more important things to worry about at the time than "design." They had to somehow get by. But that's the very reason why there is so much to say on the subject: people got by, and the business of doing so gave birth, in a broader sense, to the word "design." Apart from the architecture and interior design of these years, we also must consider Swiss toolmaking and mechanical engineering, the railway car industry, major producers of commercial vehicles, and important auto body makers.[1] And of course there was the precision engineering industry that produced world-class clocks and watches, sewing machines and typewriters, turntables and film cameras; there were numerous manufacturers of household appliances (stoves, electric shavers, lamps, electric and gas water heaters); there were weapons makers (neutral for friend and enemy), furniture factories and cabinetmakers; there was the ceramics industry, manufacturers of lighting fixtures, and producers of textiles for the garment industry and the home. Crisis and war did not bring the production of these goods to a standstill. Most of them were still made domestically at the time.[2]

It is nevertheless difficult to pinpoint objects typical of the forties that spontaneously come to mind, such as the tubular steel furniture of the thirties or the cone-shaped lamps of the fifties. Their prominence or even popularity was due in part to their internationality. This effect is absent for the forties. Like all the other European countries—but perhaps even more so because of the way it was wedged in between the Axis powers early on (from 1933)—Switzerland was largely left to its own devices culturally. There were hardly any exhibitions or publications anymore with international participation. This period therefore took on a very different tone from the time before the world economic and political crises. There was no point on the horizon anymore outside the objects themselves that they could look to for orientation. Things no longer had to lay claim to universal validity. They had lost their character of reference points within an all-encompassing progressiveness. The crisis had pulled the rug out from under the dream of lifetime happiness brought by serial and mass production; people were grateful to at least be able to make things by hand.[3] International intellectual exchange ground to a halt; the gaze turned inward. The national borders once again became a system boundary. This situation may have narrowed horizons, but above all it changed them. Things went back to being very objective. The *Rex* economy peeler and the *Flex-Sil* pressure cooker were children of their time, their purpose already apparent in the product name.[4] The here and now of the specific case replaced the utopia of the "world devoid of objects" (Kazimir Malevich). The enemies of modernism triumphed and spoke of a *retour à l'ordre.* But the result was not regression. The actual modernization of life continued, slower for sure and with some twists, but as lived reality in all its eclectic variety—with the growth of cities, the decline of the rural peasantry, secularization, increasing mobility—and less in terms of form or style. A different kind of modernism came about than that of the International Style.[5] Is the notion of *pragmatism* apt for this era? Only when the word is used to refer to actual practice, to concrete action (and not to approximation). With its measured solutions and contributions, the concept of pragmatism was surely more congenial to most of the Swiss protagonists of the period than the generalized and abstract programs of the twenties.[6]

There is yet another reason why it is not easy to succinctly capture the decade of the forties. Design is not identical with the domestic environment. There was also the design of technical equipment, machines, and vehicles—all of them of great economic importance.[7] Between about 1925 and 1935, the two spheres had converged somewhat semantically, but since then they had grown further and further apart. Technological design continued to evolve in technical and engineering fields, but in their living environments people sought instead warmth and comfort. Design was technicity *or* a compensation for it, depending on what you mean by it.

The First Stirrings of "Product Design"

There is thus no clear answer to the question: How did *things* look back then?[8] From today's vantage point, the objects of the forties seem surprisingly *non-simultaneous.* Capital goods such as a floor cleaning machine were still undesigned apparatuses, while a consumer product like a vacuum cleaner already displayed a certain stylistic expressiveness. Product aesthetics were still not an explicit issue, but they did start to become an intrinsic factor to be considered.[9] One example of a transitional object is the Therma electric iron, which in the forties was equipped with a handle made of plastic instead of the usual rounded wood. This improved both handling and protection from burns, but it only serves to make us more aware today of the difficulty that was once involved in harmonizing formal and useful properties.

The difference between capital goods and consumer goods at the time is evident wherever the aim was to induce an audience reaction, for example in the step from truck to *travel experience* in the PostBus, both using a chassis of the same make. Radios were status symbols with deluxe polished surfaces emphasizing the wood grain.[10] Typewriters traded in their shiny black surfaces for matte housings of gray-green wrinkle-finish paint to prevent reflections that might strain the eyes. A new generation of tram cars with pneumatic folding doors (popularly known as "sardine cans") left the factory in Zurich (1940) to make the rounds through Switzerland and the world.[11] At home, however, lampshades were still covered with brownish parchment or flowered fabric. As a matter of fact, shades of brown were widespread in clothing as well as in wallpaper, carpets, and upholstery fabrics. It was taken for granted that an office was dominated by wood, with a beech or oak desk, file cabinets with roller shutters, a post-and-rung chair with felt seat pad, index card boxes, patterned wallpaper, and undemanding indoor plants.[12] The interiors that were invariably depicted in ads may look to us today like collections of clichés, but they were then par for the course and for that reason are so intriguing to us as a cultural indicator.

From the distance of time, we notice how many things were still made of wood after the war: laundry racks, step ladders, wheelbarrows, strollers, scooters, wash tubs... Woven willow was used to craft shopping, picnic, and waste baskets, garden furniture and bassinets.[13] Wood was also still a key construction material in the vehicle industry: for rolling stock, railcars, elevator cabins. In home furnishings, wood was both an important material and a medium of expression. The year 1926 was already far in the past when Germaine Krull had published the portfolio *Métal,* her photographic tribute to the genuine material of the great awakening. However, wood too had already been substantially updated under the gaze of the avant-garde. Beginning in 1928, the Bauhaus in Dessau under the direction of Hannes Meyer almost programmatically distanced itself from the metal furniture vogue and developed new ways of using wood, and Alvar and Aino Aalto presented an entire collection of plywood furniture in 1933.[14] During this expansion of modernism, people rediscovered the material appeal of the visible wood grain, which after World War I had often been hidden under a (dematerialized or homogenizing) coat of paint. Now, the material shone forth again in all its fresh naturalness.

Wood as Material: Expanded Technical Possibilities or Trusty Old Tradition?

The forties reinforced this tendency to emphasize material properties, betraying an obvious penchant for *authenticity.* There were two sides to this coin, though. Both the advocates of modernism and also its denigrators now took recourse to wood, but they each meant something different by it. The former camp treated wood as an eminently constructive material, as a *contemporary* material with high technical performance capabilities.[15] For them, wood's authenticity could be found in the ready traceability of the path from raw material to object. Their opponents used wood instead as a means of paying homage to pre-modern Switzerland and the vernacular *Heimatstil.*[16]

Hermann Kienzle, director of the Gewerbemuseum Basel and older brother of the designer Wilhelm Kienzle, criticized the Heimatstil by saying, "With its artificial primitiveness, [it] is antagonistic in two ways to the natural progress of the home and of technical equipment: for one, by partially ignoring the realities of today's technology, and secondly, because it can only fulfill the needs of today's lifestyle artificially, if at all.... In addition to the radical *Heimatstil* furniture made of native solid wood using exaggerated

359

360

361

362

handicraft techniques, there are also factory-made compromise items in which only the raw wood and bulging panels qualify them as *Heimatstil.*"[17]

As opposed to the "artificial naturalness" of *Heimatstil,* other approaches emerged that have for some time been subsumed under the label *regionalism.* The difference can be found in the object itself and in the attitude of its designer toward authenticity and closeness to nature. Some *believe in it,* while the others *make use of it.* For one group, it is a worldview; for the others, a cultural technique. This distinction may seem somewhat Manichaean, but it can be demonstrated based on concrete examples.[18] The distanced, non-ideologically motivated attitude was clearly superior in terms of design.

In 1939, Hannes Meyer designed for his last building, the children's home in Mümliswil, a large ring-shaped table for the dining room, constructed of massive pine and with cruciform feet. The material made sense both objectively and constructively and yet was symptomatic of its time as a decision favoring elementary simplicity. At around the same time, the colonial armchair—an anonymous design from the nineteenth century for the colonies of the British Empire—was discovered for Swiss living rooms, so to speak as a "protest chair."[19] And in 1947, Hans Bellmann used its elementary simplicity, together with the three-legged draftsman's stool, as inspiration for his circular colonial table, with three holes where the legs were inserted. Bellmann called the table, in retrospect, "a momentary flash of inspiration."[20] But in order for this to happen, it takes the right intellectual milieu.

The Pestalozzi Children's Village in Trogen, inaugurated in 1946, and the furnishings for its classrooms, lounges, and bedrooms are among the best examples of regionalism. The village was built in response to an appeal made in the magazine *Du* in 1944 by its deputy editor in chief, Walter Robert Corti, calling for a home for war orphans.[21] As a building ensemble, this is contemporary architecture that displays a high degree of empathy with local and regional house types: respect for local traditions but without blatant sycophancy. Dictated by the need for the greatest possible economy, a type of modern architecture emerged here that had emancipated itself from the rhetoric of the International Style.[22] The architect, Hans Fischli, had become aware of the Werkgenossenschaft Wohnhilfe (founded in 1945)—a cooperative association of cabinetmakers that was based on the principle of the competitive division of labor as a way of securing their joint livelihoods—and commissioned it to supply a substantial amount of furniture for the project.[23]

The designer of this first Wohnhilfe collection was Jacob Müller (1905–1998), an undogmatic, anthroposophical master carpenter with a conspicuous talent for writing whose spectrum ranged from daring experiments with materials to the rustic solitary piece of furniture.[24] Müller, for whom wood was a uniquely versatile and inspiring material, was the creative mastermind behind Wohnhilfe.[25] The furniture for the Children's Village was made of solid pine, simple but lovingly detailed, cleanly crafted, and congenial. For the bedrooms, Wohnhilfe supplied single and bunk beds. The chairs were a new take on the traditional board chair type (with a pincer design for the legs and backrests). The school desk consisted of a trestle topped by a table leaf whose tilt could be adjusted, with a storage compartment underneath. Cross bracing was provided by round bars at the back. When the furniture was delivered to the Children's Village, Fischli is said to have exclaimed that for the first time at such a moment he was not disappointed.[26] He called the village a "small league of nations of orphans from eight countries speaking seven different languages."[27] The architecture of the village and the furnishings for its houses are utterly free of the eschatological notions of the prewar avant-garde. Or, to put it more simply, and in Swiss terms: this is modernism not as a worldview but made for real life.

Max Ernst Haefeli wrote in 1950, "In recent years, we have shunned the term 'functional form.' It has been interpreted in a manner that is too unilaterally rationalistic. We wanted to use it instead to characterize a kind of design that also pays heed to psychological factors."[28] As early as 1936, Haefeli had already fended off the accusation that the devotees of Neues Bauen (New Building) overemphasized *sobriety* and thus provoked a reactionary backlash with the sentence: "We all carry within ourselves a considerable amount of healthy 'romanticism': more and more, a certain *versatility* in form, material, color, lighting, the relation to nature, etc. is coming to light within the framework of the 'objective'—resulting in the *lively, nuanced* fulfillment of each construction task."[29]

363

364

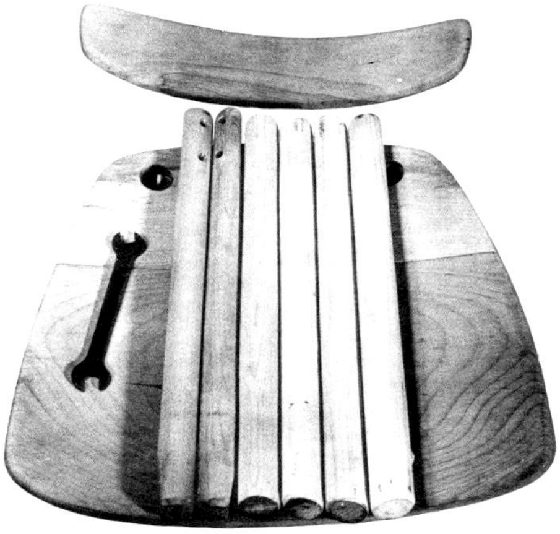

365

359 Four-axle light railcar used by the Zurich tram lines, 1944. First tram with pneumatic folding doors. Electrical engineering: Brown Boveri, car body: Schweizerische Industriegesellschaft, Neuhausen
360 Hannes Meyer, Cooperative children's holiday home in Mümliswil, 1938–39. Dining hall with "democratic seating arrangement." Food is distributed from the center of the ring-shaped solid wood table. Photo: Hannes Meyer
361 Building an auto body atop the Rover chassis, 1950s, Carrosserie Hermann Graber, Wichtrach
362 Auto body atop the Alvis chassis, 1946, Carrosserie Hermann Graber, Wichtrach. Sleek and elegant, ornament-free custom auto body on an English chassis
363 Hans Fischli, Pestalozzi Children's Village, Trogen, 1945–49. Wooden architecture in a regionalist design idiom
364 Pestalozzi Children's Village, Trogen. Classroom with collapsible solid wood furniture by Jacob Müller, Werkgenossenschaft Wohnhilfe, 1947 Photo: Anita Niesz
365 Jacob Müller, Wohnhilfe Chair ("Stabelle," or "board chair"), 1945, first Wohnhilfe program, Werkgenossenschaft Wohnhilfe. The chair was sold as a kit to be assembled by the customer.

Both statements, divided by an interval of fifteen years, apply to architecture as well as design.

Rematerialization of Furnishings, with Patterns Everywhere

In 1946, the book *Möbel und Wohnraum* (Furniture and Living Space) was published by Hans Guyer and Ernst Kettiger on behalf of the Vereinigung Schweizer Innenarchitekten (VSI, Association of Swiss Interior Designers).[30] This exemplary document captures what was typically *Swiss* at the time: interior design that is German Swiss, French Swiss, a few examples from Ticino, and the careful response by designers to the living conditions of their customers. None of the objects or interiors pictured had the character of a manifesto. Only upon closer study of the featured examples do the various positions on the scale between modern and traditional, urban and rural, independent and conformist become discernible; there is almost nothing that could be considered experimental here—because these are all commissioned furnishings and interiors and the clients were not enamored of the avant-garde. The solid corporeality of the furniture (almost always made of wood) is striking: cabinet doors with panels, table leafs with edge profiles, beds with corner posts (see Rüegg, "Solutions d'urgence," pp. 162–67). A whiff of "blood and soil" can be sensed in some of these items, but it rarely swells to a stench.[31] Sliding doors in furniture—before the war almost a hallmark of modernism—have almost completely disappeared, and with them the planar, disembodied, sideward-gliding effect; doors on cabinets, dressers, and credenzas once again swing out into the room, further emphasizing the physicality of the pieces.

Also conspicuous is the profusion of patterned products: curtains with stripes, flowers, diagonal patterns, or stars; upholstery fabrics and carpets with polka dots, rhombi, coats of arms, repetitive ornaments. The department store catalogues show patterns used everywhere, including on wool blankets, shopping bags, lampshades.[32] One might be tempted to explain this away as part of the product language of popular merchandise, which is normally distinct from the work of the creative elite. But, in fact, this tendency toward patterned surfaces and the use of ornament can also be found in the work of the members of the VSI and the Swiss Werkbund (SWB). The *Schweizer Warenkatalog* (Catalogue of Swiss Products), published by the SWB for the first time in 1949, features exclusively *patterned* home textiles. Although the patterns are subtler and more sophisticated than the ones in the majority of department store products, it is obvious that they have proliferated compared to the period before the war. Back in 1930, Georg Schmidt had noted that ornament in utilitarian objects was on the wane, citing cultural-anthropological reasons for the phenomenon. Ornament was apparently "the vestige of what were once full-fledged cult symbols. Just one more step and it has, for example on fabrics, no more meaning than the natural function of hiding stains! As ornament is depleted of its original magical content, it also loses, step by step, its formal and expressive power." That decorative patterns nonetheless continued to survive, for instance on the dishware in many households, was for Schmidt nothing more than an "atavistic residual anxiety about the unprotected object."[33] Contrary to his expectations, however, changing times did not bring about the final disappearance of ornament but instead its return. A rebuttal or confirmation of his theory? Hard to say, for during these years the atavistic residual anxiety turned into real fear.[34] In the aforementioned VSI book, nearly all the examples of living room, dining room, bedroom, and office decor incorporate ornament, sometimes even in unflinching combinations of stripes and flowers.[35]

A final look at the decade reveals that the various achievements were made at differing developmental tempos. Switzerland was more innovative and courageous in the technical arena than in home design, which was characterized more by what people thought was *possible.* The products thus manifested a careful and controlled air. The fifties would swiftly make up for all the delays that had accumulated during the forties. The traveling exhibition *Die gute Form* organized by Max Bill in 1949 anticipated and contributed to this resurgence. Comparing the first *Schweizer Warenkatalog* from 1949 with the second edition from 1954 already reveals significant changes.[36] Elsewhere, this transformation proceeded in different ways. The Thorens company presented a new kind of radio in 1948 whose components were divided between different modules: the control unit was portable, connected via cable with the stationary amplifier and speakers.[37] In the propectus, a little girl

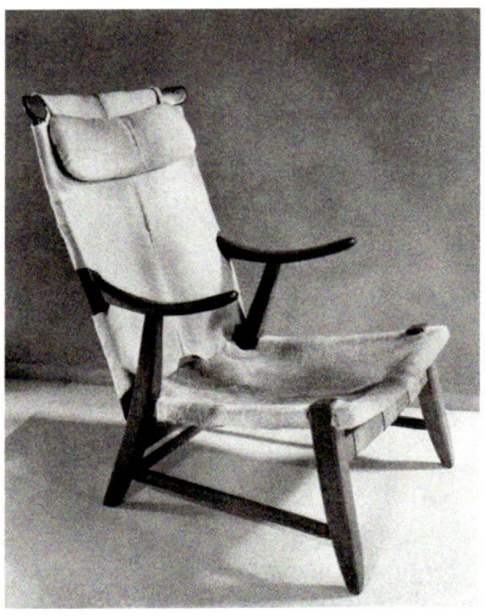

366

367

366 Walter Frey, Armchair, before 1946, solid wood with hide upholstery
367 Jacob Müller, Ascona high-back armchair, 1950, Werkgenossenschaft Wohnhilfe. Adjustable seat made of airplane plywood, horsehair cushions. Collection of Peter Tellini

brings her mother the controls so she can listen to the children's hour. A new chord is struck here, with designers heeding momentary needs and their instant gratification, the mode of consumer society. The name of the new device itself sends a strong message: *New Century.* In reality, of course, it was only mid-century. But it was as if a long-pent-up blockage was finally freed. And the English name now somehow has a different ring to it, sounding more like "America," whose era would now dawn in Switzerland as well.

1 For example, Hermann Graber (Wichtrach), Worblaufen, Beutler (Thun), Carrosserie Langenthal, Ghia (Aigle).
2 With regard to the dates defining the era under examination: Did it begin with the stock market crash in 1929? Or with Hitler? With the outbreak of war in 1939? Did it end in 1945 or later? It is usually easier to pinpoint the beginning of an era than its end. The "forties" begin in the thirties and end in the fifties.
3 The traveling exhibition organized by the Swiss Werkbund (SWB) in 1943, *Unsere Wohnung* (Our Apartment), for example, tried to show that the goal of contemporary living could also be achieved with handmade furniture. See Streiff 1943.
4 The pressure cooker, for example, was interesting at the time due to cost savings through reduced cooking time. See *Unbekannt – Vertraut* 1987.
5 The International Style (Philip Johnson, Henry Russell Hitchcock, US) is a blanket term used to refer to modernism in building and interior design between the wars—a problematic term in some respects because it leans toward formalism.
6 For example, Paul Artaria, Lux Guyer, Max Ernst Haefeli, Wilhelm Kienzle, and Werner M. Moser. The same applies to the younger generation: Hans Bellmann, Camille Graeser, Willy Guhl, Lisbeth Sachs, and Erika Thöne.
7 The term "designer" was as yet unknown at the time in Switzerland.
8 And for that matter: What did they sound like? Different: motorcycles rumbled, vacuum cleaners howled, riding in a bus was accompanied by the creaking of the wooden body.
9 Max Bill spoke of this in his 1948 lecture "Schönheit aus Funktion und als Funktion" ("Beauty from Function and as a Function"), Bill (1948) 2010.
10 There was also a manufacturer that used molded resin housings from an external supplier (Owa). It would still be several years before the age of plastics really got off the ground—with thermoplastics, then called "soft plastic," after 1955.
11 Motor trams (steel box, 1940) and light motor trams (aluminum, 1941) with trailers were developed in Switzerland and produced by the Schweizerische Wagons- und Aufzügefabrik AG in Schlieren-Zürich with technical equipment from the Maschinenfabrik Oerlikon. They were also produced by other manufacturers starting in 1944 as the Swiss Standard Car (Tramway standard suisse). The motive behind this development was business-related: less staff was needed with unidirectional *passenger flow* (getting on and paying at the back, getting off at the front).
12 One need only think of the movie *Wachtmeister Studer* (1939).
13 The semiannual general catalogues of the Grands Magasins Jelmoli contained countless products that were probably supplied by small family businesses and manufactories.
14 Did the development of this furniture perhaps reveal a desire to avoid the physiologically disadvantageous heat absorption—i.e., loss of body heat—of tubular steel during the long Finnish winters, e.g. at Sanatorium Paimio? I suspect this was the case.
15 Besides the aforementioned Haefeli, this group also included—in Switzerland—Wilhelm Kienzle, Moser, Steiger, and—abroad—the Aaltos, Marcel Breuer, Gerrit Rietveld, Charlotte Perriand, and even Eileen Gray.
16 The term "Heimatstil" incidentally originated in the nineteenth century and had positive connotations at first, but the style was mostly viewed critically in the period under discussion. See Crettaz-Stürzel 2013.
17 Kienzle 1943.
18 For example, Charlotte Perriand's hotel interiors in Méribel-les-Allues, ca. 1946.
19 1928 is cited as the year when Wilhelm Kienzle rediscovered the colonial armchair. In Eckstein 1934, the armchair is shown next to Kienzle's Embru shelf (p. 117).
20 Hans Bellmann, in conversation with the author, October 1989.
21 The magazine *Du,* founded in 1941 by Conzett & Huber publishers and created by Arnold Kübler, is another telling example of the possibly unique Swiss fusion of humanism, originality, and thoughtful preservation of traditional values that does not see itself as an enemy of modernism.
22 See Maurizio 1952, pp. 23–24, 63–66.
23 Wohnhilfe was an association of cabinetmakers that divided up the various production steps between them and thus put into practice the economic theory of Emil Anderegg, who in the forties emphasized the importance of the "creative factor" found in progressive handcraftsmanship as an economic entity between agriculture and industry. J. Müller revived the idea. Anderegg was long-standing president of the Wohnhilfe cooperative, even while still mayor of St. Gallen. The name "Wohnhilfe" ("living help") refers to the original idea of humanitarian aid in the war-ravaged country; see also note 24.
24 On the history of Wohnhilfe, see Lichtenstein 1988 and Schilder Bär 1988.
25 See his folding furniture set *Plio,* described in this publication on p. 155, and in *Jacob Müller* 1988.
26 Jacob Müller, in conversation with the author, Summer 1988.
27 Hans Fischli, in *SBZ* 45–47 (1949), cited in Maurizio 1952, pp. 23–24.
28 Haefeli 1950.
29 Haefeli 1936 (italics in the original).
30 Guyer/Kettiger 1946. On the inside cover is the dedication: "Dedicated to Wilhelm Kienzle, our revered friend and teacher."
31 Meant here is "Swiss blood and soil," in deliberate contrast to "Greater German" Blood and Soil, even if the Swiss version sometimes displayed an irritatingly similar formal repertoire.
32 The frequency of patterns in the pictured products can be explained in part by the single-color printing of the catalogues, since the patterns provide an element of information that is missing from a unicolor product.
33 Schmidt 1932.
34 For Schmidt's opponent Peter Meyer, lack of ornament did not connote originality and authenticity; decoration was for him an anthropological constant: "It is therefore usually not the case that a specific object of use is decorated by a particular user, but vice versa: an urge for expression seeks a field where it can operate." See Meyer 1937, p. 55. Elsewhere he stated, "In any case, the ornamented surface appears to be activated, to have a more intense presence than an unornamented one, which offers the eye no interest and thus, so to speak, silently eludes attention." See Meyer 1944, p. 13.
35 The fifties would continue to witness a great deal of ornament, especially on curtains, but different kinds of patterns came onto the scene, more lively in expression than those of the forties, with dynamic free-form shapes and bold color combinations, the whole field betraying the strong influence of Art Informel.
36 The *Schweizer Warenkatalog* was published by the SWB to provide the public with orientation in choosing well-designed items for the home. The first edition appeared in 1949, the second in 1954, and others followed.
37 This appliance was mentioned by Max Ernst Haefeli; see Haefeli 1950.

368

369

370

368 Willy Guhl, Pavatex Wardrobe, 1942,
Frame of natural wood, painted Pavatex interior shelves, Schreinerei Guhl. From the SWB picture collection of well-designed consumer products.

369 Wilhelm Kienzle, Stackable Shoe Rack, 1951, firwood model, Werkgenossenschaft Wohnhilfe. From the SWB picture collection of well-designed consumer products.

370 Thorens New Century radio with separate receiver, controls, and speaker. Front and back of prospectus, 1949, Thorens S.A., Sainte-Croix

371

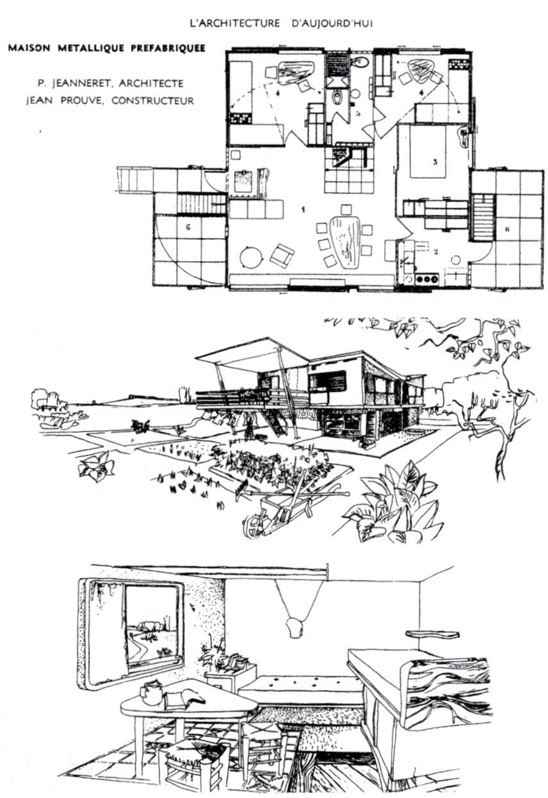

372

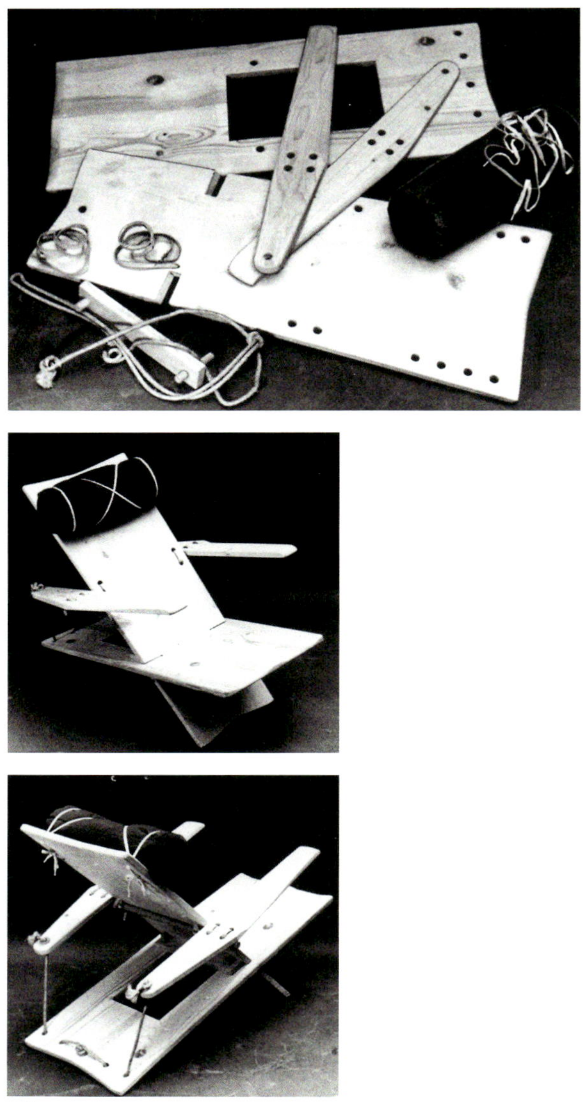

373

371 Le Corbusier, *Les constructions "Murondins,"* Chiron, Paris/Clermont-Ferrand, 1942. Cover. Emergency housing made of rammed earth and round timber

372 Pierre Jeanneret and Jean Prouvé, Maison métallique préfabriquée, 1940–42, project drawing. Prefabricated minimal houses, planned as metal structures, executed only a few times in wood (Maison F 8 × 8, BCC Grenoble, ca. 1942).

373 Pierre Jeanneret, Emergency furniture (Mobilier d'urgence), wood, cord, fabric, prototype, ca. 1942. Experimental collapsible seating furniture made of interlocking boards.

World War II was, according to Jean-Louis Cohen, "un moment fondamental dans la modernisation tant théorique que pratique dans l'architecture."[1] The modernization of architecture through war? What Cohen is referring to is the astounding scope of the economic and industrial reorganization of entire nations and the resulting boom in the construction of military facilities, industrial complexes (such as the aircraft factories designed by Albert Kahn in the US and Herbert Rimpl in the German Reich), and residential developments on a scale never before seen. The Pentagon, erected in just a few months,[2] supplied 32,000 jobs and the nuclear facilities in Oak Ridge as many as 75,000. Prominent architects—from Le Corbusier to Ernst Neufert—experimented with radically different approaches to planning the reconstruction of the destroyed cities. The result presents us with a paradox: the 1920s dream of industrial production and large series seems to have finally been realized only in the wake of a destruction-bent "sursaut surhumain"[3]—the prophetic formula already arrived at by Le Corbusier in 1931.

Design in the Economy of Scarcity

What does all this have to do with Switzerland and its design problems? At first glance, not much. It's true that Switzerland had accomplished its own epic feat by constructing the extensive underground fortifications within the *Réduit,* or National Redoubt, which consumed a tremendous amount of building materials and labor. In contrast to the Atlantic Wall, however, which the French philosopher and architect Paul Virilio made the subject of a momentous technical and—most importantly—aesthetic analysis in the 1970s,[4] the invisibility of the Swiss bunker system was an essential part of the program. Only the cloaks of invisibility in the form of chalets, barns, or boulders hiding the aboveground gun turrets offered enough visual potential for a later rediscovery (for example in Christian Schwager's photo series *Falsche Chalets,* published in 2004).[5] Like the nostalgic, Swiss folkloric furnishings of the underground troop caverns, these camouflage structures are not exactly symbolic of Switzerland's definitive modernist breakthrough; on the contrary, they demonstrate how the country withdrew into its shell—a withdrawal that seems an inevitable knee-jerk reaction to any threat to Swiss security, stability, or autonomy.

Few alternatives were proffered to this omnipresent "symbolism relating to the national past and the rural environment."[6] Former standard-bearers of Neues Bauen (New Building) like Werner Max Moser and Max Ernst Haefeli did, however, succeed in finding a contemporary organic form for the simple Swiss mood of their early interiors without completely succumbing to the traditional rural *Heimatstil.* Moser's Reformed Church in Zürich-Altstetten (1939–41) boasts a host of inventive details that do not in the least betray the material shortages prevailing at the time; here, the specially developed plywood seats mounted on concrete pedestals and wide-spanning tubular steel frames anticipate by several years the stadium seating designed by the Eames Office, ubiquitous today the world over.[7]

In view of the bloodletting then taking place concurrently in war-torn Europe, such an evocation of normality seems downright touching. In the countries all around Switzerland, "solutions d'urgence" born of necessity instead became the dominant design problem. As André Bruyère remarked in the magazine *L'Architecture d'aujourd'hui,* this problem could not be surmounted by applying textbook theories, but might indeed serve to mobilize the imagination of "a latter-day Leonardo."[8] Le Corbusier, for example, responded in 1940 to the devastation in Belgium and Holland with an unusually primitive—"Mesopotamian"[9]— architecture of rammed earth, logs, and grass tiles. In 1942, he drafted a guide for distribution among French youth clubs for how to build his archaic-looking "murondins" as emergency housing; however, he neglected to supply any plans for furnishing this alternative to the "military barracks with neither soul nor proportion."[10] The opposite approach was pursued by his former working partner Pierre Jeanneret (1896–1967) of Geneva, who was now active as an architect for the Bureau central de construction (BCC) in Lyon, and by the engineer Jean Prouvé, who designed prefabricated eight-by-eight-meter minimal houses that could be transported on trucks and set atop foundations made of rubble.[11] Jeanneret proved with detailed furnishing proposals that these six-person dwellings could function as precisely as clockwork. The simple sliding-door cabinets with their distinctive handle moldings, the organically shaped tables, and the rustic stools with woven rush seats were taken from a line of wood furniture developed at the outbreak of the war in conjunction with Charlotte Perriand;[12] beds that could be stacked at an angle were designed especially for the project. The BCC already began producing some of the models during the war. For both Jeanneret and Perriand—who worked in the Far East from 1940 to 1946—these independent interpretations of traditional construction methods were the springboard for important contributions to the domestic culture of the postwar years. Jeanneret, a gifted tinkerer, also experimented with "sièges d'urgence" made from boards that were slotted together or held in balance by cords and whose volume could thus be reduced for transport to just two or three layers of boards[13]—but his efforts did not meet with immediate success.

Furniture Kits

If it is possible to speak of a pioneering development in Swiss product design, we should not look to the country's military armament, and even less to its vernacular architecture. Such a development can be found instead in a special situation that affected Switzerland only indirectly, namely the postwar reconstruction of Europe.

Under the protection of neutrality, the architect, artist, graphic designer, and journalist Max Bill (1903–1994) succeeded while the war was still raging in gathering together, mainly from the United Kingdom, Sweden, and the US, the materials needed for the initial phase of European reconstruction, while at the same time taking stock of "new forms of living." His compendium on reconstruction, *Wiederaufbau—Dokumente über Zerstörungen, Planungen, Konstruktionen,* was published in 1945 by the export department of the Swiss Trade Association, tellingly with the note "with respect to commercial export to the war-damaged regions."[14] Commercial and humanitarian motives were of equal weight here. Bill himself was interested mainly in the "aesthetic aspects of industrialization methods," as he confessed in the book's blurb. He demonstrated what he meant by this in his Villiger House in Bremgarten, constructed already in 1942 with Durisol. In contrast to the barracks built to house the armaments industry and labor offices in Bern, likewise made of these well-insulated panels, he took advantage of the Villiger House as an opportunity to lend the support structure with its modular character an aesthetic reminiscent of Japanese architecture—using logs resting on large boulders, fascine walls, and stone walls and paths to create a memorable "Gestalt."[15]

Bill's documentation focused on the building systems that were being developed in large numbers in the countries at war. He devoted only two pages to a local innovation that would mark the beginning of a second heyday for Swiss design—the *Aermo Wohncolis* (Aermo furniture kit) for emergency housing.[16] Unlike the construction industry, which had not yet undergone the industrial breakthrough described by Cohen, many Swiss designers and manufacturers were already experimenting at the time with modularity, diversified manufacturing, and new materials. The "Aermo emergency furniture kit" was developed by Mauritius Ehrlich (1903–1988), a native Ukrainian who had already been a refugee himself twice in his life.[17] Between 1944 and 1948, he and Huldreich Altorfer produced around 30,000 furniture kits under the company name Aermo GmbH.[18] Inspired by the compactness of a cigarette case, the invention comprised "an entire set of furnishings including cookware and tableware for four persons in as space-saving and rational a package as possible for ease of transport."[19] "Just like in a construction kit,"[20] the two couch frames held the easy-to-assemble components for four stools, a table with a drawer, a closet, and two thin mattresses, along with the twenty-seven-part set of aluminum cookware and tableware. Fifty of the 166-kilogram Aermo kits could be transported in a single railway car. Similar to Pierre Jeanneret, Ehrlich approached the scarcity situation as a key design parameter. He revisited the concept of multifunctionality that had been so central to Neues Bauen and now varied it under the altered conditions of his time.

Reconstruction aid aimed not only at standardization and typification, but once again took up the topic of "minimum subsistence level."[21] For the interior design class taught in Basel by Paul Artaria (1892–1959)—a former representative of the most radical faction

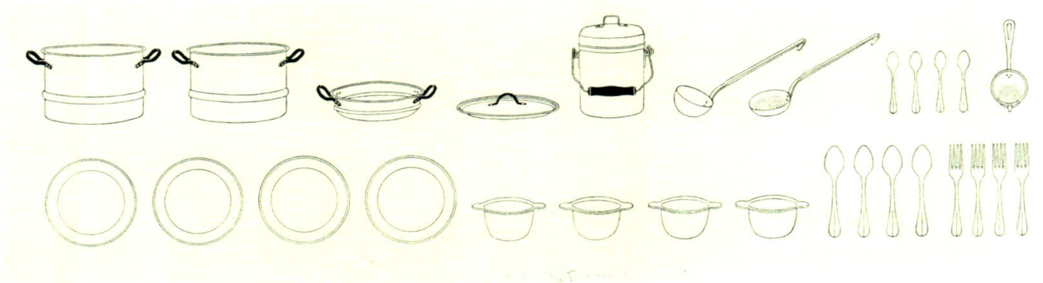

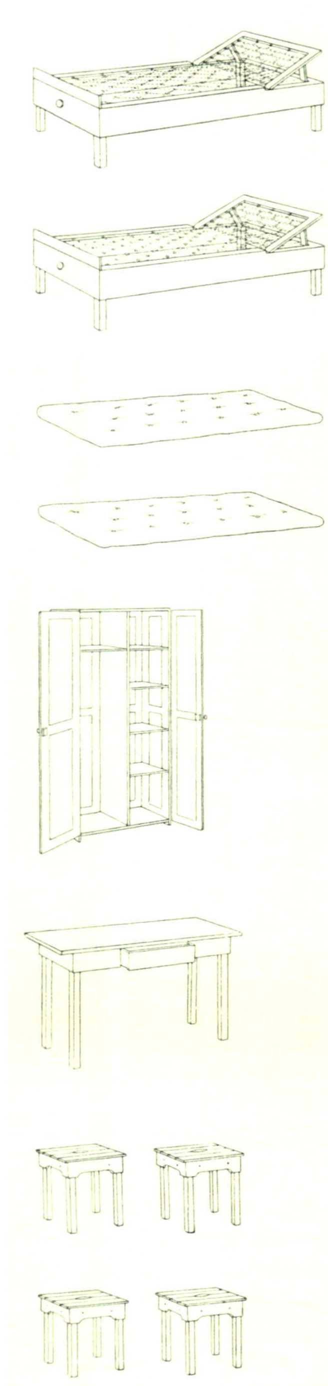

374

375

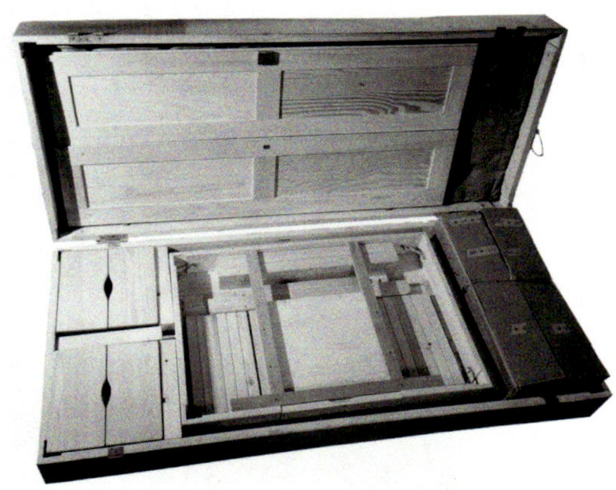

376

377

Mauritius Ehrlich, Aermo furniture kit, 1944

374 The contents of the Aermo furniture kit: two upper and lower mattresses, a wardrobe, a table with drawer, four stools, two pots, one pan, four table settings comprising plates, cups, and cutlery, two ladles, and one sieve.

375 Mauritius Ehrlich, Aermo furniture kit, complete furniture set for emergency housing, Aermo GmbH, Zurich. 30,000 sets were produced from 1944 to 1948.

376 The Aermo furniture kit, when folded out. The lower mattresses are part of the packaging.

377 Aermo furniture kit prospectus, ca. 1945. The title page shows the home kit ready to be shipped.

378

379

380

378 Max Bill, Villiger House, Bremgarten, 1942. Photo: Binia Bill. Interior. This minimalist house was the practical basis for Bill's book on reconstruction, *Wiederaufbau.*

379 Max Bill, *Wiederaufbau. Dokumente über Zerstörungen, Planungen, Konstruktionen,* Verlag für Architektur, Erlenbach-Zürich, 1945. Cover

380 Max Bill, Villiger House, Bremgarten, 1942. Photo: Binia Bill. Building with Durisol lightweight construction elements.

of Neues Bauen—such premises inevitably resulted in do-it-yourself furniture. The idea was to supply packages of prefabricated standardized fir wood boards complete with fittings and screws for assembly by the user into benches, tables, bookshelves, chests, and cabinets.[22] Willy Guhl (1915–2004) as well, who had studied at the Kunstgewerbeschule Zürich shortly before the war, now created a series of exemplary export and kit furniture, based initially on the rustic aesthetic of wooden basement partitions and transport crates. He, too, expected customers to put together by themselves the cheap *Crate Cupboard* he designed in 1940—many years before IKEA would reap enormous profits from the same concept. Unlike this elementary box, the two-door clothes and linen closet he developed in 1942 was presentable enough for use in the contemporary Swiss home. It consisted of a take-apart skeleton made of hardwood with inserted walls of colored Pavatex panels; the internal arrangement was fully adjustable. The cabinetmakers Schreinerei Guhl in Stein am Rhein produced this slightly more deluxe version up until the sixties.

Perhaps the most virtuosic contribution to the topic of early furniture systems was made by the doyen Wilhelm Kienzle (1886–1958), who had been teaching at the Kunstgewerbeschule Zürich since 1918. His "Collapsible frame for goods"—the title of the 1944 patent application[23]—was a furniture system that could truly be disassembled again at any time. X-shaped ash wood support elements were inserted laterally into lightweight shelves and secured with split pins. Available in a number of different dimensions, the elements enabled the construction of shelf or cabinet units of different depths, which could be added to laterally and even made into desks. Although the prospectuses advertising the "Strub-Gestelle" (Strub Frames)—the name comes from the co-developer and producer Robert Strub—suggest primarily office or warehouse use, the lightweight and fresh-looking furniture system also found its way into the homes of the Swiss intelligentsia.

On the Way to *Gute Form*

The formal unpretentiousness, clean design, and visually comprehensible "idea" behind the kit furniture no doubt signal a turn toward design aspects that would shape the "nomadic living" of the postwar period. The early designs Jacob Müller (1905–1998) created for the Werkgenossenschaft Wohnhilfe, a cooperative initiated by around ten cabinetmakers in 1946, were however still indebted to the "principles for postwar assistance."[24] Benches, stools, chairs, and tables were equipped with easy-to-remove legs or backs that were inserted into slotted crossbars and braced using screws. Like the one- or two-door highboy cabinets and the novel Lättlicouch slatted bed frame, they were designed for "economical transport and self-assembly."[25] The basic program of the Wohnhilfe cooperative, which was active until 1980, combines the constructive expressiveness of furniture kits with the organic *Heimatstil* idiom and the teachings of anthroposophy. It was implemented, for example, in the Pestalozzi Children's Village and it helped industrial enterprises to rapidly furnish foreign workers' accommodations. But it also made its way via the Zurich salesroom on Oberdorfstrasse—along with designs by Max Bill, Willy Guhl, Wilhelm Kienzle, and Urs Roesch—into the reduced spaces of the homes and holiday cottages of the postwar period. The *Plio* program developed in 1948 introduced kit furniture into a world in which leisure time spent in the garden or on vacation was beginning to play a major role in an increasingly affluent society. "It works," was the headline of the *Neue Zürcher Zeitung*'s article on the new trend: "What's special about this light-footed, almost weightless furniture (which calls itself Plio) is the use of plywood like that found in aircraft construction.... When you take one of these folding chairs out of your briefcase, all you have to do is insert the two pairs of legs into the slots, snap the joints into place, and your stool is ready." Two such stools produced by Aermo GmbH could be combined with a collapsible table into a package weighing only 7.8 kilograms: "Motorists will reach for this furniture to settle in for a cozy Sunday afternoon in a particularly beautiful spot, in the shade at the edge of the woods for example."[26]

It was once again Max Bill who stepped back to review these developments and to find cogent concepts in which to couch his conclusions. Shortly before the end of the war, he had kept a lookout for the latest trends in the commodity industry—not only in Switzerland but also abroad, where the end of wartime production would set free enormous energies, appearing to offer a realistic chance "for coordinated strategies for reconstruction and industrial

381

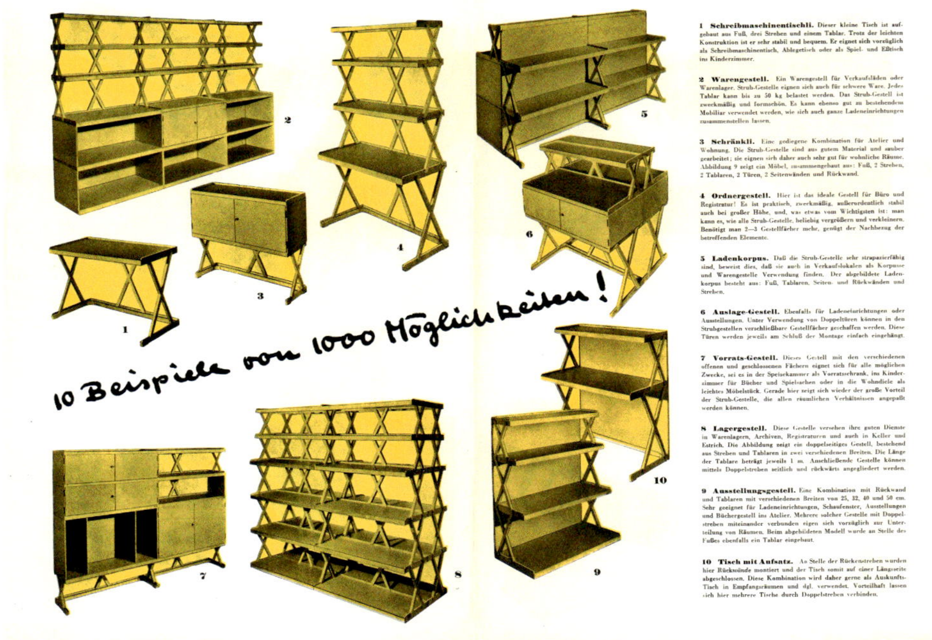

10 Beispiele von 1000 Möglichkeiten!

1 Schreibmaschinentischli. Dieser kleine Tisch ist aufgebaut aus Fuß, drei Streben und einem Tablar. Trotz der leichten Konstruktion ist er sehr stabil und bequem. Er eignet sich vorzüglich als Schreibmaschinentisch, Ablegetisch oder als Spiel- und Eßtisch ins Kinderzimmer.

2 Warengestell. Ein Warengestell für Verkaufsläden oder Warenlager. Strub-Gestelle eignen sich auch für schwere Ware. Jedes Tablar kann bis zu 50 kg belastet werden. Das Strub-Gestell ist zweckmäßig und formschön. Es kann ebenso gut zu bestehendem Mobiliar verwendet werden, wie sich auch ganze Ladeneinrichtungen zusammenstellen lassen.

3 Schränkli. Eine gediegene Kombination für Atelier und Wohnung. Die Strub-Gestelle sind aus gutem Material und sauber gearbeitet; sie eignen sich daher auch sehr gut für wohnliche Räume. Abbildung 9 zeigt ein Möbel, zusammengebaut aus: Fuß, 2 Streben, 2 Tablaren, 2 Türen, 2 Seitenwänden und Rückwand.

4 Ordnergestell. Hier ist das ideale Gestell für Büro und Registratur! Es ist praktisch, zweckmäßig, außerordentlich stabil auch bei großer Höhe, und, was etwas vom Wichtigsten ist: man kann es, wie alle Strub-Gestelle, beliebig vergrößern und verkleinern. Benötigt man 2—3 Gestellfächer mehr, genügt der Nachbezug der betreffenden Elemente.

5 Ladenkorpus. Daß die Strub-Gestelle sehr strapazierfähig sind, beweist dies, daß sie auch in Verkaufslokalen als Korpusse und Warengestelle Verwendung finden. Der abgebildete Ladenkorpus besteht aus: Fuß, Tablaren, Seiten- und Rückwänden und Streben.

6 Auslage-Gestell. Ebenfalls für Ladeneinrichtungen oder Ausstellungen. Unter Verwendung von Doppeltüren können in den Strubgestellen verschließbare Gestellfächer geschaffen werden. Diese Türen werden jeweils am Schluß der Montage einfach eingehängt.

7 Vorrats-Gestell. Dieses Gestell mit den verschiedenen offenen und geschlossenen Fächern eignet sich für alle möglichen Zwecke, sei es in der Speisekammer als Vorratsschrank, ins Kinderzimmer für Bücher und Spielsachen oder in die Wohndiele als leichtes Möbelstück. Gerade hier zeigt sich wieder der große Vorteil der Strub-Gestelle, die allen räumlichen Verhältnissen angepaßt werden können.

8 Lagergestell. Diese Gestelle versehen ihre guten Dienste in Warenlagern, Archiven, Registraturen und auch in Keller und Estrich. Die Abbildung zeigt ein doppelseitiges Gestell, bestehend aus Streben und Tablaren in zwei verschiedenen Breiten. Die Länge der Tablare beträgt jeweils 1 m. Anschließende Gestelle können mittels Doppelstreben seitlich und rückwärts angegliedert werden.

9 Ausstellungsgestell. Eine Kombination mit Rückwand und Tablaren mit verschiedenen Breiten von 25, 32, 40 und 50 cm. Sehr geeignet für Ladeneinrichtungen, Schaufenster, Ausstellungen und Büchergestell ins Atelier. Mehrere solcher Gestelle mit Doppelstreben miteinander verbunden eigen sich vorzüglich zur Unterteilung von Räumen. Beim abgebildeten Modell wurde an Stelle des Fußes ebenfalls ein Tablar eingebaut.

10 Tisch mit Aufsatz. An Stelle der Rückenstreben wurden hier Rückwände montiert und der Tisch somit auf einer Längsseite abgeschlossen. Diese Kombination wird daher gerne als Auskunfts-Tisch in Empfangsräumen und dgl. verwendet. Vorteilhaft lassen sich hier mehrere Tische durch Doppelstreben verbinden.

382

381 Wilhelm Kienzle, Strub-Gestelle (Strub Frames), 1944, collapsible shelving and furniture system made of ashwood and particle board. Standardized system components and detail of frame secured with splints.

382 Prospectus for Strub-Gestelle for shops, offices, storerooms, homes, n.d. The inside pages illustrate "10 Examples out of 1,000 Possibilities!"

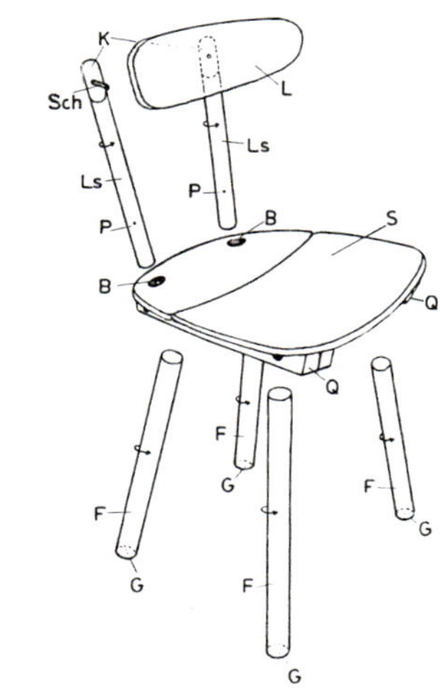

383

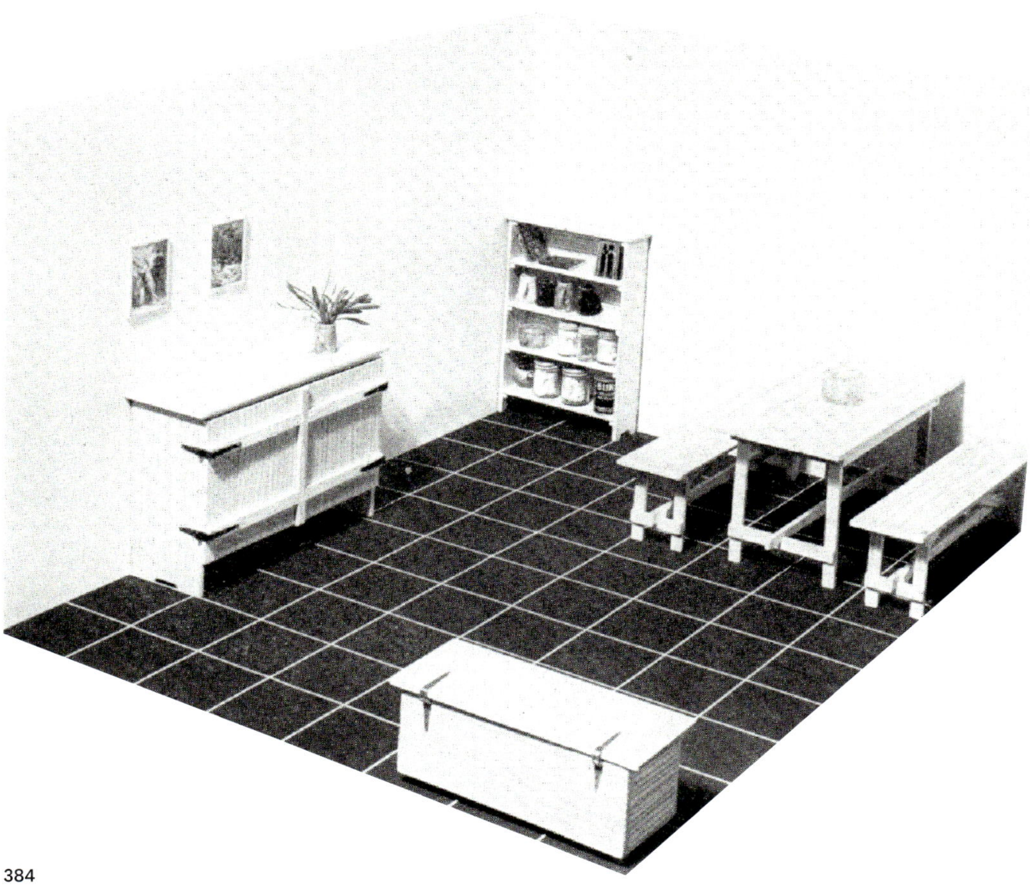

384

383 Jacob Müller, Wohnhilfe Chair, patent drawing, 1944. Sold as building kit to be assembled by the customer.

384 Interior design class at the Gewerbeschule Basel, supervised by Paul Artaria, Furnishings for emergency housing, ca. 1944. Customers were supplied with finished firwood boards and hardware from which to put together the furniture.

385 386 Willy Guhl, Pavatex Wardrobe, 1942, fir, maple, particle board. Furniture kit consisting of support frame and slide-in colored Pavatex panels. Museum für Gestaltung

387 Willy Guhl, Crate Cupboard, ca. 1940, fir. Customers could nail together the inexpensive cupboard themselves using the standard elements provided.

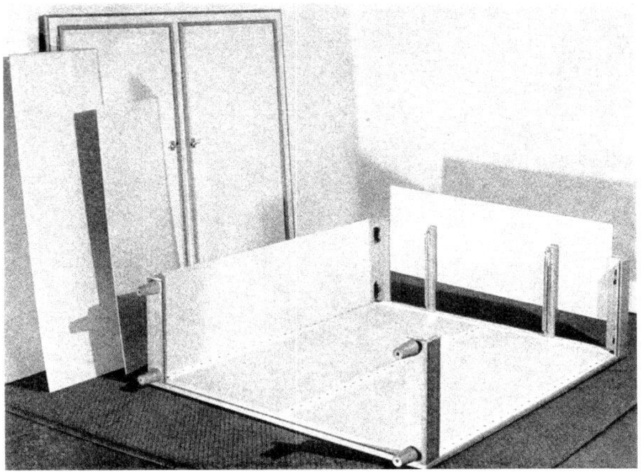

385

386

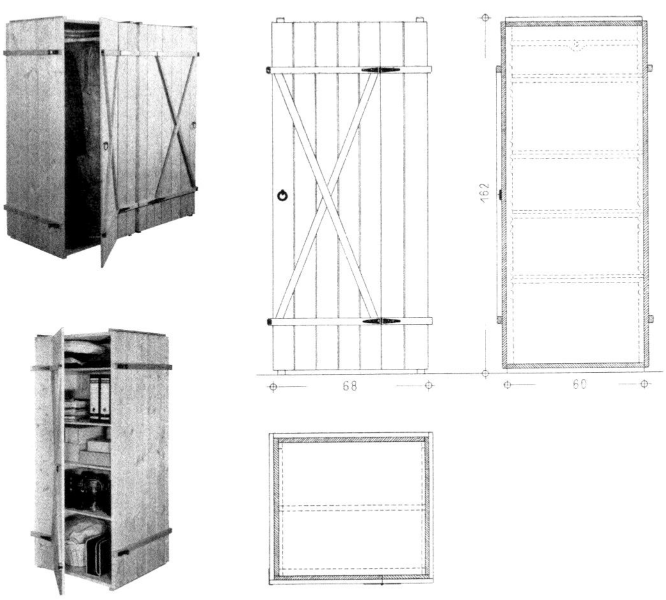

387

commodity production."[27] When Bill spoke before the Werkbund in 1948 on the subject of "Beauty from Function and as Function,"[28] he put the old familiar demands for "truth to materials" in perspective with regard to the industrial design of the future and called instead for "an ultimate utilisation of materials, where the maximum effect is achieved with the minimum of materials."[29] He cited the Eiffel Tower as a "shining exemplar of the extreme exploitation of materials, an emblem for the technical age and the rational use of materials, as well as the germ of a new ideal of beauty." The economy of means—the key feature of the "solutions d'urgence"—was, in his view, the prerequisite for solving the aesthetic question. As harbingers of future developments, the featherweight *Plio* furniture and the combinable Strub-Gestelle then also made the grade as some of the few three-dimensional exhibits in the epoch-making trade fair pavilion *Die gute Form* in 1949, which would convey Bill's ideas to a wider audience.

1 "A key moment in the modernization of both the theory and practice of architecture." Cohen 2011, jacket blurb. See also Colomina 2006: "Modern architecture is inseparable from war" (subtitle).
2 Foundation stone laid on September 11, 1941; the first occupant moved in on April 30, 1942; completed on January 15, 1943; see Cohen 2011, p. 288.
3 Le Corbusier 1931, p. 66.
4 See Virilio (1975) 1994.
5 Schwager 2004.
6 Meyer 1943, p. 80.
7 See Schmed/Rüegg 2012, fig. p. 32.
8 Bruyère 1945, p. 3.
9 See Boesiger (1946) 1995, p. 94: "Cette construction plutôt rattachée à l'architecture mésopotamienne (en plan) ..."
10 "baraquements militaires sans âme ni proportion." Le Corbusier 1942, p. 5.
11 See "Maison métallique préfabriquée – Pierre Jeanneret, architecte, Jean Pouvé, constructeur," in "Préfabrication/Industrialisation," special issue, *L'Architecture d'aujourd'hui* 16, no. 4 (January 1946), p. 58.
12 See Barsac 2005, pp. 213–33.
13 See "Sièges d'urgence de Pierre Jeanneret" 1945. "Sièges d'urgence de Pierre Jeanneret," in "Solutions d'urgence," special issue, *L'Architecture d'aujourd'hui* 16, no. 2 (July/August 1945), p. 74.
14 Bill 1945, p. 5.
15 See Rüegg 2004: Haus Villiger, pp. 49–52.
16 Bill 1945, pp. 166–67.
17 After the Russian occupation, Ehrlich fled in 1914 from Ivano-Frankivsk to Vienna, and from there to Zurich in 1938; see Müller 2011.
18 Müller 2011.
19 Bill 1945, p. 166.
20 Ibid.
21 One of the major themes of Neues Bauen in the twenties; see *Die Wohnung für das Existenzminimum* 1930.
22 See Guyer/Kettiger 1946, pp. 162–63. The book also contains in the chapter "Wiederaufbaumöbel" entries on Willy Guhl, Mauritius Ehrlich, Wilhelm Kienzle, Otto Kolb, and Jacob Müller.
23 See *Wilhelm Kienzle* 1991, pp. 62–64, here p. 62.
24 See *Jacob Müller* 1988, p. 55.
25 Ibid.
26 H. R. S. 1949.
27 See von Moos 1995, here p. 70; see also von Moos 1992, pp. 258–63.
28 Bill (1948) 2010.
29 Ibid., p. 33.

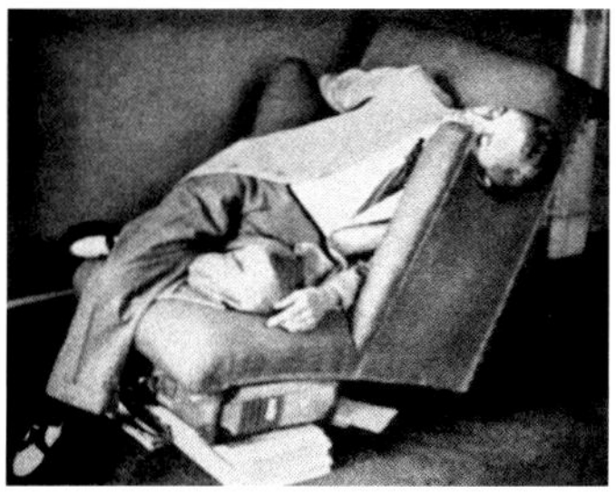

388

389

390

391

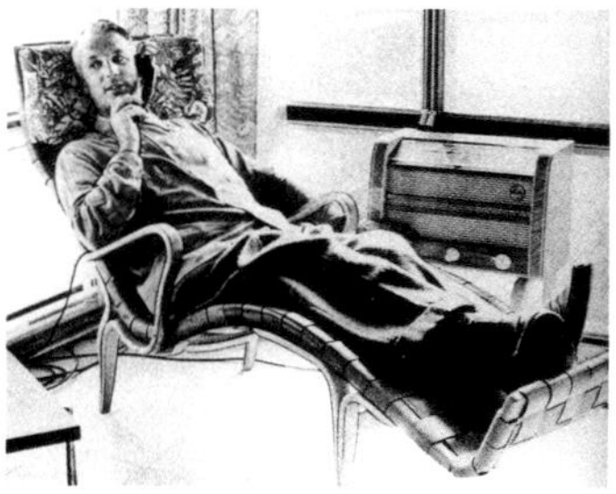

392

388 Bruno Munari, "Ricera della commodità in una poltrona scomoda," 1944
389 Willy Guhl, Basic studies on the design of seating forms, 1947. The clay shell records the imprint of the sitter and helps the designer to study various ways of sitting.
390 Charles and Ray Eames, Kazam! machine, 1942. A frame made of wooden slats, plaster, and an electric cable, set up in the Eames's home for the production of seats made of glued laminated plywood.
391 Otto Kolb, Strap Chair, 1945. The hemp straps hung between glued arcs of ashwood conform to the sitter's body. The seat can also be lifted out and used separately as a rocking chair.
392 Bruno Mathsson sitting in his Eva Chair (1934) at his home in Värnamo, 1950.

As though the soldiers returning home from the war were utterly unable to sit back down at their tables, as though sitting itself, as well as chairs, had to be reinvented from scratch, designers in the 1940s experimented feverishly with new materials and manufacturing techniques. The iconic modernist furniture pieces, as well as the heavy furnishings typical of the middle class, were not particularly accommodating to the human body. The Italian artist and industrial designer Bruno Munari provided a rather theatrical demonstration of this in 1944, using an upholstered armchair. He asked, "How many different armchairs have you seen in your life?... Everyone wants different furniture and so the true function of a chair, for example, comfort, goes to hell."[1]

Forms of Seating

In the summer of 1947, the brothers Willy (1915–2004) and Emil (1918–2013) Guhl built three wooden seat frames and upholstered them using a thick layer of clay: one positioned to sit upright, one for leaning back, and one for a reclining position. They invited the guests at a nearby river bathing facility to sit down on them. The test subjects were not informed of the brothers' intentions. They sat on the clay "upholstery" for between thirty and sixty minutes, changing their posture as needed and then leaving the seats to the next volunteer. This experiment will not have taken longer than one afternoon. Afterward, Willy and Emil Guhl recorded the results by casting them in plaster.[2]

Elevated atop wooden legs, these three seating studies were presented to the broad public for the first time in 1950 at a Schweizerische Werkbund (SWB) exhibition held at the Kunstgewerbemuseum Zürich. Willy Guhl describes them in the journal *Werk:* "We began by considering the different requirements a well-upholstered chair might meet, but we also wondered whether it was possible to create chairs with forms that were suited to certain ways of sitting. We asked ourselves, 'Is it possible to design a chair form that accommodates many different personal ways of sitting?' We launched our studies by forgoing any sort of comparison to existing models as well as any formal, fixed idea about construction, instead focusing simply on sitting itself."[3] In the same article, Guhl introduced several chair designs for the home and garden as well as other Swiss seating furniture. These examples included the fabric-covered, lightweight tubular steel armchair by the Kim company, the stackable *Aluflex* chair, made of airplane plywood, as well as the *Tripod Chair* by Max Bill. They all met Guhl's criteria for new forms of seating: "Cases in which new uses, new materials, new production methods, new notions of form, or new approaches to the fundamental aspects of seating lead to novel forms."[4] A year after this exhibition, Willy Guhl, who taught at the Kunstgewerbeschule Zürich from 1941 onward, took over as teacher of Wilhelm Kienzle's interior design class, and in 1965 he introduced Switzerland's first training course in product design.

The Body as a Design Factor

The armchair *Eva,* designed in 1934 by the Swedish cabinetmaker Bruno Mathsson, is based on a natural seating posture. This wavy chair, made from curved laminated wood, was used as an example in Willy Guhl's lessons and must have been so strongly identified with his own design ideas that it was described in the photography class as "Mr. Guhl's Swedish chair."[5] Similarly, the furniture designed by Alvar Aalto for the Paimio Sanatorium and sold by Wohnbedarf in Switzerland was groundbreaking in its human-oriented approach to design, as expressed through its organic language of form. Scandinavian domestic culture, as well as American industrial design, was highly influential for Swiss designers in the forties. As in Switzerland, in Scandinavia functionalist design was combined with traditional craftsmanship, leading to locally inflected, livable postwar modernism. In the US, renowned European Bauhaus figures such as Walter Gropius, Ludwig Mies van der Rohe, and Marcel Breuer imparted their knowledge at universities, and thus a young generation of architects continued to develop the International Style. After the war, the US furniture industry was able to take advantage of arms-industry expertise, enabling the implementation of new designs at a time when comparable production facilities were still lacking in Switzerland.

Like many Swiss architects, Otto Kolb (1921–1996), an associate in Alfred Roth's Zurich architecture firm, devoted himself to the development of new types of furniture for the needs of the reconstruction era: "It should ideally be lightweight, practical, and organic, featuring aesthetically clean shapes."[6] For his studio house in Brüttisellen, he designed in 1945 a variable *Strap Chair,* with different frames used to create a normal chair, easy chair, or rocking lounger. Other features of this visionary interior design were the *Book Tree,* a tension-loaded steel pole reaching from floor to ceiling from which books, magazines, or houseplants "grew," held by metal clasps, as well as a free-hanging lamp balanced by two counterweights. Like Alexander Calder's mobiles, this lamp can be adjusted to reach any corner of the room. Appointed to the Chicago Institute of Design (ID) on the recommendation of Sigfried Giedion, Otto Kolb directed the product design department there from 1948 to 1951, where he taught his scientific approach and designed further pieces of furniture based on the constructive principles of nature, as well as studying the human musculoskeletal system.

Although at first glance Kolb's *Strap Chair* does not immediately appear to share any formal similarities with Willy Guhl's plaster shells, both designs are based on the exploration of different ways of sitting. While the wooden construction dictates a certain seating angle and height, the surfaces covering them react to individual bodies. In the case of the Guhl brothers, the position of the body is preserved as a clay imprint, and in Otto Kolb's chair the straps give in response to the body's weight and then snap back to their initial positions.

Series Seat Shells

In order to offer as many people as possible good and affordable chairs, the American architect couple Charles and Ray Eames experimented beginning in 1941 with shaping laminated plywood into three-dimensional forms. After achieving success with their *Organic Chair,*[7] they attempted to recreate its shape in punched sheet metal and later in fiberglass-reinforced plastic—a new material invented during the war. Their *Arm Chair on Cross-Rod Base* won second prize in the international Competition for Low-Cost Furniture held in 1948 by the Museum of Modern Art in New York (MoMA). An example of the chair was crafted by hand for this exhibition, and starting in 1950 the plastic chairs were manufactured in series by Zenith Plastics and sold by Herman Miller. The Guhl brothers also took part in the competition with their three seating studies. In the publication accompanying the MoMA exhibition, Edgar Kaufmann, Jr., director of the Department of Industrial Design, said of the Guhl-like plaster shell *La Chaise* by Ray and Charles Eames. "An interesting example of parallel thinking on both sides of the Atlantic is presented by the reclining chair by Willy and Emil Guhl of Switzerland and that by Charles Eames, both developing the theme of a large molded shell."[8] Although their forms may appear similar, the histories of their production differ greatly: *La Chaise* was not researched through movement studies but designed as a freehand form and produced by Vitra from 1990 as a seating sculpture. The reclining shell seat from Switzerland featured in the competition was never serially produced, but Willy Guhl did find a manufacturer to make his upright molded seating shell in 1951. The company Scott Bader & Co. in Zurich had developed the weatherproof plastic Scobalit and produced Europe's first molded plastic shell seat under this name.[9] Unlike the *Plastic Side Chair,* the *Scobalit* does not feature a high seat back and is slightly convex in the middle like a tractor seat.

Provided we count the *Landi Chair* by Hans Coray (1906–1991) as a shell seat—the extrusion-molded aluminum shell is not freestanding but hung into the frame that forms the legs—it is the very first shell seat ever designed, and must therefore have been a useful reference for Charles and Ray Eames,[10] as well as for Willy Guhl. Using annealed aluminum, the MEWA Blattmann, Wädenswil, was able to create a lightweight shell stabilized by its bent edges as early as 1939, producing a series of 1,500 of the chairs for that year's Swiss National Exhibition (Schweizerische Landesausstellung, or "Landi" for short)[11]

A chair with a molded rattan seat designed by Gian Franco Legler (b. 1922), who had studied architecture at the ETH Zürich, was sold widely in the US. In order to become an industrial designer,

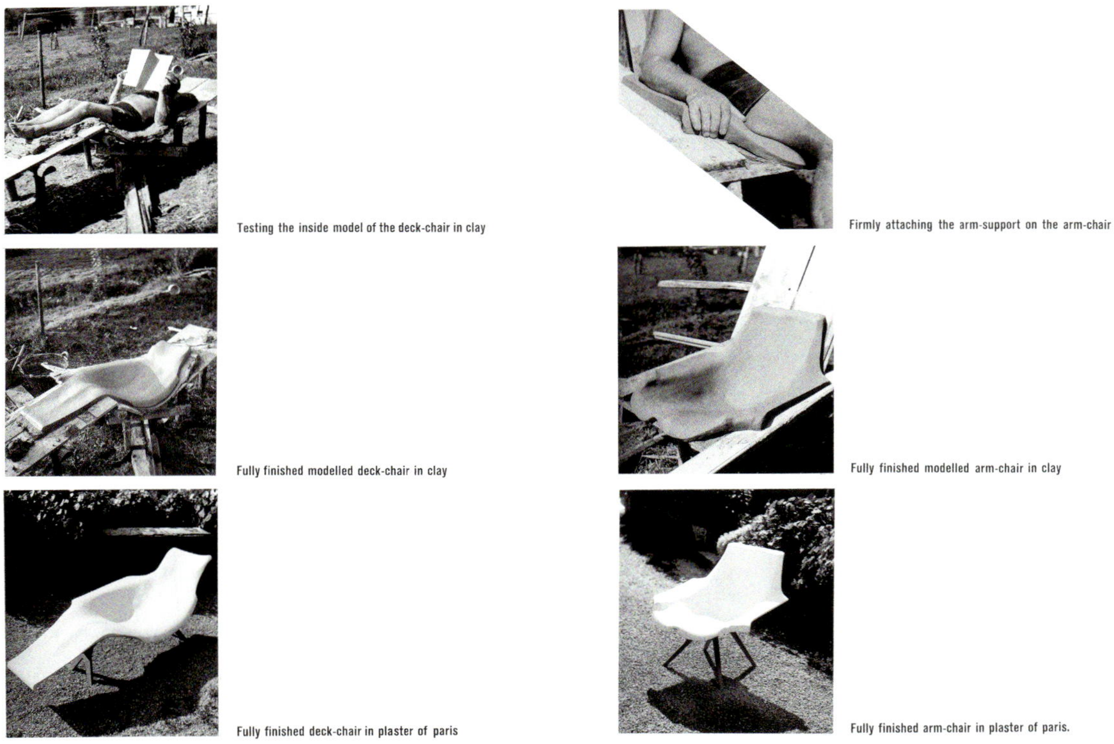

393

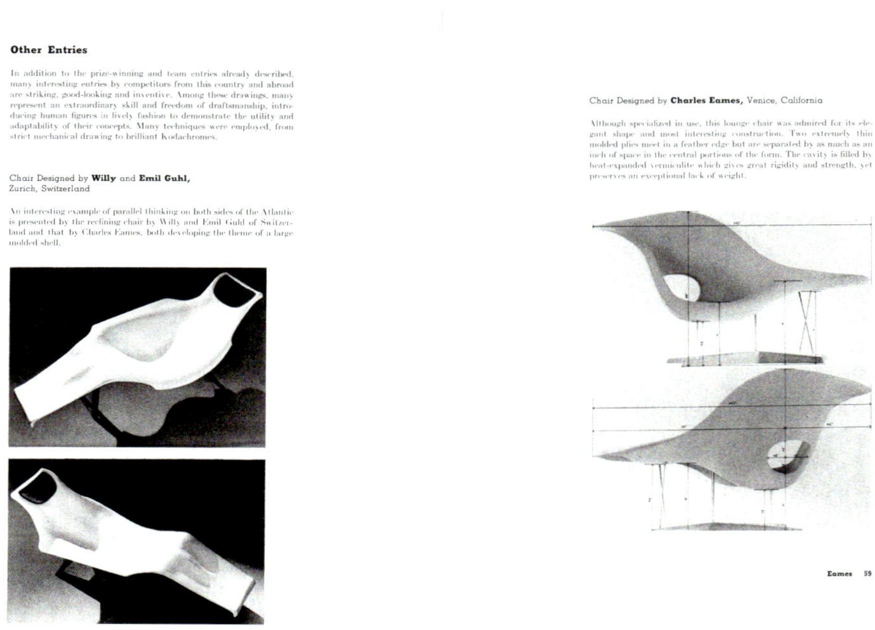

Other Entries

In addition to the prize-winning and team entries already described, many interesting entries by competitors from this country and abroad are striking, good-looking and inventive. Among these drawings, many represent an extraordinary skill and freedom of draftsmanship, introducing human figures in lively fashion to demonstrate the utility and adaptability of their concepts. Many techniques were employed, from strict mechanical drawing to brilliant Kodachromes.

Chair Designed by **Willy** and **Emil Guhl,** Zurich, Switzerland

An interesting example of parallel thinking on both sides of the Atlantic is presented by the reclining chair by Willy and Emil Guhl of Switzerland and that by Charles Eames, both developing the theme of a large molded shell.

Chair Designed by **Charles Eames,** Venice, California

Although specialized in use, this lounge chair was admired for its elegant shape and most interesting construction. Two extremely thin molded plies meet in a feather edge but are separated by as much as an inch of space in the central portions of the form. The cavity is filled by heat-expanded vermiculite which gives great rigidity and strength, yet preserves an exceptional lack of weight.

394

393 Willy and Emil Guhl, Presentation board, 1947–48. Part of the entry for the Museum of Modern Art's International Competition for Low-Cost Furniture Design, 1948.

394 Edgar Kaufmann, Jr., *Prize Designs for Modern Furniture from the International Competition for Low-Cost Furniture Design,* Museum of Modern Art, New York, 1950. Cover and double-page spread

395 Willy Guhl, Three plaster seat form studies on trestles, 1947. Presented at the SWB exhibition at the Kunstgewerbemuseum Zürich, 1950.

396 Prospectus for the Scobalit shell chair, Scott Bader & Co. AG, Zurich, n.d. (1st production run in 1951, 120 pieces)

397 Richard Paul Lohse, Double-page spread from the magazine *Bauen + Wohnen,* no. 8 (1950). On the left, the Contour Delineator developed by Rudolph Jegrt at the Chicago Institute of Design; and on the right, Willy and Emil Guhl's seat form studies.

395

396

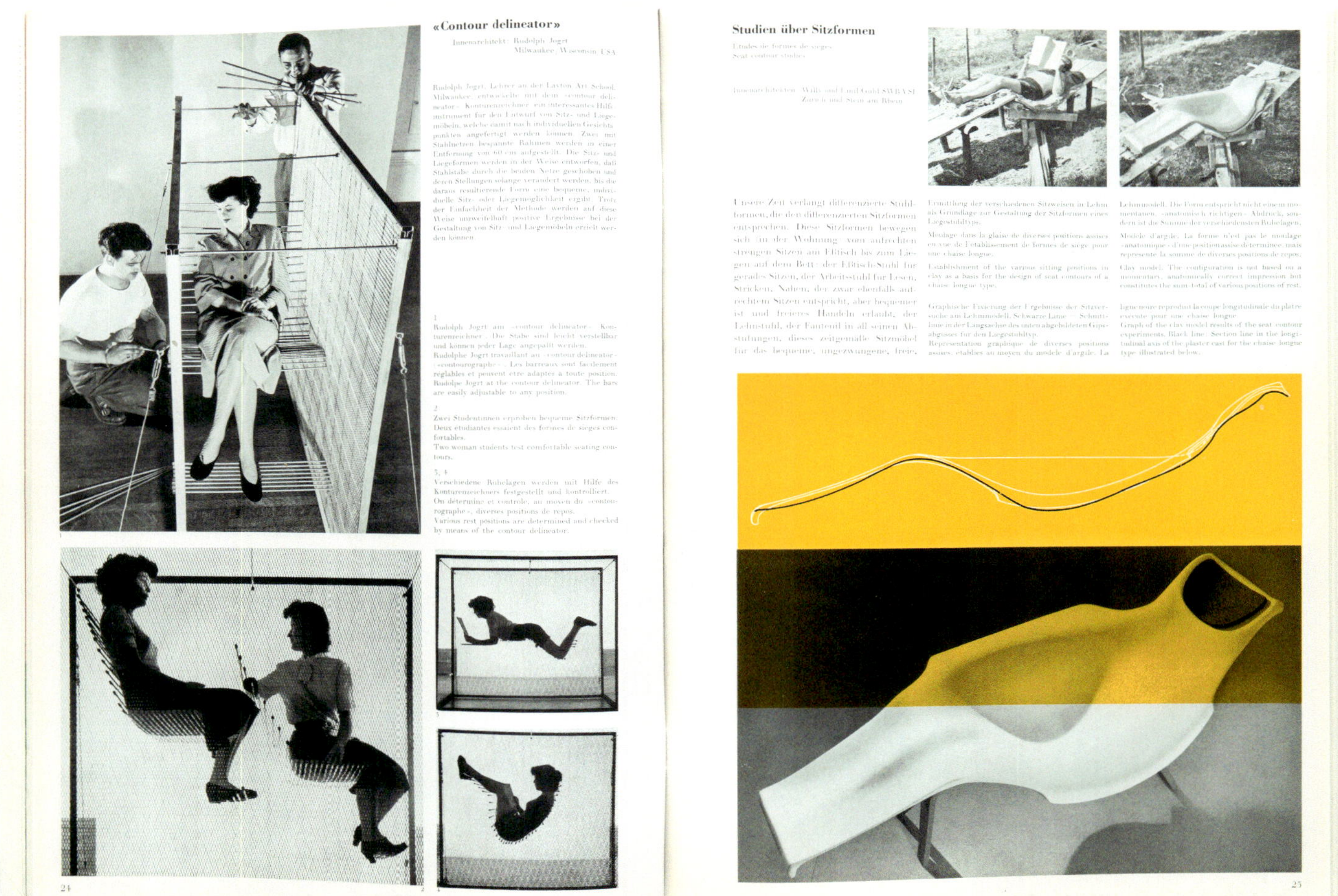

«Contour delineator»

Innenarchitekt: Rudolph Jogrt
Milwaukee, Wisconsin USA

Rudolph Jogrt, Lehrer an der Layton Art School, Milwaukee, entwickelte mit dem «contour delineator» Konturenzeichner ein interessantes Hilfsinstrument für den Entwurf von Sitz- und Liegemöbeln, welche damit nach individuellen Gesichtspunkten angefertigt werden können. Zwei mit Stahlnetzen bespannte Rahmen werden in einer Entfernung von 60 cm aufgestellt. Die Sitz- und Liegeformen werden in der Weise entworfen, daß Stahlstäbe durch die beiden Netze geschoben und deren Stellungen solange verändert werden, bis die daraus resultierende Form eine bequeme, individuelle Sitz- oder Liegemöglichkeit ergibt. Trotz der Einfachheit der Methode werden auf diese Weise unzweifelhaft positive Ergebnisse bei der Gestaltung von Sitz- und Liegemöbeln erzielt werden können.

1
Rudolph Jogrt am «contour delineator» Konturenzeichner. Die Stäbe sind leicht verstellbar und können jeder Lage angepaßt werden.
Rudolphe Jogrt travaillant au «contour delineator» («contourographe»). Les barreaux sont facilement réglables et peuvent être adaptés à toute position.
Rudolpe Jogrt at the contour delineator. The bars are easily adjustable to any position.

2
Zwei Studentinnen erproben bequeme Sitzformen.
Deux étudiantes essaient des formes de sièges confortables.
Two woman students test comfortable seating contours.

3, 4
Verschiedene Ruhelagen werden mit Hilfe des Konturenzeichners festgestellt und kontrolliert.
On détermine et contrôle, au moyen du «contourographe», diverses positions de repos.
Various rest positions are determined and checked by means of the contour delineator.

Studien über Sitzformen

Etudes de formes de sièges
Seat contour studies

Innenarchitekten: Willy und Emil Guhl SWB/VSI
Zürich und Stein am Rhein

Unsere Zeit verlangt differenzierte Stuhlformen, die den differenzierten Sitzformen entsprechen. Diese Sitzformen bewegen sich in der Wohnung vom aufrechten strengen Sitzen am Eßtisch bis zum Liegen auf dem Bett: der Eßtisch-Stuhl für gerades Sitzen, der Arbeitsstuhl für Lesen, Stricken, Nähen, der zwar ebenfalls aufrechtem Sitzen entspricht, aber bequemer ist und freieres Handeln erlaubt, der Lehnstuhl, der Fauteuil in all seinen Abstufungen, dieses zeitgemäße Sitzmöbel für das bequeme, ungezwungene, freie,

Ermittlung der verschiedenen Sitzweisen in Lehm als Grundlage zur Gestaltung der Sitzformen eines Liegestuhltyps.
Moulage dans la glaise de diverses positions assises en vue de l'établissement de formes de siège pour une chaise longue.
Establishment of the various sitting positions in clay as a basis for the design of seat contours of a chaise longue type.

Lehmmodell. Die Form entspricht nicht einem momentanen, «anatomisch richtigen» Abdruck, sondern ist die Summe der verschiedensten Ruhelagen.
Modèle d'argile. La forme n'est pas le moulage «anatomique» d'une position assise déterminée, mais représente la somme de diverses positions de repos.
Clay model. The configuration is not based on a momentary, anatomically correct impression but constitutes the sum-total of various positions of rest.

Graphische Fixierung der Ergebnisse der Sitzversuche am Lehmmodell. Schwarze Linie = Schnittlinie in der Längsachse des unten abgebildeten Gipsabgusses für den Liegestuhltyp.
Représentation graphique de diverses positions assises, établies au moyen du modèle d'argile. La ligne noire reproduit la coupe longitudinale du plâtre exécuté pour une chaise longue.
Graph of the clay model results of the seat contour experiments. Black line: Section line in the longitudinal axis of the plaster cast for the chaise longue type illustrated below.

24 25

397

Legler studied plastics and fiber technology at MIT in Cambridge, Massachusetts and later founded his own studio in Chicago, the epicenter of industrial design at the time. His *Basket* chair, designed in 1951 and manufactured in Italy by local basket weavers, was displayed in the MoMA exhibition *Good Design 1953* on the same pedestal as the *Wire Chair* by Charles and Ray Eames, likewise designed in 1951.[12] The *Basket Chair* became part of MoMA's permanent collection and was produced in a series of 40,000 by Raymor in New York.

Using Hands to See

The conclusions he drew from the 1947 plaster casts first led Willy Guhl, however, to design a prototype of an armchair in which he paid close attention to the handprints that had been left in the clay. Guhl emphasized that these "are not determined optically, or based on formal considerations, but simply 'seen' by the hands."[13] In some ways, this design approach is reminiscent of Rudolf Steiner's anthroposophy-inspired furniture and interiors,[14] or Antonio Vitali's (1909–2008) pleasant-to-handle wooden toys. But it also brings to mind the pioneering work done by Friedrich Herig at the Institut für Griff-Forschung he founded in 1941 in Rechtenstein, Germany, and his guiding principle: "Anything made to be held in the hand must be crafted by hand."[15] Not only healthy ways of sitting, as propagated, for example, by the Swedish physician Dr. Bengt Akerblom at the Cologne Furniture Fair in 1954 through the demonstration of a bent backrest, but also the anatomically correct grip became a challenge for industrial designers. The SWB thus held a competition for door handle design in conjunction with the Schlossfabrik Heusser AG in Obermurgenthal, in which Willy Guhl and his students took part. The best demonstration of Guhl's ergonomic approach, though, can be seen in his revamping of Eternit planters. In order to come up with the right form, Guhl created huge plaster models, which he stacked in the bathroom of his Zurich apartment. In order to be able to stash them away more quickly when guests came, he invented a new type of bottom grip that was perfectly suited to a plant container filled with damp soil.[16] This new grip was a decided improvement in the object's usability, while also demonstrating a simplified way of handcrafting negative shapes by avoiding undercuts. Furthermore, the slightly beveled bottom edges made formal sense as well, as they added rhythm to the rows of angular Eternit planters, making them appear lighter. Although the term "ergonomics" did not come into fashion for product design until the 1980s, Guhl's students were already well-versed in user-friendly forms that did justice to their materials and were easy to produce.

Plasticity and Elasticity

When the company Eternit AG in Niederurnen approached the teachers at the Kunstgewerbeschule Zürich, Willy Guhl was the only one to show an interest in the unusual fiber cement material. In its unbound state, it was malleable like plastic and could be molded into any shape desired. In order to uncover form-defining factors in this material and its processing, the interior design class visited the Eternit factory and developed new planters under the guidance of Willy Guhl, such as the *Spindel* or *Biasca,* the so-called elephant ears planters, which made intelligent use of and showcased the material's plasticity.[17]

In advance of the Swiss horticultural exhibition G59, which took place in Zurich in 1959, Eternit AG held a seating and planter competition under the aegis of Max Bill and then produced the fifteen best designs for display at the show. It was mostly Eternit seating that lined the paths through the vast gardens and illustrated how outdoor areas could be used as living space, based on the American example.[18] Willy Guhl had already created an iconic design using fiber cement in 1954 with his chair made out of an Eternit mat fashioned into an endless loop, using the material in its usual width as it emerges from the machine in order to avoid waste. Its simplicity, elegance, and indestructibility soon won the *Garden Chair* (Strandstuhl) pride of place at many a poolside and company

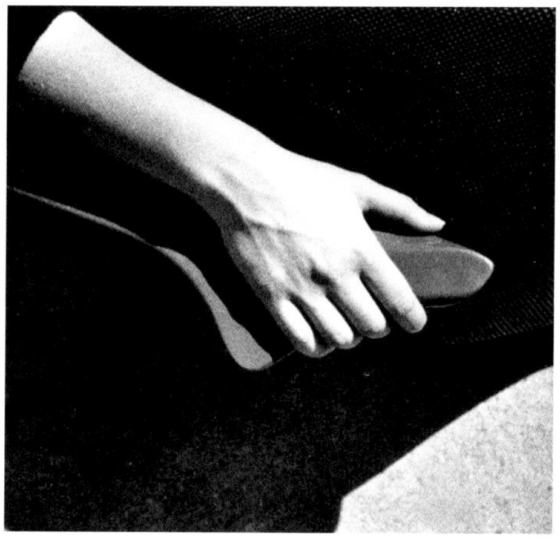
398

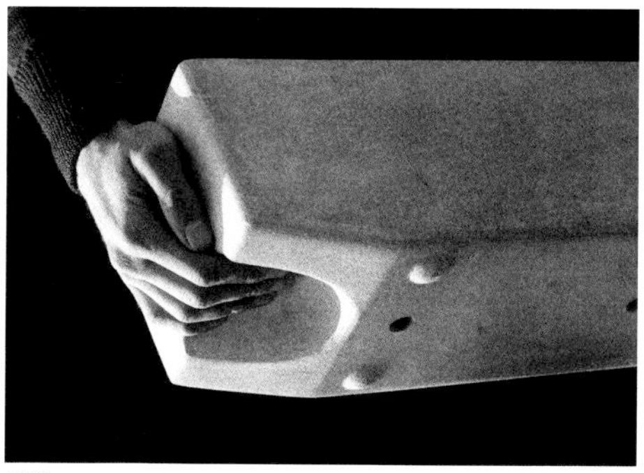
399

400

401

398 Willy Guhl, Detail of the haptic shaping of the armrest for an armchair with adjustable rattan seat, 1948
Photo: Bernard Moosbrugger
399 Willy Guhl, Detail of a flower box with recessed grips underneath, 1954, Eternit AG, Niederurnen
400 Antonio Vitali, Two foxes made of larchwood, 1944, Vitali Spielzeug GmbH, Zurich. Photo from the SWB photo card collection on "Die gute Form"; seal conferred in 1953.
401 Willy Guhl, Door handle with thumb rest, 1954, Schlossfabrik Heusser AG, Obermurgenthal
402 Eternit furniture in use: Garden Chair (Willy Guhl, 1954), Spindle Planter (Anton G. Bee, 1952), Garden Chair (Ludwig Walser, 1959) on a terrace, ca. 1960
403 Promenade at the G59 garden show in Zurich with garden chairs by Ludwig Walser
404 Charles and Ray Eames, Case Study House No. 8, Los Angeles, California, 1949. On the patio, the Landi Chair by Hans Coray, 1939.
Photo: Julius Shulman, 1950s
405 Franco Legler, Basket Chair, 1951.
Photo: Alexander von Steiger, 1965

402

404

403

405

headquarters, and it was even combined with a series of sculptures by Picasso at Documenta 2.[19]

In the fifties it was not only forms that became more fluid but also the transition between functional and freely flowing forms. Many of the utilitarian objects awarded the "Die gute Form" distinction by the SWB were notable for their sculptural presence, and "beauty as function" (see Menzi, "Designing 'Die gute Form,'" pp. 192–97) became a creative mission for Swiss product designers. It is no coincidence that their designs also fulfilled the criteria for MoMA's "Organic Design" competition. Seeing form as a "harmonious organization of the parts within the whole"[20] was intended not merely to respond to the demands of construction, material, and purpose, but also to lend expressiveness. It became crucial to explore to what extent functional forms could emancipate themselves from their standardized shapes and become sculptures. The goal was for the objects to appear neither artificial nor stylized but instead as natural as if they had grown that way.[21] The organic form is symbolic of its purpose and derives its appeal from shapes similar to those of the human body.

1 Munari 1944.
2 *Willy Guhl* 1985.
3 Guhl 1950.
4 Ibid.
5 The high esteem accorded to Mathsson was perhaps due to how he approached the manufacture of chairs as a cabinetmaker rather than as an architect, and because he managed to realize his vision of the simple life in his house by the lake (a dream also illustrated in the 1909 book *The House in the Sun* by the Swedish painter Carl Larsson). Mathsson's lake house served as a model for Willy Guhl's *Taglöhnerhaus* (Day Laborer's House) in Hemishofen.
6 Kolb 1948.
7 The plywood shell chair by Charles Eames and Eero Saarinen won first prize in the 1940 MoMA competition Organic Design in Home Furnishings. The legendary *Kazam!* machine for the manufacture of seat shells is reminiscent of the clay chairs by the Guhl brothers.
8 Kaufmann 1950, p. 58.
9 In 1954, Hans Bellmann, a former Bauhaus student, architect, and teacher at the Kunstgewerbeschule Zürich, developed together with the Strässle company a molded fiberglass shell with an outward-bending rim, which he used in the "Sitwell" seating program.
10 In any case, a photograph from the fifties shows this perforated aluminum chair on the patio of the Eames's home, built in 1949 (*Case Study House No. 8,* Los Angeles).
11 This experiment by Ray and Charles Eames failed, however, and they converted the perforated metal form into plastic instead; see Koenig 2013, p. 48.
12 Huber 2009.
13 Guhl 1950.
14 Fäth 2011.
15 Zeischegg/Auböck 1995.
16 Linus B. Fetz in a conversation with the author, August 22, 2013.
17 Grass was planted in them, and they were displayed in the lobby of the Kunstgewerbemuseums Zürich for the 1952 exhibition *Um 1900 – Art Nouveau und Jugendstil,* as well as being discussed in the journal *Werk;* see Rotzler 1952.
18 Hanak 2009.
19 Rüegg 2009b.
20 Thus did the founder of the Institute of Industrial Design at MoMA, Eliot Noyes, describe organic design in the MoMA press release, September 22, 1941.
21 Erni 1985.

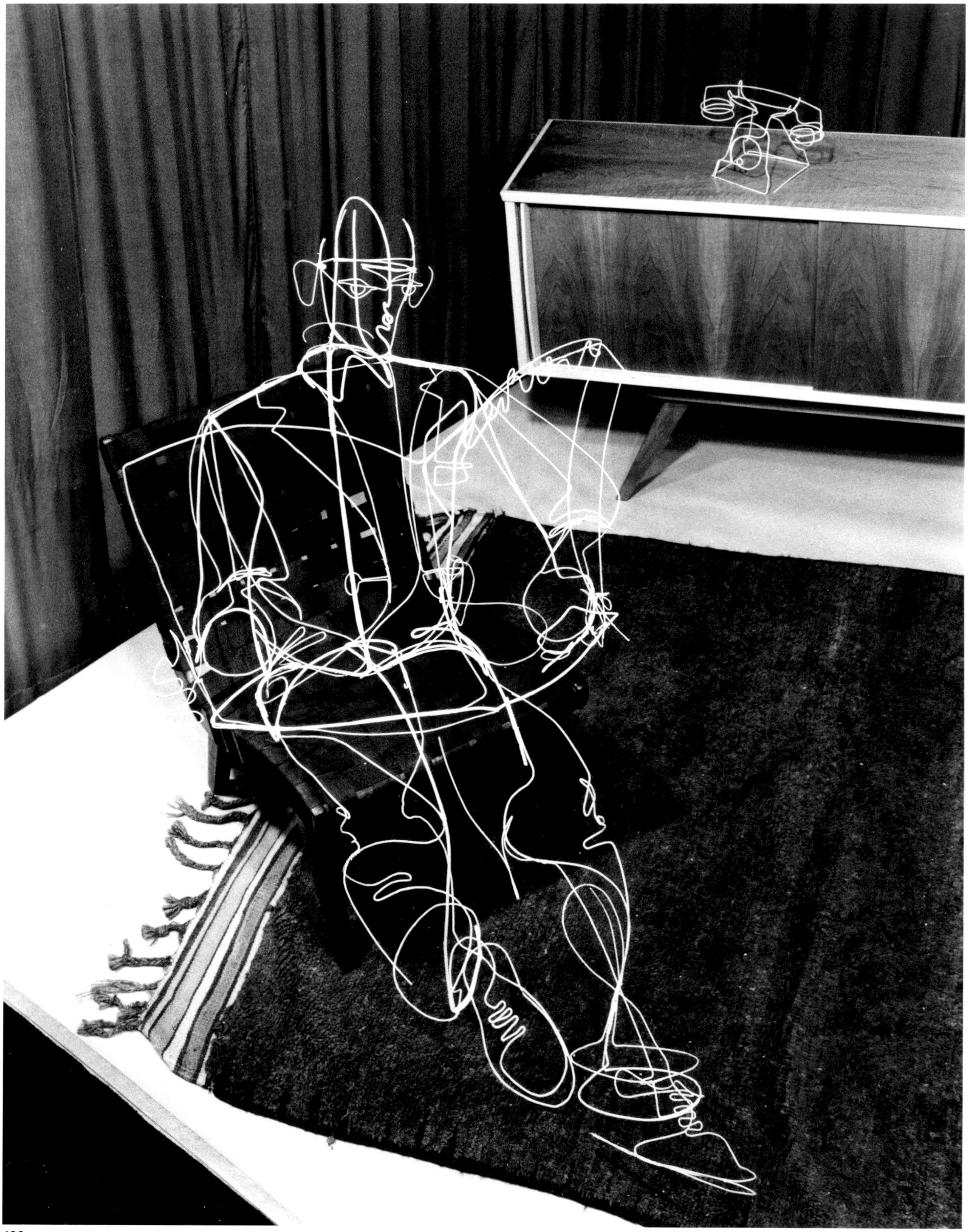

406

Jean Tinguely, Shop window decoration for Wohnbedarf AG, Basel, 1951–52. Tinguely, later famous for his kinetic sculptures, was a trained decorator. Before moving to Paris in 1952, he created on behalf of Siegfried Jehle several spectacular shop window installations that betray his Basel-brand humor in combination with his artistic sensibilities. The furniture here comes from Hans Bellmann and Jens Risom (Knoll International).

Photo: Alfred Löhndorf

1950 – Ethics

By 1950, the traditional *Heimatstil* is a thing of the past. Propagated by the consumer goods industry, *Industrieform* has arrived once and for all to mainstream society. After the hardships of war, the public is eager for new solutions that promise a freer and more comfortable way of life. When the Swiss Werkbund (SWB) inaugurates the award “Die gute Form” for exemplary design in 1952, it does so in the firm belief of laying the ethical and aesthetic groundwork for a better world. The award-winning objects with their “pretentious humility” still impress today. Conspicuous is the rise of sculptural experiments that take advantage of the malleability of plastics, fiber cement, and plywood while referencing analogue trends in art. With a variety of annual events, the SWB tries to exert a moral claim to authority in design matters among both manufacturers and the lay public.

407

Max Bill at the opening of the touring exhibition *Die gute Form* at the Basel Trade Fair (Muba), 1949. Bill himself handed out the prospectus he had designed to visitors. Two worlds collided here architecturally as well: on the left Bill's minimalist text panel with his sculpture *Rhythmus im Raum* (Rhythm in Space), 1947–48.

Photo: Ernst Scheidegger

408

Alfred Altherr jun., SWB special exhibition *Die gute Form* at the Basel Trade Fair (Muba), 1959. Altherr enlarged the round "Die gute Form" seal inaugurated in 1952 and used it to structure the exhibition space. The exemplary award-winning products were on view in display cases for visitors' perusal.

Photo: Moeschlin + Baur

409

In his satirical film *Mon oncle* (1958), Jacques Tati parodied the automated modern world. The futuristic furnishings produced by plastics specialist Arpel included the Type 600 floor lamp by the Lucerne designers Rico and Rosmarie Baltensweiler.

410

Advertising photograph for the Model 701 stacking table (designed by Hans Bellmann, 1954, produced by the AG Möbelfabrik Horgen-Glarus; sold by Wohnbedarf AG, Zurich/Basel). The photograph highlights the lightness and low-key elegance of the furniture piece, so typical for its day.

Photo: Max Buchmann

411

Hand-woven mohair and horsehair rug, 1957, photographed at the Giauque studio house in Ligerz, ca. 1960. The rug was designed by Elsi Giauque—who taught the class in textile design at the Kunstgewerbeschule Zürich from 1956 to 1966—and Käthi Wenger. The photo reflects easy outdoor living; a second photo shows Giauque's poodle at right center.

Photo: Hans Finsler

412

Advertising photo for Willy Guhl's Garden Chair (Strandstuhl), 1954, Eternit AG, Niederurnen. Guhl formed a three-dimensional loop from a standard rectangular Eternit panel without producing any waste. It was an excellent example of the sculptural trend in product design then emerging—skillfully contrasted with the relaxed pose of the female model as the embodiment of the new leisure culture of the postwar generation.

Photo: Michael Wolgensinger

Kurt Naef

Naef Building Blocks
1956

413 Kurt Naef, Naef Building Blocks, 1956
Stained maple, 18 pieces, 4.8 × 4.8 × 5 cm each,
Kurt Naef Spiele, Basel
Museum für Gestaltung

414 Kurt Naef, Catalogue, 1974, Naef Spiele AG, Zeiningen. Photo: Alfred Hablützel

415 Naef Spiele AG, Company truck in front of the company premises in Zeiningen, 1960s

413

414

415

After apprenticing as a carpenter and studying interior design at the Kunstgewerbeschule Basel, Kurt Naef (1926–2006) remained in Basel and opened a furniture and interior design store there. He received so many requests for well-crafted toys that Naef decided to produce and offer a toy of his own design in 1956. Unlike the sculptor Antonio Vitali, whose sculptural wood figures Naef added to his product range in 1958, the carpenter and interior designer took a more pragmatic approach to the material of wood and planned his toys from the outset for machine production. By making diagonal cuts into a wooden cube on two opposite sides, he created a block with multiple connection options. Children could not only stack or line up the uniquely shaped blocks but also wedge them together to build a tower. Structures could be given daringly cantilevered overhangs or could grow broader toward the top. In principle, Naef used eight saw cuts to make a simple building block into an element of a construction system. This rational approach to toy design appealed to the Swiss Werkbund, which presented the Naef blocks with the "Die gute Form" seal of approval in 1960. In 1967, the Naef company opened a production facility in Zeinigen to manufacture Naef's toys as well as those by other designers. Trading relations were established, especially with Japan, where since 1971 the popular Swiss toy brand has been distributed exclusively by the Tokyo company Atelier Niki Tiki. RM

von Büren 2006; www.naefspiele.ch.

Hans Gugelot

M 125 Furniture System
1950

416 Hans Gugelot, Bookshelf from the M 125 Furniture
417 System, 1950, produced from 1953
Core board with elm veneer, sliding doors in color-painted plywood, cast aluminum feet, 125 × 112.5 × 37.5 cm, produced in eastern Switzerland for Wohnbedarf AG, Zurich/Basel
Collection of Arthur Rüegg

418 Richard Paul Lohse, M 125 (product design by Hans Gugelot), 1954, Wohnbedarf AG.
Prospectus cover

416

417

418

Hans Gugelot (1920–1965) devoted himself early on to innovative system design. Inspired by the work of Max Bill, the ETH architect soon discovered furniture making as an ideal laboratory to try out his ideas. The most important project undertaken by the architecture firm he opened in 1950 was the development of the furniture unit system *M 125,* named after its basic unit of measurement, 125 millimeters. This was not merely a "program of boxes, cupboards, and drawers," as Gugelot himself described it, but "a system for building single pieces as well as a whole wall of cupboards." Shelves and spacer brackets made of veneered plywood, braced with metal rods, can be combined into a wide variety of units that can be used from both sides—ranging from sideboard to wall unit to room partition. Double-leaf and sliding doors, bookshelves, drawers, and clothes rails can all be integrated into the system, the units of which are mounted on cast aluminum feet. Wohnbedarf AG, Zurich/Basel had this first version produced until 1956 in elm, ash, or cherry wood, complete with sliding doors spray-coated in various colors and featuring slender grip profiles. The Zurich artist and graphic designer Richard Paul Lohse designed a beautiful prospectus for the furniture system, which was created almost simultaneously with the *Eames Storage Units* but was more comprehensive. From 1956, *M 125* was further developed with the firm Wilhelm Bofinger in Feuerbach, Germany, and produced on a large scale. AR

Wichmann 1987; Rüegg 2002d.

419

419 Jürg Bally, S.T. Table (Pull-Up Table), 1951
420 Solid maple, tabletop of linoleum with maple edge, H: 32–75 cm, D: 100 cm, Werkgenossenschaft Wohnhilfe, Zurich
Museum für Gestaltung/Donation of Jürg Bally

421 Early edition with wooden ball, produced by the designer, 1951–54

422 Detail of pull-up mechanism

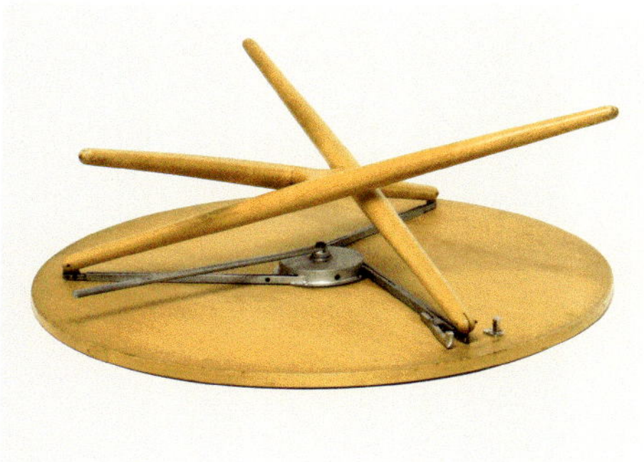

420

421

422

Jürg Bally (1923–2002), an architect and owner of a studio for product and interior design in Zurich, began a phase of intensive experimental "tinkering" in his own small workshop in the mid-fifties. Together with a carpenter at his firm, he developed domestic furniture that he began to sell on Zurich's Neumarkt in 1956. Bally designed the prototype for the round *S.T. Table* in 1951 for his wife, shortly before a two-year stay in the United States, where he also worked for Knoll. The design might be Bally's take on Hans Bellmann's also three-legged *Colonial Table,* which Knoll International began producing in 1947 as the *Tripod Table* in a low and a high version. The principal innovation in Bally's elegant, technically and formally sophisticated model is its infinitely adjustable height. The German name "S.T. Tisch" is a play on words: it can be used either as a high "Ess" (dining) table or as a low "Tee" (tea) table, and its overall quality is "äs-the-tisch," in other words, aesthetic. Three different versions are known: an early model with a wooden ball linking the legs, as well as two versions developed for series production that differ primarily in the material used for the legs. Pulling a lever concealed under the table to either the left or right winds or unwinds a steel band in a central drum. The legs then slide under the top along radially laid tracks, raising or lowering the table. AR

Bignens 2002c.

423 Rosmarie and Rico Baltensweiler, Type 600 Floor Lamp, 1950
Painted aluminum, poles of chrome-plated tubular steel, H: 180 cm, D: 40 cm (reflector); 35 cm (base), Baltensweiler AG, Ebikon
Museum für Gestaltung/
Donation of Baltensweiler AG

424 Have you read *Gigi* by Colette?, n.d.
Advertisement. The Type 600 lamp as a symbol of the modern lifestyle.

425 Type 600, Chinese hat reflector, 1950–51
Photo: Siegfried Zingg

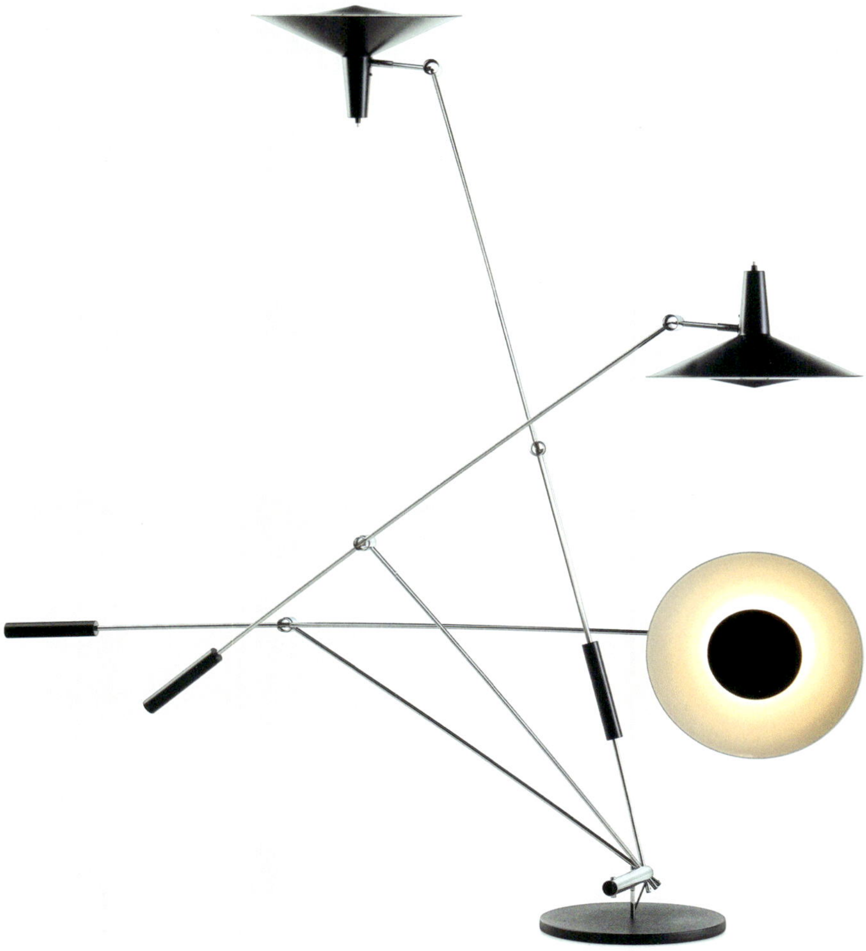

423

424

425

In 1951, Rico (1920–1987) and Rosmarie (b. 1927) Baltensweiler set up a live-in workshop in a former Lucerne bathhouse and laundry located in the middle of the Reuss River. Rico was an electrical engineer with the Swiss Federal Railways (SBB) and Rosmarie assisted Max Bill as an interior designer. Together, they developed two lamp prototypes to fit their own needs. Architect friends took an immediate interest in the elegant spring-joint lamp whose individual elements could be brought into perfect balance in virtually any position with just a light touch. A first edition of ten was produced, and then Rudolf Graber ordered a larger series for Wohnbedarf AG in Zurich. The *Type 600* floor lamp—it takes its name from the six axes of movement—soon became an icon of nomadic living. Le Corbusier considered it the ideal accessory for his sculpturally conceived Villa Shodhan, and Jacques Tati included it in his 1958 film *Mon Oncle* as part of a modern minimalist household. The spring joint patented by Rico Baltensweiler permits the lamp to lean precipitously despite its compact design—no doubt a reason for the impact the model typical of that time had on its many competitors. The first revised versions came out in 1954; Knoll International began selling the lamp in 1958 with a separate reflector and rods, the "Chinese hat" reflector was given a stabilizing articulation, and sensitive components such as the switch and socket were incorporated into a stable aluminum tube. AR

Baltensweiler AG 1996; Sonderegger 2002.

Ernst Ambühler

EA-616 Sofa Bed
1957

426

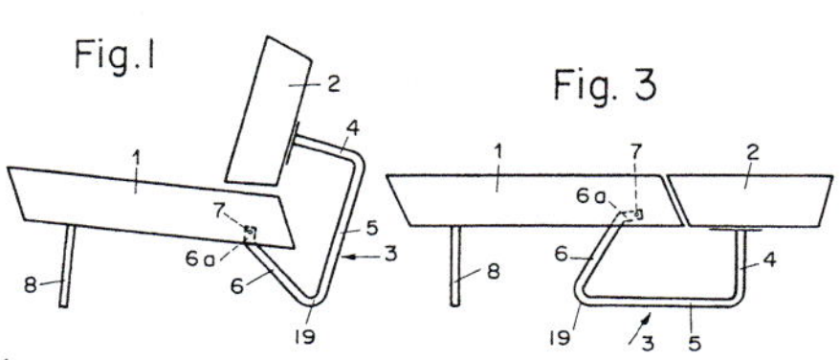

427

428

426 Ernst Ambühler, EA-616 Sofa Bed, 1957
Upholstered innerspring seat and backrest, with fabric cover, chrome-plated tubular steel frame, 75 × 200 × 95 cm, sleeping surface: 114 cm, various manufacturers for Swiss Design/ Teo Jakob Bern
Museum für Gestaltung

427 Patent drawings, 1958

428 Brochure card with sofa bed in sleeping position.

The sofa is an ingeniously simple design feat. Two seemingly free-floating upholstered sections are linked by two identical curved pieces of tubular steel. These tubes are jointed with the seat at the rear and firmly affixed to the backrest. Applying firm pressure to the backrest turns the sofa into a 114-centimeter-wide bed. When the edges of the tubes reach the floor, they raise the slightly angled sofa surface into a horizontal sleeping position. The main novelty in the invention, for which a patent was applied for in 1958, is the simple sofa-to-bed mechanism. The aesthetic appeal of the sofa bed, by contrast, lies in the levitating impression given by the two upholstered planes, precisely delimited with piped edges. With this apparent weightlessness, Ernst Ambühler (b. 1928) transferred a topos of modern postwar architecture to furniture design. His sofa was perhaps inspired by Knoll's *Model 700* by Richard Stein, featured in the 1951 exhibition *Kunst und Wohnkultur* in Lucerne and sold by Wohnbedarf AG, which has a wooden backrest that is lowered using a foot pedal. Ambühler's creation, however, is far more convincing with its coherent forms and the efficiency and economy of the design. No wonder Teo Jakob in Bern and later Charles Stendig in New York added it to their collections. Ambühler, a student of Willy Guhl, had already developed a highly interesting couch in 1951. After designing furniture for the architectural firm of Oskar Burri, he also worked as an architect. AR

Bignens 2002a; "Neue Möbel" 1956.

Wilhelm Kienzle

Shoe Rack
1951/1955

429

430

431

429 Wilhelm Kienzle, Shoe Rack, 1951, metal version: 1955
Sheet steel painted black and gray, 36 × 96 × 23 cm, MEWA Blattmann, Wädenswil
Museum für Gestaltung/Donation of MEWA

430 Wilhelm Kienzle, Shoe Rack, 1951, wood version, Werkgenossenschaft Wohnhilfe, Zurich
Museum für Gestaltung/
Donation of Hans Bosshardt

431 A MEWA product, brochure, ca. 1955, MEWA Blattmann, Wädenswil

One day at Café Select in Zurich, the president of the Wohnhilfe cooperative, Arturo Milani, described a problem he was dealing with to the retired Kunstgewerbeschule Zürich professor Wilhelm Kienzle (1886–1958): How can shoes be stored at home in a minimum of space? In just three days, Kienzle created a cardboard model on a scale of 1:10, equipped even with shoes carved out of balsa wood, in order to vividly demonstrate the principle of the tilting shoe rack and thus the solution to the problem. Kienzle had already proposed a tilting cupboard for dishes twenty years earlier for a model apartment in the Neubühl housing development in Zurich, but it was never produced. The first prototype of Kienzle's shoe rack was made out of wood by Jacob Müller and then produced in large quantities by the Wohnhilfe carpenters' cooperative (47,000 were produced between 1947 and 1965). The major advantage of the tilting cabinet was its minimal depth of twenty-three centimeters. It took up little space in the hallway of a rental apartment, and several could be stacked on top of each other. Made from inexpensive fir and color spray-painted for a surcharge, and later also available in veneered chipboard, a single unit with a length of one meter could hold five pairs of shoes. The partitions (originally made of plywood and later plastic) held the pairs of shoes together and upright when in the tilted position. In 1955, Kienzle made the shoe rack even more compact by using thin walls of sheet steel in a version produced by the metal goods factory MEWA Blattmann in Wädenswil.
RM

Lichtenstein 1991.

432

433

434

432 Hans Bellmann, Sitwell Model 101 Armchair, 1955
Fiberglass-reinforced polyester seat shell, black-painted tubular steel frame, fabric-covered foam upholstery and loose cushion, 84 × 93 × 78 cm, Strässle Söhne AG, Kirchberg
Museum für Gestaltung

433 Interior with Akari floor lamps by Isamu Noguchi, from a prospectus for Strässle, ca. 1957

434 Strässle/Sitwell. Prospectus for the Sitwell Collection, 1956, Strässle Söhne AG, Kirchberg

Europe's first industrially manufactured fiberglass shell chairs were made in 1954 in Toggenburg. Hans Bellmann (1911–1990), who had attracted attention after the war with his wooden "type furniture" for Wohnbedarf AG, Zurich/Basel, then developed a series of novel seat shells made of fiberglass-reinforced polyester ("Stracolite") in collaboration with the upholstery company Strässle Söhne in Kirchberg. Founded in 1886, the company rose to international prominence with groundbreaking seating furniture under the leadership of Alex Strässle (b. 1931), a trained upholstery specialist. Bellmann's *Sitwell* collection took up directly where Charles and Ray Eames's experiments in the United States with polyester shells in 1950 to 1953 had left off. However, in the *Sitwell* sofas, armchairs, chairs, and stools, the plastic surfaces are hidden. All models featured foam padding and—with the exception of the chairs—fabric upholstery with loose cushions. The big armchair, launched with the checked fabric typical of the time, formed the core of the collection. As in Eames's *Plastic Armchair,* the armrests form an integral part of the overall shape. The shell is couched, as it were, in a black-painted tubular steel frame with, as a signature feature, a high arc at the back. Although Bellmann later dismissed the whole series apart from this one iconic armchair, the entire *Sitwell* collection, upholstered in the latest colors, enjoyed great popularity. AR

Rüegg/Wieser 1991; Bähler 1997.

Max Bill (with Hans Gugelot and Paul Hildinger)

Ulm Stool
1954

435 Max Bill (with Hans Gugelot and Paul Hildinger), Ulm Stool, 1954
Solid spruce and beech, 45 × 40 × 30 cm, 1988, Ph. Oswald Schreinerei und Innenausbau AG, Oberglatt, small series, signed by Max Bill
Museum für Gestaltung

436 Max Bill, Stool with drawer, 1962, Werkgenossenschaft Wohnhilfe, Zurich
Museum für Gestaltung

437 Anthropology lesson on the terrace at the Hochschule für Gestaltung Ulm, 1955
Photo: Ernst Scheidegger

435

436

437

When the challenge was to achieve maximum functionality and spatial impact with minimal means, Max Bill (1908–1994) was in his element. No wonder he managed to make the most of the financial malaise that prevailed during the building of the Hochschule für Gestaltung Ulm—as architect, designer, and teacher. The Ulm Stool, made for the school out of donated fir boards and broom handles, paradigmatically embodies the added value Bill believed could be derived from the combination of an extremely prudent use of material and an aesthetically pleasing structure. Although the stool is not the result of a teacher-student relationship but rather emerged from the collaboration of Max Bill with the lecturer Hans Gugelot and the workshop master Paul Hildinger, it played a central role in instruction as a model multifunctional utilitarian object. Photographs document its use as a lectern; as seating in the dining hall, lecture hall, workshop, or student apartment; as a box for transporting books; and even as a nightstand. Whether Bill's "Platonic idea of a chair" (Bernhard Rübenach) had to be made of a specific set of materials—meticulously dovetailed fir boards and flush beech runners—was a question that was raised by Wohnhilfe in Zurich with a radically simplified 1962 particle-board version (also equipped with a drawer). Zanotta produced a birch plywood version in 1975, before the original version—by then an icon of minimalist design—was successfully revived by Wohnbedarf AG in Zurich in 1992. AR

Weinberg-Staber 1983; Rüegg 2008.

438

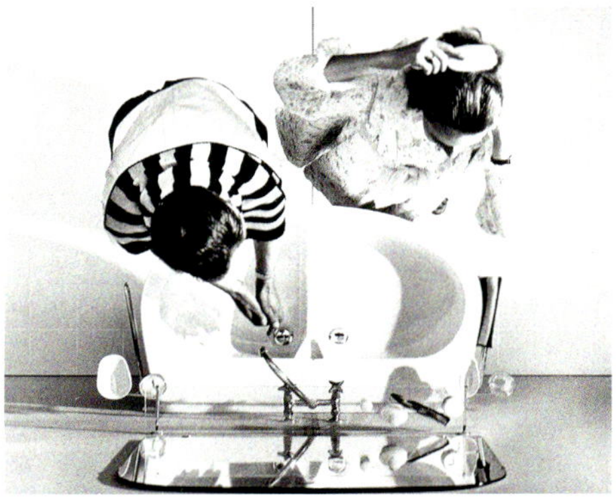

439

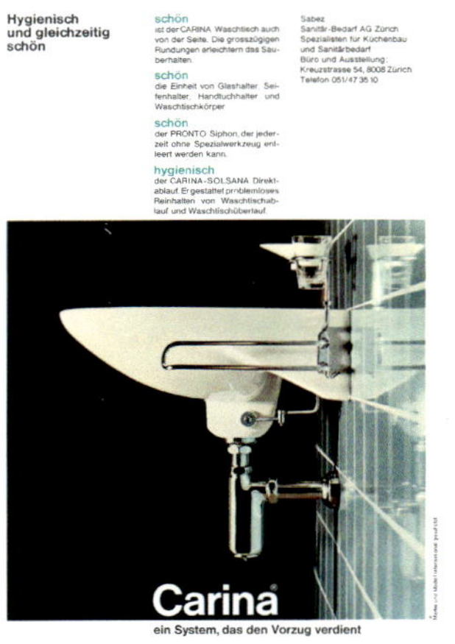

440

438 Hans Bellmann, Carina Model 5100 Bathroom Sink, 1953
Glazed white porcelain, 27 × 58 × 52 cm, Sabez Sanitär-Bedarf AG, Zurich; Kera-Werke AG, Laufenburg
Museum für Gestaltung

439 Carina Duo in use, after 1953
Photo: Max Buchmann

440 Carina, the system of choice—both hygienic and beautiful, n.d., after 1968
Sabez, Zurich. Prospectus

"Beautiful and yet practical" were the words used to describe the new *Carina* bathroom sink, as well as "at once hygienic and beautiful." Paying homage to the principles of *Gute Form,* both the front and back sides of the advertising prospectus emphasized the sink's superior aesthetic qualities. Available in white, gray, light blue, or light green, the sink was even appealing from the side, where its generous curves made it easier to clean. Also beautiful was the unity of cup holder, towel rails, and sculptural sink volume. Beautiful as well was the *Pronto* siphon, which could be emptied without special tools. Hygienic, by contrast, was the direct drain, easy to clean with a bottle brush. Practical was the large round bowl with no corners where dirt could collect. Also practical were the fittings that could be attached directly to the washstand, eliminating the need to drill into glass- or mosaic-tiled walls. The prolific furniture designer Hans Bellmann (1911–1990) expanded his field of activity after 1950 to become a design consultant to industry (for Sabez, Arwa, Therma, Homoplax, and Alusuisse). It was in the bathroom field that he was best able to realize his vision of the ideal industrial product, as early as 1952 with the *Arwa Therm*—the first pressure-independent single-lever faucet—and in 1953 with the washstand system *Carina* for Sanitär-Bedarf AG in Zurich (Sabez), a timeless design classic that is still in production today. Bellmann explicitly distanced himself from all short-lived trends, endeavoring instead to create products that people would "never get tired of as long as they were still technically serviceable." AR

Rüegg/Wieser 1991; Rüegg 1995b.

441

442

443

441 Willy Guhl, Garden Chair (Strandstuhl), 1954, produced until 1980
Wooden form for the production from cement asbestos, Eternit AG, Niederurnen
Museum für Gestaltung*

442 Willy Guhl, Product documentation: negative mold for plaster casting

443 Willy Guhl, Product documentation: plaster form (positive) cast from seating profile

When Eternit AG decided to expand its product range, Florian F. Adler, head of the advertising department, called on the Kunstgewerbeschule Zürich for help. The only teacher there to take an interest in the company's new asbestos cement material was Willy Guhl, head of the interior design class. In collaboration with Eternit, his students developed new planters that were presented to the public for the first time in 1952 in the foyer of the Kunstgewerbemuseum Zürich. In 1959, Guhl's students won the top prize in the competition held for the G59 garden show. Willy Guhl advised Eternit AG until 1960, often bringing the practical issues he dealt with there back to the classroom for further exploration. He experimented mainly with three construction techniques: molding and shaping (flower boxes), surface shell (spindle), and the loop. From the latter, Willy Guhl succeeded in creating a design icon. The *Garden Chair* (Strandstuhl), consists of a machine-made Eternit board that before hardening can be formed into a loop in a wooden mold without any scrap from trimming. To arrive at the right shape for the chair, Guhl first determined the ergonomically correct seat profile using a clay model and then cast it in plaster as negative and positive forms. Like a sculptor, he then reworked the radii of the loop, always with a mind to economy of production, as well as considering structural and material requirements. In retrospect, the constant contours of the *Garden Chair* also reflect the postwar worldview, images produced at the time within the natural sciences, and forms found in the fine arts. RM

Willy Guhl 1985.

444

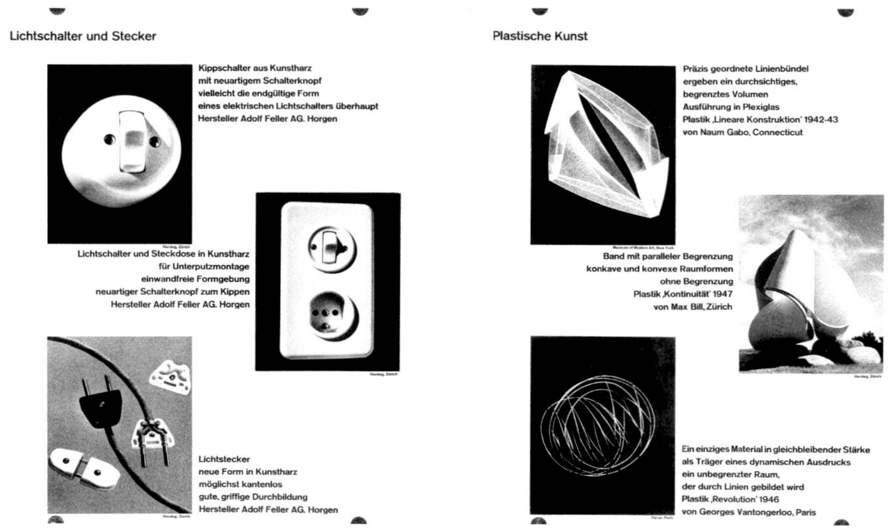

445

444 Max Bill, SWB special exhibition *Die gute Form*, Basel Trade Fair (Muba), 1949.
Photo: Ernst Scheidegger

445 Max Bill, "Light Switches and Plugs" and "Sculptural Art," photo boards from the SWB traveling exhibition *Die gute Form*, 1949–51, 84.5 × 72 cm
Museum für Gestaltung/Donation of Fritz Hobi

A design organization creates a certificate for good design and awards it each year to products from all over the country. Selected objects are shown in exhibitions at home and abroad, pictured in product catalogues and magazines, recommended by consultants, and recorded in a central photo index. With this campaign, which goes on for over twenty years, the organization becomes the national arbiter of design quality and shapes the image of Swiss design as characterized by honesty, economy, and modest beauty.

How did the Swiss Werkbund (SWB), founded in 1913, manage to instill faith in the power of good design in designers, manufacturers, vendors, and consumers, establishing a dictatorship of taste and maintaining it for over two decades?

The hope was that well-designed products would improve everyday life and ultimately also contribute to the betterment of the people using them. The SWB acted like an art director briefing the designers—not through the formulation of an aesthetic theory but by inculcating a common feel for form. This is the effect I would like to examine in this analysis of the campaign known as "Die gute Form" ("Good Form," or Good Design) in its function as visual communication through product design, photography, exhibition design, and graphics.

Beauty as Function

For its definition of *Gute Form,* the SWB relied essentially on the reform movement in the applied arts and the statutes of the Deutscher Werkbund (DWB), founded in 1907, according to which a form designated for use should fulfill its purpose through the right choice of material and construction. However, the question of whether such a utilitarian object must necessarily be beautiful was controversial.[1] The progress of modernism (and with it the further development of the functionalistic design approach) was interrupted by the Third Reich, and the Werkbund groups in Germany and Austria were dissolved. Against this backdrop, the exhibition *Die Werkbundidee* (The Werkbund Idea), mounted in 1938 by the Basel branch of the SWB under the direction of art historian Georg Schmidt (1896–1965), served as a stocktaking and defense of modernism: both positive and negative examples from all areas of design were presented and explained, and photo panels with natural forms complemented the modest installations. After 1945, the SWB felt responsible for reintroducing its version of the Werkbund idea in war-torn Europe and thus contributing to the intellectual component of reconstruction. This was also the ambition that inspired former Bauhaus student Max Bill (1908–1994), who was able at the Werkbund conference of 1948 to open the debate on functionalism with the formula "Beauty from Function and as Function."[2] The architect, designer, and artist boldly claimed to discern a personal will to design in the visual expressiveness of things—and ultimately an autonomous beauty in both industrially produced and handmade purpose-built forms. In the following years, the concept of form/design became charged with a diffuse moral and aesthetic aura, taking on almost mystical significance in Werkbund circles.[3]

Form: Visualization of the Unity of All Functions

Representatives from industry also took part in the 1948 Werkbund conference, realizing that the Swiss devotion to form was now a competitive factor. The SWB was charged in 1949 with organizing an exhibition for Switzerland's biggest trade fair, the Mustermesse Basel (Muba) samples fair, and it commissioned Max Bill with its conception. In his written exposé "zweck + konstruktion = form" (Function + Construction = Form), he defined four sections with twenty photo panels each, whereby the first is fundamental because it is composed of examples that "each represents its function as clearly as possible, but which in their totality form a stylistic unity despite their differing functions."[4] Three further sections are devoted to forms in building, in domestic needs, and in daily life. Bill's exhibition was conceived as a traveling show and it duly set off in 1949 on a tour through the German-speaking countries under the catchy title *Die gute Form,* later continuing on under the English name *Good Design* through England, Ireland, the United States, and Palestine.[5] Bill the teacher wanted to reach especially "up-and-coming creative talents" and to ensure "further training for already active designers." Reducing objects of art, nature, and everyday use to photographs, and mixing them up in terms of their functions did have the intended effect of elevating the beauty of everyday things to the level of art forms, but the exhibition also succeeded in spotlighting the design quality of Swiss products in particular. With *Form. Eine Bilanz über die Formentwicklung um die Mitte des XX. Jahrhunderts/A Balance Sheet of Mid-Twentieth-Century Trends in Design/Un bilan de l'évolution de la forme au milieu du XXe siècle,* Bill's point of view was published in book form in 1952, the same year in which the SWB began to award household and office products at the Muba with the distinction "Die gute Form/SWB." The award became the Werkbund's ultimate tool, because it could use it to directly influence businesses while at the same time addressing consumers.[6] It took Max Bill until 1957 to summarize six criteria for Gute Form. In the meantime, state-sponsored institutions in several European countries were following the lead of the Museum of Modern Art in New York (MoMA), which had instituted its "Good Design" award in 1950, and coming out with their own quality seals for design, among them Germany (1951), France (1953), Italy (1954), Belgium (1958), and Sweden.[7] Bill realized that "all this activity in the field of design" had also led to growing confusion.

The initiator of Gute Form once again invokes in his criteria the "formal correspondence" of expediency and utility value, their "aesthetic unity," along with—as the final criterion for Gute Form, the "aesthetic function as visible expression of the unity of all functions." This visualization task seems to be Bill's main concern and the unique identifying feature of Gute Form. Overall, however, the six points listed in the brochure for the exhibition *Die gute Form 1957* read like a slightly awkward translation of the definition of "Good Design" already published by Eliot Noyes, the founder and director of the MoMA's Department of Industrial Design, in 1940 on the occasion of the exhibition *Useful Objects of American Design Under $10.*[8]

Award: Leading by Example

What the SWB declared to be Gute Form and thus part of the Swiss cultural heritage obviously had to be distinguished from mere fashionably styled goods. The idea for the award, combined with a quality showcase, was borrowed from the MoMA and the Leipzig trade fair. The 1948 MoMA exhibition *Design for Use* was later on view at the Gewerbemuseum Basel in 1952, in other words, the first year the SWB put on its special show *Die gute Form* at the Muba trade fair in that city. With this annual event, the pool of objects displaying certified good design swelled each year by 250 items, presided over by a jury of Werkbund members, representatives from the Basel samples fair, and a few international experts.[9] The award-winning companies were delighted to earn the signal-red label and use it to promote sales; the logo was also available for a fee as a printing plate for use in promotional materials and thus spread rapidly.

We can only guess today at how the selection criteria were applied and verified in each case—in particular with respect to the objects' functionality. Nor do the jury reports introduced in 1964 provide much evidence. As clear as the internalized features of good design must have been to the jury members, their intuition was nevertheless subject to the effects of the contemporary zeitgeist. Although each object deemed worthy of recognition was certainly designed for longevity,[10] the appearance of the objects changed with the times: whereas in the fifties they tended to almost pantomime their purpose, like sculptures for everyday use, in the sixties the surface tension of "soft objectivity"[11] seems to increase, accompanied by a decrease in the repertoire of forms. The development toward uneventful stereometry was criticized later even by Werkbund members as formalism.

Photography: Through the Lens of New Objectivity

The image of Gute Form was conveyed mainly through photographs. The SWB had already avowed its support of the Neue Sachlichkeit, or New Objectivity, in 1932 with the exhibition *Die neue Fotografie in der Schweiz* (New Photography in Switzerland).[12] It was during this period that the SWB also started to assemble its own photo collection. The "Fotosammlung von gutem Gebrauchsgut" (Photo

Die gute Form 1953

SWB DIE GUTE FORM 1953

Anerkennungsurkunde

SCHUHKIPPE
ENTWURF KIENZLE
WOHNHILFE, ZUERICH

Ausgezeichnet durch die Jury der Schweizer Mustermesse Basel und des Schweizerischen Werkbundes

Schweizer Mustermesse Basel:

Der Direktor: Prof. Dr. Th. Brogle

Schweizerischer Werkbund SWB:

Der 1. Vorsitzende: Hans Finsler

446

447

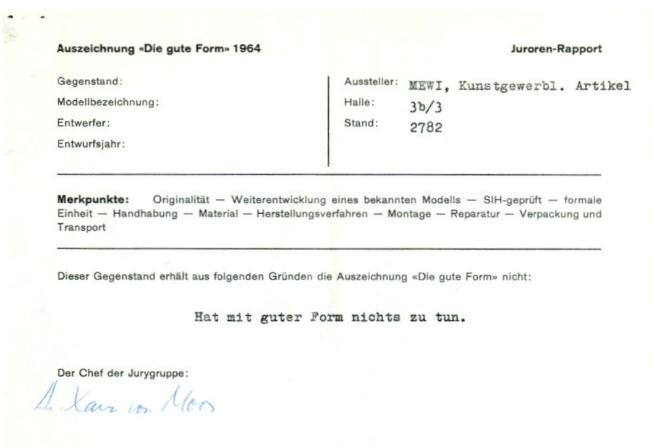
Auszeichnung «Die gute Form» 1964

Juroren-Rapport

Gegenstand:
Modellbezeichnung:
Entwerfer:
Entwurfsjahr:

Aussteller: MEWI, Kunstgewerbl. Artikel
Halle: 3b/3
Stand: 2782

Merkpunkte: Originalität — Weiterentwicklung eines bekannten Modells — SIH-geprüft — formale Einheit — Handhabung — Material — Herstellungsverfahren — Montage — Reparatur — Verpackung und Transport

Dieser Gegenstand erhält aus folgenden Gründen die Auszeichnung «Die gute Form» nicht:

Hat mit guter Form nichts zu tun.

Der Chef der Jurygruppe:

448

449

450

452

453

451

446 Armin Hofmann, SWB award certificate with the first red "Die gute Form" logo, 1952
447 Alfred Altherr jun., SWB special exhibition *Die gute Form 1959*, Basel Trade Fair (Muba), 1960. Photo: Moeschlin + Baur
448 Form for the jury's evaluation of candidates for the "Die gute Form 1964" seal. Includes points to be addressed and tautological reasoning for a negative decision.
449 Emil Ruder, "Die gute Form 1956" label, 1956. Redesign of the first logo by Armin Hofmann, 1952
450 Eliot Noyes, Criteria for "Good Design" for the exhibition *Useful Objects of American Design under $10,* The Museum of Modern Art, New York, 1940
451 View of the exhibition *Useful Objects of American Design under $10,* curated by Eliot Noyes, The Museum of Modern Art, New York, 1940
452 Morton Goldsholl, Good Design, labels and stickers, logo design from 1950 for the first award presented by The Museum of Modern Art on the occasion of the *Good Design* exhibition at the Merchandise Mart, Chicago, 1950.
453 Charles and Ray Eames, *Good Design,* exhibition at the Merchandise Mart, Chicago, 1950. Organized by the Merchandise Mart Association and The Museum of Modern Art, New York.
454 Cards from the SWB photo card collection of some 1,700 award-winning products in the categories household, office, home furnishings, and miscellaneous, 1952–67

454

collection of good consumer projects) also includes international examples of good design and is mentioned for the first time in the SWB annual report from 1936. It may also have been incorporated into Max Bill's 1949 photo panels. With the inception of the "Die gute Form" awards in 1952, the SWB then set up a photo card file as an inventory of the awarded products and their manufacturers and designers, maintained up to the last year the award was presented, 1967. With some 1,700 pictures, this file offers a complete overview of the objects of use favored by the SWB and thus a specific snapshot of Swiss product culture in the fifties and sixties. Although "Die gute Form" aimed to address all of society, it in fact reached primarily the educated middle classes to which the designers themselves also belonged.

More than just a catalogue of criteria for product design, "Die gute Form" was also influential in schooling the eye, because, unlike a mere verbal description, the photographs objectively captured the impression made by the objects. As a proxy observer, the photographer transformed his personal perceptions into a reproducible, valid view. In the service of the New Objectivity, he put the spotlight on everyday things, isolating them from the surrounding visual noise and showing them to their best advantage. There were role models for this approach, such as Karl Blossfeldt's publication *Urformen der Kunst* (Art Forms in Nature), or *Die Welt ist schön* (The World is Beautiful) by Albert Renger-Patzsch (both 1928). After the highly publicized 1929 show *Film und Foto* put on by the DWB in Stuttgart, the Swiss architectural historian Sigfried Giedion lured one of the protagonists of German New Objectivity to Zurich. Hans Finsler (1891–1972) established a photography class at the Kunstgewerbeschule Zürich in 1932, and was the first chairman of the SWB, from 1946 to 1955. He and his students were not only portraitists of Gute Form, they also shaped the image of Swiss quality with their precise product shots.

Exhibition: Staging Interiors

Similar to what happened in the photo studio, the perception of Gute Form could also be influenced by how objects were displayed in exhibitions, although the installations always included photographs and graphics as well. The Werkbund already organized in the 1910s spectacular *Raumkunst,* or interior design, exhibitions[13] and experimented with presenting model interiors.

The SWB's early presentations of industrial design, for example in the 1935 Basel exhibition *Land- und Ferienhaus* (Country and Vacation Homes), did not, however, turn practical machine-made items into rarified museum objects,[14] but instead unabashedly offered them for sale, similar to the "Tiller-Girl Principle"[15] in object photography. The Muba's stated objective was indeed to promote the sale of Swiss goods.

The annual special *Die gute Form* showcases presented at the Muba from 1952 onward, and the later SWB exhibitions at museums, made use mainly of three types of presentation: the educational interior design ensemble, the "white cube" museum setup, and typical trade-fair product displays—often side by side. Furniture, textiles, and ceramics tended to be arranged in groups as if in a domestic interior, while household appliances and office equipment were detached from their everyday context and staged like sculptures. The book *Neue Ausstellungsgestaltung/Nouvelles conceptions de l'exposition/New Design in Exhibitions* published in 1952 by the artist and graphic designer Richard Paul Lohse (1902–1988) shows an international cross section of examples starting in 1930 and declares the characteristics of good exhibition design to be systematization and limitation. The standardization and reduction of display elements facilitates transport, while the flexible modules, for example the latticework of metal tubes used by Walter Gropius in 1934 in the *Non-Ferro-Metall* exhibition or the lightweight wood frames deployed in the Swiss pavilion at the Milan Triennale in 1936, designed by Max Bill, could be adapted to suit the different exhibits while also integrating pictorial and textual documentation. Lohse's book must have been an important reference work for the scenographers of Gute Form. After the photo-panel exhibition at the Muba in 1949, Lohse mounted the following year a small sidewalk show of well-designed objects outside the Basel trade fair building; in 1951, the chairman and director of the SWB, Alfred Altherr jun., staged the exhibition *Form und Farbe* (Form and Color); and from 1952 to 1967 the annual *Die gute Form* showcase was entrusted to various scenographers, with changing themes. Very influential in this context were the exhibitions at the Kunstgewerbemuseum Zürich (its director, Alfred

1
Duroinox-Kochgeschirre aus rostfreiem Chromnickelstahl mit aufgegossenem Wärmeverteilboden aus Leichtmetall; mit blankem Chromnickelstahl- oder rotem Leichtmetalldeckel in der praktischen hochgewölbten Form.

2
Gas-Kochgeschirre aus Leichtmetall mit dem speziell für die Gasküche gedachten, praktischen Bakelit-Isolierstiel. Mit hochgewölbtem blankem oder rot eloxiertem Deckel lieferbar.

1
Casseroles Duroinox en acier au nickel-chrome avec double fond en métal léger facilitant la diffusion de la chaleur; couvercle brillant en acier nickel-chrome ou en métal léger rouge, de forme très convexe et pratique.

2
Casseroles pour le gaz, en métal léger, avec poignée isolante en bakélite, spécialement conçue pour la cuisinière à gaz. Disponible avec couvercle très bombé brillant ou rouge éloxé.

1
Duroinox kitchen ware of stainless chrome-nickel steel with light-alloy bottoming for better heat distribution, polished chrome-nickel steel or red light-alloy lids in the practical dome-shaped form.

2
Light alloy kitchen ware for gas stoves with special bakelite handle designed for the gas kitchen. Obtainable with the dome-shaped lids, polished or anodized in red.

Werksentwürfe / Créations d'usine / Works designs 1956

Heinrich Kuhn, Metallwarenfabrik AG, Rikon ZH
Telephon 052/3 52 82

455

Deutsches Warenbuch

456

457

458

459

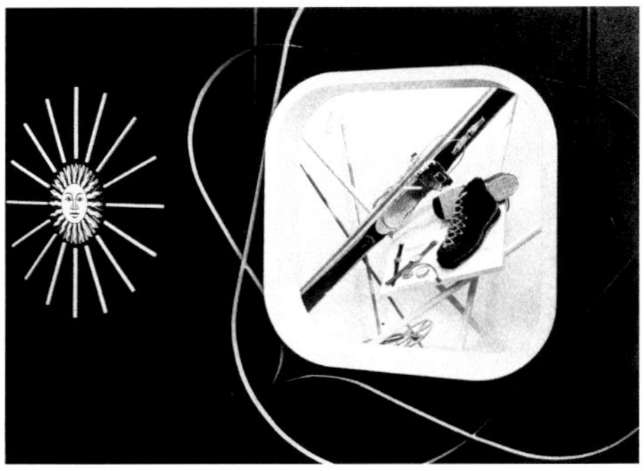

460

461

462

463

455 Richard Paul Lohse, Page from the SWB product catalogue *Wohnen heute 3. Warenkatalog des Schweizerischen Werkbundes 1960,* ed. Swiss Werkbund, Arthur Niggli, Teufen 1960
456 Pages from the German product catalogue *Deutsches Warenbuch,* ed. Dürerbund-Werkbund-Genossenschaft, Hellerau bei Dresden, 1915
457 Mia Seeger, page "Der neue Wohnbedarf: Trinken" (Today's Household: Drinking), from *Moderne Bauformen. Monatshefte für Architektur und Raumkunst* 32, 1933
458 SWB exhibition *Land und Ferienhaus* (Country and Vacation Homes), Department III: *Wohnbedarf* (Household) under the direction of Georg Schmidt and Egidius Streiff, Basel Trade Fair (Muba), 1935. Detail of product presentation according to the "Tiller-Girl principle" of object photography.
460 Ernst F. Burckhardt, Niche 1: *Bei Spiel + Sport* (Games + Sports), Swiss section at the Milan Triennale, 1940. Shop window with arranged still lifes. Photo: Atelier Eidenbenz
461 Max Bill, *Swiss Design* exhibition, London, 1959. Booth with table and Saffa chairs by Hans Eichenberger and picture by Max Bill. Presentation of art and design in the white cube. Photo: Kurt Blum
462 Richard Paul Lohse, *Trottoirschau* (Sidewalk Show), Basel Trade Fair (Muba), 1950. The SWB used the simplest of means to express its mission, with tubular steel frames and picture panels communicating the cause of *Gute Form.*
463 Alfred Altherr jun., SWB special exhibition *Die gute Form 1960,* Basel Trade Fair (Muba), 1961 Photo: Moeschlin + Baur

Altherr jun., was also head of the Kunstgewerbeschule) and those in Basel and Winterthur, as well as presentations abroad, such as *Swiss Design* in London in 1959, staged by Max Bill with museum-like partitions, and the traveling exhibition *Good Design in Switzerland,* which toured thirteen venues in the US and Canada, and was also shown in Tokyo in 1961.

Visual Communication: The Coherent Image

In addition to the articles published in the magazine *Werk,* the official mouthpiece of the SWB, the brochures accompanying the exhibitions were also important means for communicating "Die gute Form." Many of the leading figures in the Werkbund were familiar with the world of advertising from their work as photographers or graphic artists. Max Bill for example designed numerous promotional materials for Wohnbedarf AG from 1931 to 1933, and Richard Paul Lohse created both advertisements and Concrete Art, ensuring with his design of SWB exhibitions and the SWB *Warenkatalog* (product catalogue)[16] an incisive style of visual communication. Similar to the *Deutsches Warenbuch* issued by the DWB, the comprehensive SWB catalogue was distributed in home economics schools to encourage the purchase of well-designed dowry items, and in 1957 director Alfred Altherr jun., initiated a SWB domestic advice center.[17] The SWB also spread its views on Gute Form through its own uncompromisingly designed leaflets and brochures, full of clear messages for the edification of the general public. The graphics used to convey Gute Form did not obey strict guidelines but instead partook in the overall design quality of the entire campaign, while the intermingling of two- and three-dimensional communication vehicles lent the brand a multimedia look and feel.

Form Follows Market

With the social upheavals of the late sixties and the increasing criticism of functionalism, the SWB lost its monopoly on judging what constituted good product design. Critical voices were raised even in the Werkbund itself, and—after a brief but joyful period of self-discovery—attention turned toward environmental issues. An extended concept of design gradually evolved and the SWB pushed onward from material product culture to "invisible design"[18] and its political and environmental implications. It thus ceded the field of product design to a largely ideology-free marketplace.[19]

The preeminence of the Werkbund, based on a mixture of aesthetic and moral categories that are to this day difficult to digest, was by all means buoyed by its concise notions of design. Beyond all the convoluted discourses, these ideas about design have been preserved in objects, photographs, graphics, and documented interiors that boast an imposing presence. It is worthwhile—not only from the perspective of cultural history or enthusiasm for vintage goods—to regard them as design achievements and to try to learn from them, from the necessary distance of course. After all, the conflation of aesthetic with economic value that is common today, and the consequent adaptation of design to fit supposed public taste, has brought no new freedom for designers.

1 Ever since the dispute within the Deutscher Werkbund in 1914, Henry Van de Velde has been regarded as an advocate of an artistic, subjective element in the beauty of a utilitarian form.
2 This and two further presentations at the Werkbund conference of 1948 were published in the official mouthpiece of the SWB, the magazine *Werk* 36, no. 8 (1949); see Bill (1948) 2010.
3 This opinion is interesting in connection with the concept of design first applied by Johann Wolfgang von Goethe, whereby "design," as a quality of perception, cannot be analyzed and can only be, in its totality, grasped intuitively. See Simonis 2001, p. 27.
4 The typescript dated April 1948 also has an alternative title by Alfred Altherr jun.: "Nützlich und schön. Form aus Zweck und Konstruktion" (Useful and Beautiful: Form Derived from Function and Construction). Design Collection archive, Museum für Gestaltung Zurich.
5 A letter dated July 22, 1949, from Max Bill to Hans Finsler provides evidence of these expansive plans. Design Collection archive, Museum für Gestaltung Zurich.
6 As early as 1945, the SWB had already selected products at the Muba with exemplary design qualities for illustration in its magazine *Werk* 33, no. 5 (1946).
7 Max Bill mentions the Deutsches Rat für Formgebung (German Design Council, 1951) and the Design Centre London (1956) in the brochure accompanying the exhibition. See Bill 1957.
8 "Good Design (1) fulfills its function, (2) respects its materials, (3) is suited to the method of production, (4) combines these in imaginative expression."
9 The nineteen-member jury from 1954 is listed in the SWB annual report and includes big names such as Morton Shand from London, Richard Paul Lohse and Alfred Altherr jun. from Zurich, Max Bill from Zurich and Ulm, Elsi Giauque from Ligerz and Zurich, Alfred Roth from Zurich, and Otto Senn from Basel.
10 In the brochure for the exhibition *Die gute Form 1964,* the question of whether the design of the products awarded in 1953 had stood the test of time is examined based on examples that had in the meantime been developed further both technically and formally (such as the Elna sewing machine) and others whose design had remained unchanged despite technical innovation (such as the Feller light switch).
11 An expression used by Peter Erni, the profound analyst of Gute Form.
12 See also the exhibition *Bilderstreit* at the Fotostiftung Schweiz and the accompanying publication: Gasser 2007.
13 Janine Schiller examines the history of these interior design exhibitions in her unpublished master's thesis, Schiller 2012.
14 Such as in the exhibition *Machine Art* at the MoMA in 1934, curated by Philip Johnson.
15 According to Gerda Breuer, the term was coined by Erich Salomon in 1928 with reference to advertising. See Breuer 2004, p. 95 n13.
16 The extensive *Warenkatalog des Schweizerischen Werkbunds* (Product Catalogue of the Swiss Werkbund) was published in 1949 and 1954, and thereafter under the title *wohnen heute* (The Home Today) in 1960, 1962, 1965, 1967–68, and 1969–70. In an effort to stay in step with the times, it was not averse to less enduring forms of design.
17 In 1956 at the Gewerbemuseum Winterthur and from 1958 at the Kunstgewerbemuseum Zürich.
18 Burckhardt (1980) 2012.
19 Irma Noseda, the former director of the SWB Central Secretariat, regrets this retreat from the discussion on product design; see Noseda 2007.

464

465

466

464 Herbert Matter, All roads lead to Switzerland, 1935. Poster. The Schweizerische Verkehrszentrale (SVZ, Swiss National Tourist Office) promoted tourism with a collage of a car trip over the Tremola pass road along the Bernina Massif.

465 Paul Jaray, Auto body on a Ley chassis, 1923. The pioneer of streamlining in the driver's seat of his self-built car.

466 Robert Maillart, Salginatobel Bridge in Prättigau, 1930. The reinforced concrete bowstring bridge is exemplary of modernist bridge building.

Mobility Design

Since the end of World War II, greater mobility and growing prosperity have become part of everyday life for increasing numbers of people. And these daily needs beg to be served by the consumer goods industry. In the meantime, the parameters for design seem to shift back and forth between culturalist and technicist poles. The culturalist approach represents a tendency toward holistic solutions that build on familiar traditions, while the technicist side by contrast seeks to tackle new challenges on the basis of the latest technological advances and innovations. This latter pole has predominated in the field of mobility design during the last hundred years or so, primarily because there are so many new tasks that do not yet have a tradition behind them. The technicist approach coincides with the rampant industrialization of everyday life and the sweeping formal and structural changes that have resulted. More and more travelers expect to be transported with ever-increasing speed, efficiency, and convenience, and sometimes even spectacularly—all the while accompanied by values such as safety, reliability, durability, and profitability.

Since the sixties and the advent of postmodernism, culturalist strategies have been regaining currency. People are recognizing in historical precedents or traditional achievements qualities they would like to build or expand on. At the latest since the Club of Rome's report entitled *The Limits to Growth* (1972), both positions are overlaid by issues of sustainability, efficiency, and sufficiency. Moreover, many producers deliberately "charge" today's consumer goods with certain values and emotions.

Infrastructure

One reason why travel, once the sole province of the well-to-do, gradually lost its aura of privilege was the advent of the "threshold product" known as the railway.[1] In international comparison, Switzerland lagged behind at first, exhibited with a kind of "train delay." By 1890, however, percentage growth in this mode of transport exceeded that of the European railway pioneers England and Germany. The megaproject that is the Swiss rail network is being continuously expanded, including the clock-face scheduling adopted in 1982 and, in 2016, the scheduled opening of the Gotthard Base Tunnel.[2]

In parallel with the railway, major progress was made on the expansion of the road network, with newly built roadways, tunnels, and bridges. Notable projects included the construction of the first largely standardized and purpose-built route over the San Bernardino Pass (1818–23) by Giulio Pocobelli (1766–1843), the Salginatobel Bridge by Robert Maillart (1872–1940) in 1930, the Ganter Bridge at the Simplon Pass by Christian Menn (b. 1927), or most recently Jürg Conzett's (b. 1956) seven bridges over the river Flem completed in 2013, this last project part of a hiking trail that, unlike the others, has the effect of reducing speed.

The poster designed in 1935 by Herbert Matter (1907–1984) for the Schweizerische Verkehrszentrale (SVZ), the Swiss National Tourist Office, magically encapsulates the theme of the road, indeed even functioning as a general advertisement for all the infrastructure facilities built into the heroic landscape. The viewer finds him- or herself on a paved road that continues straight off into the middle ground, blurring in the distance but at the same time focusing in on the mountain pass as if through a telescope. The pass road winds uphill before clear blue skies and snow-white peaks: "All roads lead to Switzerland," or even: Switzerland for beautiful driving.

With the expansion of the national road network starting in 1963—the first section between Geneva and Lausanne—travelers experienced a new era of mobility. The Swiss divided highway par excellence would have to be the Kriens-Ennethorw-Stansstad stretch. Built in 1955, the "parkway" led comfortably through the landscape, right up to the foot of the mountain pass, so to speak, where Herbert Matter's pass road then lured the driver onward.[3]

Proceeding hand-in-hand with this development since the mid-nineteenth century was the development and conquest of the Alps for the benefit of tourism—by way of cog railway, funicular, cable car, and ski and chair lifts, ultimately reaching far beyond 2,000 meters in elevation.[4]

Streamlined

The notions of "faster" and "more efficient" soon congealed into physical forms, forever altering the appearance of trains and automobiles, which were then indeed faster and more efficient thanks to lower air resistance. Paul Jaray (1889–1974) of Vienna spearheaded many of these advances. From 1915, he worked at the company Luftschiff Zeppelin GmbH in Friedrichshafen on aerodynamic forms. After Jaray moved to Switzerland, the residents of Brunnen were surely astounded to see him coming down the street in the streamlined car he built in 1923 on a Ley chassis. The boxy shape of popular cars up to then, with running boards and fenders, gave way here to a basic form derived from the airship, sleek and low to the ground. To make sure the driver had a good view, a second volume was placed atop the body that resembled a Zeppelin gondola. All transitions, from the front grille to the headlights to the wheels, were rendered as seamlessly as possible. Only when one sees Jaray as the driver behind the side window in the illustration is it clear how narrow and stilted the design of this prototype was.[5]

Faster and more efficient also meant taking advantage of possibilities for lightweight vehicle construction. The Swiss Federal Railways (SBB) thus ordered a lightweight railcar in the early thirties that would not need a great deal of adhesive force but would be able to reach high speeds, accelerate better, and do both with the same amount of energy. In 1935, the CLe 2/4 railcar, known as the *Red Arrow,* went into operation. The *Schweizerische Bauzeitung* remarked regarding its form that "the cars do not project the image of a hypermodern *streamlined* automobile."[6] Although the *Red Arrow* certainly recalled Jaray's *Ley* from 1923 in more than just its details, the merger of cab and the box wagon looks like inadvertently hybrid design.

Walter Henne (1905–1990) came up with a more logical design for this hybrid when developing the Trans Europ Express (TEE) train type in the late fifties. He did away with the engine as a separate element at the front. The front-steered railcar was in service from 1961 on and was recently described by Stanislaus von Moos as "of almost un-Swiss brilliance."[7] After seventeen years of service, the TEE trains were put out to pasture, but two of them were renovated and rebuilt by Franz Romero in 1987. With a new coat of paint, they can now occasionally be seen as "Gray Mice" on the Swiss rail network. The foundation SBB Historic still has an original red and beige TEE in use.

But let's get back to streamlining in the thirties. Not only the SBB's *Red Arrow* but also the Gotthard locomotive of the Ae 8/14 type was subjected to a facelift for the 1939 Swiss National Exhibition in Zurich, anticipating, as it were, Max Bill's critique circa 1950, when he called for a synthesis of engineer and designer.[8] The front of the cab was now slightly slanted, the aprons pulled down, and all edges rounded, and the window of the cab along with the box windows formed an all-around band. The SBB emphasized that the body was developed not by designers but by engineers. The models for the redesign could be found circulating on the Italian rail network.

The streamlining euphoria in Switzerland soon spilled over into other areas, motivated more by considerations of form and fashion than engineering. In Arosa, for example, Martin Schmid, an auto mechanic, began tinkering in 1935 with a motorized snow sled in streamlined form. The "Schneeteufel" (Snow Devil) flamboyantly took to the snowy roads around Arosa in 1940.[9] At the same time, the company Stella S.A. in Bassecourt was at work building a lightweight, "arrow-fast" aluminum bicycle for the 1939 Landesausstellung, lending it an aerodynamic shape. As if to prove that this was the right choice of form, a speedometer is integrated into the strikingly designed "tank area." The streamlined thickening of the head tube also holds headlights and the gearshift. The bicycle frame segues dynamically without interruption via the top tube, seat, and chain stay to the pedal mechanism. A good fifty years later, Hannes Wettstein (1958–2008) designed a fitting counterpart with his *EST Highway* bicycle (1998). Now minimalistically pared down, the high-tech bike features a clever straight connection between top tube and seat stay, allowing for easy mounting, as well as a special rhythmic spoke arrangement that suggests movement.

467

468

467 Red Arrow railcar, Rae 2/4 1001, 1935, Swiss Federal Railways (SBB). A streamlined light metal railcar.
468 Walter Henne, TEE train, Rae TEE II, 1961, Swiss Federal Railways (SBB). Almost un-Swiss brilliance.
469 Nose Design, High-speed train for the Swiss Federal Railways (SBB), rendering, 2013, on behalf of Stadler Rail AG, Bussnang. International high-speed train for north-south traffic. In keeping with the SBB's brand values, the passenger should feel at home while on the go. The front cabin resembles a snake's head.

469

470

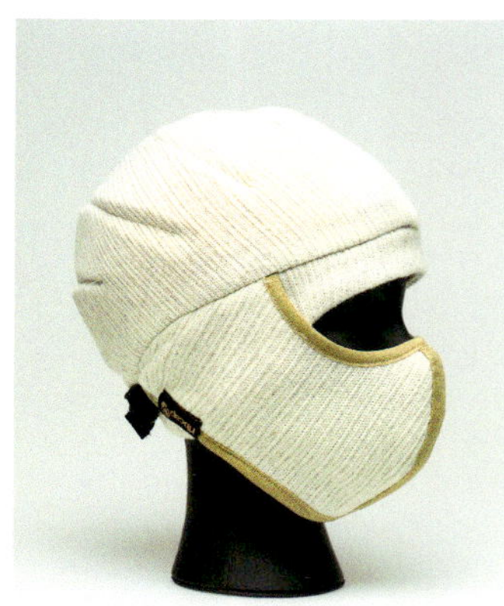

471

472

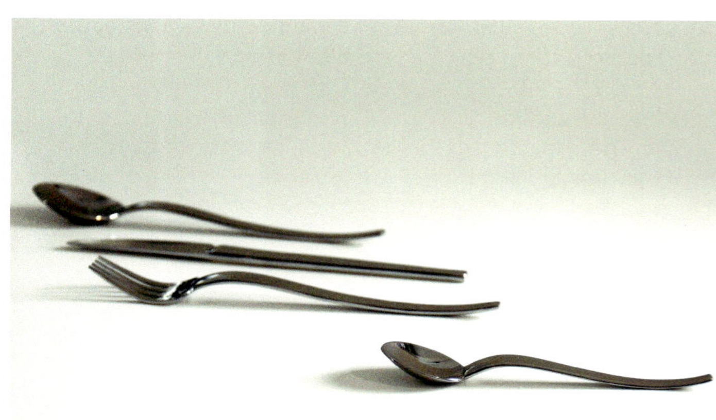

473

470 Twike three-wheeled lightweight electric vehicle with pedaling support and weather protection for two people, 1996, Twike AG, Gelterkinden. Developed with students at ETH Zürich.
471 Thomas Wüthrich and Lisa Besset, Ribcap knit cap, 2005, Ribcap AG, Münsingen. The sewn-in protector hardens upon impact and thus protects against head injuries.
Museum für Gestaltung*

472 Tobias Wülser, Monotracer, 2008, Peraves AG, Uster. Wülser redesigned the former Ecomobil (US) and was hired by Peraves for its realization. 130 horsepower enable the vehicle to reach a top speed of 250 kilometers per hour.

473 Christine Lüdeke, Poesie des Fliegens (Poetry of Flying), 1993, Berndorf Luzern AG. Onboard cutlery for first-class passengers, Swissair, Kloten
Museum für Gestaltung/
Donation of Christine Lüdeke

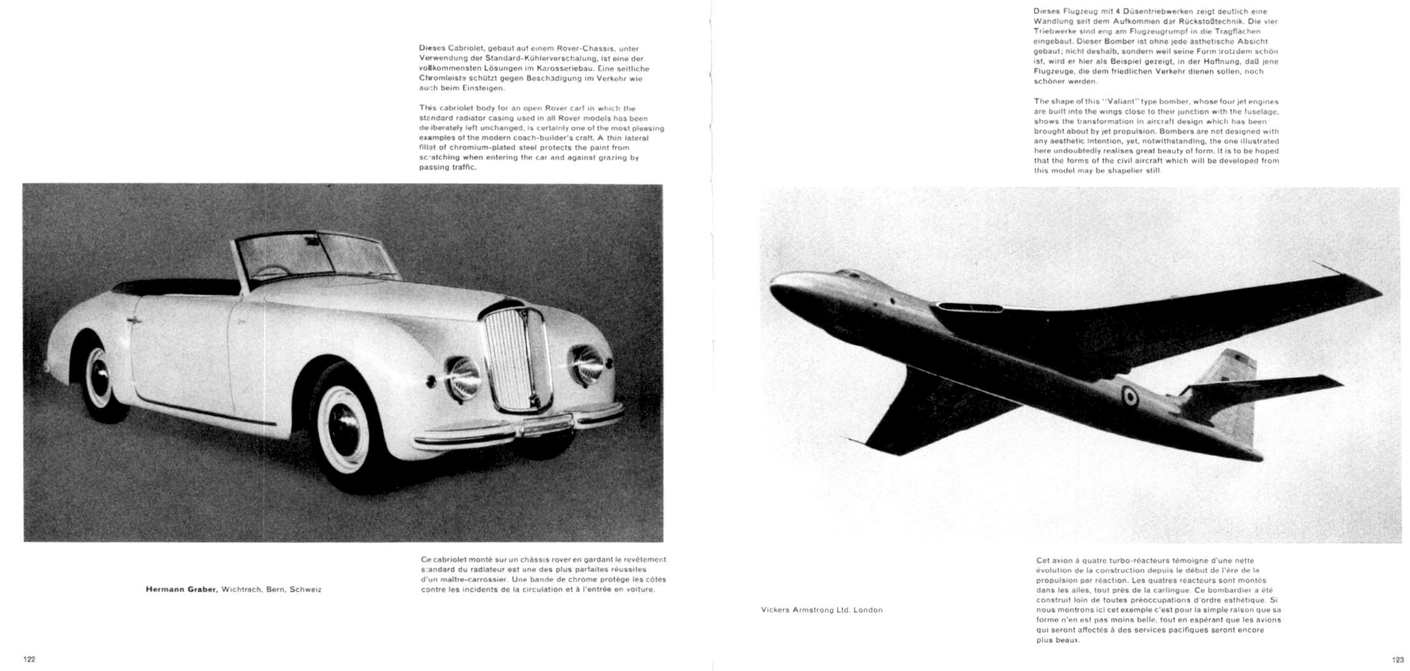
Dieses Cabriolet, gebaut auf einem Rover-Chassis, unter Verwendung der Standard-Kühlerverschalung, ist eine der vollkommensten Lösungen im Karosseriebau. Eine seitliche Chromleiste schützt gegen Beschädigung im Verkehr wie auch beim Einsteigen.

This cabriolet body for an open Rover carf in which the standard radiator casing used in all Rover models has been deliberately left unchanged, is certainly one of the most pleasing examples of the modern coach-builder's craft. A thin lateral fillat of chromium-plated steel protects the paint from scratching when entering the car and against grazing by passing traffic.

Hermann Graber, Wichtrach, Bern, Schweiz

Ce cabriolet monté sur un châssis rover en gardant le revêtement standard du radiateur est une des plus parfaites réussites d'un maître-carrossier. Une bande de chrome protège les côtés contre les incidents de la circulation et à l'entrée en voiture.

122

Dieses Flugzeug mit 4 Düsentriebwerken zeigt deutlich eine Wandlung seit dem Aufkommen dar Rückstoßtechnik. Die vier Triebwerke sind eng am Flugzeugrumpf in die Tragflächen eingebaut. Dieser Bomber ist ohne jede ästhetische Absicht gebaut; nicht deshalb, sondern weil seine Form trotzdem schön ist, wird er hier als Beispiel gezeigt, in der Hoffnung, daß jene Flugzeuge, die dem friedlichen Verkehr dienen sollen, noch schöner werden.

The shape of this "Valiant" type bomber, whose four jet engines are built into the wings close to their junction with the fuselage, shows the transformation in aircraft design which has been brought about by jet propulsion. Bombers are not designed with any aesthetic intention, yet, notwithstanding, the one illustrated here undoubtedly realises great beauty of form. It is to be hoped that the forms of the civil aircraft which will be developed from this model may be shapelier still.

Vickers Armstrong Ltd. London

Cet avion à quatre turbo-réacteurs témoigne d'une nette évolution de la construction depuis le début de l'ère de la propulsion par réaction. Les quatres réacteurs sont montés dans les ailes, tout près de la carlingue. Ce bombardier a été construit loin de toutes préoccupations d'ordre esthétique. Si nous montrons ici cet exemple c'est pour la simple raison que sa forme n'en est pas moins belle, tout en espérant que les avions qui seront affectés à des services pacifiques seront encore plus beaux.

123

474

474 Max Bill, Double-page spread from *Form,* Werner Verlag, Basel, 1952. Exemplary shapes for mobility

As a sustainable way of getting around, the bicycle is often combined today with mobility aids in the form of electric motors and made more comfortable by the integration of weather protection. Paul Jaray already developed and patented his *J-Rad* in 1921. It is a recumbent model equipped with an off-center pedal-lever drive, and the patent furthermore specifies the addition of weather protection.

A combination of muscle-powered locomotion, electric pedaling assistance, and perfect weather protection is offered by the *Twike.* The fully enclosed three-wheel lightweight electric vehicle resulted from a study by ETH students and has been in production since 1996. Around 1,000 of them are in use worldwide.

Beauty as a Function

Max Bill's (1908–1994) aforementioned critique in the years circa 1950 demanded nothing less than that design itself—that is, beauty—should be viewed as a function in its own right and not merely as the outcome of functional engineering considerations. He focused on identifying instances where both aspirations were synthesized.[10] Bill most likely would not have regarded the *Red Arrow* as embodying such a synthesis.

Evidence of what Bill considered good examples of this synthesis of beauty and function can be found in the broad selection of consumer goods presented in his 1952 book *Form.*[11] He devoted a few pages therein to mobility. In addition to international examples, he also includes a picture of Otto and Heinrich Pfister's dining car furnishings for the SBB, as well as an ambulant post office they designed. The automobile is cited several times in the discourse on *Gute Form* (see Menzi, "Designing 'Die gute Form,'" pp. 192–97). In his foreword, for example, Bill compares Ernst Neumann's car body with Umberto Boccioni's *Development of a Bottle in Space,* and the 1931 Adler sedan designed by Walter Gropius with a sculpture by the Belgian Georges Vantongerloo. Finally, Bill juxtaposes one of his own sculptures with his own 1947 Bentley Mark VI. But the epitome of contemporary modern design was in Bill's eyes Hermann Graber's 1950 convertible. It is illustrated as a triumph of good taste: "This cabriolet body ... is one of the most pleasing examples of the modern coach-builder's craft."[12] The Swiss auto body makers enjoyed an excellent international reputation due to their partly "anti-American"— in other words, restrained—style and the exquisite craftsmanship of body and interior.[13]

The work of more recent designers would surely fit in well with Max Bill's "collection" of good design. One need only think of the *Ecomobil* from 1984 and the successor model revised by Tobias Wülser (b. 1977) in 2008, the so-called *Monotracer* produced by the Peraves company. But would Christine Lüdeke's (b. 1960) cutlery set *Poetry of Flight,* designed for Swissair, also feel at home there with its sober and organic design in a scientific context?

Taking Off

In addition to cog railways and funiculars, the first cable car tramway soon went into operation in Switzerland. The Wetterhornbahn in Grindelwald carried the first guests up the mountain in 1908. At a length of just 556 meters, the tramway overcomes a difference in elevation of 420 meters. It was closed in 1915 due to a lack of visitors, and plans to expand the tramway to reach the peak at 3,692 meters were abandoned. The Gerschnialp-Trübsee connection in Engelberg, built in 1927, is thus the oldest cable car still operating today in Switzerland.[14] In 1936, Herbert Matter put the tramway with its aluminum cabins (installed in 1931) and their powerful gear mechanisms at the center of his promotional poster for Engelberg. The absence of the tramway pylons from the picture reinforces the impression of cabins silently wafting upward before a stormy sky and the starkly silhouetted mountain range. In the foreground, a couple, barefoot and in summer dress, look up in wonder at the cabins passing by. One can't help but see in the scenery compressed and synthesized here by Matter a glimmer of anticipation of the *CabriO* cable car that would open on the Stanserhorn in 2012, the world's first cable car in which guests can experience nature up close from the open upper deck.

Until the 1950s, Switzerland had developed and built, partly under license, airplanes but not jets. But then the Federal Military Department commissioned the Flug- und Fahrzeugwerken AG, Altenrhein (FFA) and Pilatus Aircraft in Stans to develop the first Swiss-made jet. The FFA's P16 took off for the first time in 1955. In 1958, the federal government decided to purchase a hundred of

475

476

475 Gangloff Cabins AG, Bern, Open-deck cabins, 2012. The world's first open-deck cable car cabin on rides from the Kälti inter-station up to the Stanserhorn summit. Photo: Kaspar Hitsch

476 Herbert Matter, Trübsee-Engelberg, 1936, Verkehrsverein Engelberg. Poster

477

014

479

481

478

480

482

477 Hansruedi Fellmann c/o Devico Design, FibreJet ski boot, ca. 1967, Raichle, Kreuzlingen. Inside the open-out plastic ski boot is a leather inner shoe that can be used for après-ski. Collection of Hansruedi Fellmann

478 Advertisement for Raichle "Red hot," ca. 1970

479 Goblin climbing shoe, 2007, Mammut Sports Group AG, Seon. The tight-fitting, pointed-toe shoe with Vibram sole gives good purchase. Museum für Gestaltung*

480 Monte Rosa Hut, 2008–09. In a study project at ETH Zürich (chair Andrea Deplazes), students developed a high-tech lodge with a crystalline structure for the Swiss Alpine Club (SAC).

481 Vintage I, boot with horse hair, 2003, Kandahar Schuhmanufaktur AG, Thun-Gwatt. Reedition of the boot with the Riri zipper and crepe sole from 1948. Museum für Gestaltung*

482 Gion A. Caminada, Extension of the Terri Hut on the Greina Plateau (built in 1925), 2007, for the Swiss Alpine Club (SAC). Adding on to the existing building fabric in the context of traditional Alpine architecture. Photo: Lucia Degonda

the jets. After the hydraulic system in a P16 failed on a test flight, however, the exercise was cancelled. Although production was already underway, the government withdrew the order. The P16 was ultimately developed further in the United States under the name *Lear Jet.* But the dream of a Swiss jet would still come true. Pilatus Aircraft in Stans has now developed and presented to the public the PC-24 jet aircraft—which can even take off and land "off road"!

The space race between the Americans and the Russians that also took off in the fifties climaxed with the American moon landing in 1969. Switzerland, too, was swept up in space fever, but only in symbolic and at times culinary form. The rocket-shaped ice cream treat brought out by Frisco the year of the moon landing is still the best-selling ice cream on a stick in Switzerland today. The water ice is subdivided into an engine section and a shield made of chocolate. In reality, Switzerland was one of the founding members of the European Space Agency (ESA) in 1975. The ESA pursues the exploration of outer space and has trained successful astronauts, including Claude Nicollier from Switzerland. Furthermore, for the *Ariane* project, which involved the conception of a rocket launcher for high payloads, RUAG Space was the developer of payload fairings that are still proving their value today.

Outdoor

Another poster by Hermann Matter, for the tourist resort of Pontresina, designed in 1936, heralded another successful chapter in Swiss design history, this time for the whole outdoor movement. The tanned face of an androgynous-looking woman dominates the image, looming out toward the viewer. She turns her gaze slightly upward to watch the goings-on in the mountain world of Pontresina. The snow goggles she has donned to protect her eyes from the blazing winter sun lend her a heroic air. Just as in Matter's Engelberg poster, the mountain world appears in the background. A skier is ready to plunge down the slope, wearing what are probably wooden skis and intricately laced boots.

In retrospect, it is clear that Switzerland was always a leader in the field of design for winter sports, pioneering a range of innovations. The successors to the *Wandervögel* hikers of yesteryear and the first climbers to conquer the Swiss peaks are today taking to the mountains outfitted with GPS, trekking poles, and high-tech textiles. The first mountain shelter, opened by the Swiss Alpine Club on the Grünhorn in 1863, has since been joined by 151 more, some of them welcoming visitors today with down duvets rather than rough wool blankets. Outdoor activities are booming, even in these days of rampant urbanization, and magazines such as *Landliebe* and *Landlust* attest to people's longing for a "genuine" encounter with nature.

It was the dedication of the ski instructor Fritz von Allmen from Mürren that led to the founding of the Kandahar-Schuhmanufaktur in 1932. It still produces high-quality sports footwear. Karl Molitor from Grindelwald, for his part, developed a solid leather ski boot with which he won the Lauberhorn ski race seven times, among other triumphs. Later, more sophisticated straps replaced the laces. The Henke company introduced the first ski boot with buckles in 1954. Raichle followed suit in 1968 with the *FibreJet*—a hard plastic shell with a soft liner. The shell has a hinge in the middle of the sole from front to back, so that the shoe can be opened outward to slide the foot in more easily. The Heierling company in Davos finally brought out ski boots with a central twist closure in 1974.

The development of skis likewise witnessed many innovations over the years. The "small" Stöckli company, for example, managed with its continual advances in high-tech racing skis to secure contracts with today's top racers. The Zai company in Disentis, by contrast, focuses on innovative sports equipment as part of the active lifestyle, producing skis that impress with their high-quality handcraftsmanship. Simon Jacomet's ski *Tila 160,* for example, has a sandwich construction that combines materials such as cedar, ultra-high-molecular-weight polyethylene fiber, carbon, and walnut wood. In emotional terms, these skis target not alpine racers but rather those attracted by the aura of "zai for Bentley."

The motto "Safer, faster, and more beautiful" is also reflected in the textiles industry. Swiss companies are known for developing innovative, high-performance textiles. Terms such as breathable, water-repellent, wind-resistant, and stain-resistant have in the meantime become part of our everyday vocabulary. Woolen men's swim trunks, which could weigh in at a good nine kilograms when wet, have thus been replaced by competition swimsuits that do not top hundred grams even when soaked.[15]

If there is something that hasn't changed since the thirties in mobility design, it is the feeling of the "liberated body" that shines out from Herbert Matter's posters. Modernism frequently connected this feeling with ice hockey, sled and bobsled races, alpine skiing, or doing gymnastics in the great outdoors. Dynamic movement and physical exercise were the main thing, with sports being hyped as a guarantor for mental as well as physical well-being.

In view of this attitude, it comes as no surprise that mobility design in particular experienced such tremendous advances, especially after World War II. In the process, the design criteria shifted from what were at first decidedly functional aspirations, onward to the more expansive interpretation by Max Bill, who ascribed an inherent function to beauty and form, all the way to today's consumer mobility products, which attempt to elicit target-group-specific emotions. Some clothing, for example, in addition to technicist or culturalist aspirations, is functional but is not supposed to look that way, while other outfits may not really be functional at all but take a form that makes them seem just that. Sporty, functional clothing has been an integral part of urban everyday life ever since the 1990s.

* The title alludes to Italo Calvino's novel *If on a Winter's Night a Traveler,* New York 1981 (orig.: *Se una notte d' inverno un viaggiatore,* Turin 1979). This essay spotlights a few aspects of Swiss "mobility design" that are meant to stand for the whole.

1 See Schivelbusch 1977.
2 Stalder 2013.
3 Heller/Volk 1999.
4 See, for example, Gugerli 1997.
5 See Lichtenstein/Engler 1993(?), pp. 16–17 and 186–89.
6 Cited in Jost 1981, p. 158.
7 von Moos 1992, p. 209.
8 See Elsasser 1997.
9 See Just/Kübler 2007, pp. 58–61.
10 Bill (1948) 2010; Erni 1983.
11 Bill 1952.
12 Ibid., p. 122.
13 See, for example, Rey 1999; see also *Schweizerische Carrosseriegeschichte* 1994.
14 See Jost 1981, pp. 156–57; see also Ganz/Rieker 2012.
15 See Karbacher 2009.

483

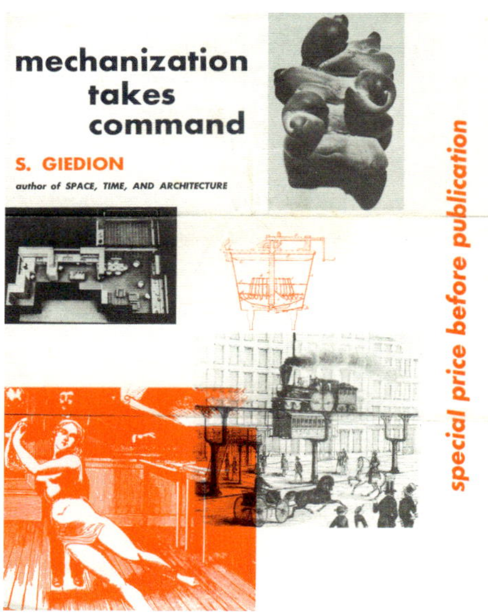

484

485

483 Advertisement for folding furniture and camp equipment from the Gold Medal Mfg. Co., Racine, Wisconsin (US), n.d. (early 1920s)

484 Herbert Bayer, Leaflet for Sigfried Giedion's book *Mechanization Takes Command: A Contribution to Anonymous History,* Oxford University Press, New York, 1948

485 Glenn Curtiss, Tent trailer, 1921, lightweight construction using fabric. American prototype for a motorhome for six people with foldout beds and a fully equipped kitchen cabinet. "At home on the go"—an early form of the *Co-op. Interieur* by Hannes Meyer?

The westward expansion of the North American continent in the course of the nineteenth century—coinciding with the Industrial Revolution—brought forth an enormous variety of lightweight, multifunctional, and mobile "patent furniture." Folding chairs, stools, benches, and beds; folding, rotating, or expanding tables; even folding bathtubs were used to facilitate moving house over long distances and for establishing an efficient economic and administrative structure in the far-flung corners of the nation.[1] In the 1940s, the Zurich art historian Sigfried Giedion traced the development of modern furniture and interior design to the "anonymous history" of this patent furniture. His book *Mechanization Takes Command: A Contribution to Anonymous History,*[2] published in 1948, compellingly captured the spirit of the times, shaping the thinking of an entire generation. Years earlier, Adolf Loos and Le Corbusier had pointed out similar lines of development for collective creativity, and in Hannes Meyer's essay "Die neue Welt" ("The New World") many of the "standard wares" listed were camping or travel products. As early as 1926, Meyer defined the individual of the future as a "semi-nomad of our modern productive system" who "because of the standardization of his needs as regards housing, food and mental sustenance ... has the benefit of freedom of movement, economies, simplification and relaxation, all of which are vitally important to him"[3] (see Gumbrecht, "Rational Impatience with the Past," pp. 82–87; Rüegg, "Arts and Crafts, Industrial Art, Machine Art," pp. 52–57).

Postwar developments only seemed to confirm Meyer's nomad metaphor. Migrations of a hitherto unknown scale anticipated the increasing mobility that would emerge in the new economic world. The refugee problem and the rebuilding of Europe brought new impulses for a renewal of product design—especially in Switzerland (see Rüegg, "Solutions d'urgence," pp. 162–67). Nomadic living implicitly or explicitly became the leitmotif of the new domestic culture. This trend is what prompted Le Corbusier to replace works of art that were permanently integrated into residential buildings with wall hangings that could be rolled up again at will. "Nous sommes devenus des 'nomades,'" he asserted in a text about the "Muralnomad" tapestries: "we change residences as our family develops, and sometimes we change our living conditions, or move to a new area."[4]

The Emergence of New Forms

Max Bill (1908–1994), the main ideologue of *Gute Form* ("good form," or good design), sought to find a common denominator that described the evolution of the contemporary world of objects. "When designers create new forms," he wrote in 1952—when the term "form" was still used with such awe as would be unfathomable nowadays[5]—"they are consciously or unconsciously responding to trends in contemporary art, because it is in art that the intellectual and spiritual currents of every epoch find their visible expression."[6] Bill was also undoubtedly right: after the long years of crisis and repression by a totalitarian regime, the public was ready for a new design world, its visual power holding the promise of a new, freer way of life.

The first *Typenmöbel,* or "type furniture," by Hans Bellmann (1911–1990), which was indeed the fruit of a philosophical somersault, would become an index fossil of the postwar period. In 1933, shortly before the closure of the Berlin Bauhaus by the National Socialists, Bellmann had received the "last"[7] architecture diploma signed by Ludwig Mies van der Rohe and Ludwig Hilberseimer. His subsequent work in Mies's studio would shape his work ethos, with Lilly Reich's statement "Furniture is cultivation to the millimeter" becoming his oft-quoted maxim. In 1944 he worked in the studio of the Bern-based Hans Brechbühler, who had himself worked for Le Corbusier and who upheld the idea of rational and yet poetic architecture even in times of war. In the studio's common room stood wooden "safari" or colonial chairs, grouped around an untreated tree stump used as a table. While this piece of nature isolated within the artificiality of the room possessed a poetic aura for Brechbühler, for Bellmann, who gave up his German citizenship in 1936 out of protest against the Nazi regime, it represented only the primitive world of "Blood and Soil." Heated discussions ensued. Brechbühler was willing to give up his poetic tree trunk if Bellmann could come up with a satisfactory alternative within an hour. Bellmann promptly drafted a "colonial table" to match the colonial chairs. Based on the idea of the folding sketching stool, the table has three lathed legs connected at a joint that spear, as it were, a round tabletop. Brechbühler was so pleased with this design that he had two samples of the table made immediately. He also used them to furnish the new timber home built for his carpenter, Ommerli.[8]

This colonial table encapsulated the essence of Bellmann's later designs, as well as those of many of his contemporaries. Their works took their cue less from the use of the latest technologies than from forms that arose from an extremely frugal use of materials and the possibility of rapid assembly and disassembly. The ease with which the furniture could be taken apart for packing, making it possible to move quickly from place to place, went hand in hand with a makeshift aesthetic, or, to stay true to our own terminology, a nomadic aesthetic. For reasons of economy, the use of wood was inescapable; nonetheless, the young designers succeeded in shaking off the specter of the provincial *Heimatstil* they so scorned by creating a decidedly modern alternative.

Furniture Types for the Postwar Era

In 1946–47, Bellmann created some of the most important furniture designs in Switzerland, if not all of Europe. A case in point is his freestanding *Spannstützengestell* (Screw-Tension Shelves), which did double duty as a room divider—to our knowledge the first shelf that could be fixed anywhere between floor and ceiling. It was conducive to a feeling of spaciousness that perfectly suited the new attitude to life; furthermore, thanks to its integrated cabinet and table elements, it can be regarded as one of the earliest examples of the "multifunctional wall unit." Through the use of two maple poles sliding in opposite directions, Bellmann had made the option of adjusting the divider to different ceiling heights the formal and technical basis of his design, while trying to reduce the number of parts to a minimum. Contemplating the elegance of his solutions, one is tempted to speak of an "intellectual economy" that is not ruled by aspects of manufacture, finances, or practicality, but focuses instead on the visualization of an idea.

Bellmann was not a lone wolf. Wilhelm Kienzle, Jacob Müller, and Willy Guhl all pursued similar goals with their semi-industrially manufactured reconstruction-era furniture. Soon, a younger generation would struggle with the typical contemporary challenge of shelves suspended between floor and ceiling. Foremost among them was the ETH architect and designer Jürg Bally (1923–2002) with his *G 15* rod shelving (1950), as well as the trained carpenter and qualified interior designer Ulrich P. Wieser (b. 1927) with his pole-mounted *UPW* bookcase (1956–57). Andreas Christen (1936–2006), a graduate of the Zurich product design class, contributed a freestanding aluminum bookshelf in 1964. With his clever *S.T. Table* in 1954, Jürg Bally was able to realize a height-adjustable version of the three-legged colonial table that had the added feature of being fit for use as either a dining table or a tea table (hence the name: *S* = "Ess," or eat; *T* = "Tee," or tea).[9]

Bellmann's most productive and most publicly active competitor was without a doubt Max Bill. While Bellmann's take-apart furniture tied in with the staunch but cheerful ethic of the *Wandervögel* movement,[10] Bill associated his works, which were often inspired by natural forms or mathematical models, more closely with the fine arts. If we compare his *Three-Circles Table* (1949–50), made up of three circles with an equilateral triangle as frame, with one of Bill's sculptures, the former would appear to be the outcome of a step-by-step metamorphosis. All the same, the slightly outward-tilting, conically tapered legs elude any evident geometrical rationale; they seem to have been taken from the arsenal of the era's more widely used furniture components rather than the repertoire of an artist. This sort of stylistic incongruence is typical for Bill, who integrated suitable finds to achieve inimitable and specific solutions.[11]

The proximity of these different positions led to at least one controversy, in this case also about the adaptation of a find. In 1951–52, Bellmann connected a wooden seating shell to a lightweight metal frame through the use of a single screw, naming the result the *One-Point Chair.* The plywood shell was a standard product, but Bellmann cut a freehand shape out of the back to give the chair elasticity; this large hole also proved to be a convincing way

486

487

488

489

490

491

486 Hans Brechbühler, Ommerli House, Köniz, 1945. Interior. Prototype for the Chasseral construction technique, furnished with Alvar Aalto's Convertible Sofa (Volkssofa), Wilhelm Kienzle's Colonial Armchair, and Hans Bellmann's Colonial Table.

487 Richard Paul Lohse, Prospectus for the Colonial Chair (product design by Wilhelm Kienzle), Wohnbedarf AG, Zurich/Basel, ca. 1947

488 Hans Bellmann, Colonial Table (Wohnbedarf Model 1015), ca. 1944, maple tabletop, stained beechwood feet, collapsible, Wohnbedarf AG, Zurich/Basel; AG Möbelfabrik Horgen-Glarus (from 1946). Photo: Hans Finsler

489 The Colonial Chair by Wilhelm Kienzle and the Colonial Table by Hans Bellmann dismantled for transport, photographed on a Berber rug. An iconic image of nomadic living.
Photo: Hans Finsler

490 Hans Bellmann, One-Point Chair, 1951–52, molded plywood seat shell, chrome-plated metal frame, Wohnbedarf AG, Zurich/Basel; AG Möbelfabrik Horgen-Glarus (from 1952)

491 Max Bill, Particle-board chair with horizontal slot in back, 1950/51–1959, molded plywood seat shell, chrome-plated metal frame. 1961 series for the Pestalozzi Children's Village in Trogen.

492

493

492 Max Bill, Exacta kitchen clock, 1951, light blue ceramic, glass, Junghans Uhren GmbH, Schramberg (DE, 1957–62)
Museum für Gestaltung

493 Max Bill, Three-Circles Table, 1949, black linoleum tabletop, maple frame, Wohnbedarf AG, Zurich/Basel; AG Möbelfabrik Horgen-Glarus (1949–ca. 1960). Design based on an equilateral triangle.
Museum für Gestaltung

to convey the chair's light weight. Bill for his part developed a chair in 1951 made of molded plywood with a horizontal slit in its back, and therefore claimed the idea as his own. While in Bellmann's opinion the elegance of connecting the shell to the frame using only one screw was central to the design, for Bill the combination of a readymade shell with a special skeleton-like frame was the main feature of the invention. Perhaps reacting oversensitively, Bellmann proceeded to construct the seat of his next important chair design—the *GA Chair* of 1955—from two identical, one-dimensionally molded plywood elements.[12]

Production and Communication

In a 1951 photograph of the skylit upper room at the Kunsthaus Luzern, furnished by Bellmann for the exhibition *Kunst und Wohnkultur* (Art and Interior Design),[13] Bellmann's *Spannstützengestell* is joined by Bill's *Tripod Chair* and Eero Saarinen's *Womb Chair*—a Knoll model—to name a few elements that explicitly reference nomadic living. They are accompanied by a selection of Berber rugs (so-called shepherds' rugs), a colorfully decorated Moroccan couscous bowl, as well as the *Hallicrafters S-38* receiver by the American industrial designer Raymond Loewy, reminiscent of a bomber radio. Bellmann's installation, which, aside from further designs by Bellmann, Bill, and Saarinen, also featured Pierre Jeanneret's *Scissor Chair* and Pierre Stein's sofa bed (both from Knoll) as well as paintings by Hans Erni and Erich Müller, exercised an impact far beyond the occasion, as a pioneering synthesis of modern, contemporary interior design culture in which Swiss and international furniture types coalesced in an open, dynamic interior layout to impressive effect.

The furniture and decorative objects for the show came from the company Wohnbedarf AG, a "clearing house for contemporary living needs"[14] founded by the art historian Sigfried Giedion, the architect Werner M. Moser, and the businessman Rudolf Graber for the purpose of furnishing the Werkbund's new Neubühl housing development in Zurich. Their first collection consisted of lightweight multifunctional pieces; an article in the *Neue Zürcher Zeitung* described the metal furniture as "a quick-change artist, a picture puzzle that through resourceful tricks could serve any number of purposes"[15] (see Lichtenstein, "Furniture for All Seasons" pp. 132–37). In 1933–34, the collection, initially dominated by Swiss designs, took a decidedly international turn through the addition of Alvar Aalto's springy wooden pieces and Marcel Breuer's aluminum furniture range—just before modern design's situation began to worsen day by day in the face of the economic crisis. Graber, who had been the sole owner of the company for several years at that point, later remembered a woman entering the store during those difficult years and asking, "Do you have any furniture with tree knots?"[16] In order to survive economically while still avoiding the trap of *Heimatstil*, Graber always complemented the as-ever modern furniture range with some antique pieces, even opening the largest restoration workshop in Zurich during the war. As soon as the political climate improved, however, he reestablished contact with the international avant-garde. By 1947, he had already secured a fruitful working relationship with Knoll Associates Inc. in New York, which was reorganized only a year before by Hans G. Knoll and Florence Schust out of the former Hans G. Knoll Furniture Company (est. 1938). Graber was granted the Swiss license to produce Knoll's domestic furniture, and Knoll in turn produced a few Swiss models (such as Bellmann's *Colonial Table* as a "tripod occasional table"). Above all, however, Graber made Wohnbedarf—and with it, Zurich—into a veritable melting pot of Swiss contemporary interior design.[17] He distributed the series of wooden furniture designed by Bellmann in 1946–47, while continuously updating his offerings with Bellman's and Bill's latest designs, as well as pieces by Hans Gugelot (1920–1965), Fred Ruf (b. 1926), and Ulrich P. Wieser. Gugelot's furniture system *M 125* anticipated future developments. Further highlights were the telephone table and *Servi* table by the old master Wilhelm Kienzle (1886–1958), a revised version of the *Landi Chair* by Hans Coray (1906–1991), as well as the stackable bed made of fiberglass-reinforced polyester designed by Andreas Christen in 1960. The AG Möbelfabrik Horgen-Glarus, a specialist in wood furniture, replaced the Embru-Werke in Rüti as Wohnbedarf's main supplier. The painter and graphic designer Richard Paul Lohse (1902–1988) took over the role of house designer, which in the beginning had belonged to Bill and the former young Bauhaus master Herbert Bayer, and later graphic designer Pierre Gauchat and artist Robert Gessner—all of them

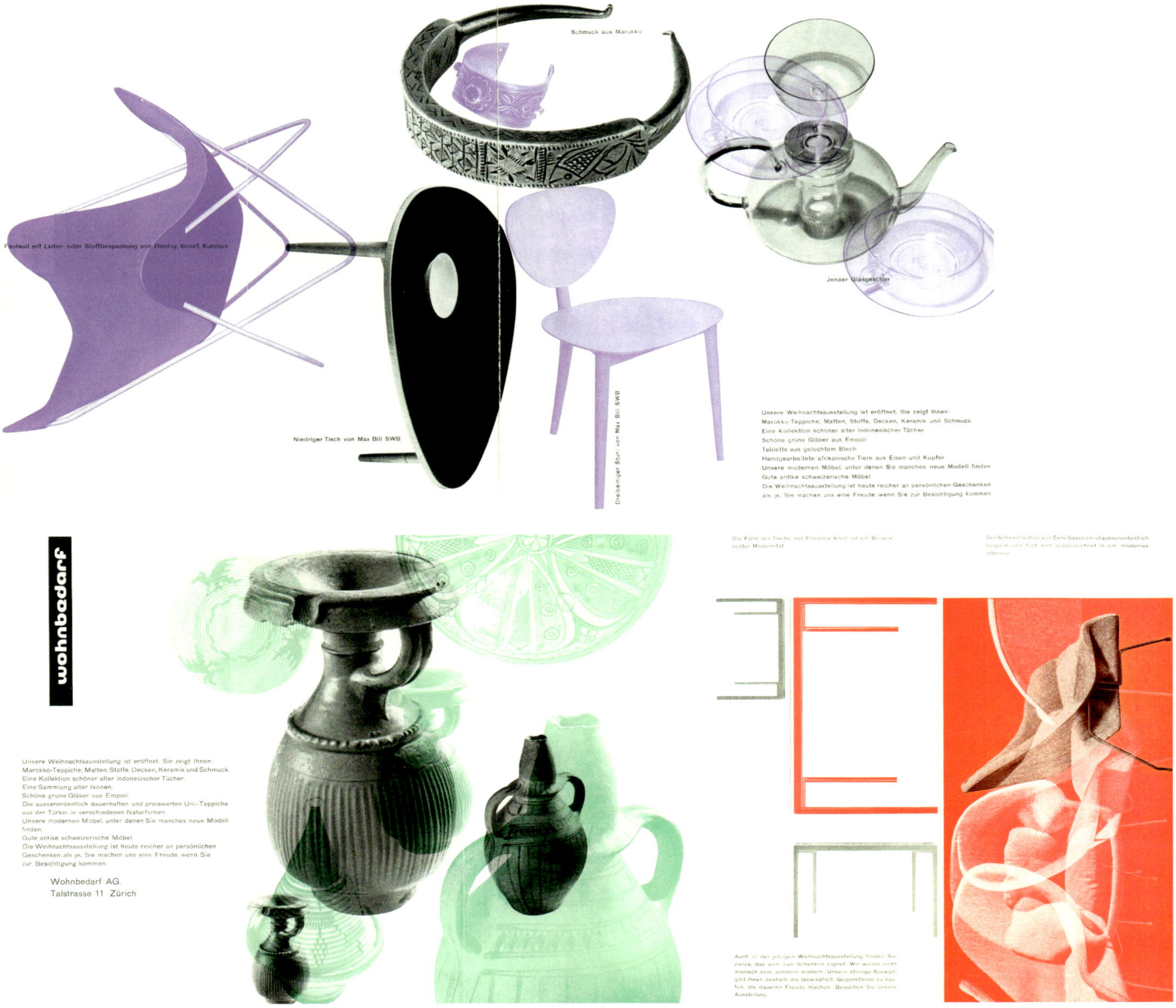

494

495

494 Richard Paul Lohse, Folding leaflets for the Christmas exhibitions put on by Wohnbedarf AG, Zurich, 1953, 1948, and 1955. Folk art meets modern design: a high point of Swiss commercial graphics.

495 Hans Bellmann, *Kunst und Wohnkultur* (Art and Home Decor) exhibition at the Kunsthaus Luzern, 1951, furnished with furniture from Wohnbedarf AG. A synthesis of the modern Swiss home decor of the time.
Photo: Otto Pfeiffer

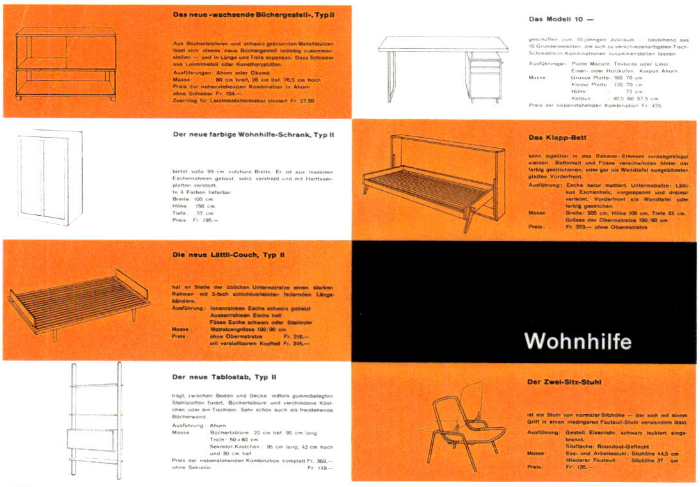

496

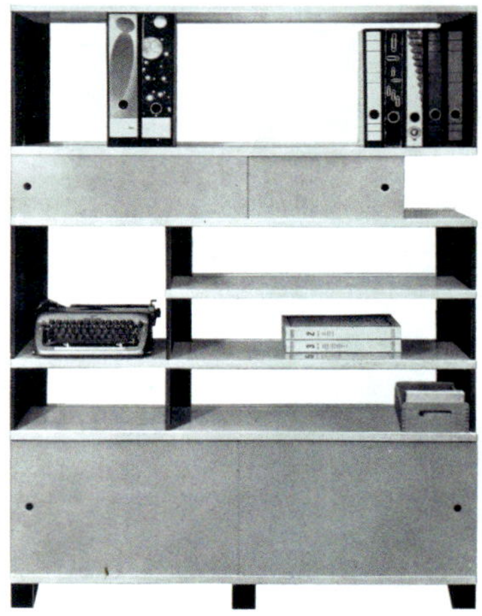

497

498

496 Werkgenossenschaft Wohnhilfe, Zurich, Prospectus on "New Products from Wohnhilfe," ca. 1957. Inside pages. Sensible, unpretentious furniture for a primarily young audience.
497 Reni Trüdinger, Type II Shelf, 1956, folded sheet metal, wooden shelves, sliding doors in synthetic resin or aluminum, Werkgenossenschaft Wohnhilfe, Zurich (1956 to 1960s)
498 SWB home furnishings advice center at the Gewerbemuseum Winterthur, 1956 to mid-1960s. A young couple obtains advice on selecting cutlery and tableware from "the home furnishings advisor in Winterthur, Miss Menzi" (from *Bauen heute,* supplement in the *Berner Tagblatt,* May 25, 1958).

masters of visual communication. Now Lohse redesigned the entire corporate identity, from exemplary product sheets to sales binders, from newspaper ads to the lettering on the delivery van; Max Bill's logo was all that remained of the original corporate design.[18] Lohse's freely composed, color-overprinted Christmas catalogues, among the greatest products of 1950s commercial graphic design, succeeded in convincingly evoking the new lifestyle; apart from furniture, they also show pottery, cutlery, rugs, textiles, and accessories.

Wohnbedarf AG had since its inception targeted the enlightened middle class. Next to the increasingly prestige-oriented furnishing items, which began to take on more sculptural forms through the use of the newly emerging molded plastics, more affordable items were also on offer, some of them direct descendants of the furniture kits of the postwar era. Cooperatives initiated by master carpenters, such as Wohnhilfe, founded in Zurich in 1945, gave rise to simply shaped furniture made of plain wood. Throughout the fifties, professional designers continued to contribute further space-saving, low-cost masterpieces, such as Jürg Bally's metal shelf and *S.T. Table,* Wilhelm Kienzle's *Shoe Rack,* Erika Schläpfer-Thöne's *Kommodenz,* Reni Trüdinger's *Type II Shelf,* and Max Bill's *Ulm Stool*—with a draw added for even greater multifunctionality. The worldview had certainly changed. The democratization of modernism was relentlessly taking hold, accompanied by a moralizing undertone taken by the housing agencies as well as Swiss Werkbund campaigns encouraging the spread of *Gute Form,* enthusiastically supported in regular columns in new women's magazines, for example in *Annabelle* from 1947 on.[19]

1 On this and many other aspects of flexible living, see Schwartz-Clauss/von Vegesack 2002.
2 Giedion 1948; see in particular the afterword by Stanislaus von Moos originally written for the German edition, Giedion (1982) 2013.
3 Meyer (1926) 1965, p. 93.
4 Le Corbusier 1960.
5 At least that was the opinion of Eugen Gomringer; see Gomringer 1967, p. 2.
6 Bill 1952, p. 11.
7 According to an interview with Hans Bellmann conducted by Arthur Rüegg and Ulrich P. Wieser on April 23, 1990; Bellmann told the Bauhaus Archive in Berlin that the diploma in question was no. 115.
8 See on this passage Rüegg 1995a.
9 See Bignens 2002c.
10 The *Wandervogelbewegung* was a youth movement in the first third of the twentieth century in Germany that hoped to develop its own outdoor way of life, setting the stage for the future naturism, *Lebensreform* ("life reform"), and progressive education movements.
11 See Rüegg 2002g.
12 See Rüegg 2002e.
13 See Schilder Bär 2002d.
14 This description can be found in a prospectus for an exhibition of the interiors at the Werkbund's Neubühl housing development in Zurich, September 1931. For an early history of Wohnbedarf AG, see Mehlau-Wiebking et al. 1989.
15 N. N. 1931.
16 Graber 1956, p. 32.
17 On the later history of the Wohnbedarf AG, see Rüegg 2011 and Rüegg 2006.
18 See Lohse Stiftung 1999.
19 See Schilder Bär 2002b, p. 142.

499

Ursula Luginbühl in Mötschwil, before 1967. The potter Ursula and the iron sculptor Bernhard Luginbühl made their farmhouse in Emmental an artistic powerhouse in the 1960s. Despite their idiosyncratic elementary forms, Luginbühl's household ceramics are undoubtedly functional.

Photo: Leonardo Bezzola

1960 — Rationality

In the decade of economic prosperity, rational thinking triumphs. The open-plan office moves into the designers' sights, and with it furniture "systems" that can be used flexibly to mark off individual territories. The interchangeability of parts dominates the office environment just as it does coffee services and the architecture itself. At Expo 64 in Lausanne, numerous artists and designers explore variations at a high aesthetic level on the topical themes of the day. At the same time, *Gute Form* loses its cachet; the new furniture designs target instead prestige and exquisite execution. Mid-decade, industrial design is introduced as a course of study. The momentum of the industrial boom leads to an uneasiness that is aired for the first time at the fiftieth anniversary of the Swiss Werkbund (SWB) in 1963 and which finds creative expression in the legendary *Chair Fun* campaign. Inspired by the counterculture movement of the hippies and the student protests, a variety of new lifestyles begin to be articulated.

500

The interior designer Rosmarie and the electrical engineer Rico Baltensweiler in their Ebikon studio, where they worked from 1965 to 1975. The couple rapidly achieved international recognition with their first two lamp models. Numerous lighting design commissions in Switzerland and abroad provided a steady stream of fresh inspiration for the development and production of exemplary light series, which they made with their self-designed tools in their self-built workshop.

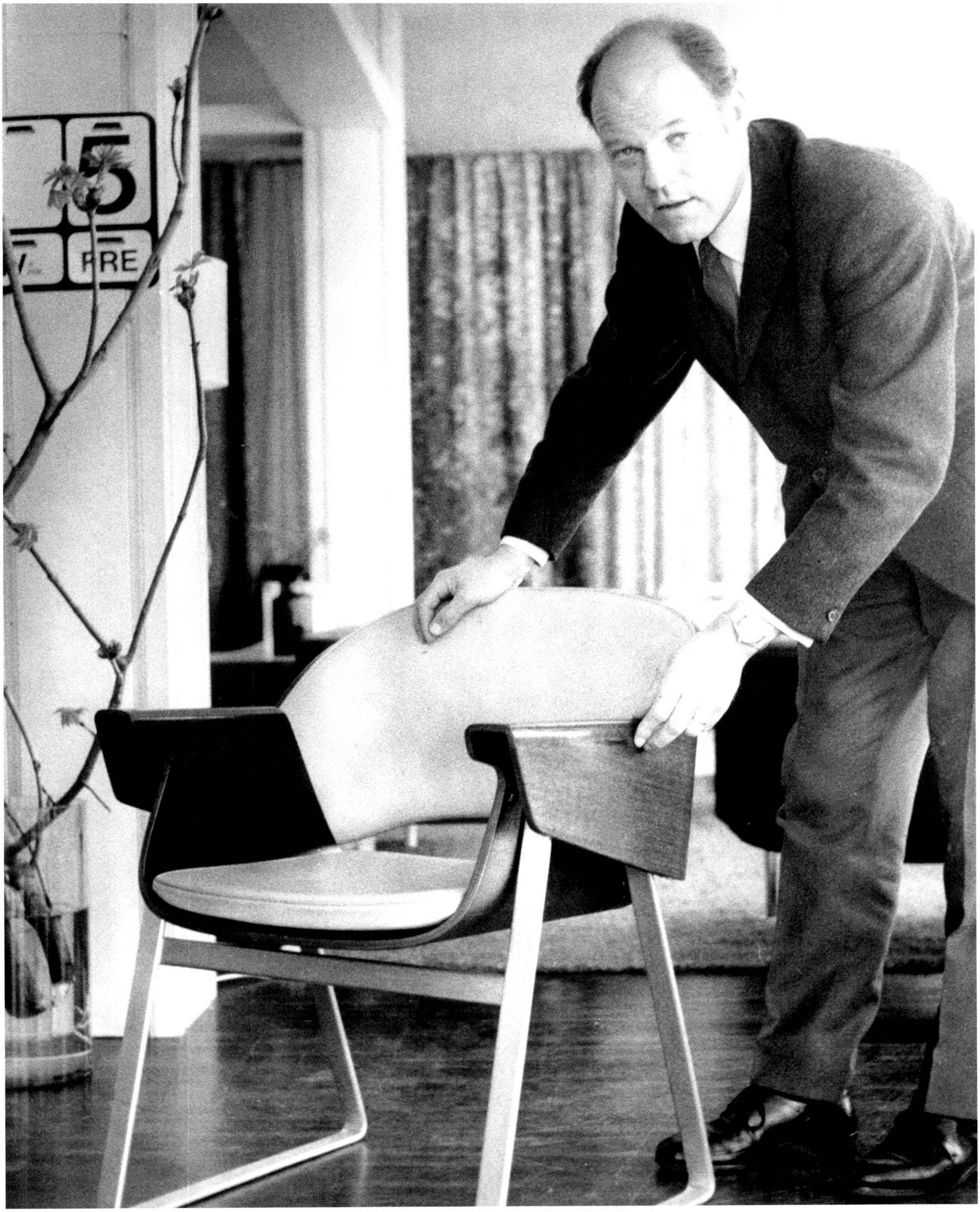

501

Jürg Bally, who trained as an architect, in the shop he opened in 1956 on Neumarkt in Zurich. Bally—who played the dual roles of interior designer and furniture designer as well as running an experimental furniture workshop and acting as salesman in his shop—presents one of his most striking pieces: the Futura 11 STA double-shell chair, produced by Girsberger AG in Bützberg.

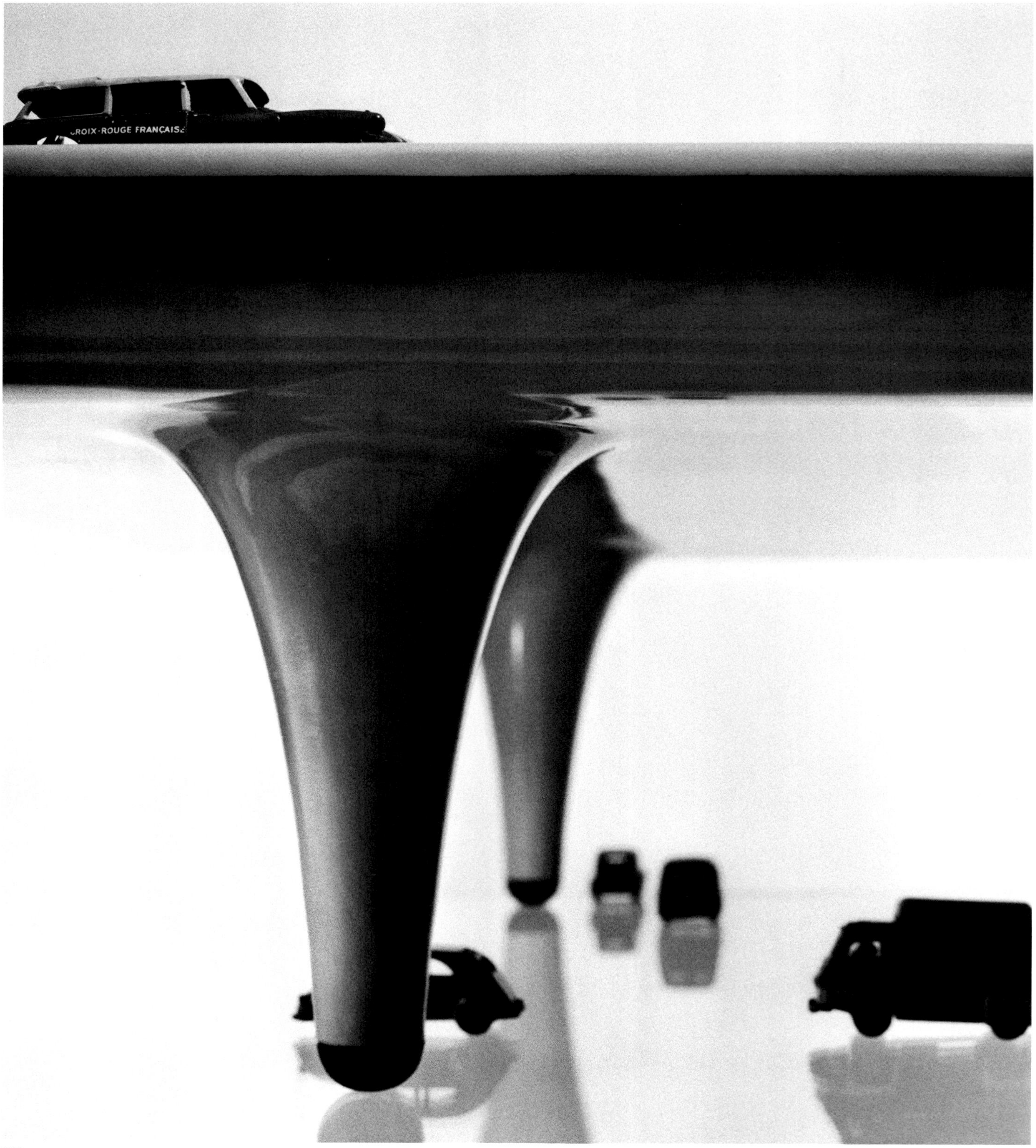

502

Experimental advertising photography using the bed designed by artist and industrial designer Andreas Christen in 1960 out of fiberglass-reinforced polyester (produced by H. P. Spengler, Rümlang; sold by Wohnbedarf AG, Zurich/Basel). The bed was molded from a single piece of material and had hollow, mushroom-shaped feet that made it stackable. Using toy cars, the photograph manipulated the scale so that the bed is perceived as a piece of modern transport infrastructure.

Photo: Alfred Hablützel

503

The faculty of the Kunstgewerbeschule Zürich in the "Valley of Youth" at the Expo 64 construction site in Lausanne. Alfred Aebersold, Walter Binder, Willy Guhl, Rudolf Bircher, Elsi Giauque, and Kristin Knell-Weber have just walked through the concrete sail, a mere twelve centimeters thick, by Geneva architect Michel Magnin.

Photo: Dölf Preisig, Photography Class, Kunstgewerbeschule Zürich

504

Nighttime mood shot taken in Pierre Zoelly's Expo Train Station, the Swiss National Exhibition Expo 64 in Lausanne, 1964. The mistreated and randomly shoved-together plastic chairs—in the meantime collector's items par excellence—tell of the throngs of people who waited by day for their special trains under the daringly cantilevered overhang.

505

Meret Oppenheim, *Läbchuechegluschti,* 1967. For their annual meeting in Bern, the Bern division of the SWB called for an uprising against functionalism and the principle of respect for materials. Along with a host of other artists, Meret Oppenheim racked her brain to think of alternative seating options. Her board chair shows a face—eyes, nose, and a big red velvet tongue—as well as sporting a cushion that paraphrases the famous Bernese *Lebkuchen* (gingerbread) with its almond decorations.

Susi and Ueli Berger

Multi-Soft Chair
1961

506 Susi and Ueli Berger, Multi-Soft Chair, 1961, produced from 1967
Vinyl cover, polyether foam upholstery, 73 × 70 × 80 cm, Victoria-Werke AG, Baar
Museum für Gestaltung

507 Susi and Ueli Berger, Soft Chair, 1967, Victoria-Werke AG, Baar
Museum für Gestaltung

508 Multi-Soft, Victoria-Werke AG catalogue, 1968

506

507

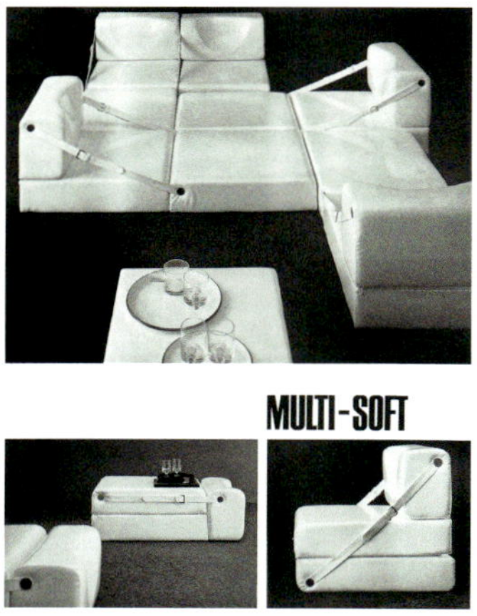

508

The increasing demand for seating to suit "young people in blue jeans" was an issue that had been raised at a Werkbund convention as early as 1955. However, American pop culture with its message of freedom and liberation was not easily reconciled with the aesthetic tenets of *Gute Form.* In 1967, the SWB therefore launched the *Chair Fun* campaign, calling for unconventional seating furniture designs. The artist and designer couple Susi Wyss (b. 1938) and Ueli Berger (1937–2008) submitted a chair devoid of any kind of inner support frame, which they named *Soft Chair.* The cartoonish seating sculpture was cut from a solid block of polyurethane foam. The chair's S-shaped back, which met flush with the floor and had five ventilation holes, served as structural reinforcement as well as accentuating the chair's dynamic, undulating form. In 1967, the Victoria-Werke in Baar began producing the *Soft Chair* as well as the *Multi-Soft* armchair designed by the Bergers in 1961 (three years prior to Joe Colombo's famous *Multi Chair*). *Multi-Soft* can thus be considered an early Swiss contribution to the concept of lounge landscapes. The upholstered elements are connected by zippers and can be folded to form an armchair or unfolded as a mattress and are held in place by side straps. Ultimately, with the release of *Terreno* in 1972, Susi and Ueli Berger abandoned the idea of designing single pieces of furniture and instead collaborated with other designers to create freely combinable foam blocks that could be arranged into seating landscapes. RM

Schilder Bär/Wild (2001) 2003, p. 102; Bignens 2002b.

Sondor M3 Baby Magnetic Film Recorder
1969–70

509

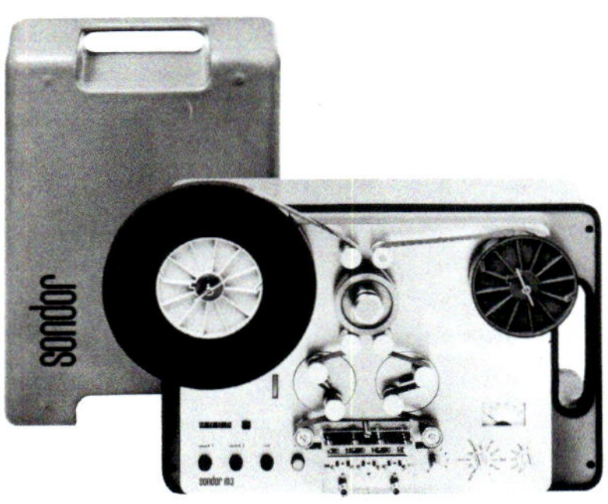

510

511

509 Ludwig Walser, Sondor M3 Baby Magnetic Film Recorder, 1969–70, produced from 1976
Fiberglass-reinforced polyester case, 24.5 × 41.5 × 28 cm, Sondor Willy Hungerbühler AG, Zollikon
Museum für Gestaltung*

510 Detail of prospectus page, 1970s, Sondor Willy Hungerbühler AG, Zollikon

511 Sondor Baby in its portable water-resistant case
Museum für Gestaltung*

The Sondor company has been developing, manufacturing, and marketing professional sound recording equipment for use in film postproduction since 1952. Their products use magnetic film—more specifically, perforated magnetic tape—for synchronous audio recording, editing, and playback. In the sixties, Sondor invented the bi-phase interlocking pulse signal to reliably sync sound to film. In his attempt to make this technological innovation visible through design, company president Willy Hungerbühler began collaborating with the industrial designer Ludwig Walser (b. 1936). Walser, who had spent some time working abroad in the United States (with Eliot Noyes, among others) after training under Willy Guhl at the Kunstgewerbeschule Zürich, founded his own industrial design studio in Dättwil in 1968. Designing primarily capital goods, he considered himself more of an engineer of human/machine interaction than a designer. Walser believed an object should represent its function, and his job was to make the design speak. For Sondor, Walser revised the hierarchy of the control elements. Along with displays, control buttons, and labels, the whole corporate design needed an overhaul. While the company's bright orange signature color ensured high brand recognition, it had presumably also been chosen as a concession to contemporary pop culture. Walser encased the internal image of precision and reliability in a soft molded shell. RM

Gantenbein 1989; Ludwig Walser Design AG, catalogue, Zurich, n.d.

512

512 Klaus Vogt, Squadra Cupboard, 1964
513 Painted plywood, 140 × 60 × 34 cm, prototype Museum für Gestaltung*

514 Detail of Squadra Cupboard with piano hinges

515 Klaus Vogt, Drawings for two prototypes of the five-piece Squadra Cupboard family; bar cabinet and secretary desk, 1964. Three other models were available, for use as a bookcase, wardrobe, or linen cabinet, 1964.

513

514

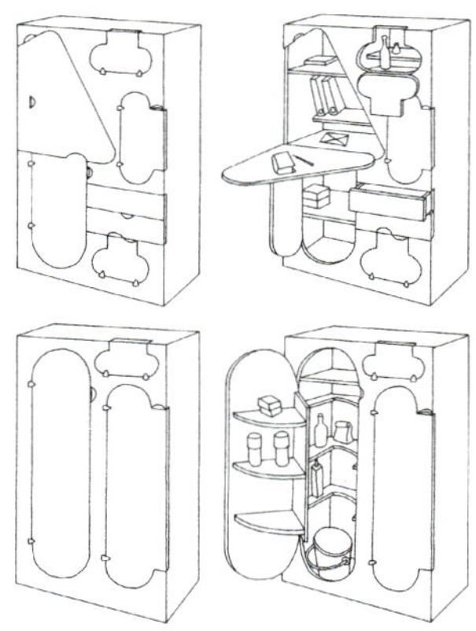

515

This colorful object's perfect rectangular shape makes the lineament cut into its front appear all the more intriguing. It is only upon second glance that it can be recognized as the outline of a set of drawers and cupboard doors. Counterintuitively, the flush-mounted doors do not open at the punched-out holes but at the edges of the cupboard. While the prototype shown here features piano hinges, for serial production Huss hinges were inserted into the presumptive finger holes. The doors, like those of antique cupboards, do not match the exact height and width of the compartment behind, but instead betray by the shape of their outlines and the placement of subtle recessed grips whether they conceal a drawer or a set of shelves, or even a fold-down tabletop. Klaus Vogt (b. 1938), who trained as a boat builder and was a graduate of Willy Guhl's interior design course, later becoming an assistant in the architecture department at the ETH, was a member of a young generation of designers who prized the joy of experimentation and playing with forms, colors, and manufacturing techniques over the strict obedience to rules. Vogt initially constructed the *Squadra* cupboards—a Helvetian form of Anti-Design—for his own personal use, endeavoring to make do with minimal means. He nailed together standard plywood boards to construct a sturdy body, out of which he then sawed the doors. Eventually, the cupboards were named after Squadra, a working group for architecture, design, and graphics that Vogt cofounded in 1967. AR

Das ideale Heim, July 1973; Schilder Bär 2002e.

516 Pierre Zoelly, Expo Chair, 1963
Fiberglass-reinforced polyester, 71 × 71 × 74 cm, MWB Metallwerke Buchs
Museum für Gestaltung/Donation of Pierre Zoelly

517 Two of the many possible sitting positions on the Expo Chair

518 Four Expo Chairs, stacked

516

517

518

The train station built for the 1964 Swiss National Exhibition in Lausanne, a bold, suspension-based construction, was the work of Pierre Zoelly (1923–2003), a young architect from Zurich who had spent a great deal of time in America. During Expo 64, visitors gathered together underneath his enormous "6,000-Person Umbrella" before being split into groups to board four trains standing ready at the station. Since waiting times were unavoidable, Zoelly also had to design station seating. He tried to create a single piece to suit all his requirements: stackable to make way for nightly cleaning and sturdy enough to carry heavier visitors. Furthermore, it was to have a backrest as well as a shelf to hold a glass of beer, and to be shaped to allow someone to lean against the front or side in a variety of positions. Zoelly chose a ribbed fiberglass construction. Having run out of time to choose a distinctive color scheme, he decided to simply use black. Due to his complete lack of experience in chair manufacturing, he partnered with a company that normally produced chemical tanks. The chair's development was so late getting started, and its manufacture so rushed, that the 200 pieces, projected to weigh six kilograms each, ended up with weights ranging from four to ten kilograms. Within a week's time, all the four-kilogram chairs had broken. Decorated with countless carvings, around a hundred of these charming all-rounders managed to survive the expo season's abuses. AR

Werk 51, no. 2 (1964); Zoelly 1992, pp. 9–10.

519

520

521

519 Pierre Renfer, City 3561 Hotel Tableware, 1965
Glazed hard porcelain, various formats,
Porzellanfabrik Langenthal AG
Museum für Gestaltung/
Donation of Porzellanfabrik Langenthal

520 Page from product catalogue *Wohnen heute 6. Warenkatalog,* 1967, Schweizerischer Werkbund

521 Prospectus for City 3561 service with bands of color, 1960s, Porzellanfabrik Langenthal AG

Since its founding in 1906, the Langenthal porcelain manufactory's product range had always been marketed to traditional, middle-class Swiss households. Advertising campaigns directed at housewives fancifully extolled the company's products; "Swissness" formed part of Langenthal's corporate identity. In 1965—when the company was at the peak of its growth, counting some 950 employees—its artistic management underwent a generational change. The ceramicist and painter Fernand Renfer (1898–1979) from the Jura region, who had been working with Langenthal since 1916 and dictating the company's artistic direction since 1940, passed his position on to his son Pierre (1925–1995), with whom he collaborated from 1949 onward. The young designer, who trained under Johannes Itten at the Kunstgewerbeschule Zürich and later in Bern, Vierzon (Cher), and Stoke-on-Trent, landed his first great success in 1965. For half a century, his hard porcelain dishware series *City 3561* proved to be one of the most popular and best-selling products of its kind. The modest elegance of *City*'s robust forms was not the only reason for its appeal—it also boasted a wide range of uses. With the release of *City,* Langenthal became the first manufacturer to put into practice the idea of a "gastronorm for hotel dishware." The standardized individual pieces can be combined in a number of different ways and are easily stackable: "*City* is suitable for use in system catering, but also shines in a traditional restaurant, ski lodge, or town hall." AR

Wohnen heute 1970a, p. 67; Schumacher 2012.

522

523

524

522 Fritz Haller and Paul Schärer, Model 500 File and
523 Supply Cabinet, 1964–65
Core board with oak veneer, legs missing,
110 × 78 × 52 cm, USM U. Schärer Söhne AG,
Münsingen
Museum für Gestaltung/Donation of Fritz Haller

524 Prospectus produced by USM U. Schärer Söhne AG, ca. 1970

This meticulously detailed oak cabinet is certainly a far cry from what we understand today as the *USM Haller* modular furniture system, with its iconic tube, ball, and sheet metal elements. It must have been a prototype or an early product made by the company U. Schärer Söhne in Münsingen (USM), which continued to market the piece throughout the seventies as a "boardroom unit." In 1964–65, the architect Fritz Haller (1924–2012) and the mechanical engineer Paul Schärer (1933–2011) developed a prefabricated office pavilion for USM, including furnishings. They discovered that files and paperwork tended to "flow" through the open-plan office in the same way that materials do on the shop floor—a revolutionary insight indeed. Their idea was to store the palletized files either in open stands or lockable boxes. "The wooden boxes," Haller recounts, "were made to be stackable units. One, two, or even three can be placed on top of a substructure similar to a table frame." The floors and fold-down doors of these file and supply cabinets were equipped with flush-mounted plastic rails to hold the specially developed wheeled file stands. In this sense, they became part of a cohesive construction kit that also included tubular steel frames designed to hold the same file stands. In a further iteration, these frames became the highly versatile *USM Haller* modular furniture system. AR

Bauen + Wohnen 19, no. 8 (1965); USM company brochures (ca. 1970); Haller 1978.

525

525 Rosmarie and Rico Baltensweiler, Guggerli L2 Floor Lamp, 1966
Tubular stand made of PVC, opal glass reflector, H: 40 cm, D: 17.5 cm, Baltensweiler AG, Ebikon
Museum für Gestaltung

526 Rosmarie and Rico Baltensweiler, Guggerli L4 Floor Lamp, 1966, Baltensweiler AG, Ebikon
Museum für Gestaltung

527 Guggerli model series, page from the product catalogue *Wohnen heute 7. Warenkatalog, 1969/70,* Schweizerischer Werkbund
Photo: Bigler + Ramuz

526

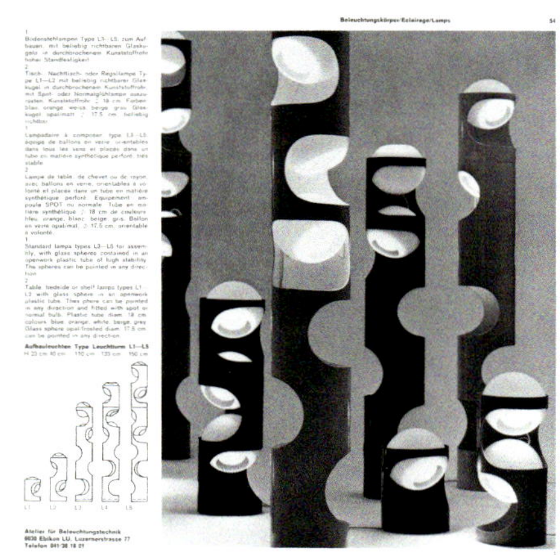

527

In the fifteen years following the design of their *Type 600* floor lamp, Rosmarie (b. 1927) and Rico (1920–1987) Baltensweiler created an entire family of visually delicate balanced-arm lamps "in the field of tension between utility and manufacturability." Their design work was characterized as much by its high level of machine-assisted handcraftsmanship as it was by its quality. The Baltensweilers made their own tools and even built the roof of their workshop themselves. The use of high-grade semifinished components allowed for the efficient manufacture of medium-sized series. Once established, these components reappeared time and again in modified form, defining the products' identity. Even before they were inspired by advances in lightbulbs to start experimenting in the early seventies with new kinds of lamps for the workplace, in 1996 the studio released the astonishingly different *Guggerli* series. The lamps consisted of white or colored standing plastic tubes with round cutouts and freely movable glass spheres. The spheres, which could be pointed in any direction, could be fitted with spotlights or standard bulbs. The series encompassed a selection of models ranging from a small bedside lamp with just one sphere to a tall floor lamp with room for up to eight bulbs. The colorful, sculptural *Guggerli* embodied the spirit of pop culture just as vividly as the earlier *Type 600* had epitomized nomadic living. No wonder, then, that is was soon being sold on New York's famous Madison Avenue. AR

Wohnen heute 1970b, pp. 54–55; *Baltensweiler AG* 1996; Sonderegger 2002.

528

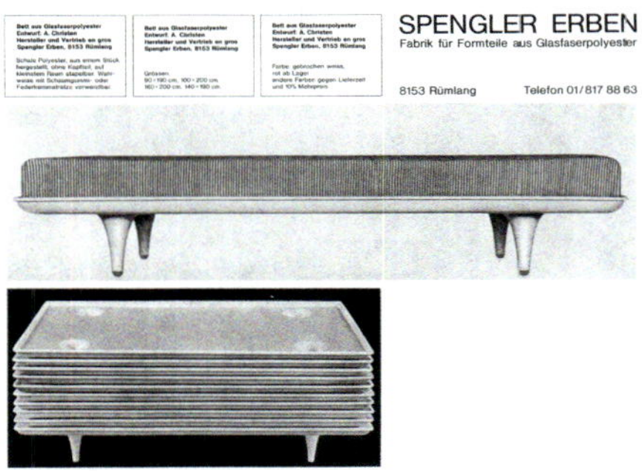

529

530

531

528 Andreas Christen, Polyester Bed, 1960
Fiberglass-reinforced polyester with a colored gelcoat finish, 100 × 200 × 28 cm, Wohnbedarf AG, Zurich/Basel; H. P. Spengler Erben, Rümlang
Museum für Gestaltung/Donation of Ruth and Damian Fink

529 The catalogue shows the fifteen-kilogram bed as a practical innovation, stackable like egg cartons, 1960s, H. P. Spengler Erben, Rümlang

530 Ensemble with Floor Lamp (1958) and Polyester Bed (1964) designed by Andreas Christen
Photo: Alfred Hablützel

531 "Wonderful system for perfect order," 1960s, J. F. Pfeiffer AG, Zurich. Page from prospectus

The advent of plastics not only led to more fluid forms but also to the dissolution of the design dogma of *Materialgerechtigkeit,* "doing justice to the material." The artist and product designer Andreas Christen (1936–2006) described plastic as a medium that allows for "a complete negation of the material." As the first and only student in the newly created trial course in product design at the Kunstgewerbeschule Zürich, Christen was able to spend three years doing cross-disciplinary training under Willy Guhl, designing both furniture and interiors. Industrial designers such as Charles Eames and advances made in the US in technical processes for producing new forms were influential for his work. Immediately following his training, Christen opened his own studio and designed the *Polyester Bed,* a plastic shell molded out of a single piece of fiberglass-reinforced polyester and thus free of any support framework. With a sheet of honeycomb cardboard incorporated to lend the necessary stability, the bed was produced through a technically complex process that Christen developed in collaboration with H. P. Spengler. Advertised as a practical novelty, the lightweight, fifteen-kilogram *Polyester Bed* was sold by Wohnbedarf AG in Zurich in the colors white, light gray, red, and black. When Knoll International took an interest in the bed in 1967, the manufacturer and rights holder, the German company Bofinger, refused to license it, thus preventing its industrial production in large quantities. RM

Hablützel 1963; Bähler 1997.

532

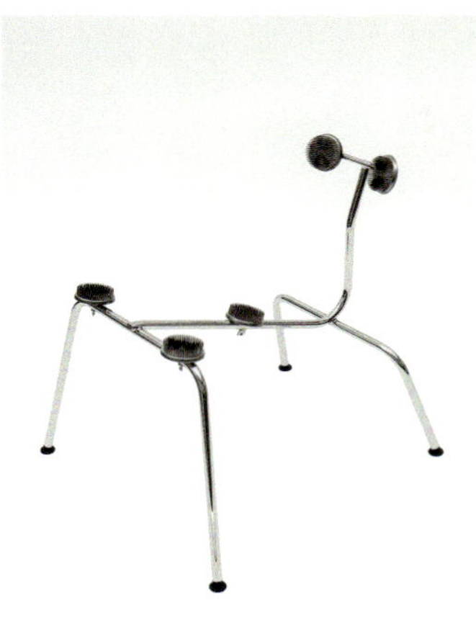

533

534

535

532 Trix and Robert Haussmann, Neon Chair, 1967
Neon tubes, seat cushion covered in synthetic leopard skin, 100 × 45 × 45 cm, custom piece
Poster: Atelier Hablützel und Jaquet, Alfred Hablützel (photo), "De Sede is Latin for sitting," 1971

533 Trix and Robert Haussmann, Maso Chair, 1967
Photo: Alfred Hablützel
Collection of Trix and Robert Haussmann

534 Trix and Robert Haussmann, Choco Chair, 1967
Photo: Alfred Hablützel
Teo Jakob Art Collection, Bern

535 Trix and Robert Haussmann, Trio, 1967
Photo: Alfred Hablützel
Collection of Trix and Robert Haussmann

In November 1967—six months before the Paris student riots—things were fermenting within the Swiss Werkbund. In a subversive move at the annual meeting in Bern, the sacred cows of "functionalism" and "faithfulness to materials" were figuratively slaughtered. For once, artists were being asked to rack their brains for an alternative species of seating: "Demonstrate that a chair can be not only a seating opportunity but also a seating possibility!" Thirty objects were gathered together for presentation at a meeting for the chair auction *Chair Fun,* including the legendary *Läbchuechegluschti* by Meret Oppenheim and the precursor to Susi and Ueli Berger's *Soft Chair.* "Chairs can be manifestos, too": for Trix (b. 1933) and Robert (b. 1931) Haussmann, the seemingly melting *Choco Chair,* Eames's prickly *Maso Chair,* and the *Trio,* consisting of three entwined Thonet chairs painted in primary colors, marked the beginning of a systematic critique of the utilitarian function and appearance of furniture. The *Neon Chair,* constructed from fluorescent tubes and with a seat covered in synthetic leopard skin, was forced to stay home due to its fragility. Its magical glow and lack of functionality as a seat make this ephemeral chair-object—which later broke during a photo shoot—the absolute epitome of Anti-Design. AR

Die Allgemeine Entwurfsanstalt 2011; Gnägi et al. 2013, pp. 326–27.

Corrugated Cardboard Chair
1969

536 Alois Rasser, Corrugated Cardboard Chair, 1969
Corrugated cardboard, H: 75 cm, D: 64 cm,
Kunstgewerbeschule Zürich
Museum für Gestaltung/Donation of Alf and Monika Aebersold

537 Making a cardboard chair, Kunstgewerbeschule Zürich, 1969

538 Alois Rasser in his cardboard chair. Photo from the exhibition *Magie des Papiers* (The Magic of Paper), Kunstgewerbemuseum Zürich, 1969

536

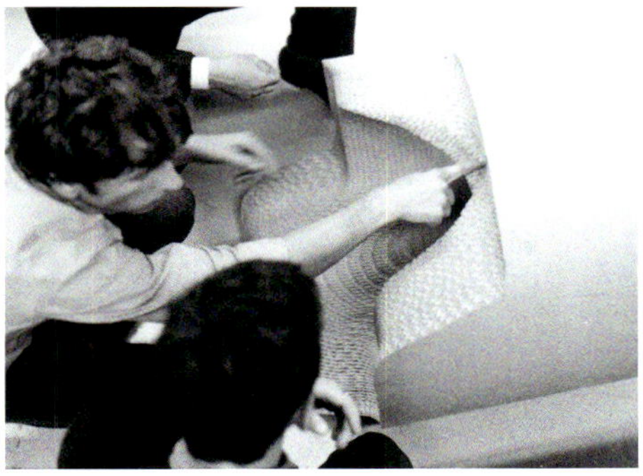

537

538

For the café and cinema furnishings that were to form part of the exhibition *Magie des Papiers* (The Magic of Paper) at the Kunstgewerbemuseum Zürich, Willy Guhl's interior design class at the Kunstgewerbeschule held a student competition. Unlike many of the other competing students, Alois Rasser (b. 1948) chose not to stabilize the cardboard using folding techniques; instead, he was fascinated by the rolls of corrugated cardboard with one-sided coating just as they were delivered. By beating down the core of a cut-off cardboard cylinder on one side and then removing the cone this created on the opposite side, he created a stool with a comfortable hollow to sit in. Rasser's next step was to develop an optimal cutting pattern for the production of chairs following the same principle. He found that by sawing a roll of corrugated cardboard into two identical halves with a band saw on a flat, S-shaped curve, he could create two chairs at once. He then only needed to beat down both cut surfaces to form seats and then remove the cones emerging at the other ends. The jury chose Rasser's chairs to furnish the café. He produced around fifty additional units using the eighty-centimeter band saw available at the company Model in Winterthur and delivered orders himself by truck. As these cardboard furnishings were not, after all, intended for longevity, only a few of the chairs have survived the last forty-five years. Rasser's ingenious invention could be recreated at any time, if only there were a suitable band saw still in existence. RM

"Papiermöbel und Kinderspielsachen" 1970.

539

540

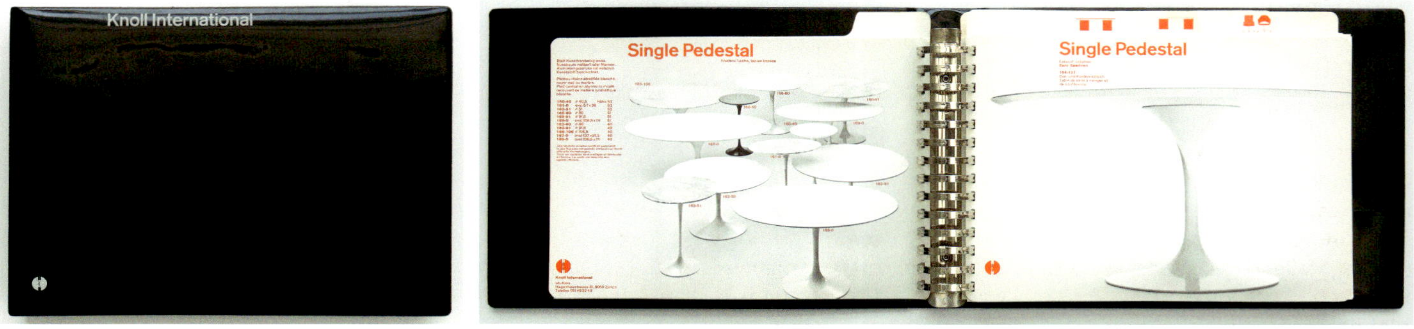

541

542

539 Otto Rudolf Salvisberg, Deutsche Krankenversicherungs-AG, Berlin, 1929–30, Accounting/filing department. An "open-plan office" avant la lettre.

540 Advertising leaflet for the Cube office furniture series from Knoll International, n.d. The Cube series was developed and produced completely "in Switzerland, to suit Swiss needs."

541 Ernst Hiestand, Product catalogue for the Wohnbedarf subsidiary wb form showing the office furnishings produced under license by Knoll International, New York, ca. 1966

542 Robert Probst, Action Office II, 1968. Individual "territories" in an open-plan office are separated here with the help of flexible, personalizable partitions.

In the 1960s the tertiary sector began to visibly expand. The rise of administration buildings, company headquarters, major banks, and office buildings changed the social infrastructure at the heart of cities. Unsurprisingly, the office building became a central topic in the architectural debate, and its inner workings a multifaceted field for experimentation by business consultants, designers, and furniture makers.

Individual Offices versus Open Plan

Open-plan offices, which were the subject of much discussion, were not an invention of the economic boom period. As early as 1906, employees in Frank Lloyd Wright's Larkin Administration Building in Buffalo handled at least 10,000 orders each day in a five-story, naturally lit, artificially air-conditioned, and clearly structured "atrium court." However, the metal furniture designed specifically for this building was still arranged in orderly rows as was customary at the time. In 1970, surveys conducted by the management consultant Dr. Robert Rosenkranz showed that only 8 percent of positions warranted their own, separate offices, while the open-plan office was ideal for 9 percent and for 83 percent of jobs either work environment was deemed equally acceptable.[1] Therefore, there was no reason not to adopt and improve these American ways of building, with their artificially air-conditioned and evenly lit "office landscapes," in Switzerland. Although it seemed initially that this might lead to monotonous mass standardization, the system was "humanized" through the differentiated arrangement of individual workspaces, as well as the option of creating different zones by using partitions, planters, and artworks. In this sense, teamwork would not only lead to the dissolution of rigid hierarchies but would also increase fashion-consciousness in the office, as Verena Huber remarked in the journal *Werk*.[2] On top of that, the integrated relaxation and communication zones helped to nip common stress symptoms in the bud.

The USM Haller Modular Furniture System

According to a (difficult to verify) statement by the architect Fritz Haller (1924–2012) from Solothurn, "the first open-plan office in Switzerland" was built in 1964, in the open countryside.[3] Haller, who had designed the HTL Brugg-Windisch technical school (now a university), had three years before he lost interest in site-specific building and instead turned to the development of steel construction systems, work he undertook in close cooperation with the steel construction company Ulrich Schärers Söhne in Münsingen (USM). The first modular construction they developed together was a new headquarters for USM, followed immediately by a sprawling steel and glass pavilion to house the company's administration offices. Since there was nothing suitable to be found on the office furniture market, Haller developed step by step, with the help of the engineer Paul Schärer (1933–2011), a modular system, which remains a global standard to this day.[4]

Haller and Schärer embarked on this task in "a very childlike manner, starting from the question of what employees really needed to have at their workplaces, and wondering especially what things they would normally store inside their beautiful old desks," Haller recalls.[5] All they found there were personal utensils; the company files and paperwork, on the other hand, tended to "flow" through the office, just like the materials of production did on the factory floor. The role played by standard pallets in production could thus be taken over by document trays in the office. "A workspace is today nothing but a table," Haller concluded;[6] the trays could be stored instead either in open frames or in lockable boxes.

Document trays were therefore the key to this solution for office spaces. The other components would have to be improvised. An iconic photograph shows the entrance area of the USM office pavilion shortly after the employees had moved in (1965). Next to a voluminous reception desk equipped with a control panel for the air conditioning, there are partitions covered in oak veneer and Eternit planters that serve to subdivide the open space, as well as modular, open document storage elements made of metal grids. What we do not see are the numerous closed, stackable filing and supply cabinets. Therefore, these initial office furnishings no doubt comprised a simple construction kit with components of different types and forms, like the ones being offered internationally by a number of companies, such as Knoll International.

The real "invention" came only later. It was a chrome-plated brass sphere with six threaded holes to which heavy-duty steel tubes could be attached in every direction of Cartesian space with the aid of an inserted connecting piece.[7] Thanks to this universally employable connecting joint, a collection of more or less modular elements became a closed construction "system" that can be used to build open or closed filing and equipment cabinets, credenzas, trolleys, reception counters, planters, wall coverings, and ceiling panels—objects that can be dismantled at any time and reassembled into different objects. Only the desks, acoustic walls, and electrical installations continued to lead their own independent existence.[8]

Apart from the system's functional performance, its reduction to just three basic visible elements—chromed spheres and tubes as well as powder-coated colored metal plates—likely accounts for the ongoing fascination with the USM Haller modular furniture system. The sphere soon came to represent the entire system. In an early advertising portfolio it peers through the window of the glossy black cover, bathed in blue underwater lighting, like a Mexican death mask. With the help of these spheres, the prestigious office landscape of the economic boom era became a colorful, nonhierarchical workshop. The freely arranged filing elements are, according to the prospectus, "the pivotal point of the work team and can define the borders of its space."[9] Nowadays, this clearly structured modular design has attained cult status and social cachet. Thanks to its certified artistic value, it has also found its way into top-class collections as well as the lofts of wealthy urbanites. Like a message in a bottle, it preserves the memory of the high point of systematic thinking in late modernism—a time when designers shifted their focus from industrially produced *Typenmöbel,* or "type furniture," to modular systems of furnishings.

Swiss Alternatives

One look at the trade journals that were being published at the time confirms that Haller and Schärer's showpiece was not created inside a vacuum. In the pivotal year of 1964, the sculptor, teacher, and inventor Robert Probst revolutionized American office spaces.[10] His *Action Office I,* designed for Herman Miller Inc., abandoned schematic rows of identical desks in favor of freely arrangeable working spaces that suited both seated and standing jobs. As a second step, Probst concentrated on two- or three-sided dividers between different "territories," using flexible, personalizable wall units packed with work surfaces, trays, and shelves. Thanks to these stylish, modifiable components, the *Action Office II* became an inspirational as well as commercial success. Futuristic office spaces with adjustable shell chairs and integrated work and communication tools were designed by Luigi Colani for Comforto Ahlen in Germany and by Odo Klose for Giroflex in Koblenz, Switzerland; however, they did not enjoy the same sweeping success. The standard for this trend was set by the progressive modular furniture developed by Knoll International, which was manufactured in Switzerland by Wohnbedarf AG in Zurich, and sold from 1958 to 1975 through a specially created contract department.[11] The designer Kurt (Koni) Ochsner adjusted the schemes designed by Florence Knoll, Eero Saarinen, Vincent Cafiero, and Richard Schultz to Swiss needs, upgraded them to current norms, and enhanced them with a number of details. The *Cube* series was even developed entirely "in Switzerland, to suit Swiss needs."[12] It consisted of different types of desks, with and without drawer units, filing cabinets of different heights and widths that could be assembled to form whole walls, as well as chairs from the Knoll collection. The strictly cubic form of these pieces and the combination of chrome with wood, or with black or white plastic veneers, were their most characteristic features. The wooden components were made largely by the Röthlisberger furniture and carpentry company in Gümligen, and the metal components by Ulrich Schärers Söhne in Münsingen, which had originally specialized in building fittings.[13]

The office furnishing industry's primary interest was in sales. Under these circumstances, most of the Swiss attempts to challenge multinational manufacturers would remain a pipe dream. The USM Haller modular furniture system is one of the exceptions that prove the rule (although the mass manufacture of this ingenious

543

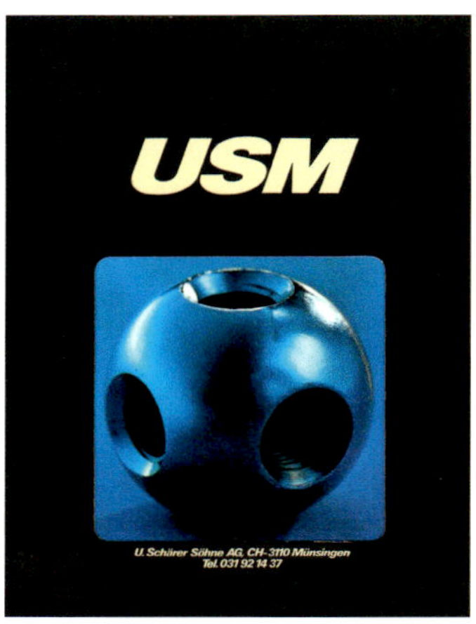

544

546

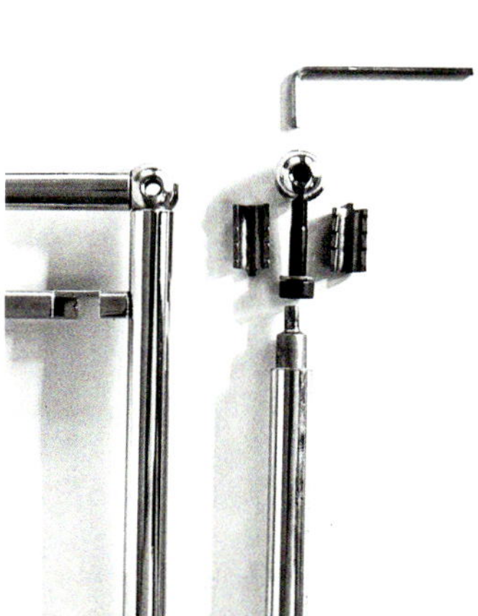

545

543 Fritz Haller, Offices of U. Schärer Söhne in Münsingen, 1965. Reception area, seating group, and reception desk with control panel for air conditioning, partitions, snap-together file frames.
544 Advertising portfolio produced by USM U. Schärer Söhne AG for the USM Haller building systems, 1970s. Cover
545 Steel tubes, ball joints, and slotted-in connecting piece—the elements constituting the USM Haller furniture system.
546 Complete workplace, constructed with the USM Haller furniture system. Photo: Alfred Hablützel
547 Richard Paul Lohse, M 125 (product design by Hans Gugelot), produced by Wilhelm Bofinger, Ilsfeld (DE), 1960, Wohnbedarf AG. Brochure. The first version of the M 125 furniture system produced by Wohnbedarf AG dates to 1950.
548 Office furnished with the M 125 furniture system and additional pieces produced by Wilhelm Bofinger, Ilsfeld (DE), from 1958
549 M 125 furniture system produced by Wilhelm Bofinger, Ilsfeld (DE), 1956–73. Standardized elements for combining various types of cabinets that can be used from both sides.

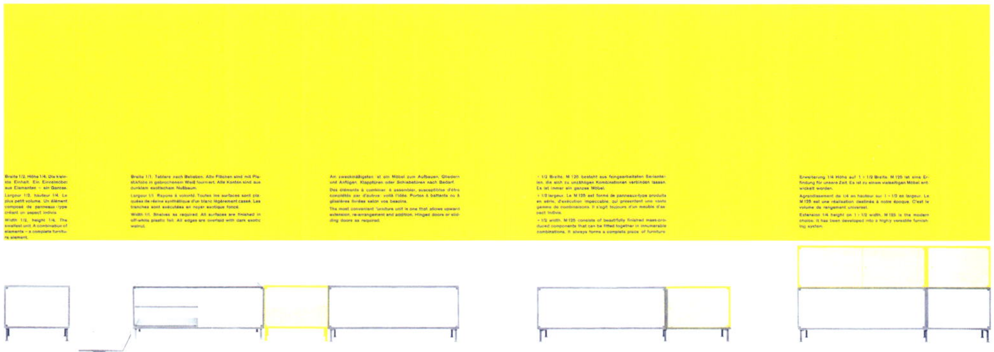

547

548

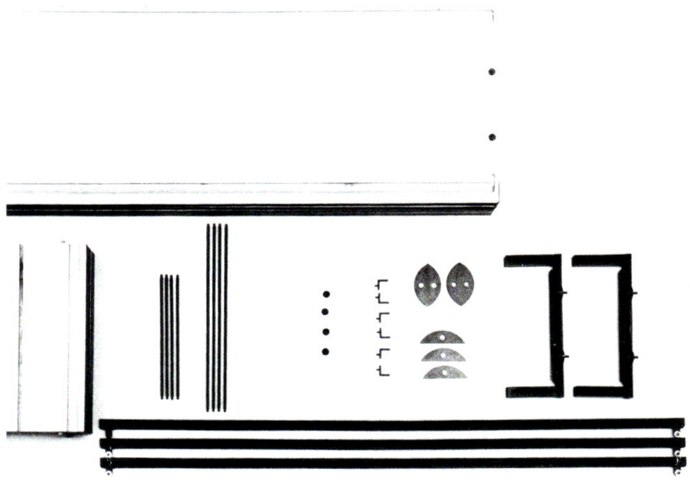

549

invention was not initiated until 1969, through an important commission by French bankers). In any case, the national scene offered several opportunities for sophisticated design ideas, in particular for modular, adjustable storage units and shelves. Wilhelm Kienzle's shelf system *Model 2722,* manufactured by Embru since 1931, is undoubtedly the prototype for this sort of concept. Steel sheets with folded edges are slotted into the shelves' milled notches at the side and behind, enabling the shelf to be extended in height or, using extra connecting sheets, in depth as well.[14] The majority of clever Swiss ideas were not designed for the office, however, but instead focused on living or studio spaces: for example, Hans Bellman's freestanding *Spannstützengestell* (Screw-Tension Shelves) of 1946, Jürg Bally's locking bar shelving *G 15* of 1950, Ulrich P. Wieser's pole-mounted *UPW* bookshelves of 1956–57, and Andreas Christen's 1964 *Lehni* aluminum shelves—just to name the bookshelves alone. The only designer to succeed in developing a modular furnishing system in the context of *Gute Form* that was suited to both living and office spaces was Dutch architect Hans Gugelot, who had studied at the ETH Zürich. He brought his system to market in 1950 in cooperation with the Wohnbedarf AG, Zurich/Basel. It was described as "a system with which ... single pieces of furniture as well as entire wall units can be built."[15] Standardized horizontal and vertical panels of veneered plywood are braced by means of metal rods. The elements, which come in three different depths, can be combined in countless ways. Revised and simplified in 1956 for the German manufacturer Wilhelm Bofinger, the furniture system *M 125,* named after its basic unit of measurement of 125 centimeters, is early evidence of the philosopher Heinrich Rickert's belief that we can only change "from theoretical chaos to theoretical cosmos by way of a system."[16]

Benedikt Rohner

The traditional Swiss sense of order has long generated revolutionary accomplishments in measuring time, in the organization of work, and in the design of everyday objects. In terms of offices, until very recently the greatest concern was probably how to deal with the constantly increasing flow of paperwork. The trained carpenter Benedikt Rohner (1924–2000), who studied at the Kunstgewerbeschule Zürich and also taught there from 1951 to 1954, dedicated half of his life to developing concepts and products for optimizing office organization. Right at the beginning of his career he focused on designing the perfect plan cabinet, a filing cabinet for oversize plans and documents and thus a piece he would use on a daily basis.[17] Created following an internship in Denmark around 1950, his solution, which was manufactured at first by his own workshop and from 1960 on by Ph. Oswald AG in Oberglatt, is today considered a design classic. The drawers set into a blockboard box emphasize the modularity of the composition with their protruding plywood bottom panels, which serve as handles, and their dark fronts. In 1965 Rohner incorporated this clearly articulated piece into a set of tables and storage units whose sharp-edged elements can be combined in a multitude of ways. The tabletops can be supported either by two easily removable legs or by a stack of drawers that has neither side nor back walls, making the perfect gliding of the drawers, despite their lack of traditional fittings, a design topic in its own right. The aesthetic of these masterful studies is reminiscent of the plain and frugal "reconstruction furniture" of the immediate postwar period; it perfectly reflected Rohner's modest personality.

The oak *Selecta* office scheme, manufactured and marketed by the company Witzig AG in Frauenfeld, is similarly cubic and angular. It was designed less for the so-called creatives than for the "normal office" that was also being targeted by the competition. A particularly remarkable feature of this scheme is the open niches furnished with plastic drawers that are integrated into the wooden unit next to the shelves. But these drawers are not only evidence of Rohner's attempt, beginning in 1970, to create a full range of "simple office furnishing systems"—the detailed workspace fittings sold by many but manufactured by few. Rohner also sought to give simple, everyday utensils a natural, "universal" form—from small desk organizers to custom-made archiving cabinets. For the company Biella-Neher in Biel, he expanded on the idea of modular plastic containers, which had been introduced by Charlotte Perriand in the form of plastic drawers between 1951 and 1954,[18] and ended up offering every conceivable variation. The stackable *Parat-Plast* letter tray from Biella-Neher is perhaps the best-known product in the entire colorful polystyrol line, which also included the desk

550

551

552

553

554

555

550 Benedikt Rohner, Plan cabinet, combined with Oswald workplace elements, Schreinerei Ph. Oswald, Oberglatt
551 Benedikt Rohner, Plan cabinet, 1950, core board, plywood, synthetic resin finish, produced by the designer, from 1960 by Schreinerei Ph. Oswald, Oberglatt. Photo: Alexander Troehler
552 Benedikt Rohner, Table and cabinet program, 1965, oak, metal cabinets, Schreinerei Ph. Oswald, Oberglatt. Precision craftsmanship
553 Benedikt Rohner, Selecta Norm, n.d., furniture program in oak with plastic drawers, Witzig AG, Zurich. "Particularly well-suited for the normal everyday office."
554 Benedikt Rohner, Styro drawer system, patented in 1981, snap-together elements of injection-molded plastic, Spichtig AG, Steinen. Universally combinable filing tray system, for use at the desk, as wall unit, or as mobile container.
555 Benedikt Rohner, Poly-Bac, n.d., filing system with letter boxes and drawers, colored polystyrol, Biella-Neher, Biel

556

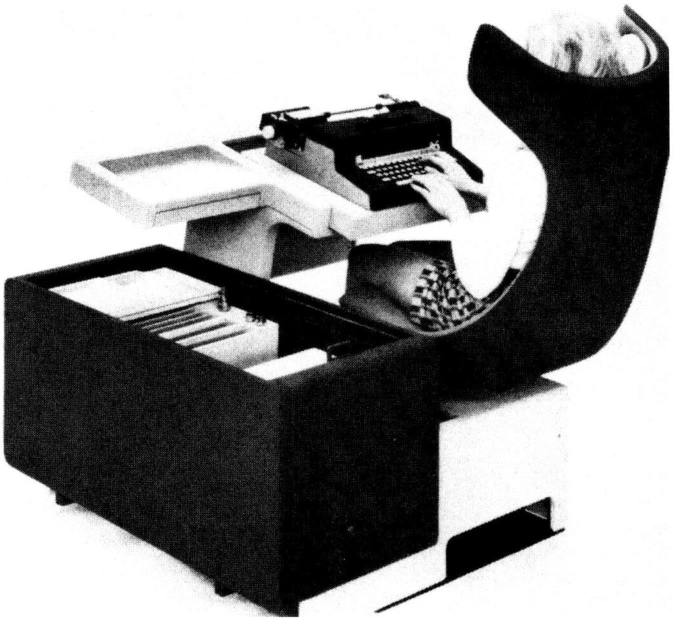

557

556 Albert Stoll jun., "Federdreh" (swivel chair with suspension), 1926, wood, metal, Stuhlfabrik Albert Stoll, Koblenz. The first swivel chair with a suspension system.
557 Odo Klose, Workspace for stenotypists, ca. 1970, Giroflex Entwicklungs-AG, Koblenz. All communication needs are integrated into a single unit with a swiveling shell seat.

calendar *Socal* as well as bookends, magazine boxes, floppy disk cases, filing containers with fold-back lids, drawers, wall-mounted trays, reusable boxes to facilitate the sending of packages, and much more. For the company Spichtig AG in Steinen, Rohner designed cubic shelving and drawer units made of plastic or recyclate, which could be assembled by customers into file trolleys or even entire walls of shelves, and reintroduced back into the production system after use.[19] These unpretentious utilitarian objects impress not only due to their straightforward functionality and the timelessness of their forms but also as a pioneering step toward the sustainable use of resources.

1 See Huber 1973, p. 921.
2 Ibid.
3 See Haller 1965.
4 For a more detailed account, see Arthur Rüegg, "Das definierte Nichts. Auf dem Weg zum perfekten Möbelsystem," in *Paul Haller. Leben und Werk,* ed. Georg Vrachliotis (planned for publication by gta Verlag in 2015).
5 Haller 1978, p. 1.
6 Ibid, p. 2.
7 The Swiss patent for a "Pipe connector for frame construction" was applied for on August 16, 1965, and registered on January 31, 1967, under the number 429 317. Named as the "inventors" are Paul Schärer jun. and Hans Steudler, both in Münsingen.
8 The tables are from the first USM Haller program; at the time, the same construction system was used for the frameworks of the filing and supply cabinets.
9 Prospectus *USM Bausysteme Haller,* vertical A4 format, including the supplement "USM Haller-Programm" and a pricelist, n.d. (1970s), Collection of Arthur Rüegg.
10 See Probst 1968.
11 See Rüegg 2006, pp. 24–25.
12 Prospectus *Knoll International Büromöbel. Serie Cube: Schreibtische, Winkelpulte, Daktylotische, Aktenschränke, Schrankwände, Stühle,* n.d. (ca. 1970), unpaginated.
13 See Rüegg, "Das definierte Nichts" (note 4). The metal parts in the Knoll programs were adapted in a collaboration between Koni Ochsner and Ulrich Schärer (according to Ulrich P. Wieser).
14 See Lepel/Spies 2001, pp. 138–41.
15 Gugelot 1987, p. 78.
16 Rickert 1920, ch. 9: "Der Kampf gegen das System," roughly cited in Wichmann 1987, p. 12; translated in Rüegg 2002, p. 164.
17 Rüegg 2002f.
18 See Barsac 2005, pp. 299, 303 (sheet metal drawers, 1949), and 330–31 (drawers of transparent and colored plastic, 1952).
19 See *styro, Katalog 1995/96* of Spichtig AG, Steinen, p. 2: "That's why we can offer you a guaranteed free return of used products ... out of the old styro products new, beautifully designed containers of the same high quality are made."

558

559

560

558 Erna Meyer, *Der neue Haushalt. Ein Wegweiser zu wirtschaftlicher Hausführung* (The New Household: A Guide to Efficient Household Management), Stuttgart, 1926. A bestseller reprinted several times.
559 Margarete Schütte-Lihotzky, Frankfurt Kitchen, 1926–27. Photo: Hermann Collischonn
560 Furnishings and fittings in a kitchen in the Neubühl housing development, Zürich-Wollishofen, 1928–32. ERFO kitchen sink with attached side sink and drain, warm and cold water tap with soap dish at the side: unconcealed, functional, and with pleasing forms. Photo shows state in 2000. Photo: Eckert & Gasser

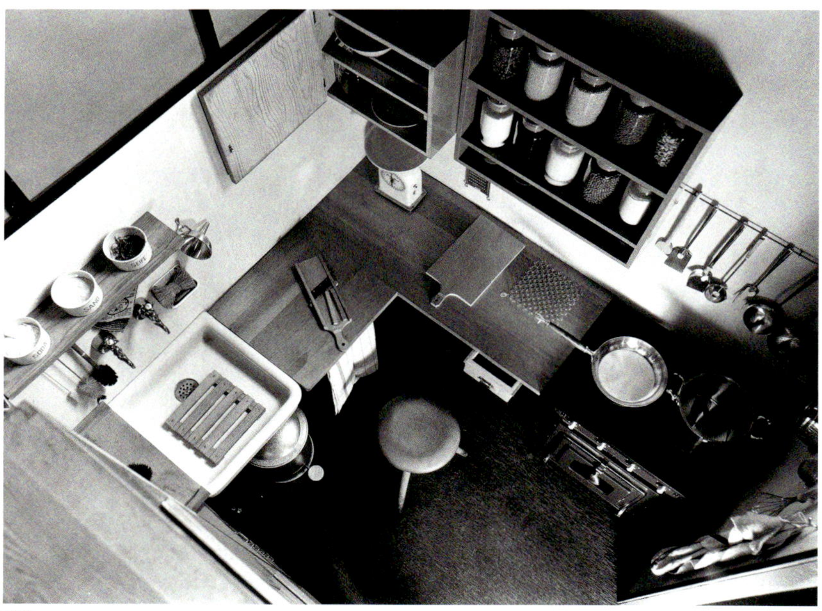

561

562

561 Rudolf Preiswerk, Model kitchen (3.5 m^2) in the exhibition *Die praktische Küche* (The Practical Kitchen), Gewerbemuseum Basel, 1930. Photo: Ochs-Walde. The new kitchen as a still life featuring individual well-designed objects.
562 Ernst Mumenthaler and Otto Meier, Eat-in kitchen in the model apartment in Block 6, Woba Eglisee housing development, Basel, 1930. Photo: Ochs-Walde. Gas stove from Solvis, boiler from Prometheus, furniture from the 3m program. The new kitchen with clearly defined areas and furnishings.

At the beginning of the twentieth century, the design of kitchens and their furnishings and fittings began to take on a whole new significance all across Europe. In Switzerland there were some truly remarkable examples of efforts to design a modern kitchen: the carefully devised kitchen spaces from the Neues Bauen (New Building) era and the groundbreaking invention of modular fitted kitchens toward mid-century, as well as design concepts utilizing high-quality materials and modern appliances that provided a stage for the act of cooking as a social and communicative event.

The "New" Kitchen

The attempts at streamlining household furnishings that developed in the United States in the mid-nineteenth century—influenced by the workflow studies of Frederick Winslow Taylor—began gaining ground in Europe at the latest during the era of Neues Bauen.[1] The economist Erna Meyer (1890–1975) from Munich was one of the most prominent and influential proponents of this more rational approach. She was often called upon as a consultant and was able to put her own ideas into practice with her *Münchner Küche* (Munich Kitchen, 1928). Her book *Der neue Haushalt* (The New Household), first published in 1926, quickly became a bestseller. It managed both to raise the ideological status of women's domestic work and to compel open-minded architects and designers to take an interest in kitchen design as an essential project of the modern era. Here, in particular, it seemed like the new maxims of design—functionality and simplicity—could be achieved almost effortlessly. From time to time, kitchen design was even apostrophized as being the true "bridgehead of modernism."

The Viennese architect Margarete Schütte-Lihotzky (1897–2000) made what is arguably the most consequential attempt at creating a "new kitchen" with her Frankfurt Kitchen, which was installed in a variety of different versions as part of the New Frankfurt housing project beginning in 1927. With its minimization of workspace, prudent use of handles, and its emphasis on ergonomics and hygiene, the "classic" version of this kitchen, made to fit a space measuring barely 6.5 square meters, was far superior to other designs of the time. Its specific details were equally innovative: Schütte-Lihotzky succeeded in combining the central elements of stove, sink, storage cabinets, and countertop workspace into a single, compact unit in a uniform color. In this sense, the Frankfurt Kitchen was not only an exemplary working or laboratory kitchen but in many ways could be considered a precursor to later, more coherently designed fitted kitchens.

Toward the end of the twenties, several exhibitions on the latest in interior design in Switzerland gave interested members of the public the opportunity to see a modern kitchen designed according to rational and functional criteria: in Zurich at the Rotach Houses (exhibition *Das Neue Heim II,* or The New Home II, at the Kunstgewerbemuseum) and at the Werkbund's Neubühl housing development (1928–32); and in Basel at the Eglisee residential development as part of the first Schweizerische Wohnungsausstellung (Woba, Swiss Housing Exhibition, 1930), or at the exhibition *Die praktische Küche* (The Practical Kitchen) at the Gewerbemuseum (1930). The kitchens presented on these occasions varied in concept and dimensions, covering a spectrum ranging from eat-in kitchens to minimalist workspaces. What they all had in common, however, was fastidious planning, execution, and furnishing, with special attention paid to ergonomics (through the use of work surfaces, appliances, and cabinets that were optimized and varied in height). When it came to choosing appliances and fittings, the architects usually selected series products by domestic or foreign manufacturers that were notable for their functional design. These "type objects," which were further optimized over the years, conveyed their functions plainly and openly; with time, they took on an almost iconic character in kitchen designs as the sculpturally formed objects of modern technological society. And yet, there was also an archaic aspect to the appliances and workspaces of these "new" kitchens: due to their functional straightforwardness and simplicity, they were evocative of ancient constants such as the hearth, watering hole, and preparation area.

Compared to the compactness of the Frankfurt Kitchen, these kitchens only marginally display the tendency toward coherent design. In this case, the architects were more the arrangers of individually built appliances, fixtures, and boxes. It was not until the thirties that there were increasing efforts in this country to create uniformly structured kitchens, designed to appear as though cast from a single mold. In most cases, these were custom interiors equipped to meet a sophisticated standard of living. And in general, only few attempts were made at the time to develop a multifunctional system made of standardized and industrially produced components. A glance across the border confirms this. Le Corbusier began to work on his modular "casiers" system around 1925, but it never went into series production. The first kitchen system based on a modular concept to make it past the prototype stage was likely *CUBEX,* designed by the Belgian architect Louis Herman De Koninck in 1930, which remained a successful top-selling model on into the fifties.

The Standard Kitchen from the Glarus Region

The first precursors of the modern fitted kitchen came onto the market in the late forties in the form of so-called sink combinations. A boiler, cupboards, and a refrigerator were combined with a sink underneath a continuous chrome steel countertop, later even including a stove, which had first simply been added on at the end. However, what was touted throughout the country as "every housewife's dream" was simply a static model, which, unlike later fitted kitchens with their interchangeable parts, could never be modified after installation.

Perhaps the most convincing and influential chapter in the history of modern Swiss kitchen design was subsequently written in the Glarus region. Those responsible, aside from the company Franke AG as supplier of the chrome steel countertops, were the company Therma AG based in Schwanden—the country's largest manufacturer of electrothermal household appliances in the mid-twentieth century—and its managing director and head designer from 1958 to 1968, Hans Hilfiker (1901–1993).[2] Under his leadership, Therma's product range—which had been rather mixed to that point—was consolidated and formally unified, and the manufacturing processes were streamlined. As a result, Therma products displayed an exceptional sense of unity in their design, construction, function, and graphical appearance in advertising.[3] Systems and modules were the main components that led in practice to the development of the fitted kitchen. They were based on uniform dimensions: the magic formula 55-60-90 (width-depth-height). Hilfiker consistently promoted these measurements, until by the mid-sixties they had become the industry-wide standard: from Therma-Norm, to Schweizer Küchennorm (Swiss kitchen standard), to the later SINK-Norm, and, since 1996, the Schweizer Mass-System (SMS, Swiss measurement system).

At the time when Hilfiker began his work with Therma, the idea of standardizing different kitchen elements, and the system of modular construction that followed from this idea, was already in the air, so to speak. The reasons were mainly economic. On the one hand, the development of such systems facilitated cooperation between planners, architects, and designers, as well as between appliance and kitchen manufacturers; and on the other, a binding set of dimensions was a precondition for achieving the goal of serial production and system-based kitchen construction. Furthermore, a modular system would, at least to some extent, fulfill the need for individuality and adaptability. Hilfiker calculated the dimensions 55, 60, and 90 centimeters based on typical kitchen work and statistics on the human body. The elements' vertical structure was also based on a modular system: their 12.7 centimeters was equal to a sixth of the height of the floor unit (after subtracting the base and countertop). Hilfiker always defended "his"—in many ways more sophisticated—standardization system when comparing its measurement of 55 centimeters to the ISO standard of 60 centimeters. Nowadays, both systems coexist on the Swiss market. The first step toward the envisaged production of entire kitchens was the development of a modular system. The range of stoves and refrigerators available had to be radically reduced for this purpose. The appliances were gradually harmonized in terms of their designs and dimensions. The system developed by Hilfiker was based on a standardized housing that could fulfill different functions, allowing for a number of different combinations while still ensuring a homogenous appearance. The new stove fronts

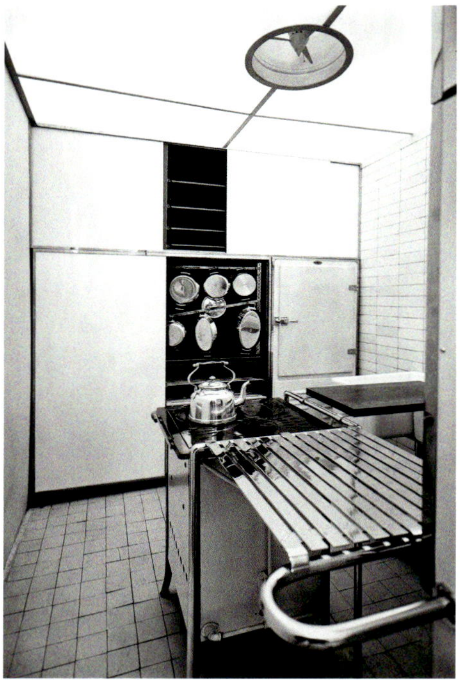

563

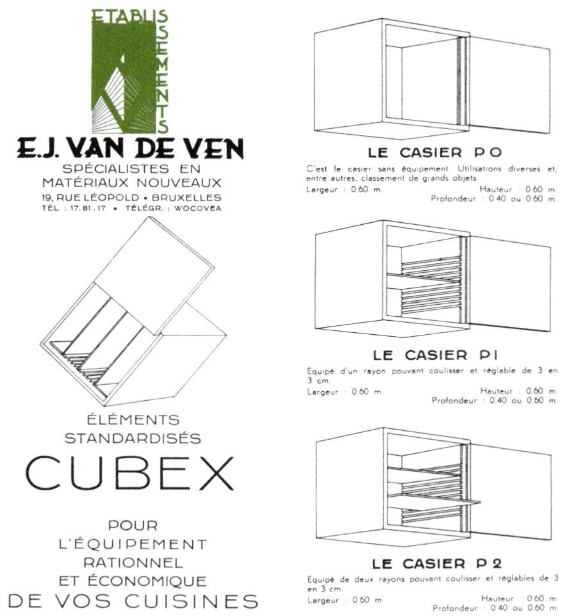

564

565

566

567

568

569

570

571

563 Le Corbusier, Pierre Jeanneret, Charlotte Perriand, Equipement intérieur d'une habitation, *Salon d'automne,* Paris, 1929. Photo: Jean Collas. Kitchen with gas stove and flip-down grates, built-in *casiers métalliques* with pan holders and refrigerator.
564 Louis Herman De Koninck, CUBEX Kitchen, 1930. Advertising leaflet for the company E. J. van de Ven, Brussels, ca. 1935. Perhaps the first modular kitchen system to go into series production.
565 "Therma modernizes the Swiss kitchen," Therma AG prospectus, 1954. Kitchen combination with stove, boiler, sink, cabinets, and pull-out counter.
566 The Therma Norm, later SINK Norm, and today SMS (Schweizer Mass-System), visualized in an advertisement, 1950s.
567 Therma one-front fitted kitchen, 1962. Photo: Friedrich Engesser. Designed to ensure unity of form, function, and construction.
568 Atelier Halpern, Therma AG prospectus, 1963 (photos: Friedrich Engesser). "Norm – System – Module: Elements of a Therma Built-In Kitchen."
569 Atelier Halpern, Campaign introducing the new Therma stove system in advance of the Basel Trade Fair (Muba), 1961. Photo: Rolf Schroeter. The Therma beta 40 model almost fully unpacked.
570 Carlo L. Vivarelli, New Therma logo, 1958
571 Freestanding Therma cooking table, 1961. Prototype. Photo: Friedrich Engesser. Multifunctional stove unit to set up anywhere in the room, with flip-up burners.

were separated into three zones of varying heights, based on the vertical module of 12.7 centimeters. This visualization of a systematic concept was "the expression of a design attitude that no longer prioritized the single piece, but the interplay of single parts within an entire complex."[4] In line with this systematic concept, many old terms such as "household range" or "work oven" were replaced by a two-part type designation: "alpha," "beta," "gamma," and "delta" denoted the comfort level, while the numbers 40 or 32 indicated the oven's interior width.

In the early sixties, Therma began producing entire kitchens. Starting in 1962, a newly built factory allowed the individual components of the kitchens and appliances to be produced on a large scale, stored in a warehouse, and fitted to become finished products. By the mid-seventies, however, Therma had stopped making full kitchens, not least due to increasing competition from other kitchen manufacturers (Therma both supplied the appliances and made the kitchen furniture). The Therma fitted kitchens were composed of a central worktable with base cabinets, lateral add-on units, and wall cabinets. These could be arranged in a variety of different combinations varying in size and complexity: kitchens with one, two, or more fronts, forming an L or U shape. The cabinets and drawers were made of enameled steel, as were the appliances—neutral white was meant to give the Therma kitchen a timelessly modern look. This did not come cheap, however, because the manufacturer relied on solid, high-quality workmanship.

In retrospect, Hilfiker's work for Therma AG is a unique chapter in Swiss design and industrial history. The products designed by him or under his guidance still captivate us today with their functionality, clean lines, and careful choice of materials. There are good reasons why practically all of the appliances and kitchens Therma produced during his tenure were awarded the "Die gute Form" seal.

Back to the Kitchen as Living Space

The years of the construction boom helped seal the success of the fitted kitchen in this country as elsewhere. Therma had set new standards in terms of the systematic approach, design, and execution, which, however, save for a few exceptions, reached only a small number of kitchen makers. Moreover, the fitted kitchen's compact size led to a further reduction in the size of kitchen spaces, often turning them into sober, unimaginative, and uninspiring workrooms. The development heralded by the Frankfurt Kitchen had made a complete about-face.

The increasing sense of unease at the negative by-products of "functional" modern design and the openness toward new lifestyles brought forth new ideas for interior design for the home. Visionary kitchen designs now emerged from the concept of the "integrated" interior. The entire living space was seen as a unified whole, with the kitchen unit freely arranged somewhere within it. In reality, however, such visions gave rise to an increasing convergence of living and work space, ultimately resulting in the now extremely common combined kitchen and living room. The ancient eat-in kitchen as the real center of human habitation thus returns to form part of the modern living environment.

Today, the design and furnishing of kitchens has become something of a first-world problem. The primary needs have long been met, the areas have been arranged in an ergonomically ideal fashion, and the appliances and gadgets have been standardized and can be replaced with others at any time. Emotion, style, and taste set the tone; when planning kitchens, costly materials and surface finishes and high-tech appliances come into play. A kitchen's layout and furnishings depend on the spatial and financial possibilities, the nature and customs of its use, and, finally, the importance its owner attaches to it: it is a heterogeneous construct of functional requirements and aesthetic lifestyle demands.

At the same time, however, critical contemporaries evidently still appreciate simple kitchen appliances and utensils that are flawless in both function and design. This is the only way to explain why the ingenious simplicity of a stove-top moka pot, for example, is able to hold its own in the face of capsule-based coffee machines and automatic latte macchiato makers, or why the simple toaster with its gull-wing mechanism—a classic Jura design—has been on the market since 1931. Several varieties of Therma, Prometheus, Le Rêve, and Bono stoves still have a place in our homes. Now that they have done their duty for such a long time, and have perhaps already reached vintage status, any shortcomings in terms of their ergonomics or energy efficiency are willingly accepted.

572

573

574

575

572 Leonardo Bezzola, *Schweinskopf* (Pigs's Head), in Bernhard Luginbühl's kitchen, Mötschwil, 1969. A renewed interest in cooking at home.
573 Alfredo Häberli, Mesa, 2007–08, Schiffini,
574 Ceparana (IT). Photo 573: Marcus Gaab. Photo 574: Luca Fregoso
The kitchen as the "soul of the house," with a cooking island staged as workplace.
575 Bonnard Woeffray, Residence, Collombey, 2005–07. Photo: Hannes Henz. Cooking in a sublime space.

576 Migros mobile shop, out and about in Canton Zurich, ca. 1950
577 Glashütte Bülach prospectus, 1926. Preserving in the Bülach bottle
578 Hero advertising leaflet, Lenzburg, ca. 1950. "Young, fresh vegetables" from the can.
579 Rolf Bangerter, Sibir refrigerators. Poster, 1955

576

577

578

579

The increasing fusion of kitchen and living space is still a trend nowadays, as well as the treatment of the kitchen as a showpiece. This can certainly be attributed to the perception of cooking and eating as culturally significant events. By now it has become second nature for us to cook with fresh, organic market produce, meat from humanely reared animals, and sustainably farmed fish, or even to make our own jam. The days of canned goods, convenience products, and frozen foods are widely considered to be over once and for all. Following this trend, architects, interior designers, and kitchen planners continue to offer convincing solutions for the convergence of cooking and living space. The kitchen island, by now very well established, has taken on special importance here, used both as a room divider between kitchen and living room, and as a focal point that encourages communication. This is indeed reminiscent of the freestanding Therma cooking table, which was built as an exhibition prototype in 1961 but never went into series production.

The current diversity in random kitchen and appliance designs, some of which try very hard to be original, makes the more sophisticated concepts stand out all the more. One of these was devised by Alfredo Häberli. With his *Mesa* program, designed for the Italian kitchen maker Schiffini, he intended to create an inspiring kitchen / living space that would be the true "soul of the house." At the center is a long table complete with benches that slide out from underneath and at its head a fireplace with a furnace-like exhaust hood. The cooking is done either on a countertop stove or a compact, freestanding butcher block, which has even been given the form of a workbench on one side. Functionality, sensuality, and cohesive design fuse these individual elements into an ensemble. Perhaps it could even be said that these spatially as well as formally precisely defined areas of function are reminiscent of the kitchen furnishings designed by the protagonists of Neues Bauen in the twenties.

1 See, for example, Corrodi 2006.
2 See Sonderegger 1995 and Sonderegger 2006.
3 See Neuburg 1963.
4 Sonderegger 2006, p. 104.

580

581

582

580 Claude Dupraz, Monorail, here in front of the US pavilion designed by R. Buckminster Fuller at Expo 1967 in Montreal. Dupraz also devised the structural concept for wheel control and power transmission in the monorail.
581 Claude Dupraz, Design for a recreational vehicle on a Citroën 2CV chassis, 1967, Namco, Thessaloniki. The vehicle body was developed based on simple sheet metal molding (bending). It never went into production.
582 Hans Hilfiker, Motor vehicle for road or rail, designed for servicing and maintaining the overhead lines of the Swiss Federal Railways (SBB), 1952
583 Uli Witzig, Standard Coach IV, SBB First Class, 1980, Schweizerische Wagons- und Aufzügefabrik AG, Schlieren-Zürich
584 Uli Witzig, Tram 2000 for Verkehrsbetriebe Zürich (VBZ, Zurich Public Transport), Schweizerische Wagons- und Aufzügefabrik AG, Schlieren-Zürich

583

584

In retrospect, the 1960s appear like a golden age of economic prosperity throughout the Western world, with Switzerland an especially privileged hotbed for the development and manufacture of quality products. At the Expo 64 Swiss National Exhibition in Lausanne, the country showcased itself as a consumer paradise and production El Dorado. Traffic planners visualized the dream of the *car-friendly city* with convenient inner-city express routes. Railroad crossings disappeared as the first stretches of highway thrilled one and all. However, all the enthusiasm was dimmed somewhat by the specter of pollution, with movies showing exhaust fumes wafting through the air and soap scum on fertilizer-contaminated waters.

Despite the economic boom, *industrial design* was not yet a common catchword, only gradually being appreciated as a subject in its own right, and it was not until the late seventies did it become firmly established as a concept.[1] Nearly all the famous and successful products of Swiss provenance until the late sixties was design without *design;* in other words, they were developed without the participation of trained design professionals: typewriters and sewing machines, turntables, radios and tape recorders, refrigerators and stoves, movie and photo cameras, household appliances and machine tools. Switzerland only belatedly introduced industrial design as a course of study, later even than Germany, and did so at the Hochschule für Gestaltung Ulm in 1955. In the United States, the pioneers of the profession who began coming to the fore around 1930 were originally illustrators or stage designers for the theater.[2] Almost all the Italian designers, from Albini to Zanuso, were originally architects. The Frenchman Jean Prouvé started his career as an ornamental blacksmith and metalworker. The beginning of something new usually comes *from outside,* as is confirmed once again in this case. In Switzerland as well, the design pioneers did not train as designers but evolved in that direction: Wilhelm Kienzle (1886–1958) started as an interior and poster designer, Hans Hilfiker (1901–1993) as an electrical engineer, Willy Guhl (1915–2004) as a cabinetmaker, and Max Bill (1908–1994) as a silversmith. In 1960, the famous Swiss models and makes were still factory-designed—outstanding achievements by mostly unknown design autodidacts: Elna, Hermes, Thorens, Bolex, Sibir, Bernina, Jura, Therma, etc.[3] In lucky cases—and there were quite a few—the technicians also had a keen sense of form.[4]

Two Meanings of *Design:* Object and Process

In this text, the word "design" is necessarily used to refer to two different things. The first is the product as a utilitarian object: consumer goods such as typewriters, refrigerators, or turntables, capital goods such as machine tools, transport devices, or agricultural machinery. Even if they are not *designed,* they still exhibit design: they have a functionality, a materiality, and a form, however satisfying those may be. And then there is the second level: the deliberately chosen form for a device and its controls. In a gradual process, professional designers have evolved beyond so-called *form-finding* to achieve recognition as professionals in construction, handling, production, operation and service concepts, and packaging and transport logistics.[5] Before that could happen, though, they had to do battle until the late seventies with the viewpoint that their work was merely putting the engineers out of business and making the final product needlessly expensive.[6] The public read about the worldwide reputation of these products in advertisements, saw them in shop windows, purchased them, and used them, but even when there was a designer behind a product no one even knew they existed. Industrial design long remained—and is in some cases still today—a hidden dimension in and behind the objects.

Even as late as 1960, the design professionals in Switzerland still came to the field from other careers. Francesco Milani (b. 1939) was a graphic designer when he founded his studio in Giubiasco in 1963. Hansruedi Fellmann (b. 1940) was a trained technical draftsman. The young Andreas Christen (1936–2006) was, from 1956 to 1959, effectively the sole private student of Hans Fischli in the "pilot class" at the Kunstgewerbeschule Zürich.[7] The first specific institutional course of training in industrial design began in Switzerland in 1965 with a multiyear pilot phase for product design at the Kunstgewerbeschule Zürich (the regular course of studies was introduced in 1971). Almost simultaneously, the same development took place at the École d'arts appliqués (today ECAL) in Lausanne, where the interior designer Claude Dupraz (b. 1933) established a study program in 1967. Willy Guhl, the design pioneer in Zurich, had taken an interest in industrial jobs at an early stage,[8] and his student Ludwig Walser (b. 1936) already knew in 1962, when he earned his diploma as an interior design draftsman, that he wanted to do technical design—he, too, was an isolated case.[9] The two areas branched off from one another and at time converged again; the ground was still for the most part unprepared.

Some designers of the younger generation of trailblazers studied at the Hochschule für Gestaltung Ulm and only came to Switzerland by way of a detour through Italy (Milan) and then gradually established themselves as designers here.[10]

Resistance to "Unnecessary" Design

It certainly can't be claimed that the Swiss consumer and capital goods industries welcomed the industrial designers with open arms. Precisely because business was already doing so well, manufacturers saw no reason to pay special attention to design. They suffered for a long time under the delusion that the quality of their products was virtually an unassailable natural law. In 1948, Max Bill's acknowledgement that the fundamentally conservative approach of some manufacturers was indeed a problem was still accompanied by confidence that things were changing: "The 'industrial designer' will arrive here too: just as the commercial graphic designer evolved from the painter who engaged with graphics into the independent profession of today."[11] But this development was much slower than expected. It was a long time before industrial design was accepted as a criterion for technical manufacturing considerations, for enhanced user-friendliness, and for convincing form. And just as manufacturers were finally becoming aware of its importance, a series of prominent companies began to experience a decline in business. The era of precision engineering was segueing into the electronic age, which then would eventually itself be replaced by digitization. For Swiss industry, with its self-assured and elitist precision mechanics culture, these two technological revolutions represented two massive cliffs to be scaled. Scarcely had the first been surmounted—and not everyone reached the top—before businesses found themselves in the vortex of Far Eastern competition, after which they then faced the second, even greater challenge posed by the digital age. The clock ran out for many well-known companies before they had even embarked on a first rendezvous with professional designers. Some firms, such as Kudelski S.A. in Lausanne, were able to save themselves by opening up new business fields in time to make the switch.

Among the first design studios to be founded in Switzerland were Devico Design AG (from Design Victor Cohen) in Zürich-Gockhausen in 1962, the Studio Francesco Milani in Giubiasco in 1963, and that same year DOG Cie. in Geneva (Claude Dupraz, André Oberson, Jacques Gallay). They did work for the western (French-speaking) Swiss precision engineering industry (on projects including the first electric typewriter for Hermes-Paillard), but also for Sigg in Frauenfeld, Lenco in Burgdorf, and the galvanizing shop Verzinkerei Zug, and they were furthermore entrusted with international tasks such as the establishment of a design school in Karachi (Pakistan) and the scenographic design of various national pavilions at Expo 67 in Montreal.[12] Devico Design was a subsidiary of Advico (Advertising Victor Cohen) and had divisions for industrial design and packaging design. Andreas Christen already opened his one-man product design studio in 1959. He made a name for himself above all with his furniture in metal (Lehni, Dübendorf) and plastic (bed and cabinet for H. P. Spengler, Rümlang), but he also designed a mailbox system (Ernst Schweizer Metallbau, Hedingen) and a dictation machine.[13] In 1967, Ernest Muchenberger (b. 1934) and Walter Eichenberger (b. 1936) founded M&E in Zug, later in Cham, and in 1968 Ludwig Walser opened his studio in Zurich (later Baden-Dättwil); the list could go on and on.

Walter Eichenberger, Claude Dupraz, Hansruedi Fellmann, and Ludwig Walser unanimously recall that they had to toil for many years until external designers began to be consulted on a regular basis. The in-house design engineers often saw them as "artists," not taking them seriously, and entrepreneurs often viewed them as merely an additional cost driver. Exceptions were rare. Ludwig Walser's first assignment, in 1968, was the design of a Precisa

585

586

587

588

589

590

585 Andreas Christen, Ultravox dictating machine, Type U-401, ca. 1967, Dictaphone International AG, Killwangen
Museum für Gestaltung

586 Andreas Christen, Modular building system for bathrooms, ca. 1972, MS Modules Sanitaires S.A., Saint-Sulpice. Prospectus

587 Ludwig Walser, Ibico-0842 pocket calculator, 1969, Ibico AG, Zurich

588 Ludwig Walser, Precisa 322 electronic calculating machine, 1971, Precisa AG, Zurich. Prospectus

589 Ludwig Walser, Aebi TP 50 transporter, 1969–70, Aebi & Co AG, Burgdorf

590 Willy Guhl, Aebi AM 60 combi motor mower, 1980, Aebi & Co AG Burgdorf. Aebi was the first company in Switzerland to hire external designers like Willy Guhl and Ludwig Walser for the development of new products.

591 Francesco Milani, System 3000 photo lab, model, 1968, Gretag AG, Regensdorf

592 Ludwig Walser, OMA 3 tape recorder, model, ca. 1970, Sondor Willy Hungerbühler AG, Zollikon

593 Werner Zemp, Playground equipment, study sketch, 1967, Burri Public Elements, Glattbrugg. The jungle gym made of fiberglass-reinforced polyester is based on two combinable elements.

594 Werner Zemp, Garbage Shark, study sketch, 2002, Brüco Swiss AG, Rümlang. A wide variety of models of the public waste receptacle have since been developed.

591

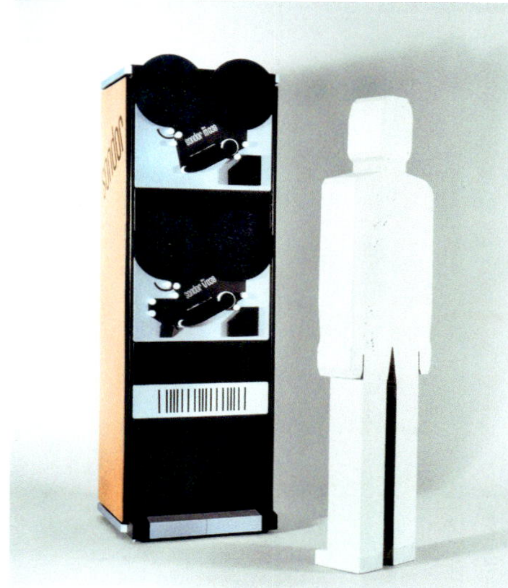

592

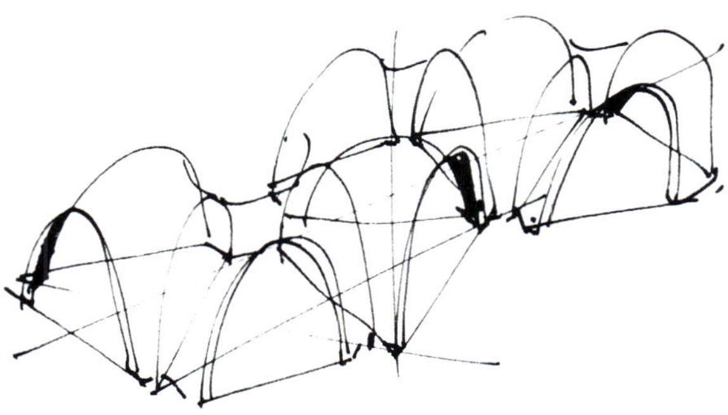

593

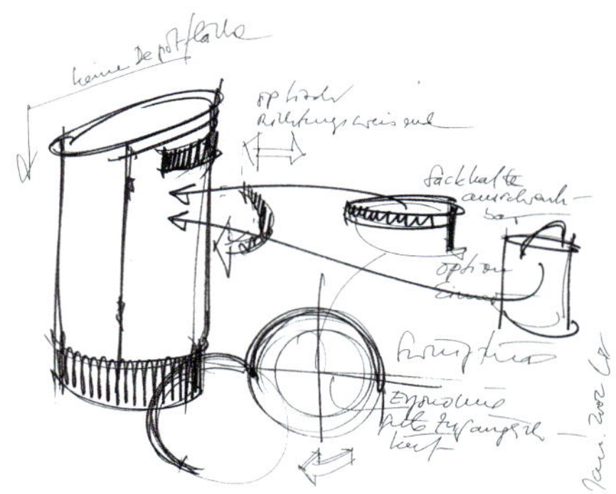

594

electronic calculator, and the firm L. Kellenberger in St. Gallen already enlisted the young studio M&E in 1970 to redesign a precision cylindrical grinding machine—the beginning of a long collaboration.[14]

Professionalized Design: Ease of Operation and Enhanced Quality

Where, in rare cases, the owners of a company were also designers, there was of course no such resistance. As a prime example of the modern attitude to design as a way of thinking, as a production method, and as a formal language, we can cite the lamps made by the Baltensweiler company in Ebikon. In the very first model, the *Type 600* floor lamp (1951), Rico (1920–1987) and Rosmarie (b. 1927) Baltensweiler used unconventional semifinished products such as pipe parts and solid bars (as counterweights) to lend the lamps a fresh, discreetly industrial spirit. In 1966, they took the same approach in their model *Guggerli,* this time using large-diameter PVC tubes in bright industrial hues to support movable light globes.[15]

Another example of a company that integrated design expertise into its workflow at a relatively early date is Studer Revox in Regensdorf.[16] The first DynaVox tape recorders by Willi Studer, a highly talented self-taught electrical engineer, still had in 1949 a brown-painted body with a handmade walnut housing. Even when the company had already attained international recognition, the subsequent Revox models were still designed nonprofessionally in-house, sporting until 1963 the reseda green body typical of those years, in a case with rounded corners, and thereafter a gray-speckled body in an angular housing.[17] The design certainly expressed technical quality, but also conventionality. Then, in 1965, the young, US-trained German designer Manfred Meinzer arrived in Regensdorf from Berlin, introduced himself to Studer, convinced him of the importance of design, and was hired as in-house designer.[18] For the next twenty-five years, Studer gave him his full support. In 1967, his design of the *Revox A77* hi-fi system as a modular and visually uniform set of tape recorder, amplifier, and tuner represented a quantum leap. The system was a worldwide bestseller and an early Swiss example of how high quality could be expressed through design without the need for bluffing.

A characteristic instance of the contribution made to a product by an external design studio, something new at the time, was the contract awarded circa 1963 to Devico Design by the company A. Sutter in Münchwilen to redesign the *Taski* floor polisher. In the old model, the multiple controls and handle adjustments were found in different places on the machine and were thus impractical to operate. (And the machine got its coat of paint from the village painter!) This was a typical engineer's product: designed for optimal *manufacture* (cast parts, metal machining process) but with little consideration as to actual consumer convenience.[19] Hansruedi Fellmann, who worked for Devico Design at the time, put the focus on the interaction between user and device, developing for the new model an operating head with centralized functions, in other words, conveniently placed ergonomic handles and body parts made of plastic.[20] Behind the redesign was the idea of *human engineering*—a new approach from the United States. Etienne Grandjean, professor of occupational physiology at the ETH Zürich, was a key figure in spreading this concept in Switzerland.[21] Not only operation but also production could be simplified through this approach, reducing costs in the process.

The Designer: Seducer or Midwife?

Producers were only slowly becoming aware of these new developments. The independently working designers had to overcome tough pushback, and often enough did not succeed. "The engineers knew how something was always done, and as a designer, I saw how it could be done," notes Ludwig Walser.[22] This statement is strikingly similar to a sentence in Sigfried Giedion's incomparable book *Mechanization Takes Command* (1948), but from the opposite vantage point. Giedion writes, "On the one hand the businessman trusts the engineer who knows how a thing should be built. On the other, he lends a willing ear to the industrial designer."[23] Giedion takes the side of the engineer here, painting the designer as an almost diabolical operator. Some Swiss manufacturers shared this critical mindset regarding styling, an attitude that generally had a negative impact on the work of the designers. They were wrong to hang back, because the fact is that the traditional certainties they relied on were undergoing dynamic changes, with new possibilities

595

596

597

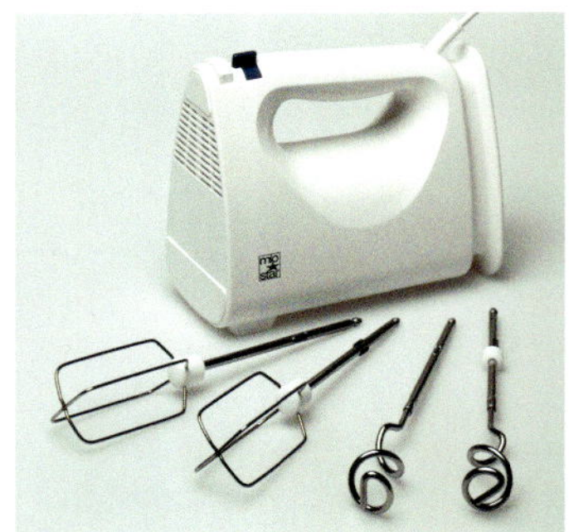

598

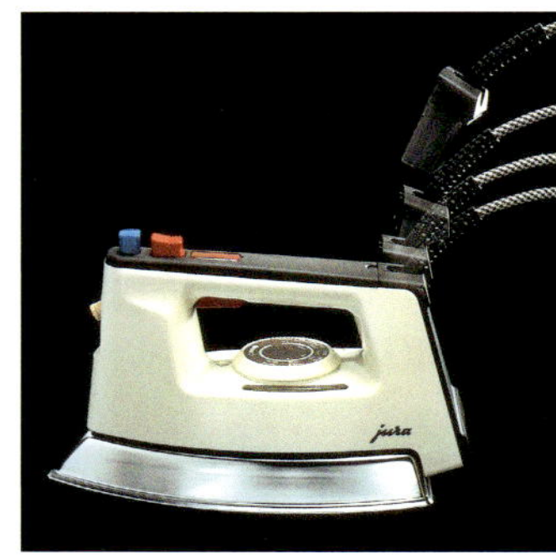

599

600

601

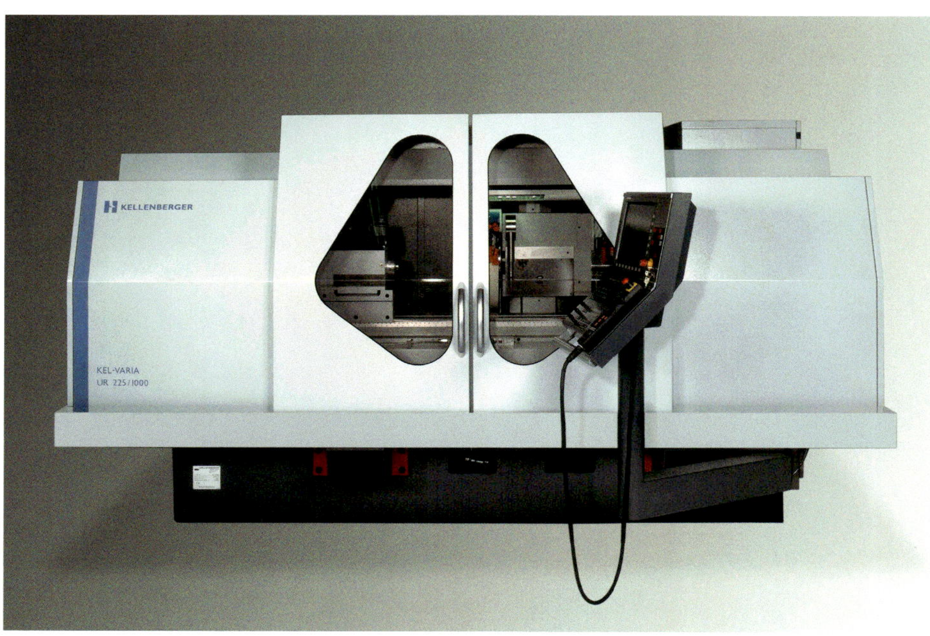

602

595 Manfred Meinzer, Revox A77 tape recorder, produced 1967–77, Studer Revox, Regensdorf
Museum für Gestaltung

596 Manfred Meinzer, Revox B790 record player, produced 1977–82, Studer Revox, Regensdorf. World's first record player with tangential tonearm.

597 Xaver Bürgi, Swissair stacking cups, 1969, Riwisa AG, Hägglingen
Museum für Gestaltung

598 M&E Design (Ernst Muchenberger, Walter Eichenberger), Mio-Star hand mixer, from 1978, Spemot AG, Dulliken. Produced for and distributed by Migros.
Museum für Gestaltung

599 M&E Design (Ernst Muchenberger, Walter Eichenberger), Steam iron, 1977, Jura AG, Niederbuchsiten

600 Hansruedi Fellmann c/o Devico Design, Atlanta telephone, before 1977, Gfeller Communications, Bümpliz. With redial, used in Singapore, Saudi Arabia, and the UK, among others. The Swiss answer to Ericsson's Ericofon.
Museum für Gestaltung

601 Fellmann Design (Hansruedi Fellmann, Christophe Apothéloz), Tritel-Sils telephone, from the Tritel program "1985," 1985

602 M&E Design (Ernst Muchenberger, Walter Eichenberger), KEL-VARIA precision cylindrical grinding machine, 1995, Kellenberger & Co., St. Gallen. The characteristic shape of the viewing window has functional reasons and is also a USP.

opening up on many fronts. This gave designers a chance to show what they could do, and sometimes they were able to seize it (and just as often they were not). Unlike the in-house specialists, their position outside of all conventions brought them into contact with new materials and production methods that gave them novel design options.[24] This proved to be a growing advantage, as manufacturers only gradually realized. Werner Zemp's summary of his activity as a designer makes the situation clear: "You always have to deal with top management."[25]

It was not until the late seventies that the M&E studio obtained a contract to redesign some of the models of the Mio-Star brand (Migros) in order to bring a consistent line into the very heterogeneous-looking range: hair dryer, hand mixer, coffee grinder, and so on. However, the contract and the fee surprisingly did not come from Migros; the retail giant only gave its consent to the products being given a visually coherent style. The supplier of the products was the electrical appliance factory Spemot in Dulliken, whose chief designer and engineering director, Jörg Pfändler, had up to then also been in charge of formal design. The boost in sales brought about by the new product line profited all those involved: the marketer, the manufacturer, the supplier, and also finally the designers, who were paid by the hour.[26]

The fact that the importance of professional design finally did gain recognition in Switzerland is documented by the development of the first plastic ski boot by the Raichle company. Raichle approached Devico Design in 1967 with a prototype in hand. It consisted of two longitudinal boot halves made of hand-laminated fiberglass polyester that were connected with hinges and clasps and enclosed a separate liner. Devico Design was commissioned to refine the formal and color concept and to explore options for cost-effective manufacture. The model went into production in 1968 under the name *FibreJet* and was successfully exported to the US as well as being adopted in alpine ski racing.[27]

Swiss companies at last generally acknowledged the importance of the designer in the course of the seventies. Uli Witzig even received a direct commission from the railcar manufacturer Schweizerische Wagons- und Aufzügefabrik in 1974 to design the *Tram 2000* for Zurich, and somewhat later was also enlisted by Baselland Transport AG.[28] His *SBB Einheitswagen IV* (standard railcar for the Swiss Railways, with Ueli Thalmann and Uli Huber) is likewise a durable design that probably only few people are consciously aware of but which is used daily by thousands of passengers. Those designers who manage to procure major contracts for the capital goods industry find themselves working in a world far away from the glamour of department stores, but they have a different kind of everyday presence. Making a milling or grinding machine vibration-free and quieter, and using safety devices behind viewing windows to make it safer as well as easier to clean, maintain, and repair, or giving an elevator cabin more congenial lighting, was and still is extremely satisfying work, even if the public automatically thinks more of furniture or cars when they hear the word "design." Ludwig Walser raves about his industrial valves for Georg Fischer: the names of the various types—ball valves, diaphragm valves, gate valves, and flaps—are like music to his ears and to all those who do not expect design to always pack an emotional charge. Such products are in use worldwide, but the less you hear about them, the better. This is design far away from the overkill of the stars with their fifteen minutes of fame.

1 The Swiss magazine *Werk* speaks until 1945 of "arts & crafts," from 1946 to 1966 of "form-finding" and "industrial form-finding," and only in 1967 for the first time of "design." The German magazine *Form* addressed in the sixties Max Bill's annoyance with the catchword "industrial form."
2 The pioneer generation in the US after 1930 included Norman Bel Geddes, Raymond Loewy, Henry Dreyfuss, Walter D. Teague, Donald Deskey, and Harold van Doren.
3 See *Unbekannt – Vertraut* 1987.
4 See Hans Hilfiker, "Wie ein Ingenieur die Schönheit sucht," address at the year-end ceremony of the Kunstgewerbeschule Zürich, December 20, 1957.
5 Ludwig Walser says that for this reason he did not include the word "designer" on his letterhead in the early years of his career but instead called himself a "Gestalter für die Industrie" (creator for industry). Interview with the author, February 11, 2014.
6 Similar statements were made by Walter Eichenberger, Hansruedi Fellmann, Ludwig Walser, and Werner Zemp, February 2014.
7 See Strauss 2008, p. 123.
8 For example, in his collaborations with Eternit, with the Aebi agricultural machinery company, and others.
9 Walser then relocated to the US to work for Latham, Tyler, Jensen (LTJ) in Chicago, and then joined Eliot Noyes in New Canaan, Connecticut; he returned to Switzerland in 1968.
10 Walter Eichenberger (from Rodolfo Bonetto), Werner Zemp (from La Rinascente). Franco Clivio returned directly to Switzerland from Ulm, as did Uli Witzig, who experienced only the final phase of the Hochschule für Gestaltung Ulm.
11 Bill (1948) 2010, p. 36.
12 DOG's work also included the interiors of the excursion boats *Général Guisan* and *Henri Dunant* on Lake Geneva, a car design on the technical basis of the *Citroën 2CV* for Greece (not realized), as well as the technically and aesthetically impressive monorail at Expo 67 in Montreal, his work on the latter also including the technical design concept. Claude Dupraz, interview with the author, Geneva, March 24, 2014.
13 Christen also made waves in 1964 with the modular, reusable Swissair trade fair booth.
14 See *Easy! Easy?* 2011, pp. 46–47.
15 See *Baltensweiler Licht und Leuchten,* company brochure, Ebikon 2012, pp. 7–9.
16 Studer is the name of the company division for studio facilities and professional equipment, while Revox is the home appliance brand.
17 See Holenstein 1998, pp. 156–59, 306–7.
18 Manfred Meinzer (b. 1943) graduated from the Art Center College of Design, Los Angeles. See "Interview mit Manfred Meinzer," *Studer und Revox,* n.d., http://www.studerundrevox.de/index.php?option=com_content&view=article&id=131&itemid=199 (accessed February 15, 2014).
19 This manufacturing-centered orientation so typical of engineers was also the basis for the development of the *Swatch* in 1982. The watch only became popular once the technical concept was overlaid with the idea of different stylistic lines for various target groups. See Lichtenstein 2014.
20 Hansruedi Fellmann, interview with the author, Gordevio, February 19, 2014.
21 Ludwig Walser, interview with the author, Zurich, February 11, 2014.
22 Fellmann, interview, February 19, 1014.
23 Giedion 1948, p. 658.
24 Lichtenstein 2014.
25 Werner Zemp, interview with the author, Amden, February 13, 2014. See also *Werner Zemp* 2010.
26 Walter Eichenberger, interview with the author, Knonau, February 6, 2014.
27 Hansruedi Fellmann, letter to the author, dated February 22, 2014.
28 Information provided by Uli Witzig, March 11, 2014.

603

The artist Rosmarie and the architect and designer Klaus Vogt with their children in Scherz, ca. 1973. In 1970 Vogt built a "comfortably livable building shell" of sand-lime brick as a stage for a forward-looking experiment in relaxed working and living in community with the neighboring families. The multicolored "psychedelic" rug was made by Rosmarie, while Klaus, a trained boat builder, designed and built the furniture with his own hands.

Photo: Elsbeth Leisinger

1970 — Pop

The modernist formal repertoire exhausts itself in the boom years. The informal lifestyle of the flower children inspires a new kind of interior decorating based on the dissolution of conventional production and behavior patterns. Inflatable seating, beanbag chairs, and lounge landscapes reflect the trend of “living on the floor.” Equally nonconformist are the instructions supplied with the do-it-yourself furnishings. Collapsible “furniture to go” experiences a resurrection. Plastics technology and Pop Art leave their mark on the world of objects. In the second half of the decade, with some delay, an alternative response to modernist functionalism appears on the scene. On various scales, in different contexts, and using a wide range of materials, designers challenge with wit and irony the social and functional meanings of traditional forms. The resulting objects—in contrast to the “take-away” products—call for consummate traditional craftsmanship.

604

Susi and Ueli Berger with their children in their living room in Bern, 1968. The two artist-designers lived in a ground-floor apartment in the Weissenbühl district from 1963 to 1970, which they furnished in their very own "nonconformist" style. In the living room, three of their mod Soft Chairs of brightly colored vinyl are pushed together to form a sofa. Hanging on the wall is a homemade cabinet and above it is Ueli Berger's object *Softy* of 1967.

Photo: Walter Studer

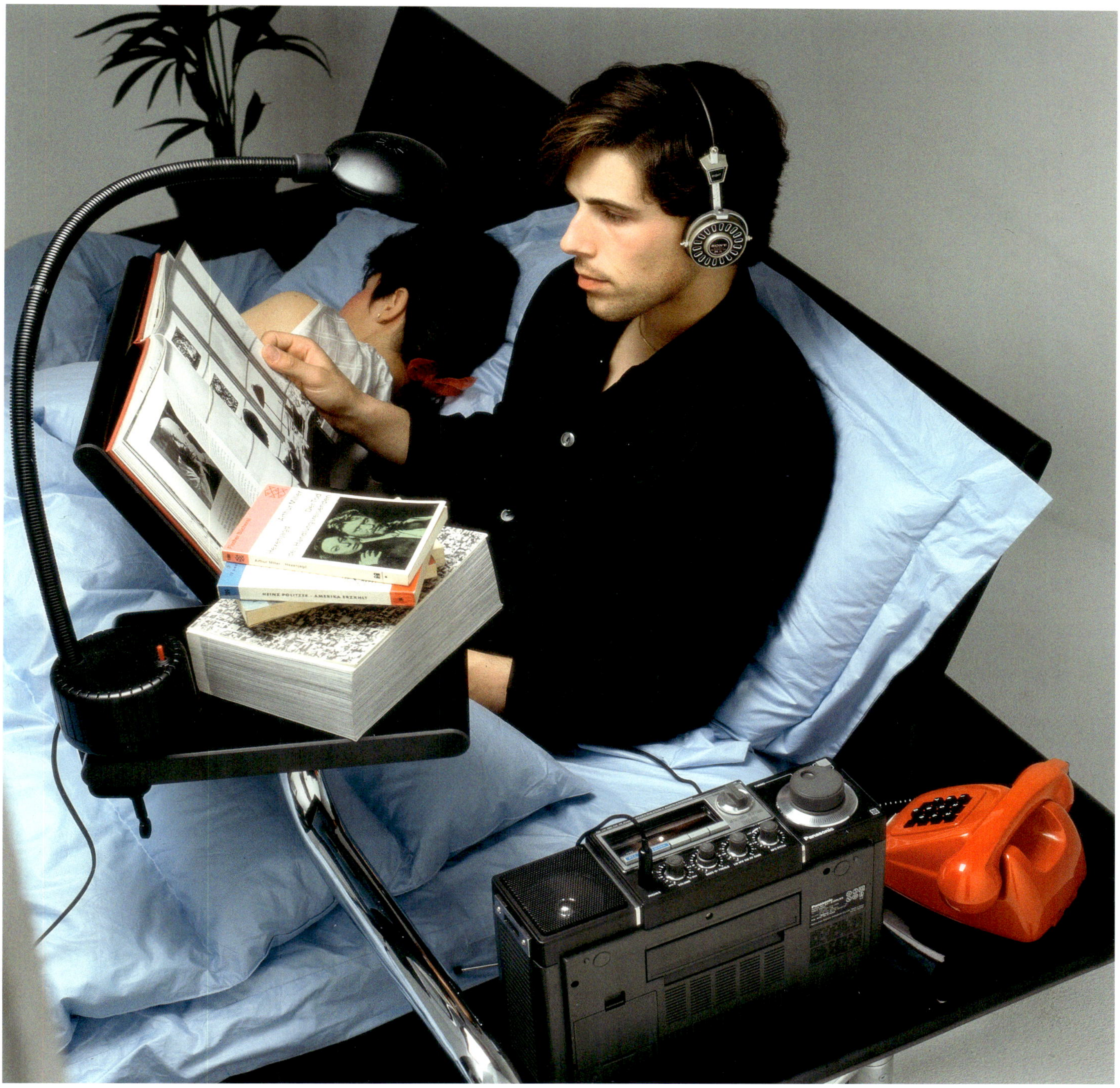

605

Advertising photo for the Wogg 3 multifunctional bed, 1983. Designed by Ubald Klug and produced by Willy Gläser Innenausbau, the bed system has a wraparound shelf as well as several swivel arms holding various consoles. The model in the photo, equipped with headphones and a red telephone, can read, listen to music, watch television, work, and eat breakfast—all without getting out of bed. The idea of exchangeable functional elements was in the air at the time—and not just in Switzerland.

Photo: Alfred Hablützel

Normbrunnen für die Notwasserversorgung

Die aus einem Gestaltungswettbewerb hervorgegangenen Normbrunnen sind dem Quellwassernetz angeschlossen und gehören zum System der Notwasserversorgung. Mit wenigen Handgriffen ist der bronzefarbene Brunnen geöffnet und der Schlauch zur Verteilröhre montiert. Bisher hat die Wasserversorgung vier solche Zapfstellen eingerichtet, nächstes Jahr kommen weitere 25 dazu; im Endausbau werden über 100 solche Wasserabgabestellen über die Stadt verteilt funktionsbereit sein.

606

Alfred (Alf) Aebersold, Emergency Fountain for the City of Zurich, 1973. The congenially formed bronze drinking fountain was the result of a competition by Zurich's public water authority. Besides this simpler version, there is also one with a drinking trough for dogs. Stored within is an apparatus that enables people to access the public water supply in an emergency.

From: *Neue Zürcher Zeitung,* October 14, 1976

607

Page from the prospectus "Plastic Items for the Household," 1970s, OWO-Presswerke, Mümliswil. After starting out as a comb factory, Otto Walter Obrecht's company responded to financial pressure by launching a new product line in 1932: plastic tableware and various molded items out of bakelite for the electrical industry. Later, OWO produced popular tableware made of indestructible polyethylene or melamine. The arrival of plastics to the home brought with it a new color spectrum.

608

Sample office, advertising photo for the USM Haller furniture system, 1970s, USM U. Schärer Söhne, Münsingen. With the advent of the universal modular furniture system, developed beginning in 1964, the prestigious office landscape of the economic boom years became a colorful, nonhierarchical workplace.

609

Presentation of the Styromodul office filing system from Spichtig AG Kunststoffwerk, Steinen, ca. 1971. Benedikt Rohner also devoted his energies to designing the workstations and organizational systems that are indispensable to any office. The colorful Styro container systems produced by Spichtig consist of snap-together plastic modules that can be used to build anything from a rolling cabinet to a complete wall unit.

Garden Chair 1703
ca. 1965

610 Walter Frey, Garden Chair 1703, ca. 1965
Fiberglass-reinforced polyester, 78 × 65 × 63 cm,
Stella Werke AG, Bassecourt
Museum für Gestaltung

611 Pages from a prospectus, ca. 1966,
Stella Werke AG, Bassecourt

610

611

Using the heading "Kunst am Swimmingpool" (Art by the Swimming Pool), the company Stella, founded in 1951, advertised a stackable, three-legged chair (No. 1703) and an accompanying footstool (No. 1743) made of fiberglass-reinforced plastic with a polyester resin coating. The material, which was new to furniture making (but already being used to build sailboats and the roof of the Citroën DS), was first used by Stella Werke, located in western (French-speaking) Switzerland, in 1955 for garden furniture under the name "Stellafort." It was touted for its non-fading, weatherproof, and long-lasting qualities, making it ideal for use in extreme temperatures. The chairs were largely machine made. The fiberglass was soaked in polyester, dyed, and formed into seat shells in minutes by hot-press molding. The chair's function and material determine its form. But the architect Walter Frey (1911–2002) from Basel had more than functionality in mind when he designed the lightweight and easy-to-clean *Garden Chair 1703.* The design was intended to appeal to a wide audience, its simplicity providing a pleasant contrast to nature's profusion of forms. Aside from the durability of the chair's material, the product catalogue also touts the aesthetic longevity of its design: "Even in a decade's time, the form and appearance of Stella garden furniture will remain as fresh and beautiful as if it was its first summer season in your garden or on your patio." Indeed, this visionary plastic chair could undoubtedly have been designed ten years later. RM

612 Bruno Rey, Model 3300 Chair, 1970
Varnished beechwood, cast aluminum,
79 × 44 × 50 cm, Dietiker AG, Stein am Rhein
Museum für Gestaltung / Donation of Dietiker AG

613 Color-painted limited edition, 2005
Museum für Gestaltung*

614 Model 3300 Chair, disassembled

612

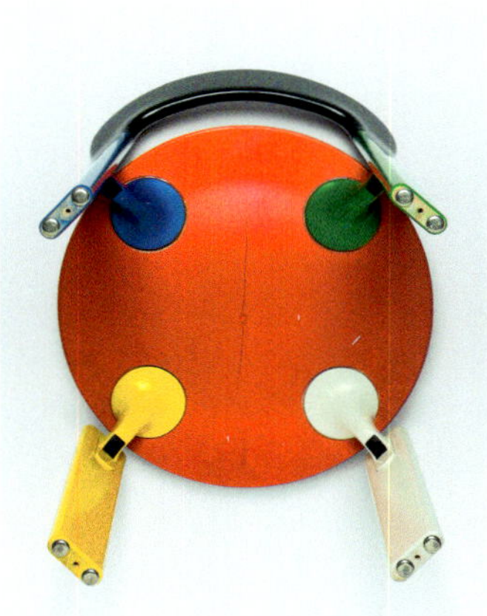

613

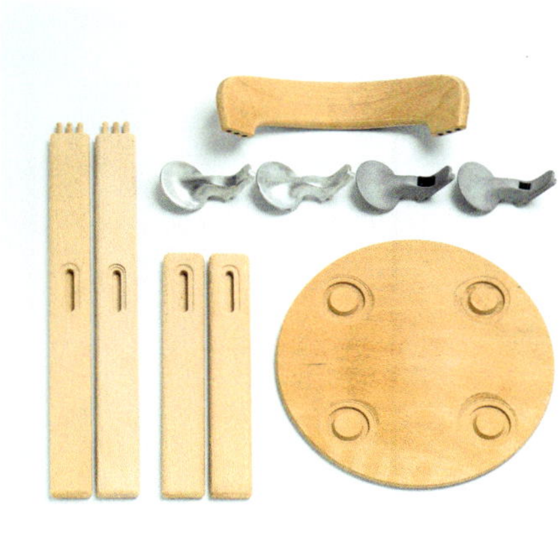

614

The most successful piece of Swiss seating furniture to date is based on an invention patented in 1972. When Bruno Rey (b. 1935), a student of Willy Guhl, began his collaborative work with the chair and table manufacturer Dietiker & Co. toward the end of the sixties, he was already working on a design that presented an alternative to the conventional wood-frame chair. Rey's design concept prioritized questions of technology, construction, and fabrication. Through the use of bonded aluminum brackets, he was finally able to connect a chair's legs to a self-supporting seat without the use of a frame. Company president Edlef Bandixen—a successful designer himself—soon recognized the idea's potential for industrial manufacture. Following several preproduction series, Dietiker invested in the plant and machinery necessary for production. The constantly evolving aluminum brackets were sand cast by hand until 1970–71, and later manufactured using permanent mold casting. However, it was not until 1989 that the die-cast aluminum process allowed for the efficient fabrication of large series. The chair was now entirely machine made, right down to the finishing—including bonding the brackets to the wood with a rotary bonder. *Model 3300,* made of native beech, is exceptionally sturdy, stackable, and, upon request, available in up to four colors. Free of any visual extravagance, the chair feels light and stands out by virtue of its practical, modern style. It is no surprise, then, that over a million models have been sold worldwide. AR

Schilder Bär 2002a.

Wisa-Gloria
Factory design

Tricycle
1970

615

616

617

615 Anonymous, Tricycle, 1970
Painted or chrome-plated metal, plastic, 62.5 × 48 × 75 cm, Wisa-Gloria AG, Lenzburg
Museum für Gestaltung/
Donation of Wisa-Gloria AG

616 Tricycle, earlier model, 1961, Wisa-Gloria AG, Lenzburg
Photo: Beat Derungs, Photography Class, Kunstgewerbeschule Zürich

617 Photo from the Wisa-Gloria catalogue, 1972

At one time, there was barely a household in Switzerland without a stroller, children's furniture, or toys by the Lenzburg-based manufacturer Wisa-Gloria. Founded in 1882, the company was able to expand its distribution to France, Greece, and South America beginning in 1915 thanks to an international trade network. Up until the sixties, Wisa-Gloria produced streamlined tricycles with curved tubular steel or sheet-metal-clad frames. In 1970, one of the company's own engineers developed a more affordable model, touted in the 1974 catalogue as an outstanding bargain at seventy-five Swiss francs. The tricycle's simple construction consists of a single steel tube, both ends of which are compressed to form flat ovals, one vertical and one horizontal. This reduction to a single structural engineering principle, along with clarity of form and a red lacquer finish, made this tricycle a timeless icon and favorite model in designer circles. Only the upward-curving handlebar is reminiscent of the two-gear motorcycles of the time. With 1,500 units produced annually, the tricycle was Wisa-Gloria's last successful product. Soon after its release, the company was forced to gradually curtail its once wide selection of products, before finally discontinuing production entirely in 1992. Thanks to this emblematic tricycle from 1970, Haderer, a small Austrian manufacturer of children's vehicles, took over the brand. The company relocated to Switzerland and has been manufacturing retro reproductions of the tricycle since 2010. RM

Unbekannt – Vertraut 1987, pp. 146–47; www.wisa-gloria.ch.

618 Heinrich Bucher, Airline Trolley, 1979–80
Anodized aluminum, powder-coated walls and doors, 103 × 30.5 × 50 cm, 16 kg, Bucher Leichtbau AG, Fällanden
Museum für Gestaltung/
Donation of Heinrich Bucher

619 Exploded assembly drawing of a trolley, ca. 1979, Bucher Leichtbau AG, Fällanden

620 View of the rear galley on a Swissair MD-81 aircraft, 1973–74

618

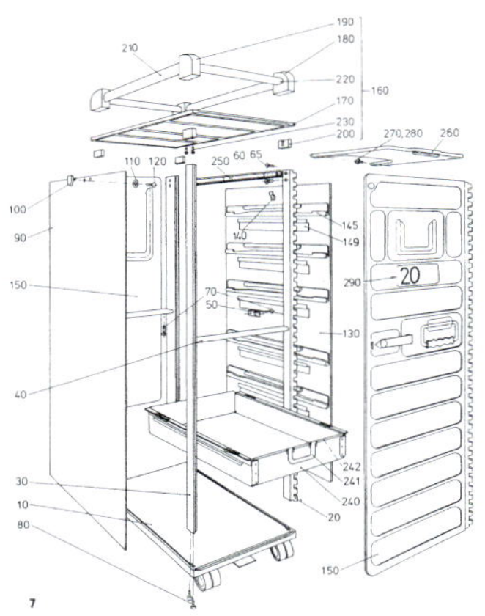

619

620

Heinrich Bucher (1918–1992), a trained sheet metal specialist, began working in Swissair's repair service in 1943. One of his responsibilities was to overhaul the on-board kitchens—so-called galleys, which store the packaged food prepared on ground, keeping it at the correct temperature. Like all airplane installations, they need to be as lightweight as possible while still withstanding heavy use as well as the forces of deceleration and acceleration. Bucher drew on his experience to begin independently manufacturing his own on-board kitchens in 1953, developing a new series of "trolleys" in 1973–74 that could be used as service carts, waste containers, or to sell duty-free products. Like the standard models of the time, these trolleys were modular aggregates that could be fitted and replaced without the use of any special tools, and also took into account the ergonomic aspects of the airplane cabin as workplace. What was new, however, was that he used "monocoque" construction, which draws its stability from the skin itself, in this case solid aluminum panels, rather than from welded, pressed, or cast elements. These "integral structures" minimize the amount of material used, reducing the trolley's overall weight while still guaranteeing a high level of sturdiness. Thanks to the CNC process, they can be easily adjusted to customer specifications. The corrosion-proof panels are powder-coated in the respective corporate color. These trolleys have a lifespan of ten to fifteen years, and although they are more expensive than rival products, their maintenance costs are lower. RM

Unbekannt – Vertraut 1987, pp. 152–55; www.bucher-group.com.

621

621 Trix and Robert Haussmann, Lehrstück II: Säulenstumpf (Truncated Column), 1978
Multiplex beech with a grained olive ash veneer, glass drawer bottoms, mirrored rear wall, 134 × 43 cm, base 46.5 × 46.5 cm, Röthlisberger Schreinerei AG, Gümligen. Limited collection
Museum für Gestaltung*

622 Trix and Robert Haussmann, Lehrstück II as skyscraper, 1978. Drawing
Archive of Trix and Robert Haussmann

623 Trix and Robert Haussmann, Model for Lehrstück II: Störung der Form durch Funktion (The Disruption of Form by Function), 1978, painted wood, 1:3 scale model by Dumeng Raffainer, Zurich.
Museum für Gestaltung/
Donation of Trix and Robert Haussmann

622

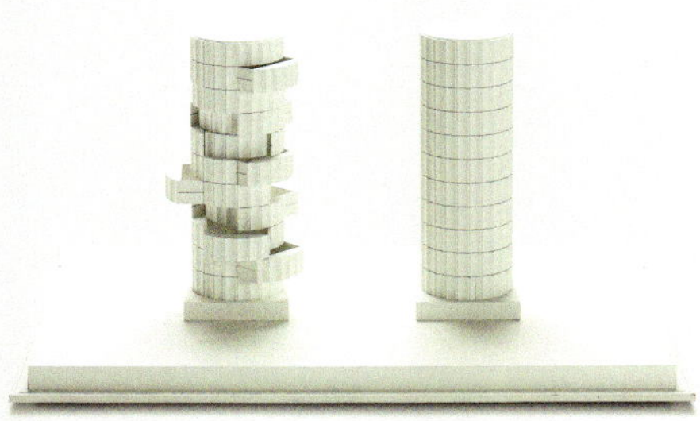

623

In 1922, Adolf Loos made his design for the Chicago Tribune Tower truly unique by giving it the form of a Doric column. Trix (b. 1933) and Robert (b. 1931) Haussmann ironically paraphrased this design when they depicted their storage unit *Säulenstumpf* (Truncated Column)—its drawers half open—as a skyscraper. Almost fifty years later, they too were attempting to challenge and convey to the senses the social and functional meanings of symbolic forms in different scales, contexts, and materials. In 1981, they coined the term *Manierismo critico,* or Critical Mannerism, to describe their research into the transmogrification of materials and the transformation of rooms; into complexity, ambiguity, multiple coding, metaphor, allegory, paraphrasing, and quotation; and into the use of contradiction, disruption, and destruction. They then chose individual topics and presented them as *Lehrstücke* ("Learning Pieces"), taking them to their logical extremes. *Lehrstück II: Störung der Form durch Funktion* (The Disruption of Form by Function), confronts the hermetic form of a fluted column with the maxim "Form follows function." In its final version, the truncated column is divided into eleven shallow cylinders that swivel around an offset perpendicular axis, functioning as glass-bottom drawers. The piece is fascinating not only due to its technical finesse and nearly unlimited possibilities for rearrangement, but also because it is reminiscent of the similarly shifted and disarrayed columns of ancient ruins. AR

Schilder Bär 2002c; *Die Allgemeine Entwurfsanstalt* 2011; Haussmann 2012.

Susi and Ueli Berger

5-Minute Chair
1970

624 Susi and Ueli Berger, 5-Minute Chair, 1970
Galvanized metal mesh, 80 × 55 cm. Prototype
Museum für Gestaltung

625 Instructions for making the 5-Minute Chair, 1971
Museum für Gestaltung/
Donation of Susi and Ueli Berger

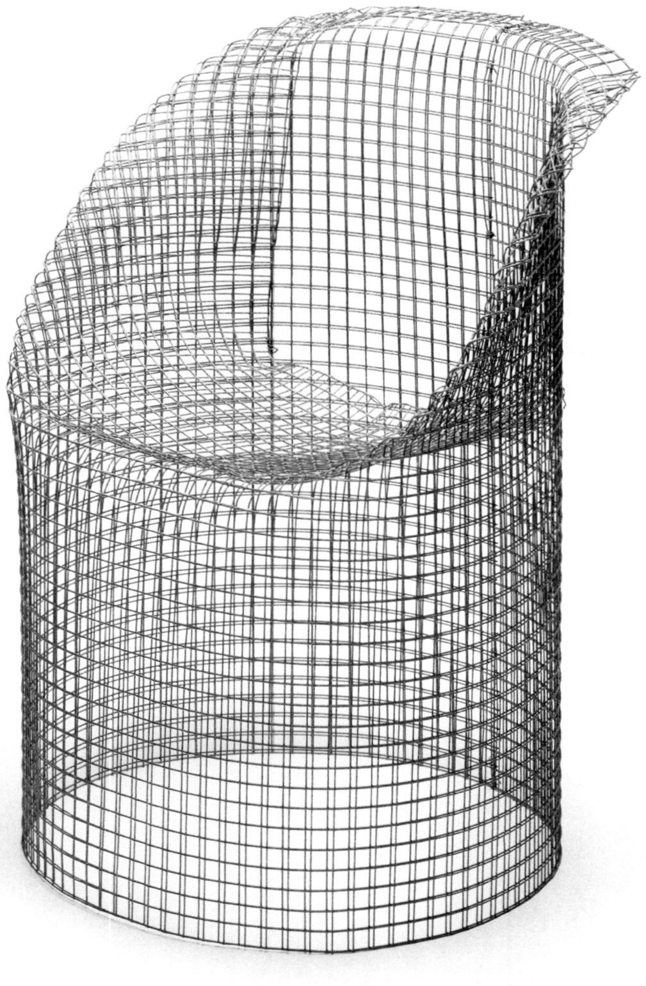

624

625

Although by the end of the sixties plastic garden furniture was already in production, and Susi (b. 1938) and Ueli (1937–2008) Berger had already contributed to the nonconformist practice of sitting on the floor and the concept of the lounge landscape with their foam furniture pieces *Multi-Soft, Soft Chair,* and *Terreno,* they still needed a chair for their own personal use to sit outside in the sun. The *5-Minute Chair* is a striking example of DIY furniture in Switzerland. The Bergers were inspired by home construction manuals born from the hippie movement (in the US magazine *Shelter,* for instance). Using a 314-centimeter-long piece of welded mesh, commonly known as chicken wire, they built a weight-bearing, comfortable *5-Minute Chair.* The design's assembly instructions would certainly have taken more time to write than to execute. Their straightforwardness shows that the Bergers, for all their free thinking, were committed to the functionality of their furniture: "(1) Roll the mesh into a double cylinder, (2) Secure both ends using galvanized wire, (3) Push wire backward and inward, respectively, at a height of approximately forty-two centimeters (seat height), (4) Use wire to fasten the far corners of the seat to the outer cylinder, (5) Round off the edges by bending them backward slightly." Unfortunately, these instructions were never published. RM

Susi und Ueli Berger 1999.

626

627

628

626 Ubald Klug, Terrazza DS 1025 Sofa, 1973
Upholstered elements on wooden base, covered in brown leather, 71 × 15 × 84 cm, De Sede AG, Klingnau
Museum für Gestaltung

627 Prospectus for the Terrazza lounge landscape, 1970s, De Sede AG, Klingnau

628 Mick Jagger on the Terrazza Sofa, autograph card, n.d.

Ubald Klug (b. 1932), a student of Willy Guhl, has proven himself successful at finding creative solutions to a multitude of different challenges—from designs for new bathroom fittings (*Bellino* sink, 1964) to concepts for urban installations in Paris (1972), from the cockpit of the Airbus A320 (1984) to a prototype for a modular apartment house (1970). Paris has been his sphere of activity since 1966. From time to time, the design consultant Alfred Hablützel has also brought him together with various Swiss manufacturers, such as in the seventies with the leather furniture maker De Sede. Klug—a nonconformist when it comes to interior design—presented the company's general manager at the time, Urs Felber, with an intriguing sketch of a sofa model based on the idea of a heap of sand deformed by sitting. De Sede agreed to produce the model as long as it was viable to do so using leather. Using rectangular panels as a foundation, Klug proceeded to construct an upholstery landscape that looked inviting to either sit or lay down on. He combined diverse elements into true "lounge landscapes." The sofa's similarity to layered topographical models or terraced landscapes is hard to miss; this is probably where the model got its name, *Terrazza.* Mick Jagger's pose illustrates the extent to which Klug's concept captured the contemporary zeitgeist. The sofa's curvy, malleable upholstered volumes featured all the characteristics of a living style then prevalent in young people's collective housing—except for affordability. AR

Hablützel 1989; Hablützel 2002b.

Cloud Lamp
1970

629

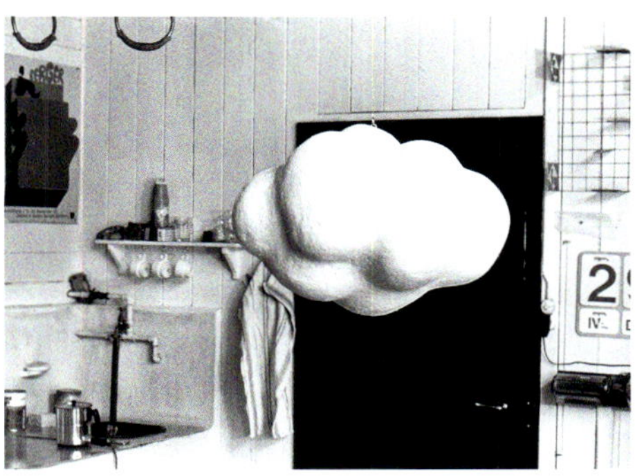

630

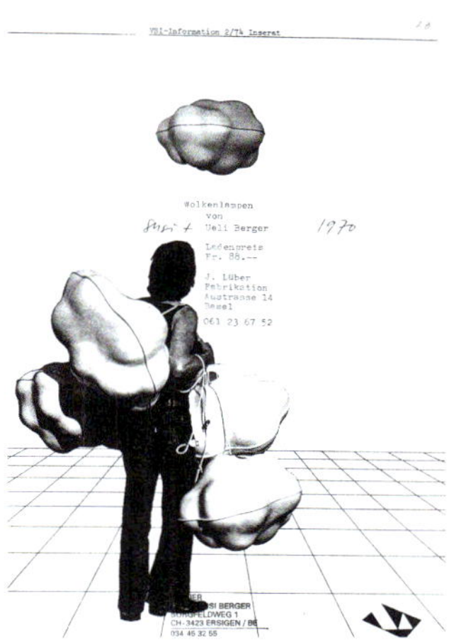

631

629 Susi and Ueli Berger, Cloud Lamp, 1970, produced 1976–79
Polystyrol, 45 × 70 × 55 cm, J. Lüber AG, Basel
Museum für Gestaltung

630 Styrofoam model in the Bergers' kitchen
Museum für Gestaltung

631 Cloud Lamps by [Susi +] Ueli Berger, advertisement in VSI-Information 2/74, J. Lüber AG, Basel

Polystyrene materials are typically stiff and hard, but when heated the thermoplastic becomes rubbery and pliable, allowing it to be formed like a membrane. Through the vacuum-forming process, small series of shells or hollow bodies can be produced with a minimum of effort. The designer and artist couple Susi (b. 1938) and Ueli (1937–2008) Berger used this novel process to create an amorphous cloud, first dividing the mold horizontally into two halves, and then using a vacuum to cause the hot polystyrene membrane to cling to the molds. Once cooled, the shells retain their form and can be effortlessly joined back together thanks to their congruent edges. Unlike the turned or pressed rotary forms abundant in the world of lamps, symmetry is alien to this cloud—it presents itself differently from every angle. Hung freely from the ceiling like a lantern, it shines brightly as an antithesis to *Gute Form.* It is not a tool with which light can be aimed or directed but simply a glowing body, a cloud in a cartoon sky. The Basel-based manufacturer J. Lüber, which was looking for small, innovative pieces of furniture to add to its product range, launched serial production of the *Cloud Lamp* in 1976. It was made by J. Lüber until 1979, then in 1999 it was produced in a limited edition by the Design Collection of the Museum für Gestaltung Zürich, and since 2002 it has been sold by Wohnbedarf. RM

Bähler 1997; *Susi und Ueli Berger* 1999; Textual information from Susi and Ueli Berger.

632

633

634

632 Anonymous, Bolex 233 Compact S Super 8 Film Camera, 1972
Plastic housing, 10 × 8.5 × 16.5 cm,
Bolex International S.A., Yverdon-les-Bains
Museum für Gestaltung

633 René Burri shoots a film with a Bolex, 1956
Photo: Photography Class, Kunstgewerbeschule Zürich

634 Bolex H16 film camera, early model from 1935, produced 1946–58, Paillard S.A., Sainte-Croix; Kern & Co. AG, Aarau (lenses)
Museum für Gestaltung/Donation of Max Zuber

Specializing in precision mechanics, the Paillard company had a flair for technological innovation, contracting aspiring young engineers, and buying out smaller businesses with growth potential. Paillard's acquisition of the Bolex brand—founded by Geneva-based Ukrainian exile Jacques Bogopolsky (1895–1962)—allowed the company to get involved in the development of "cinématographs." The company's first film projectors were launched in 1935, along with the 16mm Bolex H-16 film camera. This camera was used by personages such as Marlene Dietrich, Mahatma Gandhi, and Aga Khan. The H-16's most outstanding features included its automatic film threading mechanism, a coupling allowing the film to be wound forward or backward using a crank handle, and the option of adjusting the frame rate from eight to twenty-four frames per second. This would explain the camera's popularity among the American avant-garde: Andy Warhol, Brian de Palma, and David Lynch all shot films using the Bolex Model H. Moreover, the camera's sleek design, glinting metal, and faux leather perfectly echoed the look and feel of contemporary watches and handbags. In 1942, the Bolex 8mm became the first pocket-sized camera on the market. While Bolex met their first competitor in the form of Kodak, they were still able to keep pace in the sixties thanks mainly to their Super 8 cameras. Their Bolex 233 Compact S, designed for amateurs, was one of the smallest film cameras ever. In 1974, Bolex was acquired by the Austrian company Eumig, which went bankrupt in 1981. The electronic revolution had finally caught up with the precision mechanics of Paillard-Bolex from Yverdon. RM

Perret/Cosandey 2013.

Werner Müller

Relaxer Lounge Chair
1970

635

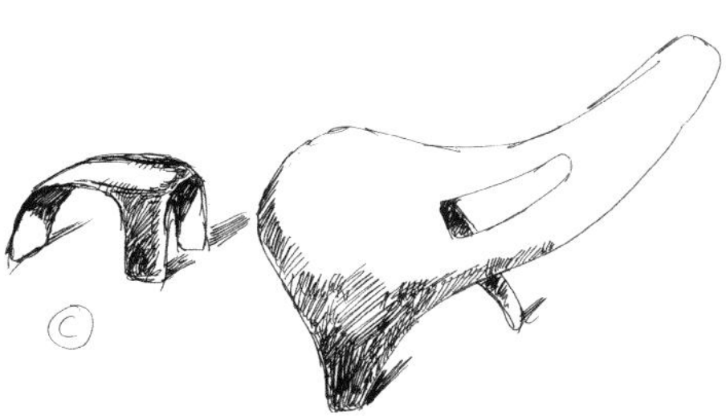

636

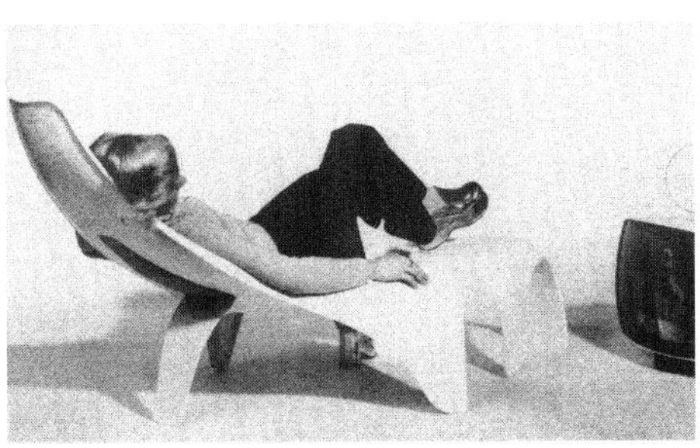

637

635 Werner Müller, Relaxer Lounge Chair, 1970, produced until 1978
Fiberglass-reinforced polyester with a colored gelcoat finish, chair 80 × 86 × 108 cm, footstool 33 × 49 × 54 cm, H. P. Spengler Erben, Rümlang
Museum für Gestaltung

636 Werner Müller, Relaxer Lounge Chair and Footstool, ca. 1970. Drawing
Museum für Gestaltung

637 Photo from Relaxer prospectus, ca. 1970, H. P. Spengler Erben, Rümlang

Werner Müller (1924–1995), who is also known as "Seepark-Müller" or by his pseudonym W. M. Anderwerd, was an architect, designer, inventor, and journalist. He studied architecture at ETH Zürich, with additional training in landscape architecture. In 1956, he quickly made a name for himself with his building project for the Lake Zurich basin. A more reduced 1974 version of the project, which originally included a car-free park and various superstructures, was, however, defeated in a referendum, as was Müller's plan for the Summer Olympic Games in Zurich. Apart from his ideas for the use of technological innovations in urban planning, such as the Compact-Bahn, an urban monorail (1973), Müller made his mark with his fountain pen for the blind, cubed milk, and nonslip underwear. His furniture is less well known but is similarly characterized by a uncompromising attitude. Müller describes the *Relaxer* as the deck chair's domesticated brother, allowing lounging to be relocated from the terrace or balcony indoors. His goal was to improve the ergonomics of the resting position—the transitional stage between sitting and reclining— for although people spend much of their time sitting down, they often do it incorrectly. Despite being molded, the three-legged chairs seem as though they were cut and formed from a single surface. The lounge chair and accompanying footstool were produced in small series in the color variations red, blue-violet, and white. They were sold by Wohnbedarf and Jürg Bally, and were used in various settings, including an indoor swimming pool in Zurich. RM

Werner Müller, "Sind wir ein Sitzwesen?," Estate of Werner Müller, Stadtarchiv Zürich.

638

639

640

641

638 Peter Emch, Sinalco—Refreshingly fruity, 1970. Poster
Museum für Gestaltung

639 Susi and Ueli Berger, Soft Chair, 1967, produced 1970–74, Victoria-Werke AG, Baar
Museum für Gestaltung

640 Studio 65, Bocca Divan, 1970, shown in 1997 version, Edra Mazzei, Pisa (IT). Foam upholstery with elastic fabric cover
MAK – Austrian Museum of Applied Arts/Contemporary Art. Photo: Georg Mayer

641 Oliver Mourgue, Djinn upholstered furniture, 1965, in Stanley Kubrick's science fiction film *2001: A Space Odyssey,* 1968

The 1960s are generally considered to have been a decade of upheaval and political change. The year 1968 in particular has become synonymous in Europe with the "spirit of revolt," expressed most strikingly in the antiauthoritarian student protests and the reform-minded civil rights movements in France, Germany, Czechoslovakia, and the United States, but also in Mexico and Japan.[1] In addition, in many other countries entrenched value systems and traditional worldviews that had been passed down for generations were fundamentally called into question by broad sections of society—even challenging the way people related to their physical surroundings. The dominant tastes on display in middle-class homes were as much a target of criticism as the range of products available in postwar society. The search for alternative, progressive, and unconventional lifestyles made many of the prized domestic status symbols seem obsolete. The focus now shifted instead to other objects of desire in the area of everyday furnishings and accessories, ushering in a structural change in commercial mass culture. For, despite all the criticism of the established consumption habits, demand for new products by the evolving pop, youth, sub-, counter-, and alternative cultures nonetheless opened up new markets—not only in the fields of music and fashion but also in those of lifestyle accessories and design.[2]

Today, the sixties thus represent, like no other decade in the twentieth century, innovative design in furniture, objects of everyday use, and consumer goods. When we think of seating furniture from this era, for example, certain modern design classics automatically spring to mind: the famous *Panton Chair,* which as the first plastic cantilevered chair is one of the all-time design icons; the polystyrene bead-filled *Sacco* by Piero Gatti, Cesare Paolini, and Franco Teodoro, a revolutionary seating object from 1968 that wittily captured the essence of the times; the inflatable plastic chair *Blow* by Jonathan De Pas, Donato D'Urbino, Paolo Lomazzi, and Carla Scolari, likewise a product of the Italian Anti-Design movement; or the futuristic upholstered *Djinn* series by Olivier Mourgue, which Stanley Kubrick used in his 1968 science fiction film *2001: A Space Odyssey.* This was furniture that represented a certain affront to petit bourgeois taste and conformity on the home front and conveyed to the contemporary public a whiff of freedom and social ferment.[3]

When so-called Radical Design or Anti-Design began in sixties Italy to emulate Pop Art and break with the functionalist design principles of postwar modernism, absorbing in the process the formal idiom and themes of pop culture, the Swiss designers had a hard time keeping up. The marks left by the Swiss Werkbund, which had for decades preached the gospel of objectivity, sobriety, and functionality in furniture and product design, were apparently too deep. Design training that aspired to meet the quality standards propagated by seminal figures such as Sigfried Giedion and later Max Bill had culminated in the "Die gute Form" quality seal, which hardly anyone dared challenge. "I liked whatever other people thought was good," wrote author Fritz Zorn, born in 1944 in Meilen near Zurich, in his posthumously published autobiographical essay *Mars,* describing the symptomatic voluntary submission to ostensibly "good" taste in the postwar period: "I had learned that classical music was 'good,' that popular music and jazz were 'bad.' So I listened to classical music, just the way my parents did; and I thought it good. I scorned jazz because I knew it was bad, even though I had never heard any jazz and didn't have any idea what jazz was.... It didn't dawn on me that I had never compared the two, that I had not made a conscious choice between them, or that I had blindly accepted the prejudice that defined all classical music as good and all modern music as bad."[4]

Even beyond the Gold Coast of Lake Zurich, the socialization of taste in the postwar period probably often happened in a similar fashion, also extending to very decided notions of the "right" way to reside. However, "modernity" of course played an entirely different role in the domestic interior than it did in music. In contrast to "modern" jazz or pop music, moderate modernist interior design concepts were widely and almost dogmatically established in Switzerland until well into the sixties. Accordingly, wherever people aspired to interiors that were as fashionable as they were elite, or in any event where they laid undisputed claim to *Gute Form,* it must have been incredibly difficult to stand back and reflect on one's domestic lifestyle in order to develop a more self-critical awareness of varying individual needs. The decades of taste education in the name of *Gute Form*—the writer Paul Nizon spoke in this context of a "specific taste dictatorship"[5]—made it hard to break out of what had been known since the era of "classical" modernism as supposedly "liberated" living.[6]

A New Swiss Style of Landscaped Interior?

When the Swiss Werkbund discontinued its "Die gute Form" campaign in 1968, the restrictive set of domestic dictates seemed to have run their course. In November 1967, the Werkbund had already organized the *Chair Fun* auction as a kind of "Pop rebellion against the principles of good form."[7] Artists and designers such as Meret Oppenheim, Robert Haussmann, Ueli Berger, Daniel Spoerri, and Bernhard Luginbühl tried to show "that a chair can be not only a seating opportunity but also a seating possibility."[8] The unusual, playful, and humorous chair experiments demonstrated that very different solutions to the purist, functionalist design principles and restricted range of materials upheld until then in Swiss design were in fact *possible.* It would however be an exaggeration to highlight this ephemeral event, of which the public was only marginally aware, as a singular watershed in Swiss design—not least because the radical nature of the artistic models ruled out both actual use as well as the prospect of serial production.

Nevertheless, around the same time, new forms and functions, coupled with new atmospheric and experiential qualities, began to appear in the Swiss home. There, too, it was mostly the seating furniture that, due to its immediate connection with the body, was taken up as a central field for experimentation, exhibiting the new approaches and attitudes most clearly. Newly developed artificial materials—chief among them foam—opened up opportunities for innovative designs. In the fifties, classic seating upholstery using horsehair, wool, or fiberfill was gradually replaced with expanded polyurethane or polyether foam, and during the sixties, foam blocks in unusual shapes and colors began to supplant conventionally built armchairs and sofas. The usual frame was abandoned and these large, pillow-like foam objects were set directly on the floor, encouraging more relaxed and casual sitting and reclining postures. Examples already presented at the *Chair Fun* auction were the *Soft Chair* (1967) and the multifunctional furniture range *Multi-Soft* (1968), both by Susi and Ueli Berger. With their characteristic shapes and colors, their material qualities and designated function, these two foam furniture series, produced by the Victoria-Werken in Baar with vinyl, nylon, or synthetic leather covers, in some cases in neon shades, fit in seamlessly with the latest seating objects by such well-known international designers as Joe Colombo, Verner Panton, and Pierre Paulin.[9]

The artist-designer couple Susi und Ueli Berger, who have conceived bold and whimsical designs ranging from lighting objects to walkable and sittable outdoor sculptures, are among the Swiss designers who seem to have been the most in tune with the spirit of international design circa 1970.[10] The Bergers also collaborated with various other designers to develop entire seating landscapes and landscaped interiors like the ones that had been in vogue internationally since the mid-sixties. These included a set of fabric-covered modules called *Terreno* (1972) for the company Strässle Söhne & Co. in Kirchberg, and another system in which the elements could be connected with zippers: *DS-600 Organic* (1970) for De Sede AG in Klingnau, the firm for which Ubald Klug (b. 1940) designed a few years later another elaborate seating and lounging landscape (*Terrazza,* 1973). Despite the unmistakable opulence of these landscaped interiors, the Bergers were clearly interested in a socially critical questioning of the usual design approaches and the commercially available products—and above all in achieving greater flexibility in furnishings and in the use of living spaces. The goal was to enable a more informal, spontaneous, and playful way of life, even going to the extreme of "living on the ground," modeled not only by the Bergers but also by the family of designer Klaus Vogt in a 1973 feature in the magazine *Das ideale Heim.*

642

643

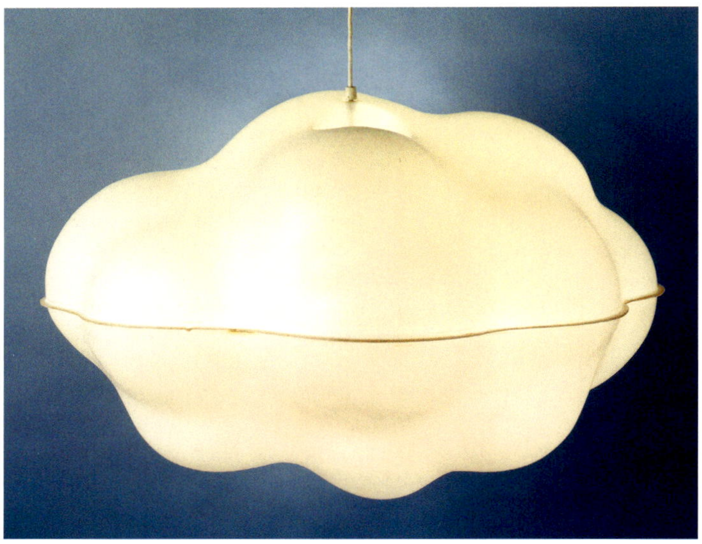

644

645

646

647

648

649

642 Photo from Ueli Berger article "Auf Polyester lässt sich auch sitzen" (Polyester is also good to sit on), in "Science-Fiction," *Die Frau* 21, no. 16 (1967). Photo: Balthasar Burkhard
643 "Dreamland Terreno," advertisement, Möbel Pfister, 1970s. Featuring the modular Terreno lounge landscape, designed by Susi and Ueli Berger and others, 1973, Strässle & Söhne, Kilchberg.
644 Susi and Ueli Berger, Cloud Lamp, 1970, produced 1976–79, J. Lüber AG, Basel
Museum für Gestaltung
645 Jonathan De Pas, Donato D'Urbino, Paolo Lomazzi, Carla Scolari, Blow Armchair, 1967, Zanotta S.p.A., Nova Milanese (IT)
Museum für Gestaltung/Donation of Peter Erni
646 Factory design, PTT telephone, Model 70 table unit, 1975, Zellweger Uster AG
Museum für Gestaltung
647 Walter Studer, Brac 2000 alarm clock, ca. 1972, Brac AG, Breitenbach
Museum für Gestaltung
648 Ubald Klug, Terrazza, 1973, catalogue, 1970s, De Sede AG, Klingnau
Sofa elements that can be combined to make a seating island or lounge landscape.
649 Susi and Ueli Berger, Sculpture to walk or sit on, Schulhaus Wiesental, Baar, 1970–71. Version in fiberglass-reinforced polyester. Photo: Leonardo Bezzola

Plastic Utopias—Perfectly Molded

In addition to foam—which the Bergers had originally obtained from the automotive industry and tested in their own living room before introducing their "low-on-the-ground" sitting objects—glass fiber-reinforced polyester, commonly known as fiberglass, also came into use in the course of the sixties as a suitable material for furniture. Designer Andreas Christen had already made a promising start in 1960 with his stackable fiberglass bed, adding in 1964 a cabinet that was likewise produced by H. P. Spengler in Rümlang, called the *Polyesterbox.* Even if plastic furniture was not quite as easy and cheap to produce as the designers and producers had initially envisioned, a few characteristic models were nonetheless created in the course of the decade whose formal vocabulary displayed vague echoes of pop culture. Examples include the oddly shaped *Expo Chair* that architect Pierre Zoelly designed for the Expo 64 Swiss National Exhibition in Lausanne, the three-legged *Garden Chair* by Walter Frey produced by the Stella Werke in Bassecourt (ca. 1965), and an armchair with footstool designed by Werner Müller for H. P. Spengler (ca. 1970). The furniture pieces made from the various synthetics certainly proved to be "faithful to their material," despite the fact that Swiss design, obsessed as it was with *Gute Form,* had long regarded their seemingly boundless malleability with some skepticism.[11] In the early seventies, these man-made materials were increasingly used for organically shaped housings for technical devices such as radios, television sets, lamps, and clocks. Plastic design had evolved into the epitome of advanced industrial production in the Space Age and was now considered trendy, sexy, and futuristic.

In contrast to Radical Design in Italy, which for example took inspiration from Marilyn Monroe's lips, a baseball glove, or a classical column capital for offbeat seating, and also contrary to international product design, which oriented itself for the design of television sets, lamps, fashion, and even whole interiors on the aesthetics of astronaut helmets and suits, rockets and satellites, Swiss design did not really engage with popular emblems or symbols. Although coming to terms with current issues such as youth culture, criticism of the consumer society, or a need for space was definitely relevant for the Swiss designers, the ubiquitous and "vital" pop imagery had virtually no effect on their work. Robert Haussmann's *Choco Chair,* presented at the *Chair Fun* art auction in 1967 as a "seating possibility," as well as the *Cloud Lamp* (1970) by Ueli and Susi Berger, are the great exceptions in this respect—at least until postmodernism took hold in the late seventies.

Even in the heyday of Pop Art and the Space Age, Swiss design essentially remained committed to *Gute Form.* The furniture pieces designed by Klaus Vogt can be taken as a prime example: with their rounded shapes and colorful painted surfaces, as well as their flexible handling, they are still regarded today as typical of their time and yet conceived for practical functionality. His *Squadra Cupboard* (1964), for instance, is considered to be one of the first pieces of Swiss furniture that expressed the influence of Pop Art on the product world. Another example is the *Roll-top Cabinet* (1969) with its pyramid-like stepped profile, fold-out desktop, and undulating laminated beechwood slats that roll up and back into the rear wall when the cabinet is opened. While designers seem to have sometimes liberated themselves during this period from the sober purist aesthetic and clear-cut lines of *Gute Form,* practical considerations evidently remained obligatory even when conceiving fun new functions. Swiss design remained straightforward, clean-lined, and sensible—and never really lost sight of the fact that furniture, lamps, and technical devices are ultimately objects of daily use.

In the late sixties, when the functions fulfilled by the domestic environment were subjected to a conceptual overhaul and were condensed into programmatic visionary functional units—cells for living, sleeping, cooking, and washing—or into variable, multifunctional interiors, Swiss design once again lagged behind what was happening in the rest of the world. Futuristic interiors like those presented during this era at the Cologne Furniture Fair by Verner Panton, Joe Colombo, Olivier Mourgue, Luigi Colani, and Rudolf Lübbens, or the contributions made by Colombo, Ettore Sottsass, Alberto Rosselli, Marco Zanuso, and Richard Sapper to the exhibition *Italy: The New Domestic Landscape* at the Museum of Modern Art in New York in 1972, were much more radical than any of the furniture systems and functional living units developed and offered around 1970 by Swiss companies and furniture shops such as USM, BASF, Zingg-Lamprecht, Wohnhilfe AG, and Idealheim AG. And yet—and this noted here only briefly in conclusion—domestic inte-

650

651

650 Verner Panton's home, Basel-Binningen, early
651 1970s. Photo: Verner Panton Design
At least as spectacular as Panton's furniture was his private residence, which he used as a showroom and laboratory for his experiments.

652

653

652 Ueli Berger, Klaus Vogt, Elena Peduzzi Riva, and Heinz Ulrich, DS-600 Sofa elements combined into an endless snake, 1971, De Sede AG, Klingnau

653 Susi and Ueli Berger, VW Beetle overgrown with grass, intervention in the cityscape, Bern, advertising photo, 1970s

riors from Switzerland better express the sixties way of life than those from almost anywhere else, and have become the epitome of living with Pop Art design. We need only think of the famous tower apartment inhabited by Gunter Sachs, in which in 1969 the German-Swiss industrial scion gathered together works by Pop artists such as Tom Wesselmann, Andy Warhol, and Roy Lichtenstein, juxtaposed with mod furniture ensembles designed by Allen Jones, François-Xavier Lalanne, and Eero Saarinen. Even more in tune with the times was the home of Verner Panton in Basel-Binningen, where the Danish architect and designer—who worked in the sixties for Vitra and fused Pop Art with design like no other—tested out his own designs and models. It must have been a domestic utopia in the truest sense of the word—a place that would have seemed extremely improbable for the Switzerland of those years.

1 In Switzerland's case, only the so-called Globus Riot in Zurich is worth mentioning in this respect; see Linke/Scharloth 2008.
2 On the turning point in 1968 and its impact on design, see Ruppert 1998 and Herding 2008.
3 On this subject, see, for example, Garner 1996; Jackson 1998; Breuer et al. 1999; Bony 2004; *Pop Art Design* 2012.
4 Zorn (1977) 1982, pp. 8–9.
5 Nizon (1970) 1990, p. 193.
6 See Bignens 2002b.
7 Margit Staber (1968), cited in Bignens 2002b, p. 193.
8 See fax from Jürg Bernhardt, Design Collection of the Museum für Gestaltung Zürich, April 11, 1994, quoted therein from the call for participation (Design Collection archives).
9 See *Formlose Möbel* 2008.
10 *Susi und Ueli Berger* 1999.
11 On the problems involved in the molding of plastics while remaining true to their material properties, see Wagner 2006. Great skepticism regarding the molding of "artificial materials" was also prompted by the Bern exhibition *When Attitudes Become Form,* which in 1969 gathered together a number of "shapeless" works of art in order to study the behavior of materials when left in their primary state, as well as the effects of gravity and entropy on the material conditions of art. However, when the exhibition organizer Harald Szeemann admitted in the catalogue that "It was inevitable that Hippie philosophy, the Rockers, and the use of drugs should eventually affect the position of a younger generation of artists," he could hope to be understood in the late sixties only by a small—albeit broad-minded—audience. See *When Attitudes Become Form* 1969, n.p.

654

655

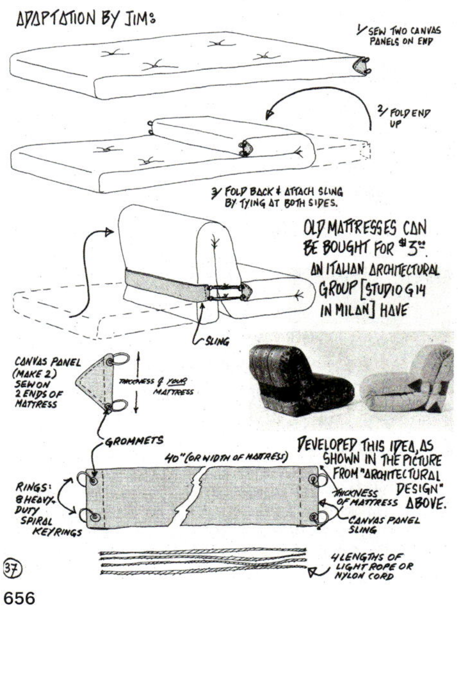

656

657

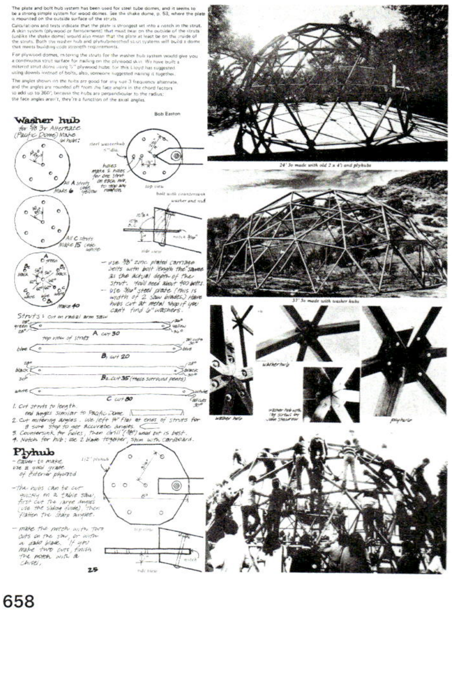

658

654 Stewart Brand, *The Last Whole Earth Catalog*, 1971. Published from 1968 to 1975 by the Portola Institute in Menlo Park, California, the *Whole Earth Catalog* provided the know-how for an alternative, environmentally conscious product culture.

655 James Hennessey and Victor Papanek, *Nomadic Furniture 1 + 2*, Atglen, Pennsylvania, 1973–74. Handbooks of the DIY movement.

656 Page from *Nomadic Furniture*

657 Page from the magazine *Shelter*, 1973

658 Page from the magazine *Domebook*, no. 2 (1971)

For some time now, "do-it-yourself" (DIY) has been a widespread phenomenon in the postindustrial societies of Europe and America. The still-growing enthusiasm for the creative possibilities and sensory experience of shaping our surroundings with our own hands has not only led to a rediscovery of craftsmanship but is also increasingly influencing the field of design through the fusion of production and consumption in the so-called "prosumer culture." It is in fact hard to think of an area of everyday life or material culture that has not been swept up in or affected in some way by the "participatory revolution" of recent years.[1] Nevertheless, today's DIY culture is an extremely contradictory fad: while on the one hand promising liberation from rampant consumerism, taste dogmas, and design norms, it on the other hand involves trying out and optimizing the market compatibility of new technologies (such as computerized 3-D printing), modes of production (including the so-called FabLabs), and sales strategies (often as a download off the Internet). Particularly in this era of the oft-invoked economic crisis, handmade products are by no means intended only to meet personal needs but are also often done on commission or represent an innovative business model.

The DIY trend has of course long since made inroads in Switzerland, and many designers have developed strategies for positioning themselves with their DIY strategies and designs on the Swiss—and international—design scenes. To what extent customers play a part in the implementation of a design varies widely. In the furniture field, the examples range from prepackaged kits for home assembly following the IKEA sales model (for example, *con.temporary furniture* by Colin Schälli, 2010), to shelving systems that can be put together from prefabricated modules in a variety of ways according to personal needs (Daniel Gafner's *1530* container system for Postfossil, 2010), or which can be combined using special connecting elements (*Indie Furniture* by Nicola Stäubli, 2008), all the way to furniture whose final form is entirely up to the consumer, with construction plans available on the Internet (*La lampada a stelo,* 2009, and *Flying Shelf,* 2010, by Kueng Caputo). Younger designers in particular often show an interest in involving consumers in the production process in order to offer sound and practical everyday items at affordable prices. Nicola Stäubli, for example, tried with his 2007 series *foldschool*—children's furniture made of corrugated cardboard that can be constructed according to free online instructions—to bypass complex furniture industry production processes and multistage distribution channels in order to "create through a smart manufacturing process and appropriate handling of materials an aesthetically pleasing, functional product at an affordable price." The designer sees his project as a "countertrend to a society characterized by its propensity for consumption." Instead of simply purchasing a finished piece of furniture, the consumer can develop through his own craftsmanship a "personal relationship" with the product.[2]

Know-How for an Autonomous Lifestyle

This consumption-critical slant to DIY is nothing new, and can be traced back to the sixties.[3] Even in the postwar period, the lack of material resources and skilled tradesmen led to a jump in the number of manuals with building instructions for every conceivable type of everyday object and recreational accessory—as well as to a proliferation of home improvement, hobby, and crafts magazines with titles like *Popular Handicrafts, Handyman,* and *Do-It-Yourself.* In the German-speaking world, a major role was played by women's and lifestyle magazines such as *Schöner Wohnen, Brigitte,* and the Swiss *Annabelle,* all of which popularized handmade products by providing handicrafts instructions, dressmaking patterns, and creative ideas for home decoration. Pragmatic and economic considerations were the main focus. For example, readers of the *Zürcher Wochenblätter* newspaper were challenged in a special supplement in 1956 on the subject of the home ("Wohnen") to "Do it yourself – mach es selbst!" By trying out their skills as car mechanic, painter, carpenter, locksmith, or electrician, they could save money on the services of tradesmen, which had become exorbitant. On the other hand, "splendid" consumer products such as "washing machines, dishwashers, clothes dryers, and the like" were celebrated as indispensable "necessities of life."[4] It was not until the late sixties that the subculture and counterculture put an anticapitalist and consumption-critical spin on the idea of autonomous production by the consumer. Under the motto "access to tools," the legendary *Whole Earth Catalog* (1968–75) led the way in the United States to amassing and transferring the knowledge required for a decentralized, alternative, and environmentally conscious product culture. Catalogues and books in the same vein were intended as reference works for an independent lifestyle, propagating "enlightened" consumption that tried to emancipate itself as much as possible from the merchandise market. They were of course also read in Switzerland.[5]

Historical books and magazines containing instructions for do-it-yourself items and tips for self-determined and autonomous living are of great value today not only for the insights they convey into the lifestyle of their times but above all because the homemade objects have rarely survived into the present. Even though DIY items exuded a hint of subversion that gave them a modicum of distinction, they were still regarded—mainly by younger people who did not yet want to furnish their homes "for life"—as affordable interim solutions and thus disposed of once "more solid," professionally made, and prestigious pieces were within reach. Nevertheless, two prototypes have been preserved in the Design Collection of the Museum für Gestaltung Zürich that seem characteristic of the Swiss DIY movement around 1970. The first is the *5-Minute Chair* devised by the designer couple Susi and Ueli Berger, which can be put together quickly and easily from wire mesh according to instructions that have likewise survived. This piece of furniture, which due to the material qualities of wire mesh has a rather unusual "atmospheric value," did not have much of an impact at the time since it was designed only for the Bergers' own use. The *5-Minute Chair* is notable however not only as the result of a design process consistently derived from the offbeat material; it also evokes the lifestyle experiments of its time and the need for more individual freedom, improvisation, and playful flexibility.

The second object from this period that survives in the museum's Design Collection did not get much beyond the prototype stage, even though it won an award in an internal competition at the Kunstgewerbeschule Zürich. The young design student Alois Rasser (b. 1948) developed his comfortable seating furniture made of corrugated cardboard with one-sided coating in Willy Guhl's interior design and product design class and presented it in 1969–70 in the exhibition *Magie des Papiers* (The Magic of Paper) at the Kunstgewerbemuseum Zürich. As a cheap and flexible alternative to heavy wooden furniture, pieces made of cardboard had already been widely sold in the mid-sixties, attracting a great deal of media attention. Like the snazzy "cardboard design classics" conceived by Peter Raacke and Peter Murdoch, these were mostly colorful folding furniture pieces made of coated and precut cardboard that consumers could easily assemble themselves. Rasser's design, which in the guide to the exhibition is rightly acclaimed as a "happy inspiration," took a fundamentally different creative route: "An original roll of the low-priced material, with the required diameter, is sawed on a curve into two equal halves. The desired seat form is then pushed into the cardboard from above, and the resulting overhangs at the bottom are cut away. Thus, two chairs are produced from a single roll."[6]

Although it was at least possible to produce a few of Rasser's award-winning corrugated cardboard chairs for the museum café during the Zurich show, one wonders in retrospect, in view of the success story of the cardboard furniture series *Easy Edges* (1969–73) produced around the same time by the US architect Frank Gehry, why this object failed to make a greater impact. As can already be seen in the case of Willy Guhl, who had experimented with ergonomically shaped plastic seat shells in the late forties—in other words, at the same time as US designer Charles Eames—the Swiss furniture scene was evidently still too cautious around 1970 to embrace alternative currents and blaze new trails in the development of experimental and edgy objects made of unconventional materials.[7]

As is clear from photos of the *Magie des Papiers* exhibition, Willy Guhl's class also inspired a number of other useful furniture pieces made of cardboard. In fact, ever since the Bauhaus—and especially since Josef Albers had his students playfully put the material through its paces in his preparatory course there[8]—paper and cardboard have been counted among the most important media

659

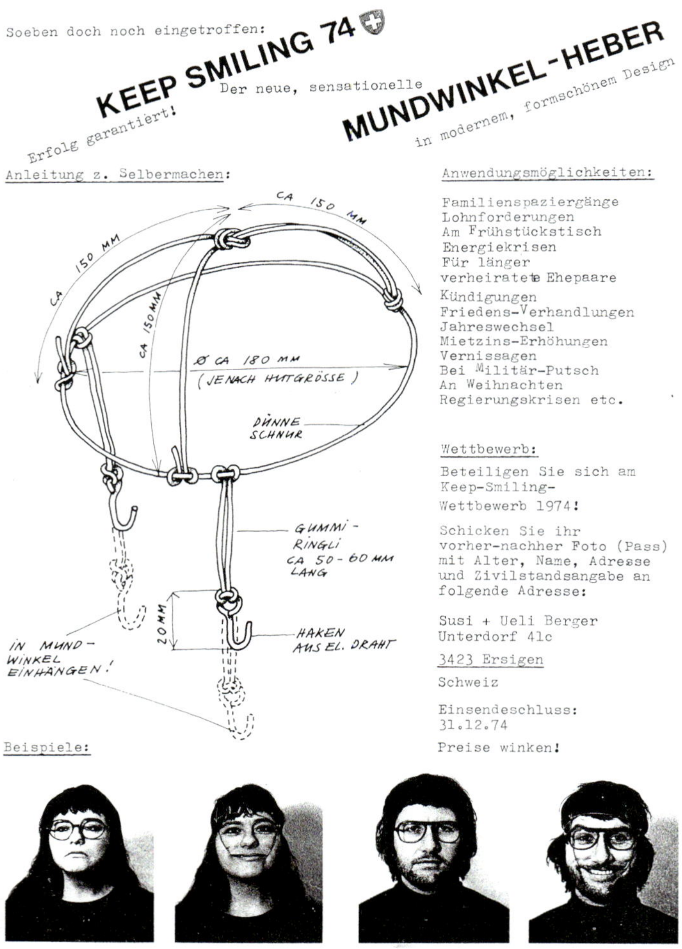

Soeben doch noch eingetroffen:

KEEP SMILING 74

Der neue, sensationelle

MUNDWINKEL-HEBER

in modernem, formschönem Design

Erfolg garantiert!

Anleitung z. Selbermachen:

Anwendungsmöglichkeiten:

Familienspaziergänge
Lohnforderungen
Am Frühstückstisch
Energiekrisen
Für länger verheiratete Ehepaare
Kündigungen
Friedens-Verhandlungen
Jahreswechsel
Mietzins-Erhöhungen
Vernissagen
Bei Militär-Putsch
An Weihnachten
Regierungskrisen etc.

Wettbewerb:

Beteiligen Sie sich am Keep-Smiling-Wettbewerb 1974!

Schicken Sie ihr vorher-nachher Foto (Pass) mit Alter, Name, Adresse und Zivilstandsangabe an folgende Adresse:

Susi + Ueli Berger
Unterdorf 41c
3423 Ersigen
Schweiz

Einsendeschluss: 31.12.74

Preise winken!

Beispiele:

660

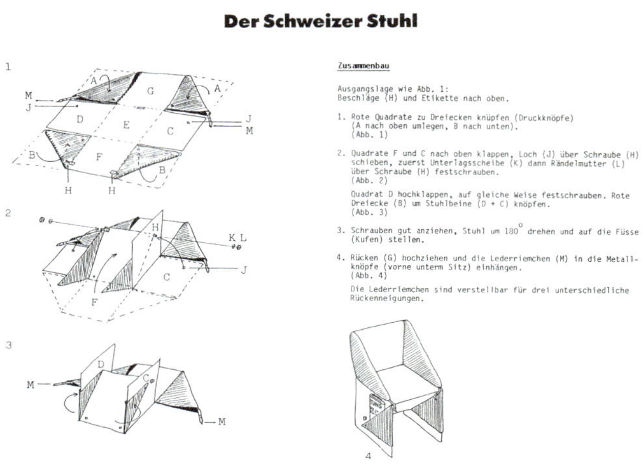

Der Schweizer Stuhl

Zusammenbau

Ausgangslage wie Abb. 1:
Beschläge (H) und Etikette nach oben.

1. Rote Quadrate zu Dreiecken knüpfen (Druckknöpfe) (A nach oben umlegen, B nach unten). (Abb. 1)

2. Quadrate F und C nach oben klappen, Loch (J) über Schraube (H) schieben, zuerst Unterlagsscheibe (K) dann Rändelmutter (L) über Schraube (H) festschrauben. (Abb. 2)

   Quadrat D hochklappen, auf gleiche Weise festschrauben. Rote Dreiecke (B) um Stuhlbeine (D + C) knöpfen. (Abb. 3)

3. Schrauben gut anziehen, Stuhl um 180° drehen und auf die Füsse (Kufen) stellen.

4. Rücken (G) hochziehen und die Lederriemchen (M) in die Metallknöpfe (vorne unterm Sitz) einhängen. (Abb. 4)

   Die Lederriemchen sind verstellbar für drei unterschiedliche Rückenneigungen.

661

659 Susi Berger, Swiss Chair, 1985, Verband der Schweizer Möbelindustrie (SEM), Lotzwil. Polaroid photos on presentation board. Instructions for folding a chair from a Swiss flag. Museum für Gestaltung/ Donation of Susi and Ueli Berger

660 Susi and Ueli Berger, Keep Smiling instructions, 1974. Photo: Leonardo Bezzola Museum für Gestaltung/ Donation of Susi and Ueli Berger

661 Susi Berger, Swiss Chair, page from press folder, Verband der Schweizer Möbelindustrie (SEM), Lotzwil, 1986

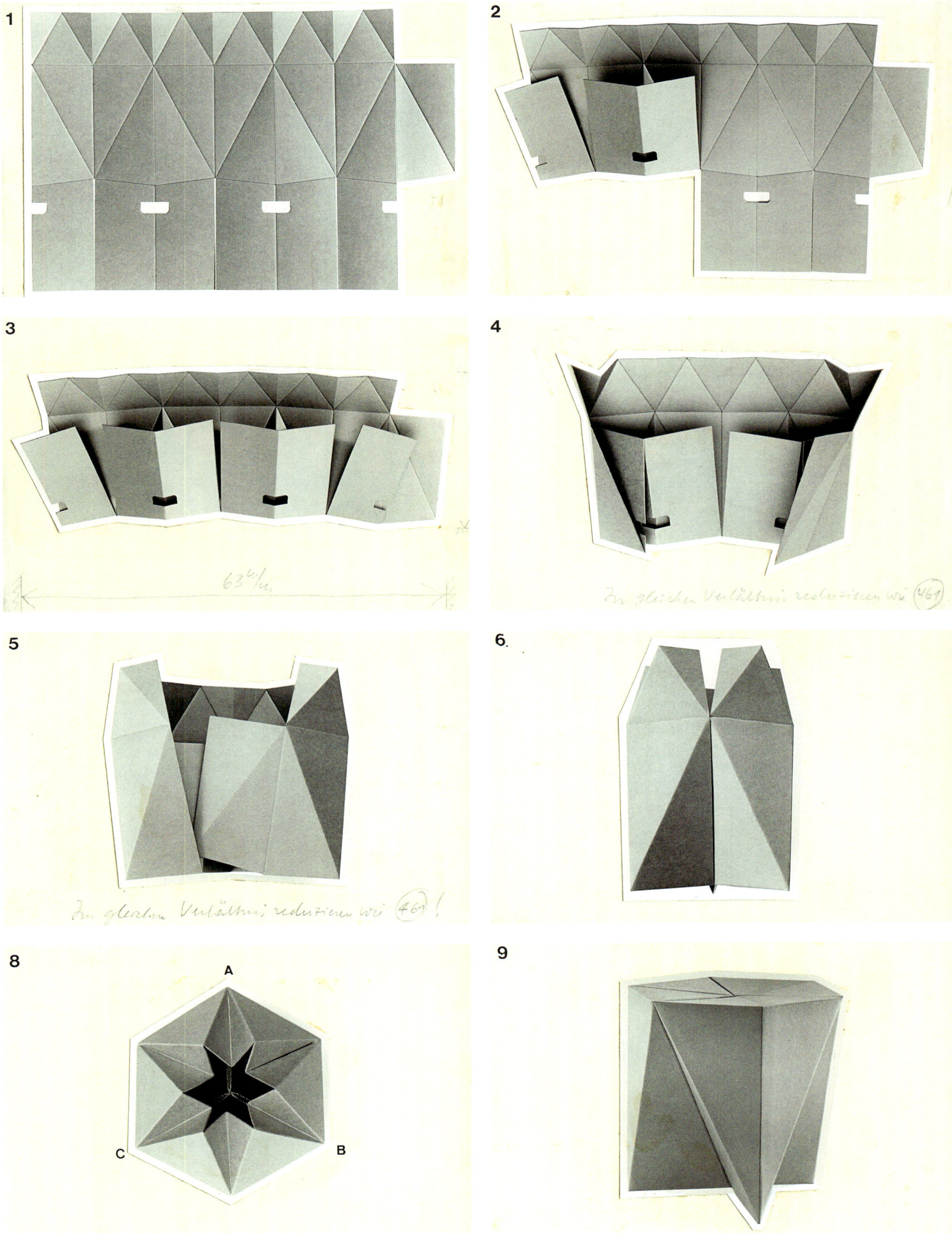

662

662 Thomas Trachsel, Instructions for folding a stool out of double-coated corrugated cardboard, 1968. Entry in a competition for exhibition furniture for *Magie des Papiers* (The Magic of Paper) at the Kunstgewerbemuseum Zürich, 1969. Designed in Willy Guhl's class, published in *Werk* 57, no. 5 (1970).

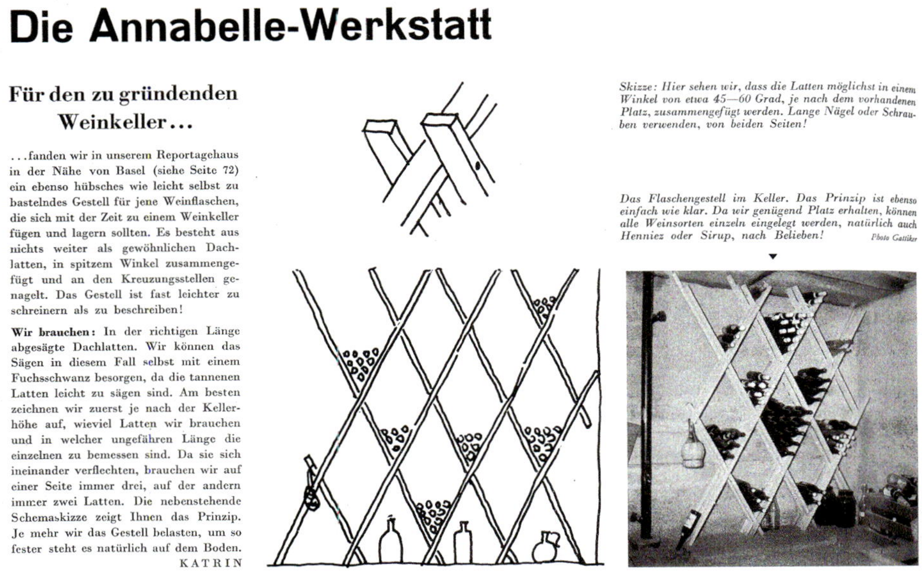

Die Annabelle-Werkstatt

Für den zu gründenden Weinkeller …

…fanden wir in unserem Reportagehaus in der Nähe von Basel (siehe Seite 72) ein ebenso hübsches wie leicht selbst zu bastelndes Gestell für jene Weinflaschen, die sich mit der Zeit zu einem Weinkeller fügen und lagern sollten. Es besteht aus nichts weiter als gewöhnlichen Dachlatten, in spitzem Winkel zusammengefügt und an den Kreuzungsstellen genagelt. Das Gestell ist fast leichter zu schreinern als zu beschreiben!

Wir brauchen: In der richtigen Länge abgesägte Dachlatten. Wir können das Sägen in diesem Fall selbst mit einem Fuchsschwanz besorgen, da die tannenen Latten leicht zu sägen sind. Am besten zeichnen wir zuerst je nach der Kellerhöhe auf, wieviel Latten wir brauchen und in welcher ungefähren Länge die einzelnen zu bemessen sind. Da sie sich ineinander verflechten, brauchen wir auf einer Seite immer drei, auf der andern immer zwei Latten. Die nebenstehende Schemaskizze zeigt Ihnen das Prinzip. Je mehr wir das Gestell belasten, um so fester steht es natürlich auf dem Boden.

KATRIN

Skizze: Hier sehen wir, dass die Latten möglichst in einem Winkel von etwa 45—60 Grad, je nach dem vorhandenen Platz, zusammengefügt werden. Lange Nägel oder Schrauben verwenden, von beiden Seiten!

Das Flaschengestell im Keller. Das Prinzip ist ebenso einfach wie klar. Da wir genügend Platz erhalten, können alle Weinsorten einzeln eingelegt werden, natürlich auch Henniez oder Sirup, nach Belieben! Photo Gattiker

663

664

665

666

667

663 The Annabelle Workshop, from *Annabelle,* no. 3 (1956). Regular magazine column with handicraft instructions.

664 Minimalist garden furniture at the SWB's housing exhibition at the "Im Höfli" housing development, Riehen (architect Hans Schmidt, 1947–54), June 1949
Photo: Photography Class, Kunstgewerbeschule Zürich

665 Nonconformist sitting on the HE 9x9 or Anti-Chair by Hans Eichenberger, 1970, Collection of Röthlisberger, Gümligen. The archaic construction consists of simple fir or oak beams firmly connected with wing screws, with woven linen seats stretched between them.
Photo: Hans Eichenberger

666 Sibylle Stoeckli and Christian Horisberger, Furniture for the Depot Basel, 2012. Reinterpretation of Enzo Mari's homemade furniture from the 1970s. Photo: Thierry Bruder

667 Yves Ebnöther – FabLab Zürich, laser-cut stool, 2010, EiABC ProtoLab Ethiopia. Beginning of the "Stool around the World" series.
Photo: Yves Ebnöther

in design teaching. In the course of similar experiments with form and material, Thomas Trachsel, a student at the Kunstgewerbeschule, conceived a lightweight yet stable, complexly folded stool made of coated corrugated cardboard that seems to anticipate by nearly forty years one of Nicola Stäubli's (b. 1978) three *foldschool* designs from 2007. While the two stools bear great formal similarities, they are nevertheless based on fundamentally different folding instructions.

From DIY Furniture to Design Classics

Many of the approaches pioneered in the sixties and seventies have come into vogue again in recent years—and sometimes today's DIY movement even gives the impression of being a retro phenomenon. For one thing, there can be found in various design blogs and Internet forums a history-conscious exchange between hobbyists and bloggers, designers and design mediators, on technical and practical as well as artistic and creative aspects of "do-it-yourself." This often leads to corrections of or new takes on historical designs—as well as freewheeling reinterpretations in the spirit of "design-it-yourself." The reinterpretation by Sibylle Stoeckli (b. 1979) and Christian Horisberger (b. 1978) of Enzo Mari's DIY furniture for the Depot Basel in 2012 is a good example. On the other hand, the sociopolitical stance of the counterculture around 1968 and thereafter, which advocated an autonomous way of life, free from the constraints of consumerism, has today been transformed in many cases into an unabashed commodity aesthetic. The ad hoc concepts for seats and recliners made out of discarded mattresses that designer Martin Schweizer presented in 1978 in an exhibition at the Design-Keller in Bern as an "inspiration for playful experimentation" can serve as an example of this tendency.[9]

In the mid-sixties, sitting on the floor had already evolved into an expression of a "posture" that was liberated from the conventions of conservative philistinism. For the protagonists of the counterculture, however, actually purchasing the ground-level designer furniture that was then in vogue was out of the question on both financial and moral grounds. Instead—as can be seen, for example, in photos of communal life in Berlin—they made do with repurposed mattresses. Gruppo DAM (Designer Associati Milano), the Milanese architect group, took up this idea in the early seventies as the epitome of casual seating, with a mattress-like piece of designer furniture that—although produced, sold, and advertised by the Busnelli company in various versions—could by all means be understood in the context of the Italian Radical Design movement as a provocation, or at least as humorous ironic understatement. For the designers James Hennessey and Victor Papanek, there was no reason to go out and buy such an easy-to-make piece of furniture. In their legendary 1973 DIY manual *Nomadic Furniture,* they adapted the design by Gruppo DAM for a homemade version that was really put together out of old mattresses.[10] Martin Schweizer's project in the late seventies can also be viewed in this context.

Mattress-like pieces very similar to what the Swiss presented back then in Bern are offered today by various furniture companies—for instance by Karup in Denmark—as upscale and high-priced luxury upholstered furniture. The whimsical creativity of improvised solutions have long been available for purchase on the furniture market as a finished product in which any hint of irony or provocation, or of immediacy and spontaneity, has vanished, replaced by a conformist, neoliberal, glossy brochure brand of chic. It therefore comes as no surprise that many a do-it-yourself design is in the meantime touted as a putative "design classic"—a perfectly engineered product that does not require the consumer to lift a finger. The *Crate* furniture designed by Gerrit Rietveld in the thirties as a way to recycle wooden pallets is a prime example, as is Ferdinand Kramer's *Knock-Down* furniture from the fifties, the folded cardboard pieces designed by Peter Raacke in the sixties, and Enzo Mari's *Autoprogettazione* projects from the seventies. Faced with such an—almost paradoxical—range of products, we can only assume that radically different intentions and motivations are at work, determining whether purchasers wind up being consumers on the commercial market or "prosumers" in an alternative world of design production.

* This essay is based in part on the deliberations on DIY culture developed and formulated in conjunction with Martina Fineder and Thomas Geisler on the occasion of the exhibition *Nomadic Furniture 3.0 – New Liberated Living?* (MAK, Vienna, June 12 – October 6, 2013); I would also like to thank Renate Menzi, Meret Ernst, and Nava Sutter for their suggestions and comments on the text.

1 See Hornung 2011.
2 See "Foldschool Cardboard Furniture," http://www.foldschool.com/pdf/press_d.pdf (accessed October 30, 2013).
3 The origins of the DIY movement itself can be traced back even further. In the late nineteenth century, magazines such as *Der Dilletant* published handicraft instructions for amateurs—for instance for working with a jigsaw or doing carvings or intarsia. The practical guides in the series *Wie baue ich mir selbst ...?,* published since 1911 in Leipzig, were also designed as handbooks for lay people rather than skilled tradesmen, featuring, for example around 1930, specific instructions for homemade furniture as well as for how to build the band saw with which to construct it. The slogan "Do it yourself" can be traced back to an article published in 1912 in the US magazine *Suburban Life* that challenged readers to paint their own walls. See Hornung 2011, p. 8.
4 M. 1956.
5 On the *Whole Earth Catalog* and other US counterculture publications as "tools" for an alternative lifestyle, see Binkley 2007. In the mid-seventies, an attempt was also made in Switzerland with the "decentrally" published *Alternativ-Katalogen* 1–3 to promote and disseminate along similar lines alternative strategies for a self-determined and more eco-friendly lifestyle.
6 Aebersold/Meyer 1969, p. 70.
7 See on this subject Schilder Bär/Wild (2001) 2003, pp. 90–93.
8 See Eimert 1994, pp. 38–41.
9 See the invitation card for the exhibition *Sitze und Liegen aus Matratzen,* Design-Keller, Bern, July 15 – September 16, 1978.
10 See Hennessey/Papanek 1973, p. 37. The two designers had discovered the "original" mattress furniture by Gruppo DAM in the magazine *Architectural Design;* see *AD,* no. 9 (September 1971), p. 575.

668

Stefan Zwicky, *Grand confort, sans confort, dommage à Corbu,* 1980. Artists held various actions in the spring of 1981 in rooms at the Kunstmuseum Bern that were scheduled for demolition to make way for a building extension, including this ironic alienation of the Fauteuil grand confort (designed by Le Corbusier, Pierre Jeanneret, and Charlotte Perriand, 1928). Constructed of concrete elements in a rebar frame, the chair has since attained museum status.

1980 — History

Postmodernism dominates the first half of the eighties. Simple consumer goods become visual manifestos. The room as a Gesamtkunstwerk is once again the focus of interest, although designers now relish in the subversive alienation of its traditional elements. There is no longer any fundamental difference between designing a chair and a house. Innovation is supplanted by remixing what already exists. The juxtaposition of old and new shares pride of place in the design world with the ennoblement of set pieces. An abrupt paradigm shift then occurs in mid-decade: away from the "thing as meaning" to the "thing as tangible experience." Designers now go after the optical sensual pleasure that can be derived directly from the impact of materials and forms. Costly, richly ornamented ingredients are no longer in demand, but rather "poor," hitherto hardly noticed materials that are carefully and subtly placed in new contexts for maximum effect.

669

Trix and Robert Haussmann, Hotel Plaza, Basel, 1984. Lobby. The few supporting columns have been supplemented with truncated columns that function as display cases and create a colonnade. The seating furniture is covered in a blue marbled fabric from the Mira-X collection: "The idea was to project architecture onto fabric, to translate hardness into softness." Conversely, the designers managed with the same imitation marble to "soften the architectural forms" and give the interior a homogeneous appearance. Critical Mannerism—a "tremendous provocation in the professional world" (Trix Haussmann).

670

Advertising photo for the Metro Lighting System designed by Hannes Wettstein and produced by Belux, Wohlen, 1982. The first low-voltage halogen lighting system suspended from contact cables got its name from Fritz Lang's film *Metropolis,* in which airships run along cables. Staged in a run-down vaulted space, the technoid lighting fixtures have a poetically exaggerated air typical of the time.

671

Peter Fischli and David Weiss, *Der Lauf der Dinge/The Way Things Go,* 16mm color film, 1985–87. A controlled happening in which everyday objects in precarious balance are set in motion in a kind of chain reaction. The film presents "history" as a stuttering, seemingly random course of events that is subject to the irrevocable laws of physics and chemistry—but does so without taking any recourse to actual incidents.

672

Bernhard Schobinger, Zerkratzes Blech – Zerbrochener Stein (Scratched Sheet Metal – Broken Stone) necklace, 1986. Portrait: Sonja with necklace, 1987. In the 1980s, the jewelry designer Bernhard Schobinger made a splash on the international scene with his provocative pieces made of garbage: artfully arranged glass fragments connected with wire, broken machine parts, pottery shards, or other worthless materials accentuated the nonconformist, individualistic look adopted by daring women.

Photo: Annelies Štrba

673

Trix and Robert Haussmann, Hauptbahnhof Zürich, hall with clock tower, model, 1987. The Swiss Railway Clock, a modern Swiss design classic (designed by Hans Hilfiker, 1944/1952), is staged as a monumental counterpoint in the historic hall built by Jakob Friedrich Wanners (1870–81)—a reversal of the principles the designer couple applied in the new underground section of the station.

674

Tilla Theus, Bar in Hotel Widder, Zurich, 1994–95. The handling of "history" as design theme. Next to the historical wooden structure a new concrete support frame has been inserted that displays thoughtful details. The crossbeam is decorated with the curving horns of the ram from which the hotel gets its name. Precious surfaces in lapis lazuli and red leather lend the simple room a noble air befitting its function as social hub within the luxury hotel, comprising several heritage buildings in Zurich's Old Town.

Photo: Heinrich Helfenstein

Trix and Robert Haussmann, Alfred Hablützel

H-Design for Mira-X 1981

675

676

677

675 Trix and Robert Haussmann, Alfred Hablützel, H-Design for Mira-X, launched in 1981
Cotton (cretonne), various dimensions, pictured designs: Mira-Lapideus, Mira-Marmoreus in various basic patterns and color schemes, Mira-X AG, Suhr
Photo: Alfred Hablützel
Museum für Gestaltung/Fabric donated by Trix and Robert Haussmann

676 Fabric remnant, Mira-Marmoreus faux uni, 1981
Museum für Gestaltung/
Donation of Eliane Schilliger

677 Mira-Marmoreus Murus with painted-on lettering, draped as soft architecture—blurring the boundaries between illusion and reality.
Photo: Alfred Hablützel

The textile manufacturer Mira-X, an offshoot of the Swiss furniture store Möbel Pfister AG, brought out its first collection in 1971. The Danish designer Verner Panton designed the *Mira-X Set* with bold colors and geometric motifs as a kind of construction kit for textiles. His Op-Art patterns virtually negated the difference between floor, walls, and ceiling. In the early eighties, Mira-X discontinued the Panton line. Living landscapes had long since been supplanted by more traditional, architecturally defined spaces. Design consultant Alfred Hablützel (b. 1931) was called in to help come up with ideas for new textiles befitting the latest in interior design. Together with Trix (b. 1933) and Robert (b. 1931) Haussmann, he succeeded in deriving a new general theme for Mira-X from the designer couple's relish for subversively alienating the reinstated architectural space. The Haussmanns had developed a "Critical Mannerism" that toyed with traditional architectural elements while reviving long-forgotten interior decoration techniques. The collection *H-Design for Mira-X,* launched in Milan and Paris in 1981, transferred illusionist means and trompe l'oeil effects to the field of fabric design. Basic architectural patterns—stone carving, coffers, fluting, paving, and slate cladding—now appeared on printed fabrics as faux marble, faux wood, or grisaille, or were woven directly into the textiles using various techniques, such as jacquard, piqué, matelassé, or damask.
AR

Hablützel 1981; Rüegg 2002c.

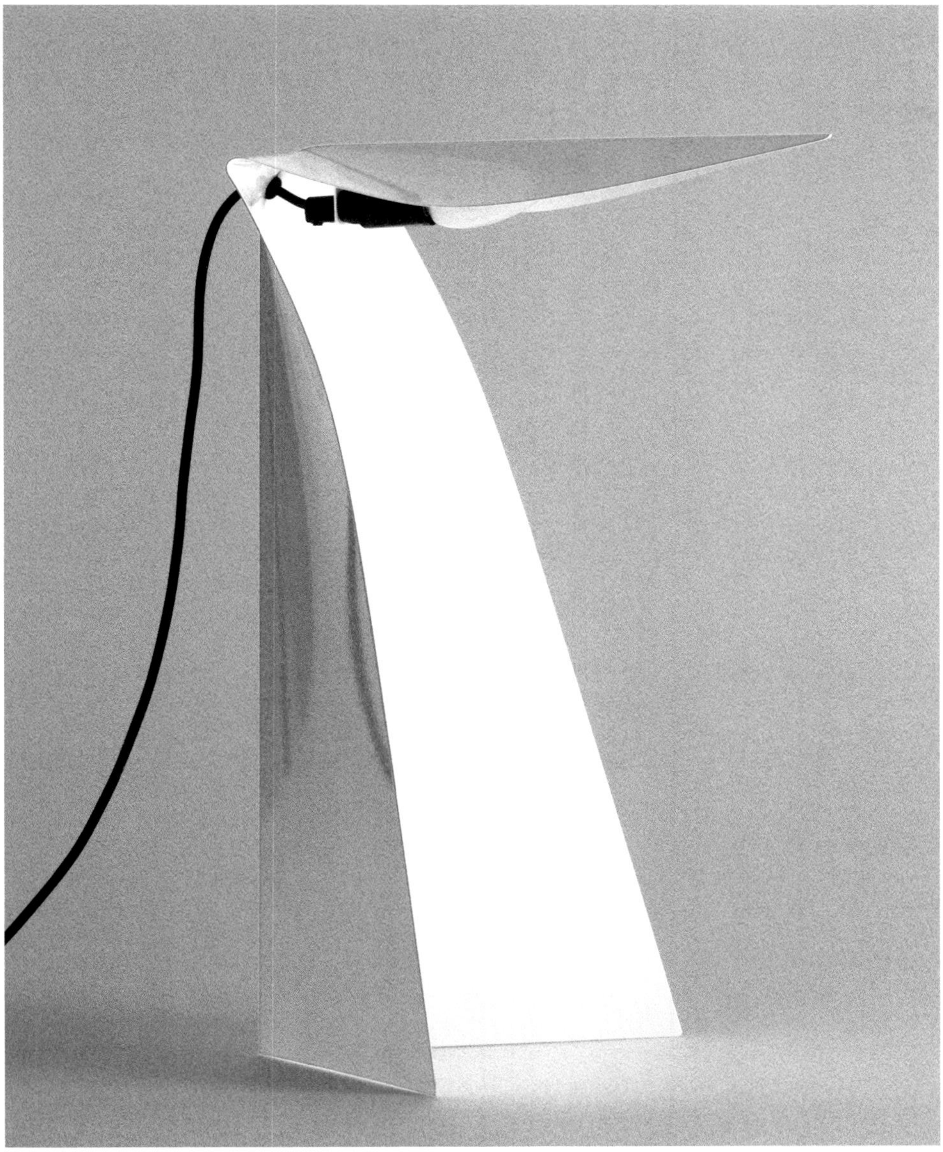

678

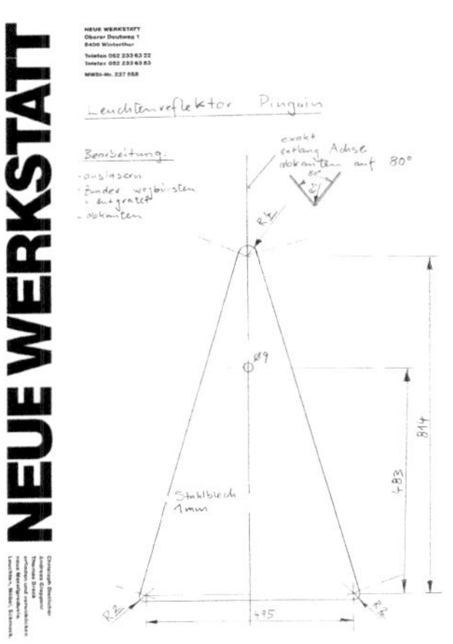

679

680

678 Neue Werkstatt, Penguin Lamp, 1989
Power-coated sheet steel, 51 × 34 × 35 cm,
Neue Werkstatt, Winterthur
Museum für Gestaltung*

679 Plan drawing for the Penguin, Neue Werkstatt, Winterthur

680 Group photo with Penguin Lamps,
Neue Werkstatt, Winterthur
Photo: Neue Werkstatt

After studying jewelry and industrial design at the Schule für Gestaltung Zürich, Christoph Dietlicher (b. 1958), Thomas Drack (b. 1962), and Andreas Giupponi (b. 1958) founded the Neue Werkstatt in Winterthur. The programmatic name (New Workshop) reflects the collective's design philosophy, combining both design and production under one roof. Design, craft, technology, and economy are not organized separately but are part of one self-contained practice. The only design requirement for the products was the feasibility of producing them in the designers' own metal workshop. They thus explored the possibilities offered by bending and shearing machines and a press to develop over the years their own product language based on reduced means of production. The collective turned out limited series of small furniture pieces, lamps, and jewelry featuring astoundingly simple construction. To mark the one-year anniversary of the Neue Werkstatt, Dietlicher, Drack, and Giupponi designed the humorous *Penguin* lamp as the perfect embodiment of their way of thinking and working. The lampshade and stand are formed from an isosceles triangle of sheet metal that is bent forward at a right angle to its central axis. Its shape arises through minimal manipulation of the material and by taking advantage of its specific properties. The upper end is simply bent forward, resulting in a zoomorphic form that is just as complex as its basic shape and process are simple. Displaced through bending, the sheet metal bulges in dynamic lines that give the stiff bird an animated and graceful air. RM

Schilder Bär/Wild (2001) 2003, pp. 131–33; Sachs 2007, p. 216.

Hannes Wettstein

Juliette Stackable Chair
1987

681 Hannes Wettstein, Juliette Stackable Chair, 1987, prototype
Power-coated sheet steel and round steel, 77 × 45 × 49 cm, Baleri Italia S.p.A., Albano Sant'Alessandro (IT)
Studio Hannes Wettstein

682 Photographic staging by Baleri Italia S.p.A.
Photo: cerruti baleri/Ezio Manciucca

683 Juliette is the result of some 200 sketches, drawing, 1986
Studio Hannes Wettstein

681

682

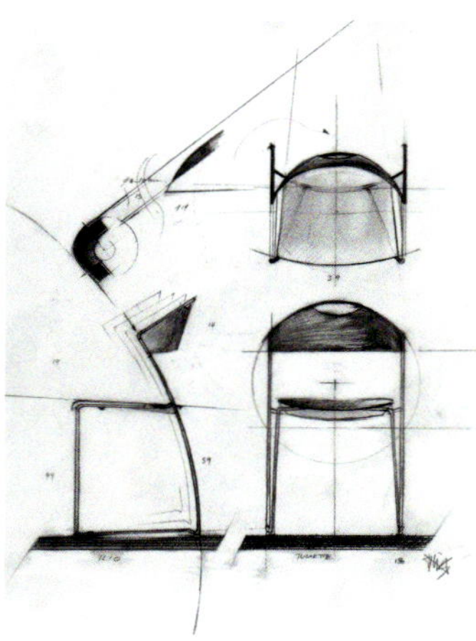

683

Hannes Wettstein (1958–2008) was a great believer in developing "things that simply maintain their presences, so that the users themselves become the main protagonists in space, and not the furniture." Creating self-referential, straightforward objects was the goal of a new generation of designers who had not witnessed the break with modernist principles in the late sixties but had instead begun their design careers in the eclectic profusion of the eighties. A trained structural engineering draftsman and technically savvy self-taught designer and architect, Wettstein produced lamps, furniture, and utilitarian objects characterized by pure, unspectacular, functional design. The *Juliette* chair was the result of 200 sketches in which he engaged in detail with the theme of sitting. *Juliette*'s rear section swings dynamically forward, offering itself for use as an armrest. A thin tubular steel frame encloses the seat and pressed sheet metal backrest. The stackable chair weighs only 5.3 kilograms and is extremely slender. Storable in high space-saving towers, it is ideally suited for temporary seating in large spaces. *Juliette* was manufactured by welding robots, a manufacturing technology borrowed from the automotive industry. Despite its practicality, the filigree metal chair has a graceful appearance. Its structural striving for immateriality gives it the elegance of a high-tech device. With *Juliette,* Hannes Wettstein rose overnight to international fame. RM

Eschbach 2006; *Hannes Wettstein* 2012, p. 49; www.studiohanneswettstein.com.

684 Mario Botta, Seconda Armchair, 1982
Tubular steel, spring steel, and perforated sheet metal with epoxy resin coating, polyurethane backrest, 72 × 52 × 58 cm, Alias S.r.l., Grumello del Monte, Bergamo (IT)
Museum für Gestaltung/Donation of Alias S.r.l.

685 Mario Botta, Sketches for Prima (upper right) and Seconda chairs, 1982

686 Color variations, 2013
Photo: Alias S.r.l.

684

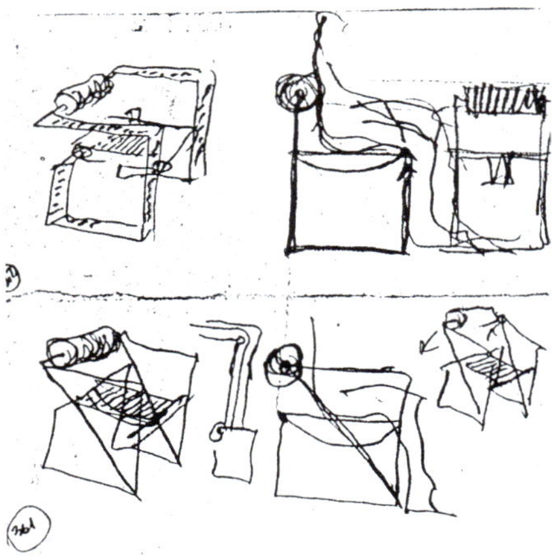

685

686

Mario Botta (b. 1943) had already developed his own distinctive architectural style when the Italian company Alias asked him to design a chair for them. A chair by a master architect who put his faith completely in the basic geometry of cubes or cylinders? The challenge evidently appealed to Botta, because he saw no fundamental difference between designing a chair and a house. And he did in fact base his first attempt, the *Prima* chair, on a cube. With consummate skill, he traced the edges of the Platonic body with flat steel and tubes in such a way that a single continuous line could carry a seat made of springy sheet metal and, as a rear extension over the seat, a cylindrical backrest. Now his creative powers were unleashed: over the next few years, Botta produced a host of designs for chairs, tables, and a sofa; for table, floor, and wall lamps; even for carafes and vases. *Seconda* is an interesting second attempt on the theme of the chair. By introducing armrests, Botta achieved a greater level of figural complexity than in the elementary *Prima.* A right-angled triangle now augments the seating cube on both sides and carries the back bar with its two hard polyurethane rolls. Elevated above the floor on plastic wheels, the formally and technically sophisticated construction is a sculpture for sitting—perfect, concise, iconic. AR

Mario Botta 1989; Loderer 2002.

Logitech S.A.
Factory design

P4 Computer Mouse
1982

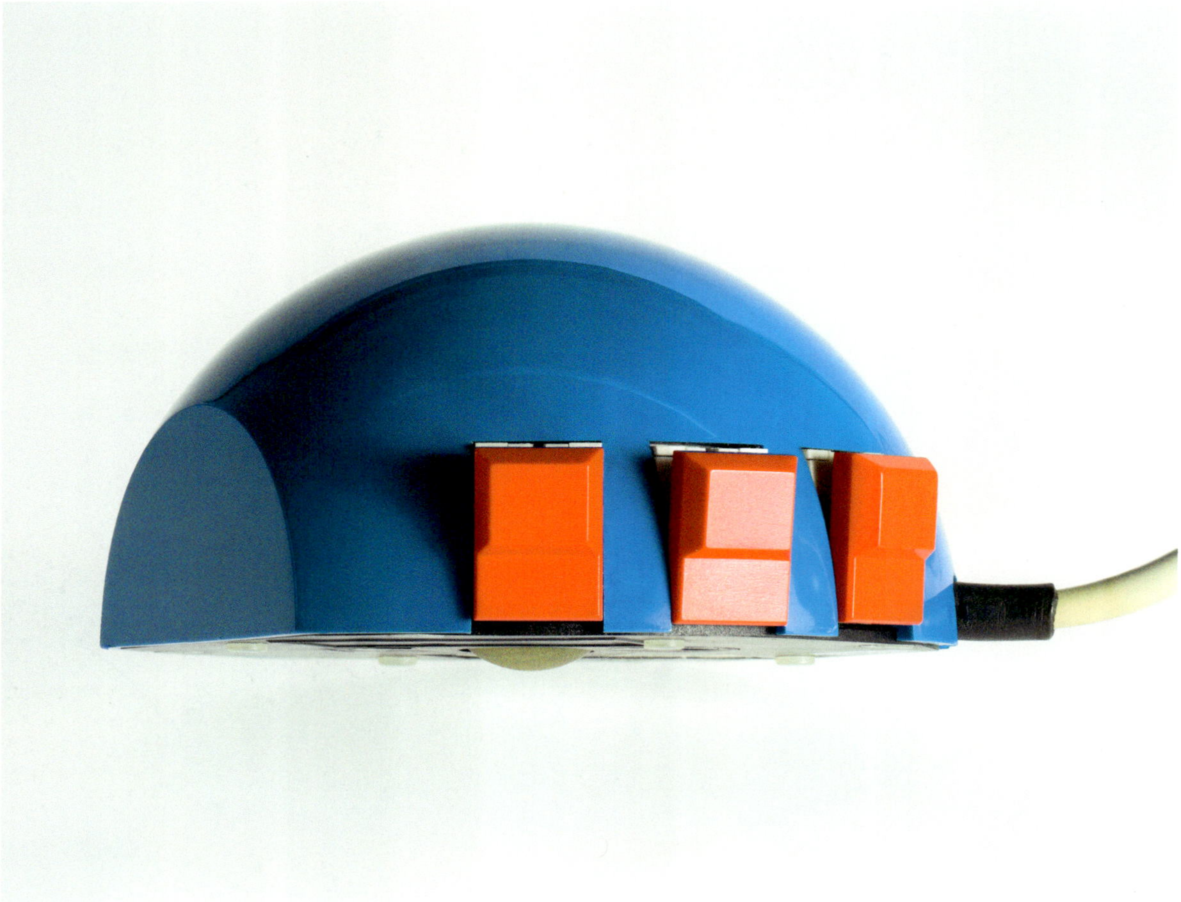

687

688

687 Anonymous, P4 Computer Mouse, 1982
Plastic, 4 × 7.5 × 9.5 cm, Logitech Europe S.A., Romanel-sur-Morges
Museum für Gestaltung/Donation of Logitech S.A.

688 P4 Computer Mouse, connected to a Lilith computer with teak housing, developed by Niklaus Wirth at ETH Zürich in 1978–79, 1984. Photo: Peter Auchli

In the eighties, the PC conquered not only offices but also made inroads into private homes. The mouse played a leading role in this success story: it eliminated the need to enter all commands via the keyboard, enabling a more direct interaction with the computer, which was particularly important for graphics programs. In the late seventies, the young Swiss physicist Daniel Borel was working at Stanford University, where he became familiar with Xerox Alto workstations, innovative interfaces, and the computer mouse—invented in 1964 by Douglas Engelbart but not as yet commercially available. In 1978–79, Jean-Daniel Nicoud at the École polytechnique fédérale de Lausanne (EPFL) improved on Douglas Engelbart's invention by developing alongside the precision engineer André Guignard from La Chaux-de-Fonds the first computer mouse "Made in Switzerland." Instead of transferring movement electrically (using potentiometers on two wheels), it used a rubber ball and introduced the opto-mechanical principle (a combination of optical technology and mechanical tracking). In 1982, Borel, back in Switzerland and looking for actionable ideas for his company Logitech (founded in 1981 with Pierluigi Zappacosta), acquired the rights to Nicoud's opto-mechanical mouse. Through the development of software, inexpensive production in Taiwan, and clever marketing, Logitech succeeded in bringing the *P4* mouse from lab to market and in distributing it worldwide as the first computer mouse. Compared to today's mice, the *P4* looks like a cute eighties-style toy. Its hemispherical shape fits nicely into the cavity of the hand and translates its motor movement to the cursor. RM

Landwehr 2001; www.logitech.com.

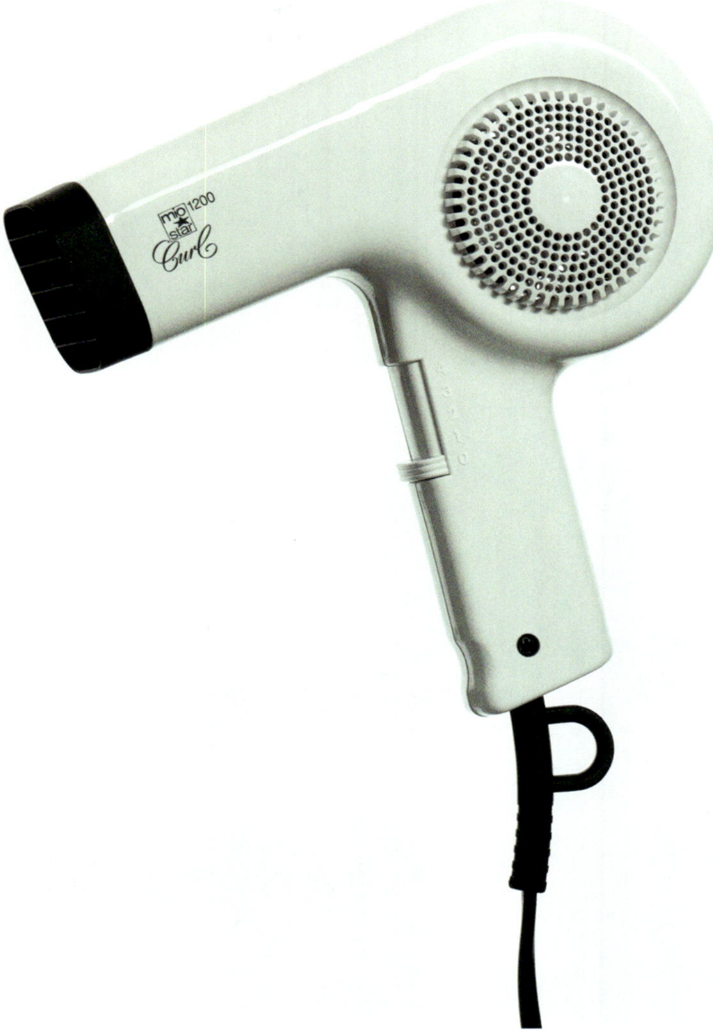

689

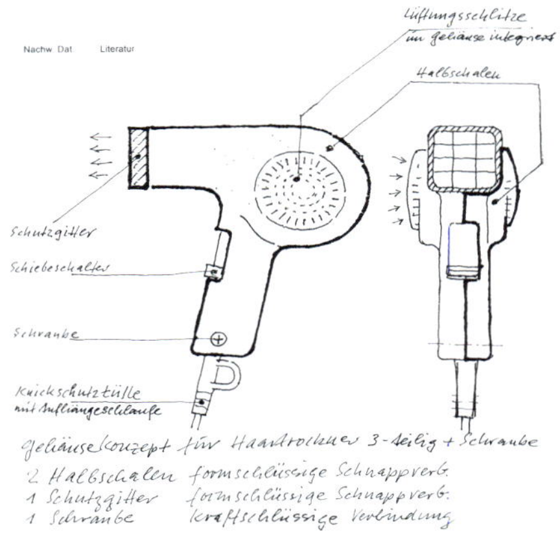

690

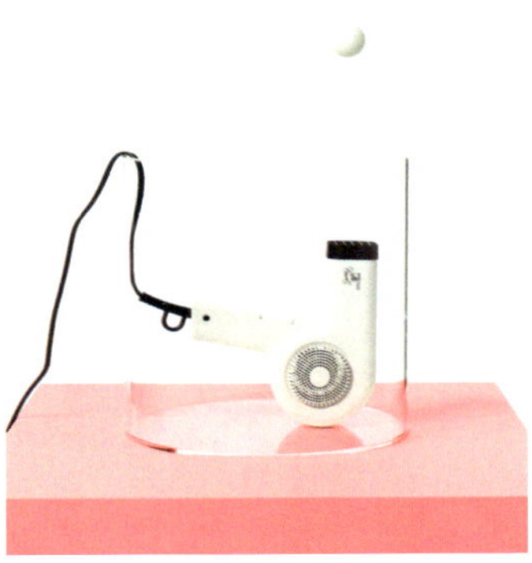

691

689 Walter Eichenberger and Ernest Muchenberger, Curl Professional Type F26 Hair Dryer, 1989
Plastic, 20 × 20 × 7.5 cm, Migros-Genossenschafts-Bund (MGB), Zurich; Spemot AG, Dulliken
Museum für Gestaltung/Donation of Migros-Genossenschafts-Bund

690 Construction drawing for the Curl Professional hair dryer, M&E Design, Cham
Archive of Walter Eichenberger

691 Damien Hirst, *What Goes Up Must Come Down*, 1994, edition for the magazine *Parkett* 40/41
Photo: Prudence Cuming Associates

Ernest Muchenberger (b. 1934) and Walter Eichenberger (b. 1936), both product designers trained at the Hochschule für Gestaltung Ulm, founded the design studio M&E Design in Cham in 1967. They designed "no-name" products for industry until 2002. For the Spemot company, a developer and manufacturer of electrotechnical equipment that was a supplier to Migros, M&E designed kitchen and household appliances, such as the hair dryer *Curl Professional Type F 26* under the Mio-Star brand. In collaboration with the design engineer Bruno von Arx, M&E developed a plastic housing for the dryer consisting of only three parts with interlocking snap connections. This saved time on assembly, reducing costs significantly, because Migros wanted to sell the blow dryer for a mere thirty Swiss francs. M&E also tried to make the shape user-friendly, balancing the center of gravity so that the wrist would not tire even when using the hair dryer for longer periods. Designed for aesthetic longevity, the basic form is reduced and the controls clearly arranged. The balanced radial perforations for ventilation manifest an attention to detail that is hard to find anymore in everyday products in this price class. M&E ensured economically efficient production and at the same time created a timeless, archetypal form. This and the regular parallel air flow emitted by the dryer must have been what moved the artist Damien Hirst to use the Mio-Star hair dryer for his installation *What Goes Up Must Come Down* (1994).
RM

Easy! Easy? 2011.

692

693

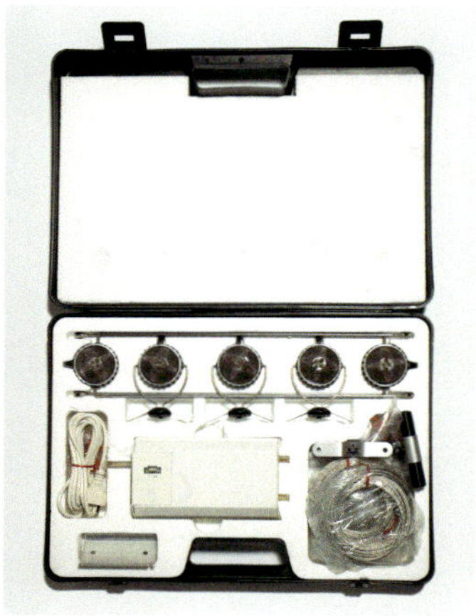

694

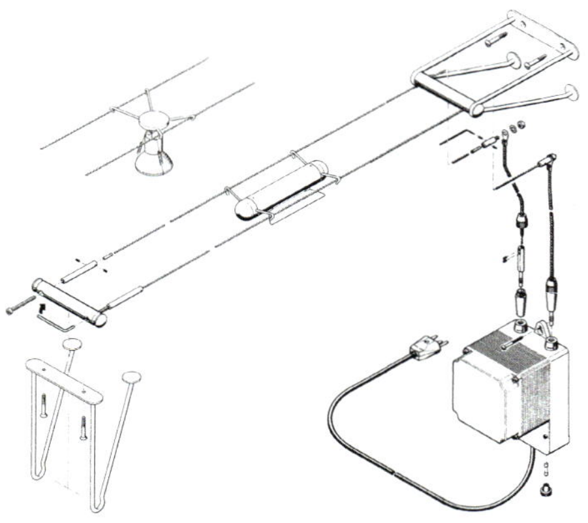

695

692 Hannes Wettstein, Metro-45 Lighting System, 1982
Injection-molded aluminum light fixtures, various reflectors, Belux AG, Wohlen
Museum für Gestaltung*

693 The first low-voltage lighting system mounted on contact cables stretched across the room.

694 Metro Lighting System in case with steel cable, light fixtures, wall and ceiling hangers, halogen bulbs, cable length (max.): 10 m
Photo: Lozza

695 Assembly plan for Metro low-voltage lighting system, 1982

In 1982, the designer and architect Hannes Wettstein (1958–2008) started his own business, and in that same year he invented a lighting system in close cooperation with the Swiss lamp company Belux that would later be copied thousands of times over. *Metro* was not only the world's first low-voltage lighting system suspended from tensile contact cables, it also inaugurated a whole new approach to lighting. Like stage lighting, each light fixture is mounted in the ideal location for a particular room and can be pointed in several different directions. Delivered in a custom case, the lighting system takes the form of a toolbox. It consists of two wires to be mounted in parallel along which small, strong light sources can be freely positioned, powered directly by the cables. The fixtures come with metal bodies in varying shapes, floating in the room like fantastical flying objects. In fact, the twenty-four-year-old Wettstein was inspired in his design for *Metro* by the zeppelins in Fritz Lang's film *Metropolis* (1927) as well as by illuminated trams and ski lifts. The difficulties inherent in this technological novelty consisted in curbing the electricity transported along the open cables, managing heat issues, and making sure the contacts functioned properly. *Metro* was meant to be a technical lighting product rather than a decorative design, and the designer at times wished he could take the lights away again from users who did not deploy them as he had envisaged. RM

Hannes Wettstein 2012; www.studiohanneswettstein.com.

696 Factory design, Swatch GN 100 Watch, launched in 1983
Plastic, 1 × 3.5 × 24 cm, Swatch AG, Biel; ETA AG, Grenchen
Museum für Gestaltung/Donation of Alfred Emy

697 Alfred Hofkunst, Vegetable Set, 1991, Swatch AG, Biel. Limited edition

698 Robert & Durrer, Swatch, 1983–88, Swatch AG Biel. Details of various dials
Photos: Robert & Durrer

696

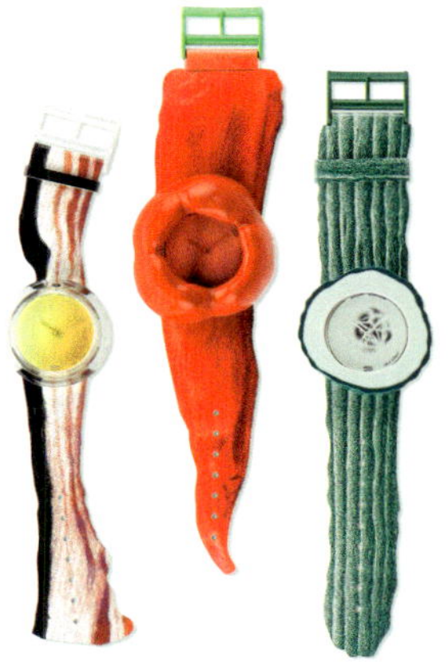

697

698

In the seventies, the Swiss watch industry found itself at an unprecedented low point. From 1977 to 1983 the volume of exports declined by half. Although the world's first electronic wristwatch with quartz crystal display was developed in Neuchâtel in 1967, Japan was quicker to bring the new technology to the mass market, and Switzerland ending up losing its leading position. The radical reinvention of the Swiss watch by Swatch (Second Watch) has an economic, a technical, and an aesthetic aspect, with all three equally determining the development and design of the Swatch. Simplifying production was central to the undertaking: the watch's plastic housing (produced by injection molding) also serves as a mounting plate for the mere fifty-one components (instead of the ninety-one in typical watch movements), which are assembled in a fully automated process with ultrasonic welding technology. Despite this pared-down construction, the Swatch, available for only fifty Swiss francs, runs on its mini-battery just as precisely as a chronometer. In 1983, the Swatch was presented in twelve colors—with plans to produce a million watches. But it was not until the plastic watch was reconceived as a fashion accessory that it experienced its true breakthrough. In the period from 1983 to 1989, the visual designers Jean Robert and Käti Robert-Durrer came up with five different product lines displaying great graphic diversity—something for every taste. Swatch thus became a pioneer of individualized surface design, a trend that would go on to shape the product culture—from sneakers to toothbrushes—in the nineties. RM

Häfliger et al. 1991; *Swatch Watch* 2000; *Swatch* 2011; Lichtenstein 2014.

699

699 Kurt Thut, Bellows Cabinet Model 480, 1997
Aluminum frame, polyurethane mesh sides, MDF shelves, 185 × 99 × 30–60 cm, Thut Möbel AG, Möriken. Photos: Alfred Hablützel
Museum für Gestaltung*

700 Bellows in folded state

700

Instead of carrying on time-honored traditions when he took over his father's carpentry shop in Möriken, the forty-five-year-old designer and architect Kurt Thut (1931–2011) proceeded to embark on a series of remarkable innovations. He used the combination design studio/production workshop for a period of inventive tinkering with new materials and techniques. In the process, he set his sights more than ever on the nomadic lifestyle of the postwar period, looking beyond the world of furniture for novel ways to construct lightweight, portable pieces. Alan Colquhoun called the analogous design procedure adopted by Le Corbusier "displacement of concepts." Thut's lightweight *Aluminum Cupboard* (1986) recalls the aluminum aircraft *Ju 52,* and the extendable *Scissor Bed* (1990) is based on a principle he derived from a trivet. Thut repeated the scissors principle for the doors of his *Folding Curtain Cupboard* (1993). And, not stopping there, in 1998 he replaced the coated particle board of this model with a bellows tunnel made of polyurethane. The *Bellows Cabinet* strikingly recalls an old folding camera, an impression accentuated by the precision aluminum construction of the front frame, equipped with a closing scissors grille holding a plastic curtain. Depending on the desired shelf size, the cabinet can be extended to a depth of sixty centimeters. Its elaborately celebrated adaptability is arguably due less to functional considerations than to Thut's interest in honing an idea that he made into a signature stylistic device. AR

Hablützel et al. 1988; Gantenbein/Hablützel 2001.

701

702

703

701 Hanspeter Weidmann, Shoe Rack, 1984
702 Galvanized sheet steel, steel springs, nylon band, 170 × 54 × 18 cm, hp Weidmann, Basel; Bemag Objekteinrichtungen AG, Sissach.
Photos: Ruedi Walti
Museum für Gestaltung*

703 Weidmann Wardrobe System, Lehni AG, Dübendorf (produced and sold since 2007). Page from prospectus

After Wilhelm Kienzle presented his legendary *Shoe Rack* to the Wohnhilfe cabinetmakers' cooperative in 1951, it would be a long time before anyone again devoted some thought to footwear proliferating in stairwells and hallways. It was not until 1984 that the versatile Basel design student Hanspeter Weidmann (b. 1958), inspired by a training project initiated by Ulrich P. Wieser (Wohnbedarf AG, Basel), focused on this problem area. Weidmann's design—formally reduced and sophisticated in functionality and construction—is now widely regarded as a prime example of "Swissness" and is considered a classic of recent design history. Its originality lies not in a spectacular appearance but in an ingenious approach that even allows it to be produced by hand. Through folding and bending, three stable base elements are shaped from rectangular sheets of metal, which then form the back and side walls, the angled shoe compartments, and the central pull bar. Two Tubtara rivets clamp each compartment to the side walls. A steel spring is attached at the lower end of the pull bar. The whole piece does without any screws. Pulling on the bar opens or closes the whole series of stacked compartments simultaneously. Since the edges of the side walls are curved slightly differently, any number of *Shoe Racks* can be lined up to create a seamless wall relief—also adding on the *Storage Units* (1986) and *Mirrored Cabinets* and *Hall Units* (2004) that Weidmann later designed. AR

Herrmann 1989; Wild 2002b.

704

705

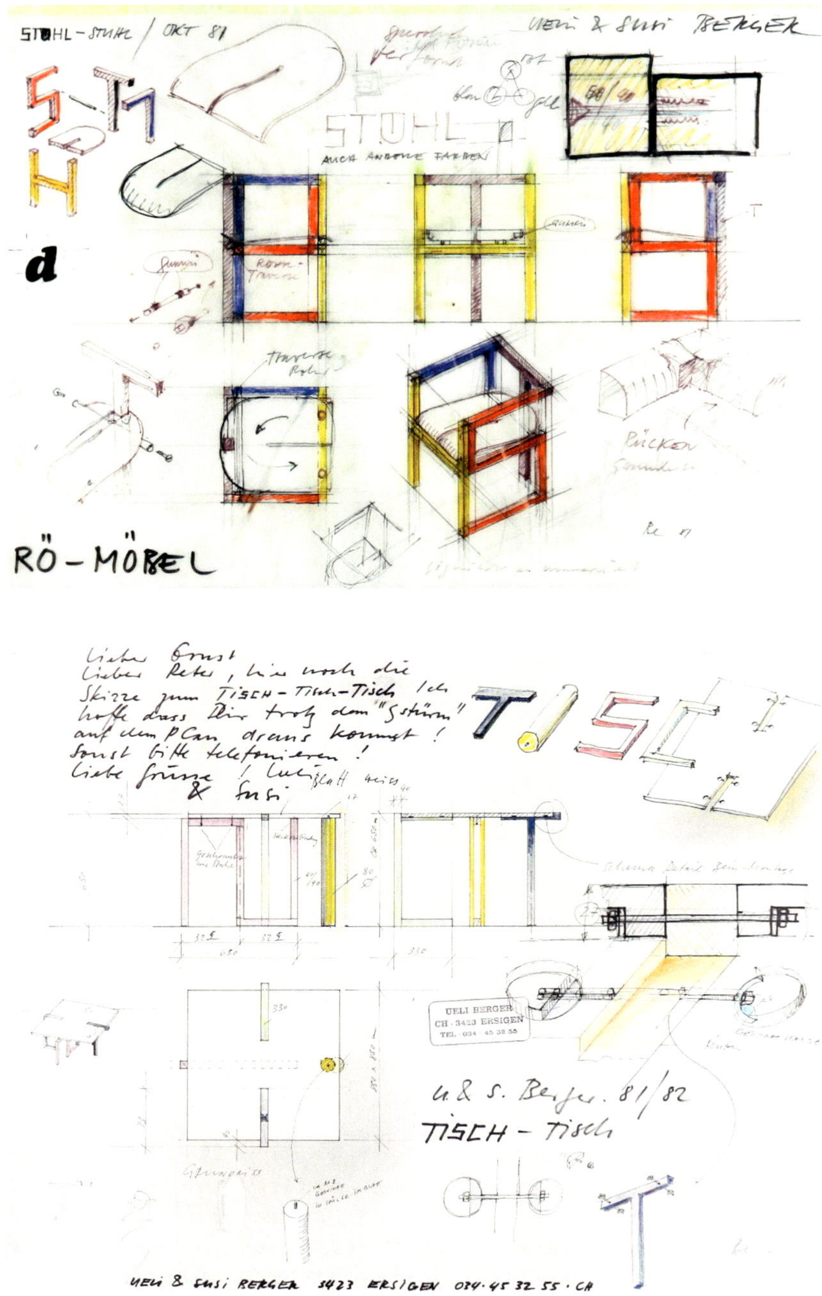

706

704 Page from Alois Martin Müller and Köbi Gantenbein, eds., *Mehrwerte. Schweiz und Design. Die 80er* (Value Added: Switzerland and Design: The 80s), Museum für Gestaltung Zürich, 1991

705 SWB Chair Fun Auction, 1967, Bern Group. Exhibition of competition entries and auction of artistically designed seating options devised by SWB members.

706 Susi and Ueli Berger, Construction drawings for the furniture pieces Stuhl-Stuhl (Chair Chair) and Tisch-Tisch (Table Table), 1981–82, Röthlisberger Schreinerei AG, Gümligen. Furniture made of linguistic signs.
Museum für Gestaltung/
Donation of Susi and Ueli Berger

The early eighties were a time of upheaval, especially for Switzerland. When in May 1980 the Zurich City Council denied funding to an autonomous youth center while at the same time approving 60 million francs for the renovation of the Opernhaus Zürich, the decision sparked prolonged youth riots. Young people demanded their freedom from the traditional high-culture mainstream. During the same years, architects and designers were also challenging the relevance of established cultural norms. A document that gives a good picture of the spirit of the times is an issue of the magazine *Werk, Bauen + Wohnen* (founded in 1980) called "Standortbestimmung" (roughly translatable as "The Lay of the Land").[1] Lucius Burckhardt (1934–2003) explains therein that the "necessity in the postwar situation to relearn the lessons of the international style"[2] had ended in the seventies. This had already been acknowledged especially by those "building on spec and by the manufacturers of building materials" while, in part as a critique of this development, leading architects had evolved in their designs a synchronous polyphony that was more attuned with the sociocultural diversity of society.

Invisible Design

Burckhardt, a sociologist, had been appointed to a post in the architecture department of the ETH Zürich after the 1968 revolts—during an "experimental phase"—and he was soon advocating expanded horizons for architecture and design. He summarized his proposals in the 1980 essay "Design ist unsichtbar" ("Design Is Invisible").[3] Burckhardt criticized an understanding of design as something that did nothing more than "acknowledge its external parameters," conceived merely to ensure that the can fits the can opener, and vice versa. This approach would only lead to a closed system of objects, which must then be overcome. Instead of equipping kitchens with still more appliances, designers should try to conceive of a kitchen that would inspire guests to help with the cooking. Burckhardt believed that design must be understood as a model expression of its respective social environment.

He explains this principle by taking recourse to the theory of *pattern language* posited in 1977 by Christopher Alexander,[4] using the example of a street corner: "For the newsstand thrives on the fact that my bus has not yet arrived, and so I buy a newspaper; and the bus happens to stop here because this is an intersection where passengers can change to other lines. 'Street corner' simply tags a phenomenon that encompasses, above and beyond the visible dimension, elements of an organizational system comprised of bus routes, timetables, magazine sales, traffic light sequences, and so on." So if you're going to design a street corner, you must change not the object itself but its rules: "What we need, perhaps, so that I won't miss my bus while scrabbling for change, or because the newsagent is serving another customer, is a simplified method of paying for a newspaper." Instead of making an electronic magazine dispenser, Burckhardt proposes intervening in the existing system by perhaps selling periodicals for a round sum. This critical-conceptual approach is paradigmatic for design circa 1980.

Conceptual Principles: Word, Letter

For incisive examples of how linguistic systems such as Alexander's pattern language influenced design, we need look no further than the furniture pieces *Stuhl-Stuhl* (Chair Chair), *Tisch-Tisch* (Table Table), and *Glas-Holz-Tisch* (Glass Wood Table) by the designers and artists Susi and Ueli Berger, all dating from 1981.[5] In the early eighties, the Bergers observed with some disillusionment that all the "good furniture" had already been built. Nevertheless, after taking up contact with the furniture maker Röthlisberger, they did perceive an opportunity to produce objects without a particular purpose. They decided to seize this chance to set off in search of something "new." Drawing on the work of the artist Saul Steinberg, who often transformed words into objects in his drawings, they wanted to produce furniture from the elements, that is to say the letters, of the words used to denominate them. They set themselves the condition of constructing furniture pieces from the letters of the German words for chair (*Stuhl*) and table (*Tisch*) as basic types for their designs. These seemingly restrictive, self-imposed criteria in fact opened up a variety of new creative possibilities for designing a unique, model piece. While exploring possible variations, the similarity to Gerrit Rietveld's *Red and Blue Chair* came to mind, so the Bergers chose pastel tones for their components instead of primary colors in order to distinguish their "differently grounded design concept." Along the same lines, instead of choosing a design that came closest to a conventional chair or table, they went for a rather bizarre-looking formal composition that would let them "try to break out of the traditionally strict design to create something more playful and emotional." The logical consistency of a chair with a lowercase *h* or a table with a round *i* is thus deliberately breached in favor of formal whimsy.

The Bergers alluded with their designs not only to the rules of construction dictated by the names of the respective types but also to the overriding conceptual principles, thus expanding the functionality of their objects.

Critical Mannerism and Formal Disruptions

The architects and designers Trix and Robert Haussmann took things a step further, ultimately in the direction of the dissolution of the object, with their 1980 work *The Log-O-Rithmic Slide Rule.*[6] On their so-called *concept slider,* adjectives and adverbs from architecture and design discourse meet randomly as the two moving scales slide in either direction. The resulting pairs of terms form conceptual proposals that are intended to supply design ideas. The piece was originally conceived for the exhibition *Manierismo Critico,* which the Haussmanns presented in 1981 at the Centro Culturale Studio Marconi in Milan. Motivated by a widespread feeling of resignation in the face of a modernism that rampant "commercialization and internationalization"[7] had emptied of ideals, they turned to the Mannerist style of the sixteenth and seventeenth centuries for inspiration. Based on that historical era, they used the term to refer to a general tendency to oppose any frozen classicism—in their case, that of classical modernism. They put a series of *Lehrstücke* ("Learning Pieces") on display in which they took their maxims to their logical extremes. Following the principle of the *Log-O-Rithmic Slide Rule,* the titles of the *Lehrstücke* recombined the modernist tenets in new ways and implemented them in material form by way of model objects. *Störung der Form durch Funktion* (The Disruption of Form by Function) was the name of one of the nine pieces. Here, a form resembling an ancient column took on the role of a chest of drawers. By opening and closing the drawers, the column could be made to dance, exposing its Mannerist underpinnings. The intended result of this revelatory design was to propose a *Critical Mannerism.* Often only produced in limited editions, the furniture pieces became concepts and metaphors and thus drew closer to the characteristics of fine-art sculpture. This development was in turn fostered by with their relationship with Röthlisberger, with which Hans Eichenberger likewise collaborated, experimenting with elaborate, illusionistic wood marquetry. His *Vorhang-Kredenz* (Curtain Credenza) has a pattern imitating an interrupted curtain or waterfall, and the *Piedestal* chest of drawers creates the illusion that the drawers are a sliding staircase or chair elements.

Susi and Ueli Berger tried their hand at exquisite wooden cabinetry as well, designing the *Schubladenstapel* (Stack of Drawers) commode. The piece creates the illusion of a tower of drawers loosely stacked at different angles. Kurt Thut had already experimented in the seventies with cushions sewn in the shape of the letters of his name, which he wanted to use to label his furniture, similar to the letter games in Trix and Robert Haussmann's *Log-O-Rithmic Slide Rule* or the *Stuhl-Stuhl* designed by Susi and Ueli Berger. A *Schweizerstuhl* (Swiss Chair) by the Bergers could be created by folding a Swiss cross shape, and their *Einmann/Einfrau* (One Man/One Woman) bookshelf had a human silhouette. Koni Ochsner referred directly to art with his *Mondrian* cabinet, dealing in a new and nondogmatic manner with the paintings of the modern master. Finally, a direct exchange between fine art and serially produced design took place when the company Lehni in Dübendorf produced a series of furniture in 1984 by the American minimal artist Donald Judd, thus breaking down the boundaries between the two disciplines of art and design.

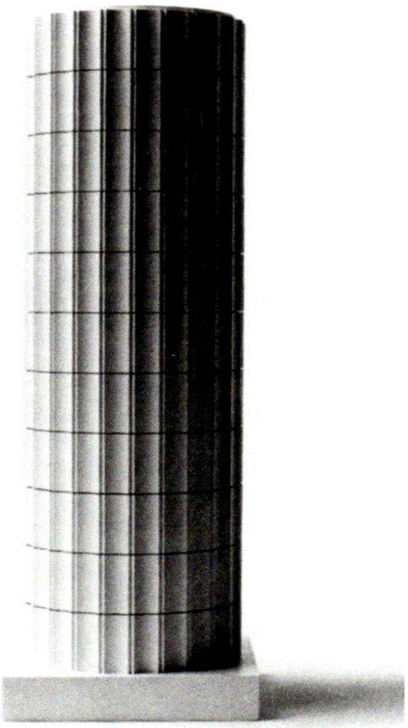

707

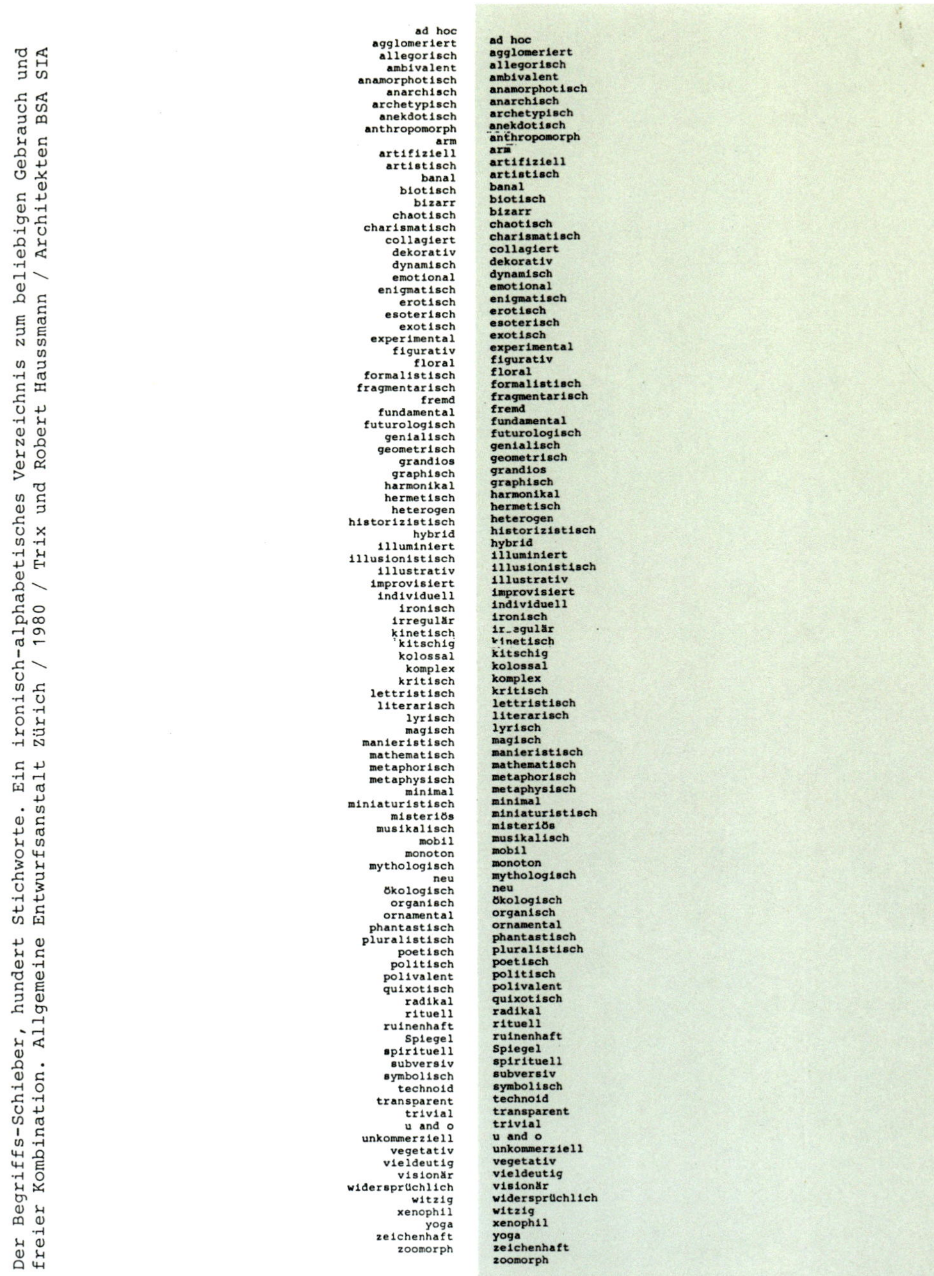

708

707 Trix and Robert Haussmann, Lehrstück II: Störung der Form durch Funktion (The Disruption of Form by Function), Säulenstumpf (Truncated Column) as stacked drawers, 1978. Model
Museum für Gestaltung/
Donation of Trix and Robert Haussmann

708 Trix and Robert Haussmann, *The Log-O-Rithmic Slide Rule:* "One hundred catchwords. An ironic alphabetic directory for free use and combination," 1980

709

710

711

709 Trix and Robert Haussmann, Lehrstück I: Möbel als Architekturzitat (Furniture as Architectural Quotation), Bridge Desk, Column Cabinet, Wall Dresser, Temple Table. Model.
Photo: Matthias Buser
Museum für Gestaltung/
Donation of Trix and Robert Haussmann

710 Atelier Roger Pfund, Alfred Hablützel (photo), poster for Erich Dardel, Aarberg, 1987. Shows the Manhattan Furniture Collection by Trix and Robert Haussmann, 1987

711 Trix and Robert Haussmann, Lehrstück V: Function follows Form, 1979. Model. One of five pseudofunctional metamorphoses of a given form.
Photo: Fred Waldvogel
Museum für Gestaltung/
Donation of Trix and Robert Haussmann

712

713

714

715

712 Trix and Robert Haussmann, UN Building in New York, 2012, alienated on the basis of Lehrstück IV: Seven Codes (Mirrored Cabinet, 1978). Digital print on polytrans (360 × 280 cm)
Museum für Gestaltung*

713 Trix and Robert Haussmann, Lehrstück III: Störung der Form durch das Ornament (The Disruption of Form by Ornament), 1978. Model
Museum für Gestaltung/
Donation of Trix and Robert Haussmann

714 Susi and Ueli Berger, Schubladenstapel (Stack of Drawers), 1981, Röthlisberger Kollektion, Gümligen
Museum für Gestaltung

715 Hans Eichenberger, Vorhang-Kredenz (Curtain Credenza), 1987, Röthlisberger Kollektion, Gümligen
Museum für Gestaltung*

716

717

718

719

716 Robert and Trix Haussmann, Boutique Lanvin, Poststrasse, Zurich, 1977. Interior with faux marble painting. Photo: Heinrich Helfenstein

717 Poster design for the *Manierismo Critico* exhibition at Studio Marconi, Milan, 1981. Designed with the fabric H-Design for Mira-X by Trix and Robert Haussmann and Alfred Hablützel, 1981.
Photo: Alfred Hablützel

718 Trix and Robert Haussmann and Alfred Hablützel, H-Design for Mira-X. Staging of the Mira-X fabrics for a prospectus produced by Mira-X AG, Suhr, 1981. Photo: Alfred Hablützel

719 Meret Oppenheim, *Table with Bird's Feet,* 1939, gilded wood tabletop, bronze feet, 65 × 70 × 50 cm
Private collection (FR)
Produced by Dino Gavina, Bologna, as the model Traccia in the Ultramobile Collection, 1972.
Currently produced by Cassina S.p.A., Meda (IT)

720 Trix and Robert Haussmann, Neon Chair, 1967 (2011 replica)
Collection of Trix and Robert Haussmann

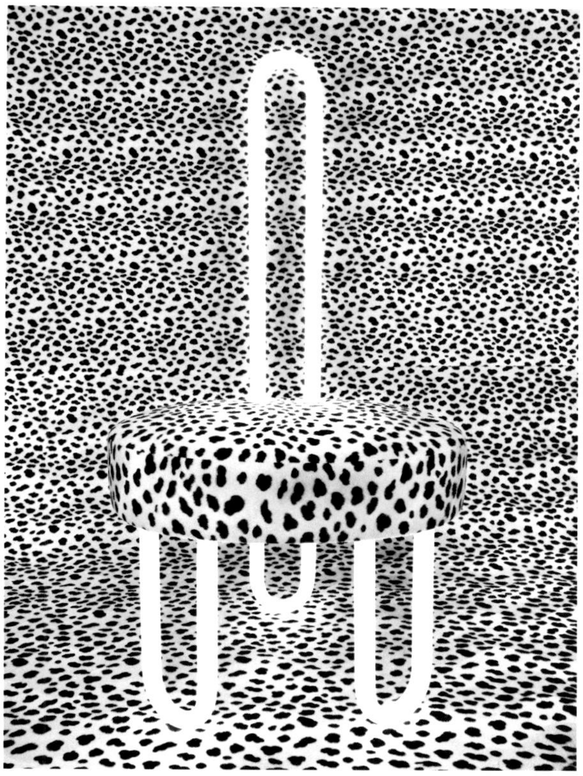

720

Furniture as Houses

The protagonists of the Ticino School, among them Mario Botta, Ivano Gianola, and Livio Vacchini, extended the boundaries of design even further, in the direction of architecture. "I approach a design with the same logic as I do a house, a church, or a bridge,"[8] Botta once said, and Vacchini as well made no distinction between disciplines (see Rüegg, "The Beginning—or the End—of History?," pp. 300–305). For him, furniture was "the true reflection of the experience" he had when building in the city;[9] it produced in him the same emotions. That was the very quality Ivano Gianola sought in the context of production: "The milieu in which I grew up in Ticino encouraged me to take advantage of the traditional craftsmanship that still exists there."[10] As in architecture, there could also be found "in the structure of furniture [the same kinds of] supports and braces and relations between them as part of a design. In a structure of solid wood, these relationships are visible in places where the parts of the frame meet. In my designs, the joints are a means of expression of the utmost importance. Taking advantage of them makes it possible to create organizing lines that, together with the wood grain, become part of the furniture's form. The piece is then completed by the decorative play of exotic wood veneers. A polished finish turns furniture into a mirror that reflects the interaction between material, structure, and form." Using a geometric vocabulary of forms that had already stood the test of time in the realm of architecture, Botta for his part began in the early eighties to produce a series of chairs he counted off as *Prima, Seconda, Tertia,* and so on. Here as well, the sculptural shape with its didactic geometric construction took priority over the chairs' primary function, especially regarding the comfort of sitting in the individual pieces. Turning the conventional dictum on its head, the function of the chair seems to follow its form.

Collages à la Duchamp

The young architects Jacques Herzog and Pierre De Meuron neatly encapsulated in their *House Bar* of 1977 what the various combinations of geometry, language, and historical formal vocabulary really constituted: collages. The title of the piece already suggests a self-reflexive handling of the play with language and form: Is it a bar inside a house or a bar in the form of a house? As a bar and simultaneously an architectural model, the design is apparently capable of taking on both tasks. "A Duchamp readymade gets its tension and sophistication from the juxtaposition of the generally familiar and well-known congruency of form and content in the readymade with its changed, unusual location. The artist's work consists in working with readymades. Duchamp was always well aware of that fact.... The same can be said of the work of the architect. Existing architectures are his starting points: the compositional elements available to him are parts of ready-to-use historical architectures that are determined by their form and content. The collage based on Duchamp's readymade *Fresh Widow* of 1920 illustrates these ideas in a simple fashion: together, the three elements—window, base, Manhattan Island—form a new architecture."[11] As early as 1983, however, Herzog and De Meuron already found grounds to criticize their own design: "While in 1977 we wanted to incorporate readymades in our designs, today we are much more interested in how we can develop our own visual metaphors out of our real-life experiences and convey them to others. The *House Bar*—made of ceramic, glass, mirrors, chrome-plated steel, and painted wood—is to be understood both as a collage and as a piece in its own right, while today we try to give each room of a building a clear identity within a sequence of spaces. This also includes choosing or designing the furnishings, the tables, benches, and chairs, in such a way that they contribute to conveying this identity."[12]

And yet it is precisely the effort to detach objects from a unique identity that characterized design in this period. In response to impulses from art, architecture, and not least the social upheavals of the day, the established paradigms were called into question without designers feeling the need to come up with an unambiguous new vision. Instead, the challenge to convention was itself the subject of the debate around 1980, as shown by the work performed by Trix and Robert Haussmann's *Log-O-Rithmic Slide Rule,* an open design tool connoting ambivalence more than clarity.

1 Burckhardt 1980.
2 Ibid., p. 19.
3 Burckhardt (1980) 2012, p. 154.
4 Alexander 1977.
5 See Berger 1983.
6 The *Log-O-Rithmic Slide Rule* was used as the cover for the publication Studio Marconi 1981.
7 Haussmann 1981, p. 29.
8 Cited in Loderer 2002, p. 250.
9 Cited in Vacchini 1983.
10 Cited in Gianola 1983.
11 Herzog/de Meuron 1977; cited in Herzog/de Meuron 1983.
12 Herzog/de Meuron 1983.

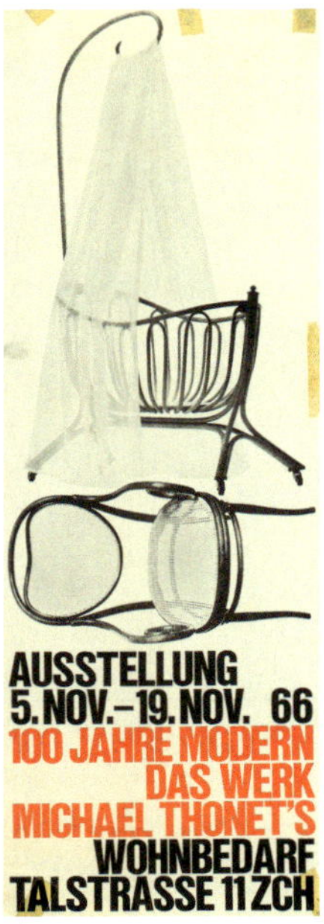

721

722

723

724

721 Poster for the exhibition *100 Jahre modern. Das Werk Michael Thonets,* Wohnbedarf AG, Zurich, 1966. The cradle was a thrift-shop find by Ulrich P. Wieser.

722 Aldo Rossi, Contribution to the project Tea and Coffee Piazza launched by the company Officina Alessi, 1980–83, coffee and tea utensils, silver, edition of 99, Officina Alessi, Omegna (IT). A small tabernacle for the celebration of tea and coffee rituals.

723 Friederike Mehlau-Wiebking, Arthur Rüegg, and Ruggero Tropeano, *Schweizer Typenmöbel 1925–1934. Sigfried Giedion und die Wohnbedarf AG,* gta Verlag, Zurich, 1989. Cover. The first overview of Swiss furniture design of the 1930s.

724 Moser Chair (1930), Roth Cocktail Trolley (1932), and Aalto Sofa Bed (1930/32) as reeditions. Prospectus produced by the company Atelier S.r.l., Mariano Comense (IT), 1989.

The milestones of design history take pride of place today in every progressive and elegant interior. Even though many of them blazed new trails in their day, after the first flush of success they often went out of production for a long period of time. This was the fate especially of the icons of the interwar period. Their later reedition was to become a crucial chapter in the history of their reception.[1] It took a special set of circumstances to bring this about—in Switzerland as elsewhere—as the dilemma of formal and technical innovation is each time a new challenge.

The Utilitarian Object as Style Icon

Between 1975 and 1985, modern "useful furniture" was elevated to the status of a semiotically charged "pictorial manifesto."[2] In the heyday of postmodernism, even coffee pots or salt shakers could be understood as architectural metaphors. The Ticino architect Livio Vacchini describes the phenomenon when he writes, "I look at furniture in a space in the same way I look at palaces and buildings in a city. There are essential pieces that set the tone, and then there are the more humble items that created more nuanced relationships.... I love this constellation and would not want to live without it."[3] He then gets more explicit: "Modern series furniture is not conceived along these lines: placed in a room, all of these pieces have equal significance and weight. They don't convey any emotion." The concept of a hierarchy of significance in architecture and urban planning had been propagated by Aldo Rossi from Milan, who taught in the early seventies at the ETH Zürich and from there made waves worldwide. His ideas were subsequently transferred with success to the design of utilitarian objects. While for Alessandro Mendini's Alessi project *Tea and Coffee Piazza* (1980) Rossi gathered together a costly ensemble of tea and coffee utensils, some of them crowned with domes, in a *tempietto,* the colored wooden chair *AR 2* (1983) recalls instead the simple industrial products of his Lombardian hometown environs. Rossi and his contemporaries weren't interested in inventing new objects for a specific purpose but instead in engaging with existing types. Their design efforts focused on studying the "images" that connect such types with human visual experience and in this way—according to Vacchini—"conveying emotions."[4]

Despite this clear stance, Vacchini hinted at an irritating paradox. On the one hand, the modernist repertoire of forms had been exhausted in the building spree of the economic boom years. On the other hand, however, the marketing of modern "useful furniture" reached its absolute high point in the eighties, not only in terms of production, as Gerda Breuer stresses, but also with regard to the awarding of licenses and contracts.[5] As a matter of fact, the invention of the modern classic—as she titled her book (*Die Erfindung des Modernen Klassikers*)—did not take place until the most important designs of the prewar era were tracked down, handpicked, and marketed as reeditions. Only then did their manufacture in the originally intended numbers become a reality—indeed, to an extent that sometimes reached epidemic proportions. The spectrum of reeditions ranged from the "series furniture" of Neues Bauen (New Building) reduced to its most elementary form (such as Marcel Breuer's stool from the canteen of the Bauhaus in Dessau) to unique pieces by Frank Lloyd Wright, Josef Hoffmann, Antoni Gaudí, Charles Rennie Mackintosh, Otto Wagner, Gerrit Rietveld, and Gunnar Asplund, finally extending to the entire realm of modern interior design.

The Swiss public proved eager to snap up these classics. However, the rediscovery of Switzerland's very own *Typenmöbel,* or "type furniture," did not lead to commercial success, let alone to an international breakthrough. In 1980, Stanislaus von Moos had initiated a discussion of Swiss design in the thirties with his essay "Wohnbedarf und Lebensform" (Wohnbedarf and Lifestyle).[6] Starting in 1982, Ulrich Paul Wieser organized a series of exhibitions at Wohnbedarf in Basel on the company's history and brought out reeditions of some of the old Wohnbedarf models in small series (such as Alfred Roth's *Cocktail Trolley* from 1932). Likewise in 1982, Ruggero Tropeano and the author of this essay launched a documentation series on Swiss design history, including, among other publications, the monograph *Schweizer Typenmöbel 1925–1935. Sigfried Giedion und die "Wohnbedarf AG"* (1989, with Friederike Mehlau).[7] Responding with "pretentious humility"[8] to international developments, the Swiss designs would, despite all their efforts, have to wait until the nineties to finally arouse greater interest and find favor with a wider public.

By contrast, the most radical manifestos expressly directed against the historicism of the nineteenth century now became prestigious must-haves, a development that surely had something to do with the eclectic exploitation of all the ways in which a piece of furniture could be viewed as a pictorial icon. Gerrit Rietveld's trademark *Zig-Zag Chair* possesses the necessary incisiveness of formulation, but so does even an eccentric piece like the surrealistically charged *Table with Bird's Feet* (1939) by the Swiss artist Meret Oppenheim. It is only natural to directly associate this elite selection, based not on functionality but iconic presence, with postmodernism. Few studies have dealt with this theme to date, but Charles Jencks, author of *The Language of Postmodern Architecture,* aptly covers with his term "radical eclecticism" the revival of the most diverse traditions in furniture as well as in architecture.[9] The German design theorist Christian Borngräber even went so far as to speak in 1996 of a "new historicism" in the "reception aesthetics of movable goods," referring also to the reedition of the modern "furniture classics."[10]

Back to the Roots

Successful marketing in the postmodern era, whether regarded as bane or boon, has forever changed the way we see the design icons of modernism. Their rediscovery in the immediate postwar period took place under completely different auspices than at their inception. After what seemed like the endless downward spiral of the Great Depression, the totalitarian ideologies that stepped into the breach, and the hardships of war, the public was hungry for a freer and more comfortable way of life—and ready to embrace it at an especially early date in Switzerland, where no reconstruction was necessary. The vernacular *Heimatstil* now looked dated, and the metal furniture from the era of "liberated living" had long since gone out of production. In Zurich, the firm Wohnbedarf AG, founded in 1931, lost no time in reviving its own traditions. Rudolf Graber, now the company's sole owner, already managed in 1947 to establish a cooperative arrangement with the recently reorganized Knoll Associates Inc. in New York and to obtain the license to produce Knoll International's home furniture line in Switzerland.[11] In the same year, Knoll in turn secured the rights to Mies van der Rohe's 1929 *Barcelona Chair.* The very epitome of modern elegance, the chair came onto the market in 1948 but didn't find its way to Switzerland until around 1956. Almost concurrently with this worldwide reedition, Graber began on his own to once again produce in a small series the *Chaise longue basculante* conceived in 1928 by Le Corbusier, Pierre Jeanneret, and Charlotte Perriand, based however on the model produced by the Embru-Werke in Rüti in 1933. Access to these witnesses of a bygone avant-garde high point was of course limited to a wealthy elite, yet even the very availability of the legendary pieces sent a clear signal. It was not the "end of history" that was up for debate but rather a reconnection with its auspicious beginnings.

At the same time, the origins of modern design became the subject of a comprehensive reassessment. Nineteenth-century bentwood furniture now found its way into both the homes of Werkbund members and the decorative arts museums. There was a new appreciation of Lobmeyr glasses and Wedgwood tableware, and a keen interest in anonymous equipment from the laboratory and mechanical workshop.[12] The Zurich branch of Wohnbedarf AG exhibited in 1948: *Gläser aus Empoli* (Glasses from Empoli), *Alte Berber: Teppiche, Stoffe und Keramik* (Old Berber: Carpets, Fabrics, and Ceramics), and later *Echt englische Schiffskommoden* (Genuine English Ship Cabinets) and *Bauernsteinzeug aus der Gegend von Autun Kunst aus der Südsee* (Peasant Stoneware from the Autun Region, and Art of the South Pacific). In 1948, the Zurich art historian Sigfried Giedion published the book *Mechanization Takes Command: A Contribution to Anonymous History,*[13] with which he "elevated the history of the development of everyday mechanical devices beyond a listing of the great inventors to chart a historiographic program."[14] In Switzerland, the colonial chair reworked by Wilhelm Kienzle in 1928 achieved renewed recognition. In Italy, some of the paradigmatic "mobili storici" (historical furniture)

725

726

725 Alfred Hablützel, Poster for Galerie Mezzanin, Zurich, 1959. Heidi Weber showed in her gallery reeditions of Le Corbusier seating furniture (the exhibition title) in the colors of Salubra's Le Corbusier wallpaper collection, which were also used for the poster.

726 Exhibition mounted by the regional group of the Swiss Werkbund at the Kunstgewerbemuseum Zürich, 1950. View from above of the garden with a group of lounge chairs. Center right: Chaise longue basculante by Le Corbusier, Pierre Jeanneret, and Charlotte Perriand, produced by Wohnbedarf AG, Zurich. Photo: Serge Libiszewski

provided the impetus for the first reeditions: the folding chair *Tripolina* was brought out in 1950 by Dino Gavina in Bologna, and Gio Ponti's wooden *Leggera* chair, a kind of tribute to the anonymous *Chiavari* models, was introduced in 1955 by Cassina in Milan.[15]

In 1958 the young Zurich interior designer Heidi Weber was responsible for the next milestone when she acquired from Le Corbusier the rights to produce the four canonical chairs he had designed in 1928 in collaboration with Pierre Jeanneret and Charlotte Perriand. The *Chaise longue basculante,* the *Fauteuil à dossier basculant,* and the two versions of the *Fauteuil grand confort* all drew—just like the *Barcelona Chair*—on carefully selected predecessors. The inimitable combination of historically and functionally legitimated furniture types with a new and innovative formal and design logic probably accounts for these models' ability to defy all passing trends and retain their timeless appeal. Heidi Weber's role as mediator is all the more significant as she brought together in her Galerie Mezzanin the "Le Corbusier seating furniture"[16] with paintings, wallpaper, drawings, and lithographs by the artist-architect. The success of this venture confirmed the astuteness of her instincts, but it also led her to contemplate around 1964 outsourcing the manufacture of her collection to a powerful license partner. Dino Gavina, who was apparently the first to be asked, had, for his part, landed a big coup in 1962 with his reedition of Marcel Breuer's *Wassily Chair.* However, the contract was ultimately awarded to Franco Cassina, who had tirelessly sought a license ever since June 1963, first appealing to Le Corbusier and then to Heidi Weber.[17] The four chairs formed the cornerstone of the Cassina series *I Maestri,* which was inaugurated in 1965. They would go on to play a major part in fueling the international reedition fever in the eighties while—posthumously—cementing Le Corbusier's reputation as one of the great designers of the twentieth century.

Swiss Design

The *Barcelona Chair* already appeared on the Swiss design scene in 1949, but only as a photograph in the exhibition *Die gute Form* conceived and organized by Max Bill.[18] Although a reedition had just come onto the market in the US, this "easy chair for discriminating tastes"—according to its caption in the exhibition—possessed an aura of inaccessibility. It is probably for this very reason that the chair struck such a chord with the young Swiss designers. With their "sense of precision, structural ingeniousness, and luxurious details,"[19] they were irresistibly tempted to try to emulate this masterful role model. Rather than merely copying the design outright, they focused on a critical examination and mustered the courage to come up with their own formulation. The young interior designer Robert Haussmann (b. 1931), who after earning his degree had worked from 1953 to 1955 as assistant to the architect Florian F. Adler at Eternit AG, presented an independent tribute to the model by the great master in 1954. His *RH 301* armchair, which won an award at the Milan Triennale in 1957, consists like the *Barcelona Chair* of a frame of flat steel with straps supporting two leather cushions, but Haussmann deftly avoided the difficult-to-manufacture junctures in Mies's flat-bar construction. The model launched an ongoing series that resembles a kind of didactic experiment. With the model series *RH 302* Haussmann then parsed from 1954 to 1956 the theme of the *Fauteuil grand confort,* replacing the handcrafted metal frame of the original with a polished construction with screw connections that is fine-tuned down to the last detail. The elimination of production difficulties evidently formed his pretense for a congenial further development of the original models: "From the beginning, we attached great importance to furniture that could be easily taken apart, because it was manufactured on a large scale and exported."[20]

Haussmann was not acting alone here. The autodidact Hans Eichenberger (b. 1926) from Bern, for his part, used angled pieces of chromed flat steel with visible screws for his round table *HE 116* in 1956, and attempted that same year with his armchair *HE 113* to achieve a delicate balance by conjoining a strap construction and tufted leather cushions borrowed from Mies, Breuer's chromed tubular runners, and the overall form of the *Fauteuil à dossier basculant.* The third member of this league, Kurt Thut (1931–2011), was, like Haussmann, a graduate of the Kunstgewerbeschule Zürich. He explicitly mixed the flat-bar furniture collection he developed in 1956 with design elements borrowed from Mies van der Rohe: "The link between the flat-bar frame and the four angles in the base corresponds to the method of fashioning steel used by Mies van der Rohe."[21] A "proportion fanatic,"[22] Thut used metal spacers in

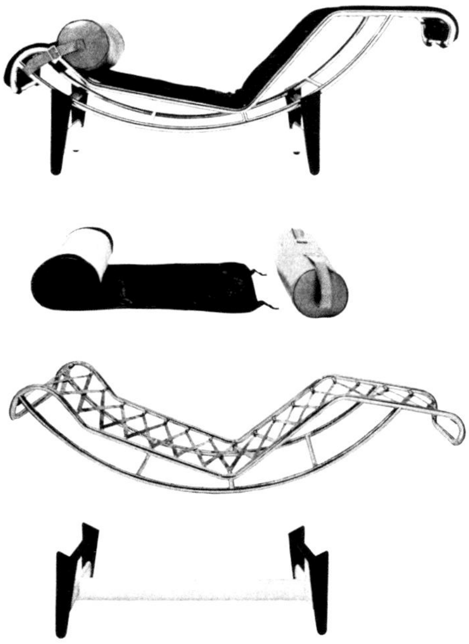

727

728

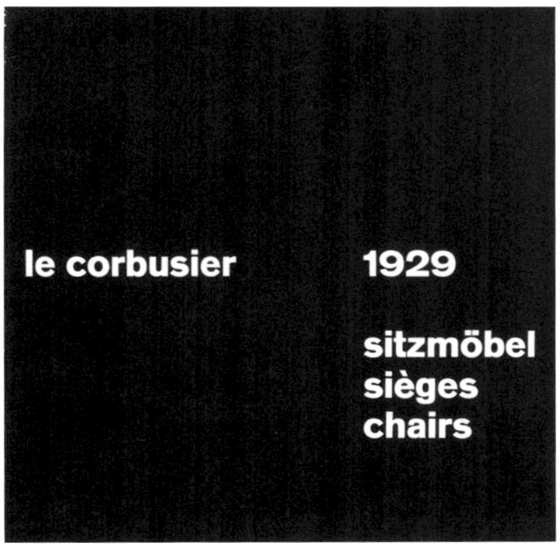

729

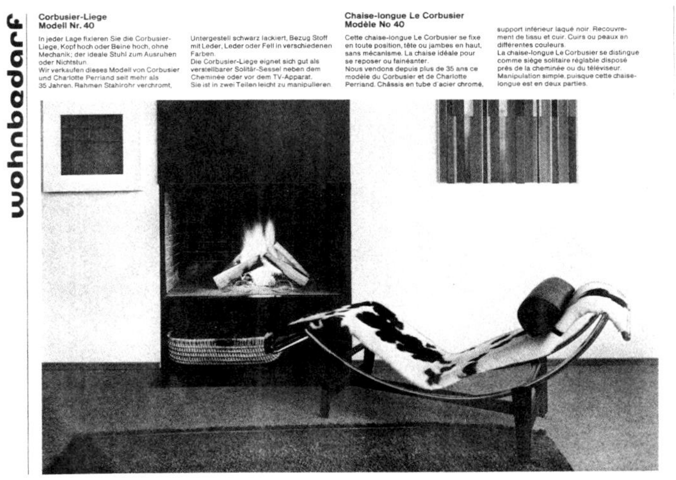

730

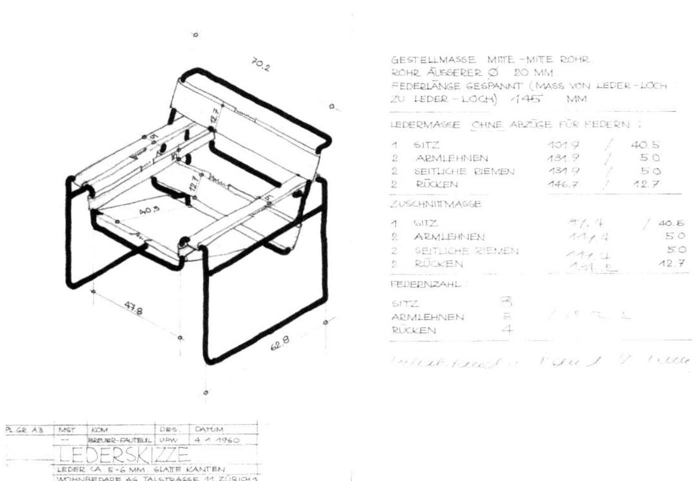

731

732

727 Le Corbusier, Pierre Jeanneret, and Charlotte Perriand, Chaise longue basculante, 1928, produced by Heidi Weber (1959–64). This is a "deluxe version" with a fur mattress atop a slatted metal frame.
728 Le Corbusier, Pierre Jeanneret, and Charlotte Perriand, Fauteuil grand confort, 1928, produced by Heidi Weber (1959–64). Characteristic is the chrome-plated frame and the replacement of the bent junctures with mitered connections. Museum für Gestaltung

729 Alfred Hablützel, Cover of the catalogue *Le Corbusier. Sitzmöbel, Sièges, Chairs, 1929,* Galerie Mezzanin (Heidi Weber), Zurich, 1959.
730 Prospectus for the Chaise longue basculante by Le Corbusier, Pierre Jeanneret, and Charlotte Perriand, 1928, from the production of Wohnbedarf AG, Zurich (ca. 1950–79). The "series model" with seat made of fabric, leather, or cowhide.

731 Ulrich P. Wieser, "Leather sketch" and material sheet for the reedition of Marcel Breuer's B3 (Wassily) Chair, which had been planned by Wohnbedarf AG since 1959; dated January 4, 1960. The reedition of the 1925 chair was finally accomplished in 1962 by Dino Gavina, Bologna.
732 Marcel Breuer, Shop remodeling for Wohnbedarf AG, Zurich, 1956. The exhibition for the company's 25th anniversary prominently features next to in-house models—an Aalto sofa and Bellmann dining table with chairs—the chairs created by Ludwig Mies van der Rohe for the Barcelona pavilion in 1929, in a 1948 reedition by Knoll International.

733

734

735

736

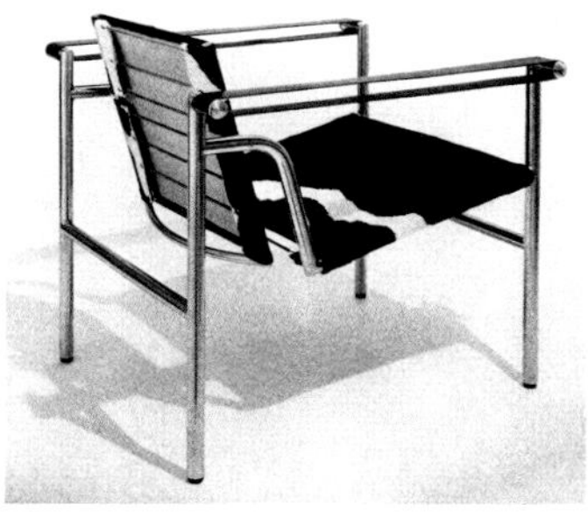

737

738

733 Robert Haussmann, RH 301 Chair "Hommage à Mies van der Rohe," 1954, chrome-plated flat steel and leather, from 1958 part of the Swiss Design Collection, produced by Heinrich Kihm, Dübendorf (1954); then by Hans Kaufeld, Brake (DE); De Sede, Klingnau, and others. The frame is collapsible.

734 Ludwig Mies van der Rohe, Barcelona Chair MR 90, 1929, chrome-plated flat steel and black leather, Berliner Metallgewerbe Josef Müller (1929–30). Inspired by historical archetypes. Photo: Werner Blaser

735 Robert Haussmann, RH 302 Sofa, chrome-plated flat steel, leather, 1954–56, from 1958 part of the Swiss Design Collection, produced by Hans Kaufeld, Brake (DE); later by De Sede AG, Klingnau. Over ten editions with various upholstery options, collapsible frame.

736 Le Corbusier, Pierre Jeanneret, and Charlotte Perriand, Fauteuil grand confort, 1928, petit modèle, first prototype, painted tubular steel, steel cable, leather. Inspired by the English club chair.

737 Le Corbusier, Pierre Jeanneret, and Charlotte Perriand, Fauteuil à dossier basculant, 1928, chrome-plated tubular steel, calves' hide. Inspired by the English colonial armchair. Photo: Sophie-Renate Gnamm

738 Hans Eichenberger, HE 113 Armchair, 1956, chrome-plated tubular steel, leather, from 1958 part of the Swiss Design Collection, produced by Hans Kaufeld, Brake (DE); later by De Sede AG, Klingnau, and others. The chair pays homage to Le Corbusier as well as Mies van der Rohe.

739

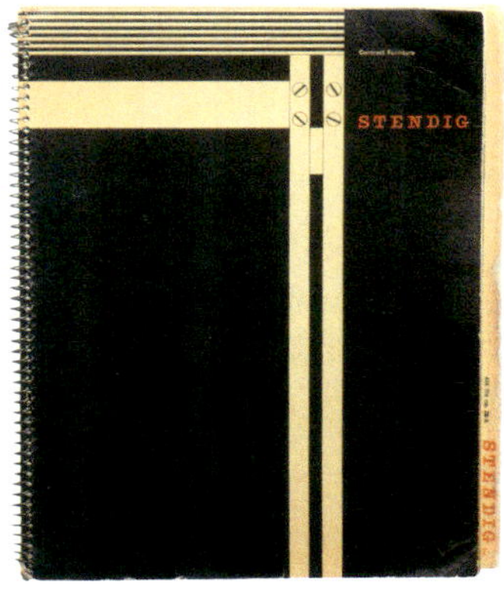

741

742

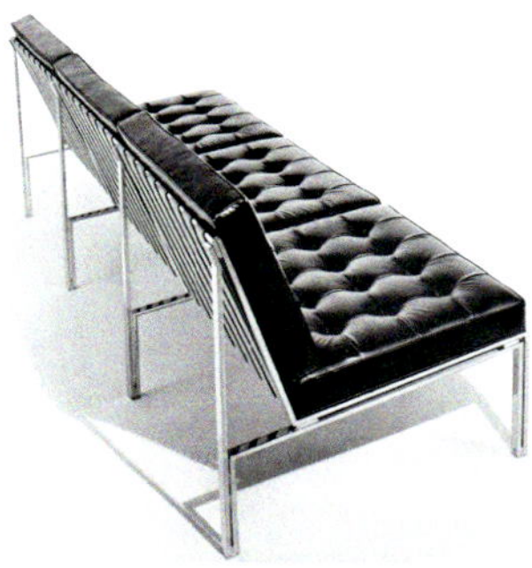

740

739 Kurt Thut, Sofa, 1956, chrome-plated flat iron,
740 leather, rubber belts, from 1958 part of the Swiss Design Collection, produced by Hans Kaufeld, Brake (DE). Detail of the backrest. Photo: Alfred Hablützel
741 Cover of the Stendig Inc. Contract Furniture catalogue, with detail of the table by Kurt Thut, New York, ca. 1959. The Swiss Design Collection was exported mainly to the US.
742 Shop for Gebrüder Haussmann und Haussmann, Oberdorfstrasse, Zurich, 1955. For the opening, designs by Robert Haussmann were presented in the display windows: on the left, the RH 301 Chair; on the right, two rope chairs. Photo: Alfred Hablützel

the form of Swiss crosses to underline the elemental nature of the construction. "We believed we had to take up where the European pioneers ... had left off fifteen years before," summed up Alfred Hablützel (b. 1931), who worked in Bern and had been a fellow student of Haussmann and Thut. He joined the trio in the role of theorist, propagandist, organizer, and photographer: "I remember how we critiqued texts and buildings by Le Corbusier, Mies van der Rohe, and Frank Lloyd Wright with a mixture of awe and youthful arrogance."[23]

The perfected easy-to-disassemble furniture models by the three designers, in the meantime working independently, were exhibited at the Kunstgewerbemuseum Zürich in fall 1958 under the simple title *neue metall-möbel* (New Metal Furniture) before, under the evocative brand name Swiss Design, being marketed both at home and abroad—such as by Stendig in New York from 1959. By representing the copyright interests of the designers, Teo Jakob turned Bern for a time into the Swiss center of modern home decor. However, the production of the line, which was considerably expanded over the following years, was entrusted to the German company Kaufeld in Brake, until De Sede in Klingnau took over production temporarily in the late seventies and then permanently in 1999. Today, almost sixty years after their creation, this company is still producing some of the timeless models designed by Eichenberger and Haussmann, with a common wooden chair paying them due reverence in an advertisement: "Swiss Design, I presume..."[24]

1 For more on this subject, see Cacciola 2008, to which I am indebted for certain information, particularly that concerning Gavina and Cassina. Ms. Cacciola worked for Dino Gavina in 2000 and conducted several interviews with him.
2 See Lichtenstein 2002.
3 Vacchini 1983.
4 See Rüegg/Steinmann 1983.
5 Breuer 2001, p. 17.
6 See von Moos 1980.
7 Mehlau-Wiebking et al. 1989.
8 The term was coined by Peter Erni; see Erni 1983, p. 11.
9 Jencks 1978, p. 127.
10 Borngräber 1989.
11 See Rüegg 2006, p. 11.
12 See Fischer 1971 and Sembach 1971.
13 Giedion 1948.
14 Harbusch 2009.
15 See Rüegg 2012b, p. 349; see also note 1.
16 Exhibition catalogue of Galerie Mezzanin, November–December 1959.
17 See Rüegg 2012b, p. 349.
18 See Bignens 2002b, pp. 185–87.
19 von Moos 1969.
20 Haussmann 1958.
21 Thut 1958.
22 Hablützel 1985, p. 10.
23 Ibid., p. 9.
24 Catalogue sheet from De Sede AG Klingnau, undated, showing the models *RH 1–3* and *5*, with the text: "Right, this is a chair, designed by Robert Haussmann for the Swiss Design Collection of De Sede."

743

With the Designers' Saturday, producers located primarily in Langenthal established in 1987 a biennial trade fair for presenting Swiss products and furniture to a wider public. Companies such as Création Baumann, Ruckstuhl, as well as Belux, Lehni, Röthlisberger, Thut, and Wogg showcased their new products there. As did the Porzellanfabrik Langenthal, who with its product line Bopla!, launched in 1993 (with ornament by Mario Grasso), tried to update its image by offering trendy, colorful tableware.

1990 – Statement

With rampant globalization and the relocation of production to foreign climes, industrial designers and producers are now geographically distant. In order to bring design, manufacturing, and sales back together again, some designers become entrepreneurs. They establish design collectives or their own companies, design their own corporate identity on the computer, and beginning in the mid-nineties their own websites as well. Subcultural scenes dominate communications. Design is appreciated as a creative economic factor and duly promoted. After semantically charged and sensually appealing objects, new and simple products appear around 1990 that offer clever solutions for everyday life. Their matter-of-factness is a program in its own right: details such as connections, locks, or hinges determine the form and proclaim their function. Rather than a signature gesture, what counts now is the designer's ability to reduce to essentials. Digitally controlled tools such as inkjet printers, CNC milling machines, and laser cutters enable cheap production in small series. Ornament is rehabilitated as a means of customization.

744

Annette Gigon and Mike Guyer, Restaurant Vinikus, Davos, 1990–92. The table settings, the furnishings, and the interior fittings all express the same design attitude. The focus is no longer on the symbolic meaning of the forms but rather on the sensual effect of the materials. Whether as millimeter-thin veneer or centimeter-thick floorboards, warm wood tones dominate the ambience of this casket-like interior.

Photo: Heinrich Helfenstein

745

Daniel and Markus Freitag cutting apart truck tarpaulins, ca. 1994. At the beginning of their career as bag producers, the brothers did the work themselves. They removed the bindings from the used truck tarps so that they could clean the fabric and cut bags out of it, which they then sewed together themselves. A friend advised them to use their family name as the brand, which invariably provided humorous advertising ideas.

746

Furniture from the Röthlisberger Kollektion in the office of a Bern textile wholesaler, 1999: Alfredo Häberli and Christophe Marchand, Zehn Hoch modular drawer unit, 1993; Carmen and Urs Greutmann, Kiki folding table, 1997; Ubald Klug, Shell cupboard, 1997, and Spiga coat stand, 1999. Assembled here are diverse furniture designs that are all dedicated to the New Simplicity.

747

Prospectus for the Swatch Twinphone—Two Phones in One, Swatch Telecom AG, Biel, 1991. Starting in 1985, private users could buy phones outside the PTT standard program. The phone thus became a personal, freely choosable accessory. In various color combinations or with modern transparent housing, the base of the phone could also be used as a receiver. The watchmaker Swatch produced its first Twinphone in 1989. Its successors bore names like Blue Lagoon (1989), Limelite (1997), and Black Venus (1997) and benefited from the lifestyle potential of the mobile phone.

748

Artistic direction for Expo.01, "Nature and Artificiality," 1998, model of the forum for Arteplage Neuenburg. The broad surfaces symbolize human freedom of movement. Five interwoven thematic bands form the shell. Architecture or design? In the concept, which was only realized for Expo.02, the goal was no longer to exude an aura of prestige through grand architectural forms or to present exemplary products, but rather to offer personal, visceral sensory experiences: "events" and "little dreams."

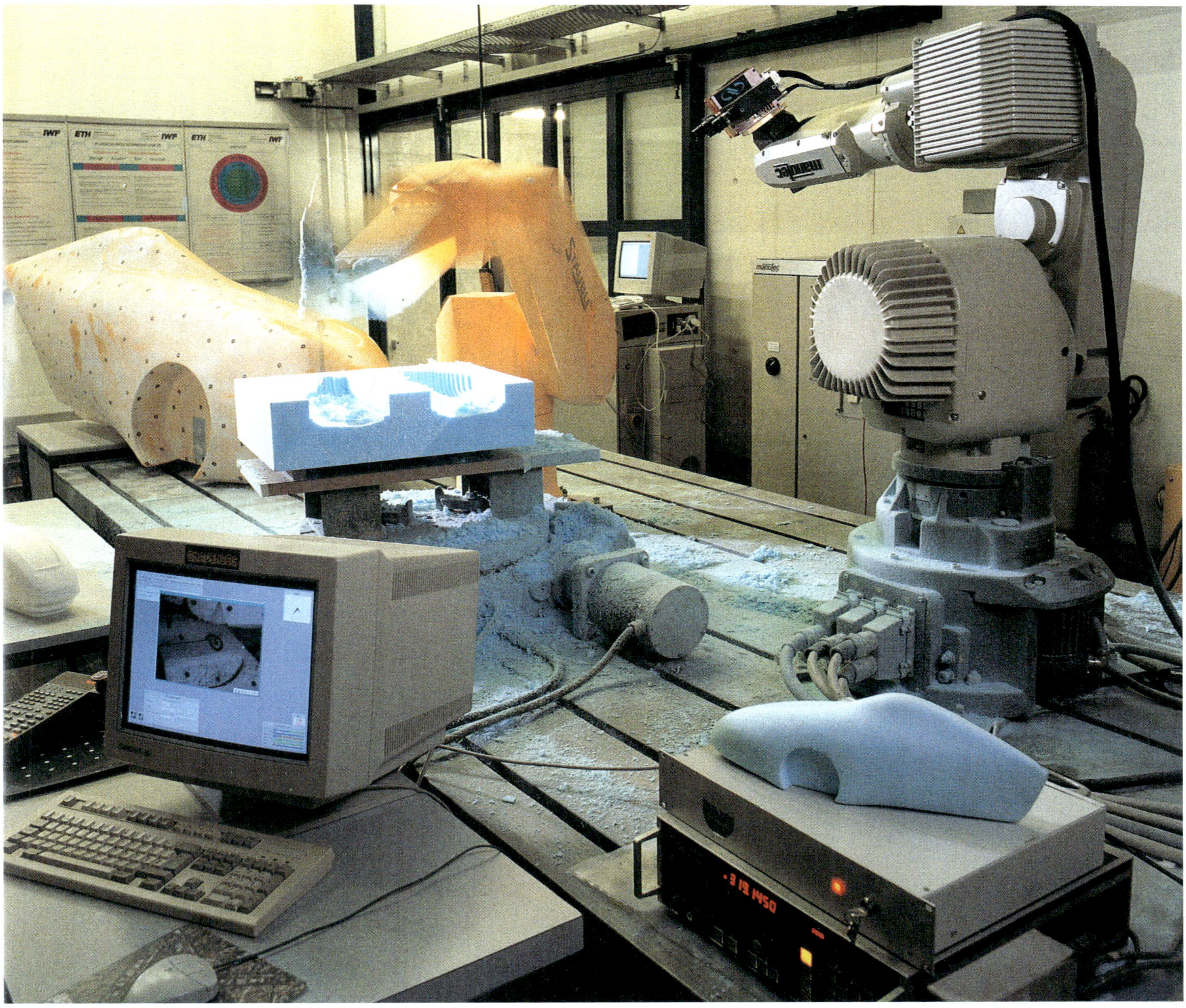

749

ETH Zürich, TCAD (tangible computer-aided design), 1999. This is the workplace where Robert Zanini and Joachim Wirth from the Institute of Machine Tools and Manufacturing tested the TCAD system. With an optical measuring technique, the curved surfaces can be translated into three-dimensional digital data, after which the physical model can be processed further by computer. In the background is the 1:1 model of a bicycle shell whose form was further developed digitally and then produced on a smaller scale using a CNC-controlled milling machine.

Photo: Peter Würmli

750 Daniel and Markus Freitag, Freitag Bag, 1993, prototype
Used truck tarpaulins made of PVC-coated polyester mesh, 38 × 53 × 15 cm
Museum für Gestaltung/Donation of Freitag

751 Freitag flagship store in a container tower, Zurich (Spillmann Echsle Architekten, 2006). The products are presented in cardboard boxes.
Photo: Roland Tännler, FREITAG lab.ag

752 Each bag is cut out individually by hand and thus unique, 2013
Photo: Joël Tettamanti, FREITAG lab.ag

750

751

752

Daniel (b. 1971) and Markus (b. 1970) Freitag sewed their first messenger bags from recycled truck tarps, bicycle tubes, and seat belts for their own personal use. Their chosen mode of urban transport was the bicycle, and they always had rolls of paper with design drawings in tow. As members of a shared apartment with views of the truck traffic on Zurich's Hardbrücke bridge, they came up with the idea of transferring the tarps from the heavy delivery vehicles to their bikes. With a design borrowed from the bags used by American bicycle messengers, the first *Freitag* bags were sold to the public in 1994 on the same bridge that inspired them. The brothers took orders for custom color combinations and then cut, sewed, and delivered the bags themselves. In principle, this business model has changed little over the past twenty years, but the bags are now a globally recognized brand. The entrepreneurially industrious and creatively talented brothers set up their own production site in Zurich, where the tarps are cleaned, taken apart, and cut into designs betraying a flair for color and typography before being sewn together and photographed. The departments for tarp acquisition, product development, sales, and marketing are all located under one roof, ideal for driving the company's iterative process. Via online shops, the company's own stores, and distributors, the unique Freitag products reach customers and fans, who then present their impressive bag collections on Facebook. Freitag is regarded today as a prime example of creative industry and proof that thinking in terms of design can yield economic success. RM

Müller 2001; Schindler 2006b; Menzi 2012; www.freitag.ch.

753 Kurt Thut, Scissor Bed 990, 1990
754 Laminated wood frame, varnished beech feet, 16 × 80–180 × 200 cm, Thut Möbel AG, Möriken
Photo: Alfred Hablützel
Museum für Gestaltung*

755 Prospectus for Scissor Bed 990. Attachable swiveling lamp by Benjamin Thut, 1984, Thut Möbel AG, Möriken

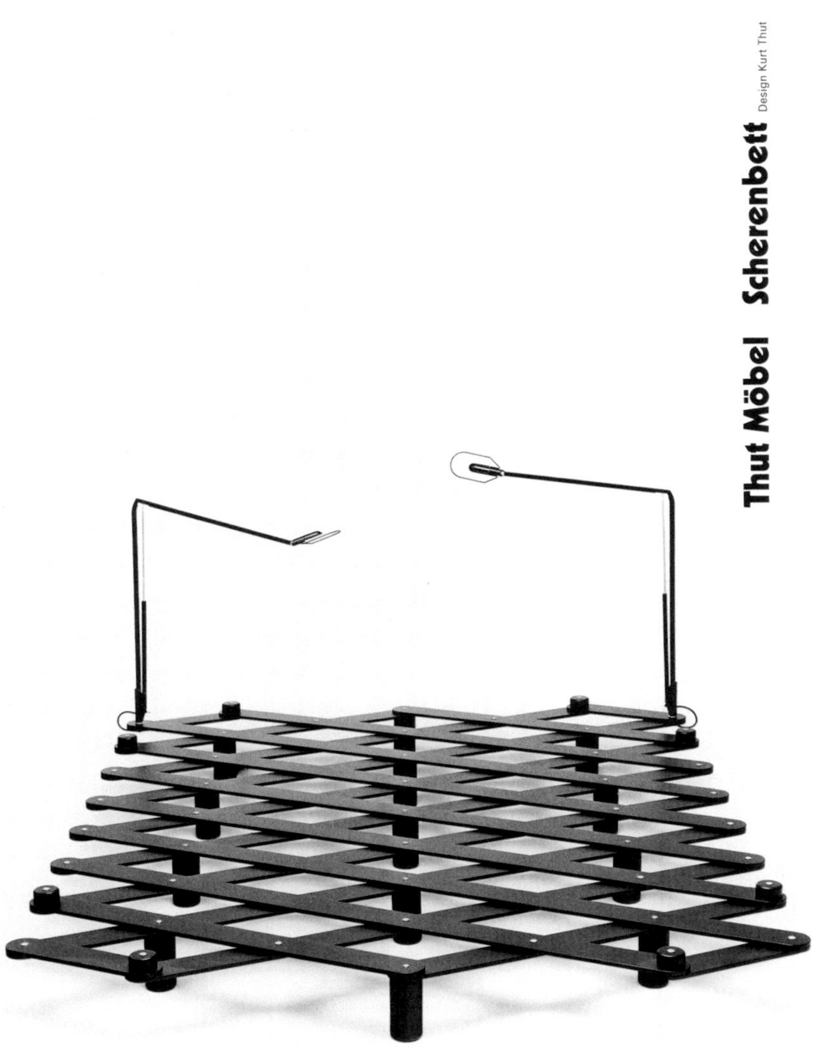

753

754

755

Taking a critical distance from postmodernism, the designer, design engineer, and entrepreneur Kurt Thut (1931–2011) virtually reinvented himself in the course of the eighties. Right after graduation from the Kunstgewerbeschule Zürich, he had attracted attention with his minimalist furnishings in the spirit of *Gute Form.* And in 1956 he was part of the devoted Swiss Design group surrounding Teo Jakob, who developed precisely engineered alternatives to the iconic seating furniture of classical modernism. After sixteen years as a freelancer, Thut took over his father's carpentry shop in Möriken in 1976, in the midst of the oil crisis. Now, in addition to streamlining the assembly of his chipboard furnishings, he also began to flout the zeitgeist with a series of groundbreaking experiments. These are inventions that spring from transferring proven principles of lightweight construction to a new context. The designer based his "Thut Bed" on the principle of accordion-style fencing, which provides a high degree of stability despite material savings of 30 percent. The protruding joints of the more or less extended grille structure can hold either one or two mattresses in place and also provide a place to attach the *Lifto* lamps designed by Benjamin Thut. The bed requires no assembly and is easy to transport in its folded state—ideal for design-conscious young urban nomads, who of course also purchase Thut's *Folding Curtain Cupboard* (1993), *Bellows Cabinet* (1998), and the *Scissor Shelf Unit* (2000), which can be raised like a hydraulic platform. AR

Gantenbein/Hablützel 2001; Hablützel 2002a.

Sigg AG
Factory design

Hot Water Bottle
ca. 1925

756

757

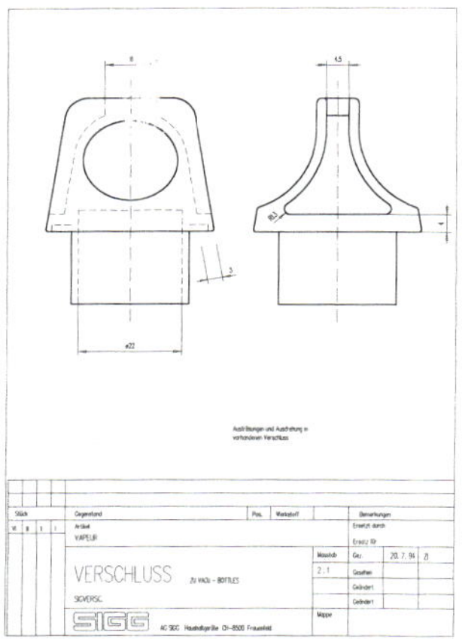

758

756 Left: Anonymous, Grooved hot water bottle, ca. 1925
Anodized aluminum, 25.5 × 8.5 cm, holds 1 liter
Right: Kurt Zimmerli, Water bottle, redesign 1983–90, with Johann Munz, screw top with finger hole, 1968
Aluminum, coated interior, powder-coated exterior, plastic screw top with finger hole, 26 × 6 cm, holds 1 liter, Sigg AG Haushaltgeräte, Frauenfeld
Museum für Gestaltung

757 Group photo
Kurt Zimmerli, Sigg bottles, redesign 1983–90
Left: Water bottle, 0.75 liters
Center: Water bottle, 0.6 liters, with active top, 1991
Right: Sport bottle, 1 liter, sport top, 1988
Museum für Gestaltung/
Donation of Sigg Switzerland AG

758 Kurt Zimmerli, Technical drawing of screw top with finger hole

Founded in Biel in 1908 by the metalworker Ferdinand Sigg, the Sigg company produced simple household goods, adding field dishes for the military to the product range after moving to Frauenfeld in 1917. A so-called pattern maker was responsible for the development of new products until, in 1965, the company hired Johann Munz (b. 1939), a graduate of the Kunstgewerbeschule Zürich, as a designer. Munz proceeded to unify the range, which had swelled in the fifties to encompass over 10,000 items that followed no discernible line. The Sigg aluminum bottle proved to be the product with the greatest longevity. Its simple extruded shape could be manufactured from a single piece—an advantage over iron and copper bedpans with corrosion-prone welding seams. Painted in plain colors, in smooth or ribbed versions, it remained popular even after plastic hot water bottles were introduced. In 1968, Munz designed a new plastic screw cap that resembled the top of the Sigg frying pan cover from the same period. The practical carrying hole in the cap may have prompted people to start using the hot water bottle as a drinking bottle. It was first sold for export as a camping bottle, and then starting in 1980 as a plain-colored water bottle for mountain climbers and athletes. From 1989 to 1994, Kurt Zimmerli (b. 1946), who had trained at the Schule für Gestaltung Basel, took the next steps in the development of water bottles for Sigg AG. Diversified to hold different volumes, outfitted with sport caps and colorful decor, the erstwhile hot water bottle blossomed in the nineties into a must-have global accessory. RM

Unbekannt – Vertraut 1987, pp. 102–4; www.sigg.com.

759 Nose Design Intelligence, Extreme Backpack, 1994, model from 1999
Polyamide, 70 × 45 × 25 cm, Mammut AG, Seon
Museum für Gestaltung/Donation of Mammut

760 Nose Design Intelligence, Extreme Backpack, 1994, launched in 1996
Museum für Gestaltung/Donation of Christian Harbeke

761 Nose Design, Bionic Carabiner, 2006, Mammut Sports Group, Seon
Museum für Gestaltung/Donation of Mammut

759

760

761

Christian Harbeke (b. 1968), Ruedi August Müller (b. 1964), and Ruedi Alexander Müller (b. 1968) all worked at the agency Zintzmeyer & Lux before founding Nose Design Intelligence in 1991. Nose covers the entire design chain from consulting to industrial and graphic design, and from interior design to exhibition stand construction, merging the competencies of the trio of engineers and designers. For the Mammut company, which originally produced primarily climbing ropes, Nose designed a new corporate identity in 1993. They gave the mammoth in the firm's logo, cuter than it was memorable, a more distinctive silhouette. This was Harbeke's first experience using the drawing program Freehand. In keeping with the company's desired brand image and the target group of serious mountaineers, Mammut next wanted to market a rugged backpack. The bag was to be based on two high-performance textiles by the Swiss textile manufacturer Schoeller. Dynatec is a nearly indestructible fabric made of polyamide fibers and designed especially for the backpack, and for its dark blue frame Harbeke applied what he had learned building aircraft. Between the two elements was the likewise tear-resistant, but lighter-weight orange nylon fabric Cordura. The contrasting colors visually accentuate the backpack's construction. The lettering, likewise designed by Nose, directly addresses the extreme athlete. The backpack was launched in the summer of 1996, and later versions were enhanced with additional elements such as ice ax holders and a lid pocket. It was a huge success and led to the systematic expansion of the Extreme Collection. RM

Locher 1994b; Pradal 2006; www.nose.ch; www.mammut.ch.

Silk Tie
1990

762

762 Cornelia Hesse-Honegger, Poisonous Frogs, 1990
Anonymous, Snail Shells, ca. 1992
Discharge printing on silk satin, Fabric Frontline, Zurich
Museum für Gestaltung/
Donation of Charles Clerc

763 Cornelia Hesse-Honegger, Eyes, 1992, discharge printing on silk satin. Fabric remnant
Museum für Gestaltung*

764 Ines Boesch, Sophie Shadow Large, 1995, Sourire en Soie AG, Zurich; Seidendruckerei Mitlödi AG. Fabric remnant
Museum für Gestaltung

765 Sonnhild Kestler, Kaleidoscope, 1995. Foulard
Museum für Gestaltung

763

764

765

By the sixties, Zurich's prominence in the global silk industry had dwindled. Swiss textile exports found themselves needing to keep up with the transformation of the international fashion industry. The triumph of synthetic fabrics and growing Asian competition necessitated a fundamental structural rethinking. Many businesses were forced to close, and investments were made in new technologies. In the field of clothing textiles, a young generation of entrepreneurs came to the fore in the eighties who put their faith in independent designs that did not slavishly follow the latest fashions. Outstanding examples of this approach are the Zurich silk brands En Soie and Fabric Frontline. Both embrace an independent attitude to style, image, and promotion, and base their strategy on integrating each individual design into a programmatically developed label. At the silk company Fabric Frontline, founded in 1980 by Andi, Elsa, and Maya Stutz, the house style features figurative motifs on a colored background. The scientific illustrator Cornelia Hesse-Honegger (b. 1944) started working for the company as a freelance designer in 1987, creating silk fabrics with the detail of scientific illustrations. These textiles with their precision drawings and high-quality printing document her ecological concerns about endangered species and animal and plant mutations in the vicinity of Swiss nuclear power plants. The fabrics make a bold statement—such as when made into ties and worn everyday by newscaster Charles Clerc on Switzerland's nightly news broadcast. RM

Schilder Bär/Wild (2001) 2003, pp. 172–80; www.fabricfrontline.ch.

766

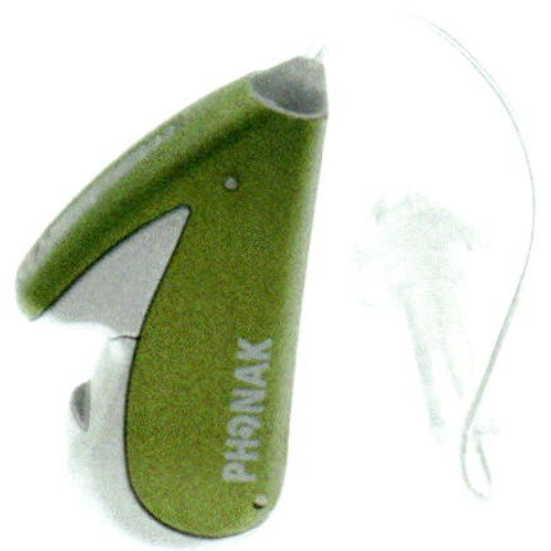

767

768

766 Kuster Design, Claro 11 CIC Hearing Aid, 1998
Plastic, 2 × 1.5 × 1 cm, Phonak AG, Stäfa
Museum für Gestaltung*

767 Milani Design & Consulting AG, Audéo Hearing
768 Aid, 2007, Phonak AG, Stäfa
Lifestyle accessory instead of prosthetic device, available in fifteen color combinations. The case also has room for replacement batteries.
Photo: Eric Schmid

What outward form should be given to a small device whose function involves being invisible? The hearing aid *Claro 11 CIC* was created by Kuster Design for the hearing aid manufacturer Phonak in 1998. "CIC" stands for "completely in the canal," meaning the hearing aid disappears in the ear canal. By the end of the twentieth century, microelectronics had advanced to the point that a fully digital hearing system including digital microphone could be developed on this small of a scale. The design houses microphone, speaker, processor, and battery in a shell of transparent plastic. A thread helps to retrieve the device from the ear canal. *Claro 11 CIC* is a technical device that reveals its inner workings like a transparent shrimp. Here, the technology reveals itself only to the user. But once the device is inserted in the ear, it is invisible. The user first has to learn how to operate the tiny hearing aid in the store or with the help of a manual, because its details are not immediately apparent from the design. Those who don't know the device even have to look for the on and off switch, and anyone unfamiliar with color coding in hearing aids (red dot for the right ear, blue dot for the left ear) is liable to get mixed up. And yet the purpose of the miniaturization is also clear: it endeavors to counteract the stigma for the wearer by making the device as small as possible. ME

Ernst 2005; Scharf 2007.

769 Wim Ouboter, Micro Scooter, 1998, launched in 1999
Aluminum, plastic, 86 × 34.5 × 66 cm, Gebrüder Ouboter AG, Küsnacht
Museum für Gestaltung/
Donation of Wim Ouboter

770 Micro Scooter, folded

771 Wim Ouboter, Micro Luggage, 2009, Micro Mobility Systems AG; Küsnacht, Samsonite AG, Muttenz
Museum für Gestaltung*

772 Child's drawing. Today the Micro is especially popular with children.
Micro Scooter

769

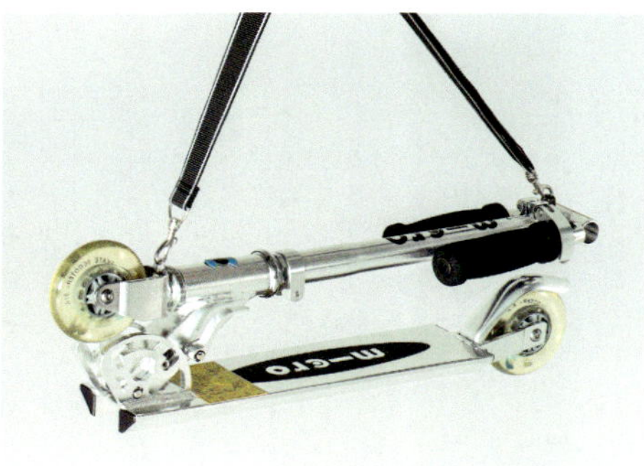

770

771

772

Legend has it that Wim Ouboter (b. 1960) built his first prototype by hand in order to conveniently reach the sausage stand on Zurich's Bellevue square. It was not really worth getting on his bicycle or in the car to cover the short distance, so the banker cobbled together a scooter with inline skate wheels in 1996. It lay around unused until 1998, when Ouboter decided to invest in micro-mobility and founded the company Micro Mobility Systems. He bought the steering rod system invented by his Berlin rival Sieghard Straka and handed the license to the US ski manufacturer K2, which used it to launch a three-wheeled kickboard. Ouboter's own design *Micro,* a two-wheeled aluminum scooter with brake, soon became a standard product sold successfully worldwide, setting off a veritable boom (in Germany alone, some four million scooters and kickboards were sold in the year 2000). To meet the rapidly rising demand, Micro relocated production to Shenzhen (China). However, it was only possible to obtain limited legal protection for the *Micro* since there were already patents for foldable scooters dating back as early as 1932. Imitators crowded onto the market and the bubble burst—simultaneously with the new, service-based economy. After this setback, Micro expanded its product range and now offers scooters for children, adolescents, adults, and extreme athletes. The company even collaborated with Samsonite to develop hand luggage that can be turned into a scooter as needed. RM

Schulz 2000.

Playtime Stool
1998

773 Frédéric Dedelley, Playtime Stool, 1998
Satin-finish stainless steel, 57 × 55 × 20 cm,
Base diameter: 30.5 cm, Burri AG, Glattbrugg
Museum für Gestaltung

774 Frédéric Dedelley, Evolution Stool, 2001.
Prospectus page

775 Frédéric Dedelley, Playtime, 1999. Drawing

773

774

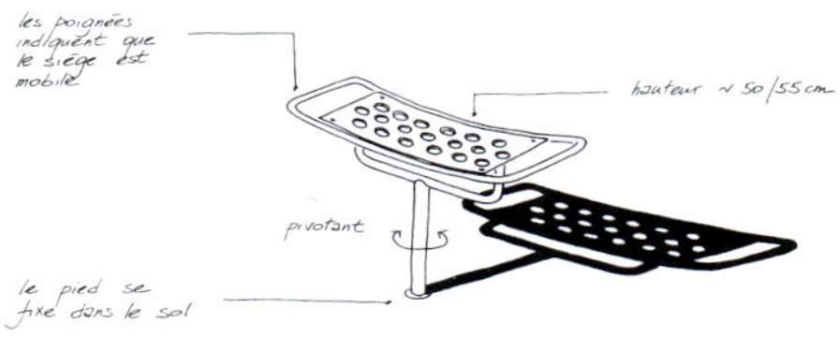

775

When Frédéric Dedelley (b. 1964) opened his own design studio in 1995, he didn't have a workshop available in which to produce models and so he experimented instead with materials that he could manipulate without machines and almost without any tools. He sketched ideas in three dimensions using aluminum wire, without drawing them on paper first. These experiments with line culminated in the design of the swivel stool *Playtime* for the Swiss garden at the 1999 World Horticultural Exposition in Kunming, China. The self-satirizing Swiss contribution to the show toyed with tourist clichés. The project by the Swiss landscape architect Rotzler Krebs, for example, consisted of a room installation of oversized picture-postcard motifs. In the context of this garden architecture, Dedelley's piece alluded to the canon of Swiss furniture classics. Recalling the design of a one-legged milking stool, *Playtime* takes up the perforated pattern of Hans Coray's *Landi Chair* (1939), while its materials and colors summon visions of the furniture adorning restaurant terraces in the fifties. In 2002, Dedelley reworked the swivel stool, optimizing its construction with a clamped plate instead of the nostalgic-looking tubular steel and thus adapting it to the contemporary vocabulary of forms. Both variants are produced by Burri, a company specializing in furniture for public spaces. RM

Pradal 2008; www.burriag.ch.

Alfredo Häberli and
Christophe Marchand

Move it Folding Table
1995

776 Alfredo Häberli and Christophe Marchand, Move it Folding Table, 1995, prototype
Coated solid core plywood tabletop, powder-coated tubular steel, 74.5 × 80 × 80 cm,
Gebrüder Thonet GmbH, Frankenberg (DE)
Museum für Gestaltung*

777 Move it tables folded up and telescoped with the help of a pneumatic spring and wheels.

778 Development of the plastic clips for the model series
Museum für Gestaltung*

776

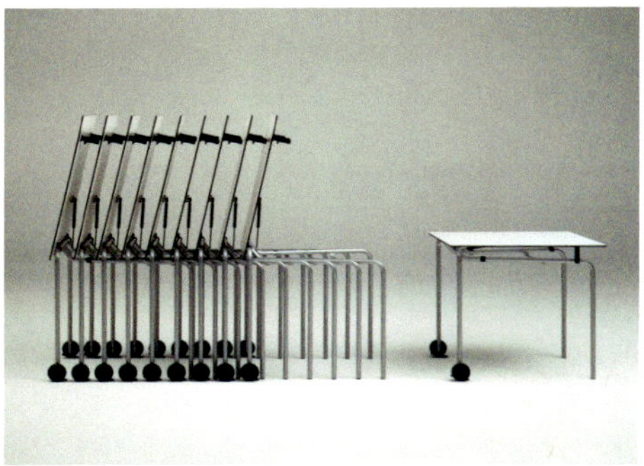

777

778

After studying at the Schule für Gestaltung Zürich, Alfredo Häberli (b. 1964) and Christophe Marchand (b. 1965) opened a studio for design and product development. Great admirers of the work of Bruno Munari, Enzo Mari, and Charles Eames, they designed useful items that could also have a touch of whimsy or wit, while lavishing attention on functional details. Häberli and Marchand had sought cooperation with international manufacturers early on. Through Heinz Ryffel (b. 1947) and his production and distribution company Sele2 they came into contact with Thonet, which was looking for a mobile table for temporary deployment. *Move it,* made of tubular steel and solid-core laminate, can be regarded as a tribute to the "Swiss tubular steel modernism" of the thirties. Similar to the folding table by Werner Max Moser (1932), it can be conveniently stored by flipping up the tabletop. *Move it* thus solves the storage problem, but not with a foldable frame; instead, it holds the tabletop in a vertical position with a pneumatic spring so that several tables can fit inside each other to save space. The *Aluflex* chair by Armin Wirth (1903–1992) was the model for this telescoping storage system. The frame derives its stability from specially designed plastic clips. Available in five colors, the table offers space for four but can also easily be combined with other tables, making it suitable for both home use and public spaces. RM

Meier 2002c.

779 Tina Grässli, Knit Shorts, 1995
Wool jacquard knit, reversible, 39 × 47 cm,
Xess + Baba, Zurich
Xess + Baba, Jacket from the Coming Home
Collection, 2011
Wool, jacquard knit, reversible, 64 × 131 cm,
Xess + Baba, Zurich
Museum für Gestaltung*

780 Xess + Baba, Sie und Er (She and He), sweater
with pants, reversible, Production I, 1998–99
Photo: Marc Straumann

779

780

During her textile design studies at the Schule für Gestaltung Zürich, Tina Grässli (b. 1969) created fabrics for specific products. She always tested her patterns first on concrete objects—contrary to the typical professional reality of a textile designer. After earning her degree, Grässli designed plaids with optical color mixing effects and with color progressions and hierarchies, and developed a reversible knitted fabric on her own knitting machine. A first series of industrially produced apparel sold well, but she needed capital to build her own collection. In 1996, Grässli and the jewelry designer Barbara Rüegg (b. 1968) took the plunge and opened their own store. Saus & Braus was the first multi-label store for clothes and accessories produced in limited series. It thus also offered a platform for other designers, enabling them to test market their products. Under the label Xess + Baba, Grässli and Rüegg began in 1997 to produce durable knitted jackets, caps, and dresses whose design quality took its cue more from the criteria of product design than from high fashion. Their plaid in many different colorways became a signature trademark. The double-face knitwear in wool or cotton is reversible and thus offers men and women two contrasting color moods in a single garment. Especially coveted are the Xess + Baba zippered cardigans from the early days, reissued in 2012 by popular demand. RM

Schilder Bär/Wild (2001) 2004, pp. 172–80; www.xessbaba.ch.

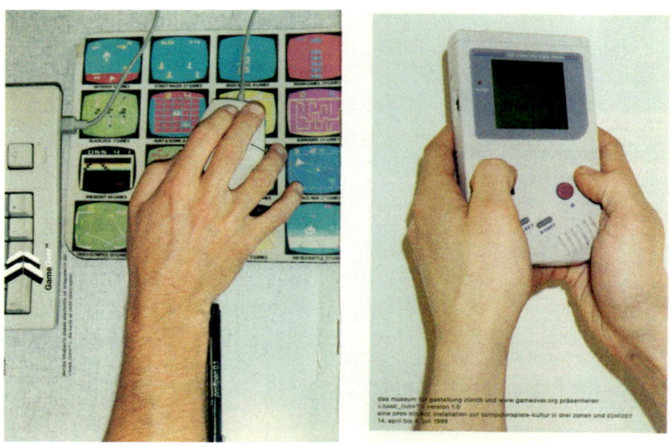

781

Who is responsible for shaping Swiss design? The designers themselves, certainly. But above all, it is the producers who for decades have not only commissioned the work but also typically manufactured it.[1] The inventors and creators who provided the designs that went on to succeed in the market have tended to disappear behind the crossbow logo.[2] Who knew that the *Rex* peeler, which is still manufactured in Switzerland, was patented in 1947 by Alfred Neweczerzal, the son of a Prussian immigrant who had settled in Davos?[3] Or that the *Recta* compass, developed in 1940 by Eric Vaucher, has ever since then been made in Biel?[4] Companies such as Jura, Sigg, Zyliss, V-Zug, and Bernina have been manufacturing consumer goods in Switzerland for decades—often at their own production plants. Designed in-house, these products have always carried the brand name, which is likewise the name of the producer / manufacturer, out into the world. The names of the designers hardly played any role at all in marketing, and no one ever spoke of "designed by."[5]

The watch, furniture, machinery, and textile industries were proud of the role they played in Switzerland's workplaces. Their self-confidence was reflected in the "Swiss Made" label.[6] However, with the outsourcing of production to southern Europe and then to East Asian countries with lower labor costs, a new situation started to emerge in the eighties. Not only the Swiss textile industry, which began outsourcing production in the mid-sixties,[7] but also the computer industry had to find ways to produce goods more cheaply. Logitech was one of the first Swiss IT companies to relocate its production facilities abroad for economic reasons, shifting computer mouse manufacture to Taiwan in 1986 and then to the People's Republic of China in 1993.[8] The company was founded in 1981, the same year IBM introduced its personal computer that would turn the computer into a mass-market product.[9] In an interview on Logitech's twenty-fifth anniversary, the cofounder and CEO at the time Daniel Borel described his fear at the end of the eighties that the relocation of production would also cause development and design work to migrate out of Switzerland: "The 1992–93 crisis in the computer industry almost killed us. With Windows 3.11, the mouse became a mass-market product overnight and prices plummeted. For us, it was like an electric shock—we couldn't hold on to our factories in America and Europe."[10] However, it was also this

781 Cornel Windlin, Invitation to the exhibition *Game Over,* front and back, Museum für Gestaltung Zürich, 1999

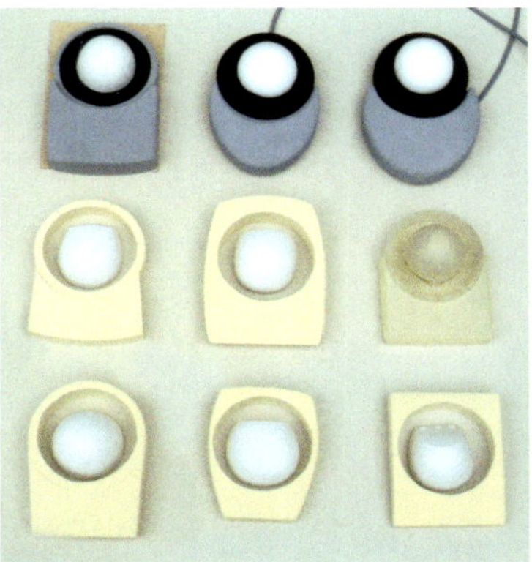

782

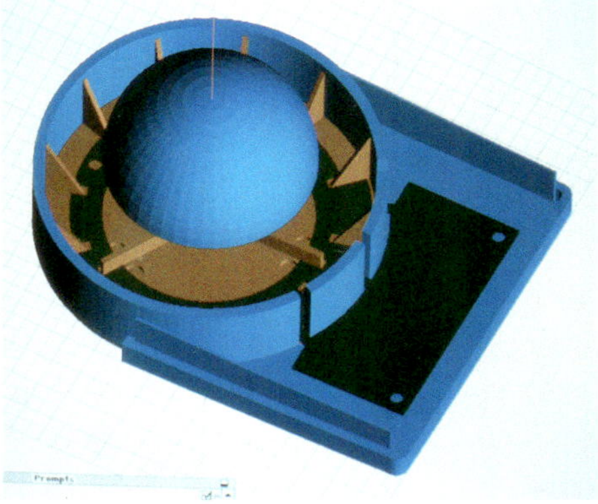

783

784

785

782 Michael Krohn, Design process for the Spacecat
783 computer mouse. Physical models made of
784 foam for the ergonomic test, CAD construction drawing and rendering, 1999, as part of a research project at the former Hochschule für Gestaltung und Kunst Zürich (HGKZ) and the ETH Zürich.
785 Bodum water kettle, ca. 1998, Agil, Stans. Manufactured using the stereolithographic 3-D printing process, which was patented in 1986 and used for the first time in Switzerland by the company Agil, in 1989. Model made of laser-hardened polymers next to the finished product in plastic. Photo: Martin Stollenwerck

outsourcing that ended up guaranteeing jobs for the remaining approximately 260 engineers and financial and marketing people in Switzerland, says Borel. Under pressure, the engineers "developed the optical trackball and the technology for wireless mice and keyboards that we still use today."[11]

Not at the Same Location: Designing and Producing

In the following, we are less interested in the economic repercussions of globalization on Switzerland than we are in the question of what impact comprehensive digitization had on design and production and how those affected saw their roles change.

The story of Swiss design from the perspective of changing production conditions has yet to be written.[12] The development of industrial production in this country can be described using the four-stage model of industrialization. The processes that are grouped together under this general concept were what made the division of labor between design and manufacturing necessary in the first place. They form the basis for industrial design and characterize many of the definitions of design. The model follows a linear narrative. It identifies progress based on the degree of automation and rationalization as well as the positive effects of economies of scale in production practices. The first industrial revolution with its mechanical production facilities began to operate in the late eighteenth century with steam and water power, followed by a second revolution around 1900 with the advent of electrically driven mass production and a far-reaching division of labor. The application of further automation in the seventies using electronics and information technology is regarded as the third phase. Currently, individualized production models are the state of the art. After the water wheel and steam engine, assembly line and computer control, this most recent development is referred to as the fourth industrial revolution and is labeled with such catch phrases as "Digital Factory,"[13] "Industry 4.0,"[14] and "The Internet of Things." It is characterized by a link between production and the Internet: work pieces, machines, and transport systems communicate via RFID chips and wireless transmission in order to autonomously mediate the most efficient means of production. The target is to produce customized products at prices equivalent to industrial production.

Since the beginning of the new millennium, this current last stage of comprehensive digitization has affected production, logistics, sales, and the market. Development cycles and production times are shorter, and producers are able to react to customer needs more quickly after identifying them through extensive market surveys.[15] Are clients thus becoming co-designers? Is design's raison d'être beginning to dissolve? Not really, but the concept of design is indeed expanding and the profession is changing. Of necessity, designers have been forced to acquire strategic knowledge in order to defend their creative freedom.[16]

Computers in the Design Process

This latest development is the result of an upheaval in the nineties that fundamentally changed both the education[17] and the practice of design. Apart from market globalization—driven by information technology—responsibility lies with the computer.[18] Its influence plays out in two ways: in design and in manufacturing.

Design theorist Gui Bonsiepe (b. 1934) already predicted the significance computers would have for design back in 1967. He envisioned that the machines would consolidate research and thus accelerate the design phase.[19] And second, that appropriate software would streamline technical drawing, leading to a broader basis for decision making.[20] The creation of several versions of a design would thus become economically feasible.

Bonsiepe formulated his prognosis with knowledge of the beginnings of computer-aided design.[21] Its forerunners date back to the mid-fifties, when the Massachusetts Institute of Technology (MIT) wrote the first programs for computer-controlled machines. In the seventies, the CAD industry grew worldwide into a billion-dollar business. But it was not until the late eighties that CAD systems spread in Swiss design studios[22] and training programs,[23] and streamlined drawing. Up to that point, one could describe the digitization of design as an unbroken extension of the drafting pencil, with which the step into the virtual had already been taken: a stroke, a second stroke, and an idea becomes an image. Whether analog or digital, whether on paper, as a cardboard model, or on the screen: the visualization of an idea is what is meant by the process of design.[24] When and with what intention the hand-drawn

786

788

787

789

790

Tools for making wooden chairs.
ag möbelfabrik horgenglarus:

786 Machine for the traditional bending of wooden frames
787 Templates
788 CNC milling machine

sketch is replaced by a digital drawing, and the drawing by a cardboard or foam model, remains an individual decision.[25]

From the mid-nineties, with increasing computer performance, three-dimensional CAD systems began to take hold: design programs that not only displayed ideas in three dimensions but could also model free-form surfaces. History-based programs, which store the sequence of design steps and the geometrical parameters of the basic elements, allow for modifications to be made. Thus, the geometry of an object can be changed after the fact without having to rebuild it from scratch. Interfaces with the programs used by the designers facilitate cooperation between formation and construction.[26] The data forms the starting point for digital mock-ups, or virtual prototypes. These join the visualizations of individual components into a whole. They also show how an object should be assembled or disassembled in production, how it moves, thermally deforms, or how it behaves under pressure.[27] Such a virtual prototype—a simulated representation of the finished product complete with a natural-looking rendered surface—can replace the construction of real prototypes.

The digitization of the design has an effect on the manufacturing process: the digital models can be converted into executable commands. The data is fed into tools and robots that precisely and quickly turn out components and products—"just in time." Data also controls rapid prototyping methods such as stereolithography,[28] which builds a prototype model, or the CNC mill, which produces the work pieces. In addition, databases that plan the work and record operational data help the manufacturer to ensure quality. If everything meshes together, the manufacturing industry has achieved the goal of computer-integrated manufacturing (CIM).

Computer Skills

During the nineties, workstations with increasingly powerful illustration and modeling programs streamlined the design process. The Internet also came into widespread use during this period. As Gui Bonsiepe had predicted in 1967, this opened up access to a global archive. The computer that found its way into offices became a research tool.[29] Internet and the corresponding universal standards also allowed the design drafts to be spatially detached from production. The data that controls the machines could now be transmitted all over the world.

How this change in the production process was handled in the nineties can be demonstrated based on the example of the Bodum design department. The Danish family business, founded in 1944, moved its headquarters to Switzerland in 1978. Two years later, the company established its own in-house design unit, Pi-Design. In the mid-nineties, Bodum produced in eight European countries and the United States and sold its household goods worldwide. This was possible because design, manufacture, and distribution were carried out with computer support, but also because the responsibility for design resided at management level and was recognized as a management task.[30] The designs came from external designers who had to engage directly with the company's internal communication system. From sketch to production decision, the team worked with two-dimensional and transformable three-dimensional data that was transmitted via modem to the toolmakers. In parallel, models were created as needed with the stereolithography process. Peter Wirz (b. 1960), part of the Bodum design team, was quoted in 1994 as saying that the competence to work with these tools was becoming increasingly important: "Designers who do not deliver their designs with at least semi-real 3-D data will have increasingly less weight within the industry."[31]

It was not only producers who were aware of the benefits of computer-aided design. Nose Design Intelligence, founded in Zurich in 1991, equipped the majority of its eight workstations at the time with CAD. Today, the agency is one of the largest of its kind in Switzerland. The three founding partners, influenced by apprenticeships at the corporate identity agency Zintzmeyer & Lux, put their faith in the Mac for their workstations. As the engineer and designer Christian Harbeke noted, it was "the ideal tool for identity-oriented industrial design."[32] Three-dimensional visualizations admittedly did not replace working models for the designers. And the transmission of the data that controlled the mills or machines did not always go smoothly. After all, that required that the manufacturers were able to offer CAD/CAM interfaces, which in the mid-nineties was not always the case.

Where it was available, computer-integrated manufacturing provided designers with new opportunities—and helped traditional

791

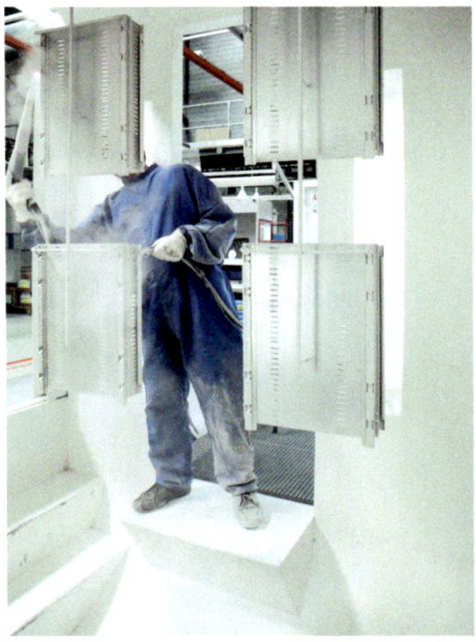

792

793

789 Hannes Wettstein, Lyra, 2008, ag möbelfabrik horgenglarus Produced with a CNC milling machine.
790 Studio Hannes Wettstein, Icon, 2012, ag möbelfabrik horgenglarus. Molded plywood 3-D seat; solid bentwood backrest, back legs, and seat frame.
791 Lista Office AG plant in Degersheim with sheet metal processing machines for punching, cutting, bending, molding, and welding. Photo: Derek Li Wan Po
792 Spraypainting of drawers in the Lista plant in Degersheim. Photo: Désirée Good
793 Damian Fopp, Merlot Waste Bin, 2012, Lista Office AG, Degersheim Museum für Gestaltung/ Donation of Lista Office

and also smaller producers who manufactured locally to take a step into the present. Since 2000, horgenglarus has used a five-axis CNC machine for the production of chair legs. Hannes Wettstein (1958–2008) was creative director of the Glarus-based furniture maker from 1999 to 2008. With his designs *Miro, Lyra* and *Epos,* among others, he carefully continued the tradition of the company's collection while taking advantage of the new production technologies. After his death, the Studio Hannes Wettstein designed *Icon,* a chair that reinterprets *Classic,* which was in production since 1998. In this design, the seat frame, rear legs, and back piece are first cut and bent on the old machines before being processed further with CNC milling. Hannes Wettstein had a life-long interest in not only form but also how to contextualize it. This led him to search for new materials and production techniques: "For us, open manufacturing processes that allow intervention are important. It's like cooking: with the exact same ingredients you can cook something great or something pretty bad."[33]

Role Play

The digitization of the design process, which began well into the nineties, can be interpreted as an extension of designers' skill sets. The design and construction programs brought designers closer together with technical developers and engineers. This led to designers having greater responsibility, but it also affected the technical side of design; now, designers are taking postgraduate courses to become design engineers.[34] The aim of such advanced training is to understand the two sides better, and also to recognize the limits of one's own competence.

Information and communication technology have also made it easier for designers to not only take responsibility for the design itself but also for bringing it to market. They can thus try their hand at assuming dual or even multiple roles as designer, consultant, producer, and marketing and salesperson.[35] Collectives such as N2, a group of seven designers that came together in 1997 for a presentation at the *Wohnsinn* exhibition at Basel's Muba trade fair, and in the same year got off to a soaring start in a Milan gallery during the furniture fair there, at first produced their own designs for furniture and lighting and offered them over the Internet.[36] Their success brought producers into the picture. The *Diaz* lamp went into production at Driade, and the *Habana* floor lamp at Pallucco Italia. A year after the studio's premiere in Milan, Jörg Boner (b. 1968) and Christian Deuber (b. 1965) designed the *Ajax* table, which won a first prize at the Interieur trade fair in Kortrijk, Belgium, and was added to the product range offered by the ClassiCon company.[37] The group, which wished to jointly design and present their work, sought producers to manufacture their own brand. But their few successes did not cover costs and they ended up parting ways, shattering the dream of a collective that was at once designer and producer.

The mid-nineties saw the exploration and teaching of role diversity, bundled under the catch phrase "cultural entrepreneurship."[38] Entrepreneurial skills built upon a design background again became a career option—just as it had once been for entrepreneurial inventors like Alfred Neweczerzal in the late thirties. This in turn allowed designers to aim for a strategic role: they began to look beyond the individual stages of the design process and to steer the entire process in the role of design manager.

The search for producers, however, remains central. They are found not only on one's own doorstep. Peter Wirz therefore decided in 2003 to move his agency at the time, Process Design, to where the production facilities were. He opened his first branch in 2006 in Taipei, and a second in 2011 in Shanghai. This also had to do with the impending loss of power for designers, because with the relocation of manufacturing, the development departments migrated as well—and thus the decisions.[39] In the meantime, labor costs have risen in Asia. So it's time to instruct the producers and manufacturers there in the value of careful brand development and design work, in other words, to help them change from manufacturers to responsible producers. For local producers, on the other hand, production costs alone no longer determine the choice of manufacturer.[40]

Finally, the traditional model of product development and production process, in which the designer plays a minor role as contractor, is crumbling. This conventional model is also called "over the wall engineering." It organizes processes sequentially and with a division of labor, as if the project were like a package passed from one area of responsibility to the next. In contrast are more recent

794

795

796

794 Gramazio & Kohler, The Programmed Wall: Architecture and Digital Fabrication, ETH Zürich, 2006. Students design brick walls using algorithmic design tools for production by an industrial robot.

795 Gramazio & Kohler, The Foam: Architecture and Digital Fabrication, ETH Zürich, 2007–08. Digitally controlled fabrication of polyurethane foam panels. The fascination lies in the impossibility of virtually modeling the foaming process.

796 Gramazio & Kohler, The Sequential Wall: Architecture and Digital Fabrication, ETH Zürich, 2008. The definition of different parameters for the functional specifications of an external timber wall merges with formal considerations.

approaches that include feedback loops and that reduce operational boundaries. Spatial proximity, integrated project management, and appropriate communication tools are necessary for this to work.[41]

To put it simply: good design requires empathy between the players. It thrives best when understanding between producer, designer, and manufacturer prevails. This in turn means nothing more than the ability to see the complex design process through the eyes of one's working partner.

1 A distinction is made in the following between the "producer," who initiates a project, commissions the designer, and organizes manufacture, sales, and marketing, versus the "manufacturer," who manufactures the product on behalf of the producer.
2 The crossbow logo is disseminated by the Society for the Promotion of Swiss Products and Services. Founded in 1917 under the name Schweizer Woche, the association was renamed Swiss Label in 1989. The approximately 600 members have the right to market their goods and services under the crossbow trademark if their value for Switzerland accounts for at least fifty percent (depending on the product group). See www.swisslabel.ch.
3 *Unbekannt – Vertraut* 1987, p. 134.
4 *Unbekannt – Vertraut* 1987, p. 120. The Recta compass is still produced today by the Biel-based company.
5 "Design" caught on as a German term in the sixties. See Lichtenstein 1997, pp. 177–79; Schilder Bär / Wild (2001) 2003, p. 11.
6 The use of the "Swiss Made" certificate of origin is regulated by articles 47 and 48 of the federal law on the protection of trademarks.
7 Scholz / Kaufmann 2003, Appendix D, p. D1.
8 Logitech, founded in 1981 in Apples (Canton Vaud), opened its first manufacturing plant in Asia in 1986 in Hsinchu, Taiwan. See http://www. www.logitech.com/lang/pdf/logitech_defining_moments.pdf (accessed May 30, 2014).
9 "In 2000, 60 percent of all households in Switzerland had at least one computer." Haber 2011. See also Tobler 2001.
10 Streitz 2006. On the high-tech industry and its importance for the development of a knowledge-based economy in Switzerland, see Bundesamt für Statistik 2004, pp. 27–29.
11 Streitz 2006.
12 For a summary, see the chapter "Design and Economy: The Link Exists," in Schilder Bär / Wild (2001) 2003, pp. 15–38.
13 *Richtlinien Digitale Fabrik* 2008, sheet 1, p. 2; Bracht et al. 2011.
14 In Germany, the Federal Ministry of Education and Research and the Federal Ministry for Economic Affairs and Energy launched the project "Industry 4.0." Within the scope of the corresponding financial planning, up to 200 million euros will be allocated. In Switzerland, the organization SwissMem champions this issue.
15 "Advanced technology creates the illusion of a unique piece as part of a comprehensively computerized apparatus." Widmer 1999.
16 The development of the profession is described as one from craftsman to applied artist to specialized designer. See, for example, Bürdek 2005. Professional sociological studies are lacking. At the University of St. Gallen, a study was conducted in 2013–14 called "Creativity as a Profession: The Sociology and Philosophy of Working Life in Visual Arts and Design." A cultural-sociological viewpoint on design is offered in Moebius / Prinze 2012.
17 See Mbiti et al. 1999.
18 Philip Ursprung points out the importance of the container as well, which to his mind takes its place alongside the computer as a key object of globalization; see Ursprung 2011.
19 "One loses too much time with manual methods of searching for information—browsing through magazines, surveying index lists, sorting printed matter." Bonsiepe 1967, n.p.
20 "Formal transitions, surface modulations, formal variants, enlargements, reductions, diversity of views and in no time at all, mass coordination could be visualized by specifically developed computer programs." Bonsiepe 1967.
21 Weisberg 2008; Porten 2008.
22 A 1994 survey of 120 members of the SID (today the SDA, the Swiss Design Association) showed that around two-thirds of design studios use the computer as a design tool and about a third use it for renderings. The investment costs were estimated at 30,000 francs. 3-D CAD systems cost around 90,000 francs. CIM is not yet common. Cited in Locher 1994a, p. 28.
23 In 1986, the first computer workshop was set up at the Hochschule für Gestaltung und Kunst Zürich. The software packages Model Maker, Picture Maker, and Auto CAD were taught. This was followed in 1989 by computer-controlled looms for the textile class, for which the software programs WB and WC were developed at the school. The first experiments with integrating the Internet were undertaken in 1995. In 1996, the terminals numbered around 130, including specialized workstations with video editing, animation and modeling, parametric CAD, and sound editing. The year 1998 saw the start of studies in New Media. Mbiti et al. 1999, pp. 2–3.
24 Ernst 2011a, p. 298.
25 See the statements by Claudia Mareis and Jörg Boner in Ernst 2013a.
26 Ernst 2011b.
27 Gausemeier 2013.
28 The process was patented in 1986; in 1988 the first production-ready stereolithography machines came onto the market; and in 1989 the company Agil in Stans started operations with the first device in Switzerland. Unlike CNC, the process can be used to produce undercut forms and to build up double and extremely thin walls. Locher 1995.
29 "Network technologies, which are cross-platform and cross-system (standard protocols such as TCP/IP), and the resulting transformation of the global network of individual stations into a global cluster allow for working communities that can collaborate on the same project independently of time and place." Matter 1999, p. 42.

797

798

799

797 Production of the Impec BMC Bike, 2010. The loom wheel is fitted with up to 128 bobbins that weave a seamless tube of carbon fibers using the load-specific weave (LSW) technique onto the positive form of the respective frame tube.

798 Production of the Impec BMC Bike, 2010. LSW, the robot-controlled, load-specific weave technique, leads to a carbon fiber and resin structure that is accurate to a tenth of a millimeter.

799 Nose Design, Impec Bike, 2010, BMC Switzerland AG, Grenchen Photos: Martin Jeker

30 Peter Wirz, who designed for the Bodum company in Lucerne from 1992 until he founded his own studio in 1998, along with Charlotte Warning and the Danish chief designer Carsten Jørgensen, took over design management tasks. Locher 1994a, p. 28.
31 Ibid.
32 Locher 1994b, p. 17.
33 Ernst/Hönig 2005, p. 22.
34 Ernst 2011b, p. 27.
35 This process is to be understood as a differentiation of role models. It is not meant to replace the classic role as a service provider/contractor as described below.
36 Wirth 1998.
37 Besides Jörg Boner and Christian Deuber, the group also included Valerie Kiock, Dave Braun, This Reber, Paolo Fasulo, and Kuno Nüssli. See Hornbogen 2003.
38 Such courses were introduced at the ZHdK in Zurich, the FHNW HGK in Basel, and elsewhere. At the same time, models are taught at art and design universities under the title transdisciplinarity, which aim to teach the students not only core competencies but also to have a wider view beyond the individual discipline.
39 Ernst 2011c.
40 Sturm 2013, p. 12.
41 Especially for complex developments, these factors can lead to more rapid success. Concurrent engineering is implemented, for example, by space agencies. The École polytechnique fédérale de Lausanne has also built a concurrent design facility at the Swiss Space Center. See http://space.epfl.ch/page-39445-en.html (accessed May 30, 2014).

A good producer is capable of entering into a close relationship with the designers. This happens when their two different roles overlap or even merge. I always assume that every designer wants to not only accomplish the task set by the producer but also realize work to personally identify with. I therefore don't view designers as service providers for the manufacturer. The art of the design producer consists in harmonizing these two sets of objectives. That is the prerequisite for success. The producer's job doesn't end there, though. A convincing way must be found to realize the design and make sure that the plans are carried out correctly. A product that is not realized according to its market potential is a missed opportunity. A good producer is capable of creating a joint work with the designer, implementing it in the appropriate quality, and seeing to it that the product lives up to its market potential.

Things get exciting when designers and manufacturers swap roles: when the manufacturer spurs on the designer, demands greater autonomy, and contributes design proposals; or when the designer questions the proposed costs and reminds the manufacturer of market requirements. A good producer must be able to step out of the narrowly defined producer role. The designer in turn is good if he or she can dispel the notion that the demands of production or the market restrict creativity. Charles Eames once said that every designer should be grateful for constraints. They are what prompt designers to really get creative. A designer who dreams of working without constraints should be wary of this wish coming true. Grappling with a rigorous set of specifications is what makes the whole thing interesting in the first place. In our product world, the requisites typically concern durability and quality. Furthermore, the price of the product must be accepted by the market. And the product should give the designer and the manufacturer the satisfaction of having given the world something new.

A design is typically achieved through a process in which the designer and producer play equal parts. This distinguishes design from a purely artistic activity. Within the producing company, there are a number of aspects to consider, such as technology, marketing, and—where large investments are involved—financing. This means that many different people join in the conversation. Nevertheless, product development can never be a democratic process for which a consensus must be reached. Producing is risky, especially if something new is to be created. Copycats know from the outset what the result will be; they risk nothing—except perhaps an accusation of plagiarism. But the more a design deviates from what already exists, the riskier the design process becomes. Therefore, it takes the will of the entrepreneur who says: "I want it, I believe in it, I'll risk it."

The whole design process is based on the principle of trial and error. You have to try out several different variations before coming up with a good solution. Usually, a number of prototypes are involved along the way, because drawings can only convey in a limited fashion the problems and opportunities presented by a draft design. In retrospect, you then wonder how you could ever have seriously considered the one concept or the other. Even in the case of giants like the Eameses, you see prototypes that make you think: How could they ever have contemplated that possibility? This process is frustrating, but in the end, if you've worked hard and are lucky, you arrive at a good solution. Because there is always room for improvement, you could go on forever with this process, but at some point you have to stop and say: "This is the best solution we can currently imagine." Failure is, of course, a commonplace event for a design producer. We at Vitra have been lucky to usually detect potential flops while they were still in the development phase, such that we didn't go on to realize the product. In other cases, we firmly believed in the project and yet it didn't work out after all because we had not properly understood the needs at hand. Those are the interesting flops that propel us forward, because we learn something. A failure due to a design that has no autonomy is uninteresting. Every new product must strive to be an improvement over what exists. And if the improvement is understood as such by the user, we have achieved our goal.

Rolf Fehlbaum (b. 1941), former CEO and a member of the Board of Directors of Vitra, explained the characteristics of a good producer in a conversation with Meret Ernst on August 6, 2013. This conversation is reproduced here in the form of a written statement. During his active days at Vitra, Rolf Fehlbaum forged contacts with many well-known designers and developed joint projects with architects such as Tadao Ando, Frank Gehry, Zaha Hadid, Nicholas Grimshaw, Álvaro Siza, Herzog & de Meuron, and SANAA, who all designed buildings on the Vitra Campus in Weil am Rhein. As an avid collector of twentieth-century design, Fehlbaum founded the Vitra Design Museum in 1989.

A good designer in Freitag's opinion is also a good co-entrepreneur. A designer can of course be permitted to be in love with the product but must also think holistically. Each project is like a small start-up within the existing company, for a brand that already exists. The designer is involved in the process early on and accompanies the product all along the way. The designer must deal with the marketing department, the head of sales, and the production plant and have the courage to question everything. Today, more designers with this kind of understanding are coming to work with us. The profession has changed since we [Daniel and his brother Markus] did our training. "Design thinking" has been declared an entrepreneurial principle. It carries more weight, or perhaps has simply made a name for itself. After all, in earlier periods there were also entrepreneurs who produced their own chairs but were actually designers or architects.

In multinational companies, designers have to become entrepreneurs again and intervene in the processes. Unfortunately, in many companies the maximum power is wielded over a product where the most value is derived from it, namely in the sales department. The perceived needs shown by the sales figures should be respected but not exaggerated, because otherwise the result will only be "more of the same." We are constantly trying to reinvent ourselves, to detach ourselves from all entrenched patterns and to rethink everything. It's a fight for innovation, even in the smallest details. Since we're all working under the same factory roof, there is a certain amount of friction. The struggle is more difficult for the lone designer, but it can lead to better quality.

Back in the nineties, it was possible to have the courage to produce your own designs. And the economic conditions in Zurich were good. I was able to make a living with a few hours of freelance work and still have some spare time. Today, the pressure to start your own business is perhaps too great. There are Internet platforms such as We Make It and other crowd-funding platforms that facilitate financing in the design business, and that's a good thing. But Markus and I have never written a business plan or calculated our finances using an Excel spreadsheet; we instead tackled everything with a desire to create and to do something. We didn't want to give up our enthusiasm under any circumstances and later we were able to bring it to bear on all areas of the company. Our first company letterhead, for example, was a photo of the letter's author, taken with a video camera installed in the computer. It was like a reward: every time we had finished writing a business letter, we were allowed to make a funny self-portrait. A normal businessman would probably never come up with such an idea.

Communication has always been important to us as graphic designers. It enabled us to convey our ideas. We did it casually and without thinking about marketing. We did our first website in 1997; that means that four years after our first bag there was already this new medium. It offers an opportunity to communicate an idea just once and in a lasting way. You don't need a budget for that. Today, you can already gain a following for an idea that hasn't even been put into practice yet. It's fantastic!

Producing in Switzerland was the right decision for us. Sure, people say that some designs can only be produced in low-wage countries because they would otherwise be too expensive. It's also a question of the specific design—and for us, for more than twenty years, it has been a question of our business philosophy. We still sell the most bags in Zurich. But that doesn't mean that they are purchased only by Swiss. This "destination shopping" is important: I also like to bring something home with me from the place where it is made. A product can be locally anchored without having a Swiss cross printed on it. Our bags are an update of the Swiss image. They pay tribute to it with well-known values such as clarity, quality, robustness, and longevity; they have something "small-clean-stark-orderly" about them. The typography, the "transit graphics" from the trucks that we transfer to our products, also fit in with the recognizable Swiss identity. On the other hand, the used look, the patina of the used truck tarps, is not necessarily something you would call Swiss. And regarding the bags' sustainability: the ecological perspective is inescapably Swiss—we can afford at least that much! But with our repurposing and recycling of tarps we have gone beyond simple conservation or collecting aluminum pull tabs and have ventured to get our hands dirty. The world won't be saved just because we recycle a few tarps, but the related message has a larger impact. We learned a lot about our product by exposing it to different places, especially in Japan: different viewpoints, a different vocabulary used to describe it, a different market, and a different culture. Reflecting these back onto ourselves and assimilating them into our own culture has taken us a step further. We don't need just more and more merchandise but rather fewer and better products that work in different situations and feature a smart service design. And that is a task in and of itself, to keep reinventing those few products that we really need.

Daniel Freitag (b. 1971), cofounder and co-owner of the bag company Freitag, explained what constitutes a good designer in an interview with Renate Menzi on December 17, 2013. It is reproduced here in the form of a written statement. Together with his brother Markus Freitag (b. 1970), Daniel Freitag designed a bike messenger bag in 1993 made of used truck tarpaulins. The Freitag brothers then went on to build their own factory, stores, and brand around this functional and individualized product. The Freitags' business model is regarded as a prime example of creative industry and became the inspiration for many other design entrepreneurs.

800

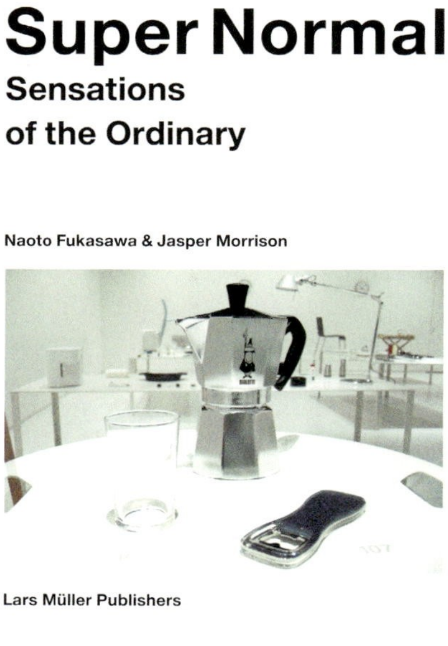

801

"The New Simplicity"
800 Jasper Morrison, *some new items for the house, part I,* exhibition view, DAAD Galerie, Berlin, 1988
801 Jasper Morrison and Naoto Fukasawa, *Super Normal: Sensations of the Ordinary,* Lars Müller Publishers, Baden, 2007

Design frees itself from its functional context
802 Philippe Starck, Juicy Salif Citrus Squeezer, 1987, produced from 1990, Alessi S.p.A., Crusinallo (IT) Museum für Gestaltung
803 Carlo Borer, Etienne Louis espresso machine, 1993 Museum für Gestaltung

Diversification as a new function
804 Prospectus for Swisstel PTT telephone, n.d. (produced 1987–90), Ascom Autophon AG, Bern
805 Martin Lotti, Nike Metal Max III athletic shoe, 1999, Nike Inc., Beaverton, Oregon (US)

802

803

804

805

Design history uses terms like "essentialism," "New Simplicity" (Neue Einfachheit), and "reduction" to label the nineties style, alluding to both formal tendencies and design attitudes that are regarded as responses to the pluralism of the eighties.[1] The critical intellectual power packed by postmodernism had in fact exhausted itself in mass-produced merchandise by around 1985. The graphic splashiness of cheap furniture à la Memphis, high-tech lighting fixtures, and pastel-colored toasters quoted the formal language of an avant-garde that had long since joined the mainstream. "Anything goes," the battle cry of the critique of functionalism,[2] had taken on a new meaning: the stylistic diversity it advocated now became a real challenge in the face of a globalized market. Producers demanded new and distinctive products tailored to a specific target group. As service providers, designers now either had to convince marketing departments to go with their ideas or to become their own brands. Taking their cue from Philippe Starck, star designers staked their claim as personalities and established an individual signature style that shaped interiors, furniture, and products. Now there was "designer furniture" whose value depended on the international renown of its author. Uniqueness and individuality became values in their own right from the perspective of both the designer and the consumer. Design as an expression of personality thus came to the forefront in the nineties as a hugely important economic and social factor.

Design as a Social Medium

As early as 1979, the French sociologist Pierre Bourdieu had investigated how individuals establish their identities in social environments.[3] The concept of lifestyles was oriented on his work, examining the attributes by which people distinguish themselves from each other or create a feeling of belonging to the group. What critics of functionalism around 1968 had lamented as missing from modern design—particularly emotions, atmosphere, and subjectivity—were aspects specifically addressed in, or even created by, advertising messages in the nineties. Design became one way of standing out in an experience-oriented society,[4] manifested in leisure preferences, forms of communal living, settings, and things.[5] In the production industry, market-based analyses became an important factor in product development. Social milieus were defined as target groups, and trend researchers were employed to study scenes or communities. New products were not only developed with a specific target group in mind, they were also offered in stylistically distinct variations for different tastes. A Candida in-house brand toothbrush sold by Migros thus reveals something about the gender, age, and lifestyle of the targeted user. So-called lifestyle products are especially notable for this customer-specific diversification. It is furthermore governed by the dictates of fashion, which—by definition committed to aesthetic obsolescence—follows the laws of distinction and transformation.[6] In product design, new technologies also promoted diversification. The least expensive option was to produce different color choices and graphic user interfaces to individualize what were actually uniform products.[7] The Swatch phenomenon, which around 1983 inaugurated systematic graphic design for the plastic watch, is representative of the "dominance of the surface" that conquered many areas of design in the nineties.[8] Digital printing, new coating processes, CNC-controlled milling, and laser cutting machines fostered a renaissance of ornament. Colored patterns and images adorned Sigg water bottles, silk ties, and porcelain dishware. Even the Swiss Federal Railways started offering in 1991 different color moods for what were once uniformly blue ticket covers, demonstrating an openness to individual customer requests.[9] Injection molding, used to meld several components into a three-dimensional form, enabled the production of an enormous variety of shapes in the world of plastics, for example in toothbrushes or sneakers.[10] In 1997, the art historian Beat Wyss described "the world as a T-shirt," succinctly characterizing the products of the nineties.[11] Colors, pictures, and ornaments were projected onto the bodies of objects, and the availability of digital data led to mass-produced, personalized everyday products while also offering a way to cost-effectively produce small series.

Simplicity as Statement

This background helps to explain the term "New Simplicity" as both a formal reaction to the sculpturally formed design icons of the international design elite as well as a countermovement to the

806

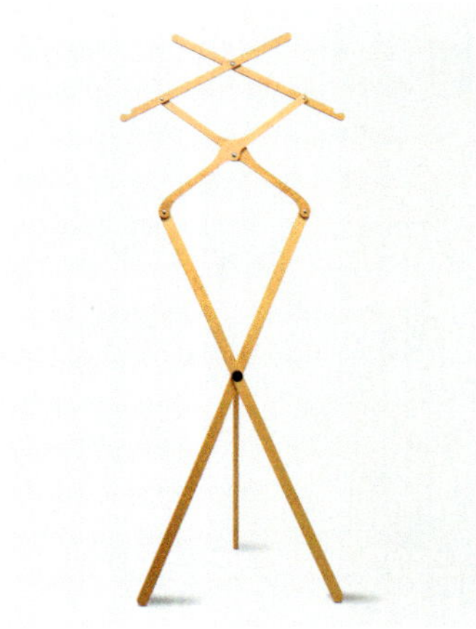

807

808

809

806 Frédéric Dedelley, Fleur flower holder, 1997, Wireworks, London
Museum für Gestaltung/
Donation of Frédéric Dedelley
807 Kurt Thut, Clothes Stand 107, painted beechwood, 1998, Thut Möbel AG, Möriken
Museum für Gestaltung*
808 Atelier Oï, Product sheet for the Torslanda Wall Hanger, 1997, Ikea. The two-dimensional PET foil only opens its hooks when mounted on the wall.
809 Alfredo Häberli, Folding table prototypes with round and square tabletops, 1995–96
Museum für Gestaltung*

commercial individualization of everyday products. Unlike perhaps in Germany or Italy, however, it is not possible to speak in Switzerland of a return to functionalist design principles, because the "minimalist tradition" had always been subliminally present in Switzerland—despite sporadic do-it-yourself, Pop, and Anti-Design projects. Since Italian postmodernism as well barely affected Switzerland, the New Simplicity cannot be understood in this country as a totally new approach to design;[12] rather, with the introduction of beauty as a function (see Menzi, "Designing 'Die gute Form,'" pp. 192–97), an "advanced functionalism" was already being implemented around 1950.[13]

Schooled in modernism and its continuous evolution, a young generation of well-trained designers continued to focus in the nineties on the core values of design reform and the maxims of the Werkbund: being true to materials, functionality, and principles of construction. The economy of means and the aesthetic durability of *Gute Form* now took on an ecologically motivated emphasis. What was new about the New Simplicity in Switzerland was a novel way of looking at ordinary things that had always been there but were now virtually rediscovered and reissued.[14] Also new around 1990 was the poetry, humor, and whimsy found in international role models. In addition to studying Swiss design classics, the younger generation looked to London and Berlin for inspiration.[15] In 1993, for example, Gijs Bakker (b. 1942) and the design historian Renny Ramakers (b. 1946) founded the Dutch design collective Droog Design (droog means "dry" in Dutch) with the intention of counteracting the prevailing design ethos centered on form with more conceptually based, pragmatic, and subjective objects.

Jasper Morrison's (b. 1959) room installation *Some new items for the house, part I*[16] is regarded as the manifesto of this New Simplicity. Producer Rolf Fehlbaum (see "What Makes a Good Producer?," p. 330) coincidentally saw a photo of this room—in which all the furniture, the walls, and the floor are made of plywood—in a magazine and persuaded Morrison to design for Vitra. Other major producers, such as Kartell, Moroso, Cappellini, and Driade, also took an interest in the new penchant for archetypal forms. Meanwhile, in Switzerland small and medium-sized companies began to manufacture designs hallmarked by the New Simplicity. They manifested the traditional quality standards typical of the country but also catered to foreign demand for unspectacular, durable, and flexible furniture.

Design Boom

Grand gestures are commonly greeted with skepticism in Switzerland, so excellent design achievements could be found instead in formally restrained products for middle-class buyers, as good democratic design. Star designers were not systematically nurtured in their own country, which is why talents like Hannes Wettstein (1958–2008), Alfredo Häberli (b. 1964), Christophe Marchand (b. 1965), and Jörg Boner (b. 1968) had to rely on cooperation with international manufacturers, seeking contacts with appropriate foreign producers early on. Wettstein made his international breakthrough with his *Juliette* chair for Baleri in 1987; Häberli/Marchand realized their table *Move it* with Thonet in 1996 before working on their furniture ideas with Swiss producers; and the N2 group positioned itself in 1997 at the Milan Furniture Fair. International success tends to draw more attention to a designer's work back in their own country. The New Simplicity as a novel form of pragmatism thus buoyed the Swiss furniture industry and boosted exports to neighboring countries.[17] The courage to embrace a new product culture and the transition from artisanal to serial production proved well worthwhile. A few traditional manufacturers started to change tracks around 1985, realigning their production and sales, and working closely with designers. The lamp maker Belux, for example, collaborated with Hannes Wettstein, and Eternit worked with Alfredo Häberli. Wogg, founded by Willi and Otto Glaeser in 1983, followed New Simplicity principles from the very beginning. Design producers such as wb form and Sele 2 began operations. Producers joined forces and appeared together at trade fairs. Design was widely perceived and celebrated as delivering added value to products. In 1986, the first Designers' Saturday took place in Langenthal, an event that features exhibitions at the production facilities of the resident companies and that has enjoyed great popularity ever since. Swiss design comes alive here, especially for interested laypersons.

In 1991, a consortium of small and medium-sized Swiss companies[18] presented the first Swiss Design Award. In addition to such

810

811

812

813

810 Kurt Thut, Scissor Shelf Unit Model 399, 2000, Thut Möbel AG, Möriken. Photo: Alfred Hablützel Museum für Gestaltung*
811 Fritz Lange, Wogg I Shelf System, 1983, Wogg AG, Baden
Photo: Alfred Hablützel
812 Jörg Boner, Dresscode Wardrobe, 2004, Nils Holger Moormann, Aschau (DE)
Simple wooden construction with fabric shell in various colors.
Museum für Gestaltung*
813 Moritz Schmid, Pile Shelf System, 2012, glass, Trösch AG, Bützberg
Simple stacking glass elements and solid wood strips in various colors and glass qualities.
Photo: Simon Opladen

private initiatives, the public promotion of design was also expanded (see Ernst, "Teaching, Research, Promotion," pp. 358–59).[19] The Museum für Gestaltung devoted numerous exhibitions and publications to "Designland Switzerland," founding its own Design Collection in 1987 (see Brändle, "Open Collections for Tomorrow," pp. 360–61).[20] In addition to the preservation, study, and communication of Swiss design, it also gears its work to fostering the contemporary design scene.[21] Founded in 1988, the magazine *Hochparterre* reports every month on current themes and projects. In 1997, the Swiss National Museum likewise developed a collection concept for the twentieth and twenty-first centuries. These efforts supported the designers' work while at the same time promoting Switzerland's international reputation as a design hotspot and breeding ground for the creative industries. And at Expo.02, the 2002 Swiss National Exhibition, interdisciplinary design achievements were ultimately presented in atmospheric rooms and in theme-based scenographic exhibitions. Especially for designers from French-speaking Switzerland, such as Atelier Oï, Expo.02 offered a platform for reaching a wider audience as well as producers in the industry.

Pioneers of the Creative Economy

The availability of computers and inkjet printers as well as a new understanding of corporate design created favorable conditions for the establishment of small design firms. Designers dared to go out on their own, setting up production facilities and also taking charge of publicizing and distributing their own products. The bag producer Freitag, founded in 1993, can be regarded as exemplary of this new business model based on artisanal production.

But even much earlier, there were already many firms founded and managed by designers, such as Baltensweiler,[22] or the multigenerational family company Thut. Interestingly enough, Thut Möbel is also considered the epitome of New Simplicity in Switzerland. The product and interior designer Kurt Thut, who had trained with Willy Guhl and Hans Fischli at the Kunstgewerbeschule Zürich, transformed the family cabinetry business in Möriken from the ground up. A new plant was built according to his plans in 1958, and he reorganized the production processes and made furniture out of novel materials and using innovative technologies. It was only on this basis that Thut achieved his international breakthrough as a designer in 1986. His *Aluminum Cupboard* has a skeleton and skin made of aluminum, a material commonly associated more with aircraft design than with chipboard cabinetry.[23] Schooled by his experience with modern architecture, his design philosophy influenced both the reorganization of the production processes as well as the resulting furniture pieces, leading to a reduction in form motivated by considerations other than merely stylistic ones. Pieces such as the *Bellows Cabinet* (1997) or the *Scissor Shelf Unit* (1990) illustrate basic principles of construction. This interplay between design and production would become a seminal theme for a younger generation of designers who, after a sound traditional training, founded their own start-ups.[24] Among them were the members of the Neue Werkstatt, a collective of three designers with training in precision mechanics and metalworking who learned in the advanced class at the Schule für Gestaltung Zürich that form could be simplified through a new approach to production, putting the principle of reduction to creative use. Christoph Dietlicher, Andreas Giupponi, and Thomas Drack experimented with rudimentary methods of forming sheet metal like bending, perforating, and folding and eventually arrived at surprising product forms and a one-of-a-kind product language.

Many of the design entrepreneurs who built their own production facilities in the nineties also took charge of publicizing, marketing, and distributing their products. Their letterhead, websites, promotions, display furniture, and business plans to submit to investors thus all came from their own drafting table. Some ran boutiques or multi-label stores and began to sell their products on the Internet.[25] In the course of the decade, many technically trained designers realized that publicizing their products was just as important as their design. While many of these nineties start-ups were short-lived, their innovative, subculture-influenced business model prevailed, and is treated in business administration textbooks under the heading of "creative industries."

814

816

815

817

New forms of production and communication
814 Daniel and Markus Freitag sewing Freitag bags, ca. 1994. The designers take production into their own hands. Photo: Jules Spinatsch
815 View of sustainable production at the woodworking school PIKA in Semarang, Java (Indonesia), 2007. Production partner of the Swiss brand INCH Furniture (Yves Raschle and Thomas Wüthrich). Photo: Thomas Wüthrich
816 Senior Design Factory, Zurich, 2010. The designers Debora Biffi and Benjamin Moser combine the local manufacture of their products with social, cross-generational engagement. Photo: Florian Kalotay
817 Depot Basel, Invitation card for the exhibition *Okolo Offline,* Basel, 2014. The exhibition showed the work of the Czech collective Okolo and explored how digital content and connections can be transferred onto the analog plane. Animal figures by Adrien Rovero, Enzo Mari, and Bertille Laguet.

1 In his essay "Die Heiligen Kühe des Funktionalismus müssen geopfert werden" (The Sacred Cows of Functionalism Must Be Sacrificed), published in 1968 in the magazine *Form,* Werner Nehls describes post-modern design as a protest: "The new forms are a painted, molded, and built protest against the past era with its mechanistic, rational, purist/utilitarian, soulless, inhumane design of our environment." Cited in Fischer/Hamilton 1999, p. 213.
2 Paul Feyerabend was a philosopher of science who advocated freedom from methodology; in the design realm, his motto "anything goes" was interpreted as a liberation from the design principles of modernism.
3 Bourdieu (1979) 1984.
4 The sociologist Gerhard Schulze aptly described Germany society in 1992 as being experience-happy, breaking down its various milieus—similarly to the milieus developed around 1980 by the Sinus Institute—into target group segments according to social situation and value orientation. Schulze (2005) 2010.
5 The cultural theorist Hartmut Böhme examines fetishism as a central category in the self-description of European culture, up to and including present-day cult forms. See Böhme (2006) 2014.
6 Elena Esposito tries to explain the fashion industry using the tools of system theory. Esposito 2004.
7 See Menzi 2010.
8 See Buck 1998.
9 An exhibition at the Museum für Gestaltung Zürich focused on this new penchant for ornamentation in everyday products: *Musterspiele – Ornamentale Gestaltungsansätze* (Pattern Plays—Ornamental Design Approaches), Designraum, 2004.
10 See Menzi 2000.
11 Wyss 1997.
12 Norbert Wild wonders whether the "New Simplicity" in Switzerland was even new at all. Wild 2002a.
13 This term was coined by Jochen Gros, who described in 1976 the "sensual functions" of design (in *Form* 75 and 76, 1976).
14 Schilder Bär/Wild (2001) 2003, p. 114.
15 Important exponents of the New Simplicity were living and teaching in London around 1990, including Jasper Morrison, Ron Arad, Konstantin Grcic, and Sebastian Bergne.
16 Following his industrial design training at the Royal College of Art in London and the Berlin University of the Arts, Morrison founded a design studio in London in 1986. He showed the installation *Some new items for the house, part I* at the DAAD Gallery in Berlin in 1988.
17 According to Norbert Wild, exports even increased in the crisis-fraught nineties. See Rüegg 2002a, p. 259.
18 The "Forum Schweizer Fabrikanten," founded in 1985, was made up of twenty-one companies that wanted to work together toward strengthening Swiss design. It was dissolved in 1993 but gave rise to "Forum 8," whose members in 1996 were Belux AG, Wohlen; Greter AG, Zurich; Lehni AG, Dübendorf; Kollektion Röthlisberger, Gümligen; J. Seilaz AG, Zurich; Sele 2, Zollikon; Walter Thut AG, Möriken; and Wogg AG, Baden-Dättwil.
19 The Swiss Federal Office of Culture (BAK) presents design awards, arranges internships at prestigious studios, and also promotes young designers through acquisitions. These objects are permanently loaned by the BAK to the Design Collection of the Museum für Gestaltung Zürich.
20 In the early eighties, this included the Schweizer Design-Pioniere (Swiss Design Pioneers) publication series and further publications on exhibitions, as well as trade publications and scholarly articles (see the bibliography of this volume).
21 A good overview of the state of Swiss design in the nineties is provided by Lotte Schilder Bär, the co-initiator and first curator of the Design Collection: Schilder Bär 1997.
22 Baltensweiler AG, Ebikon, founded in 1951 by Rosmarie and Rico Baltensweiler.
23 Gantenbein/Hablützel 2001.
24 In the nineties, production techniques were taught in the graphic design as well as the textile, jewelry, and appliance design classes at the Schule für Gestaltung Zürich and the corresponding machines were available for use by students.
25 The Museum für Gestaltung Zürich devoted an exhibition and publication to the creative and economic potential of this business model: Menzi 2012.

818

819

820

821

822

823

Current design practice straddles the fields of industrial design, art, and craft

818 Big-Game, SCOUT Watch for Lexon Design, Boulogne-Billancourt (FR), 2013. Photo: Big-Game

819 Velt – Stefan Roland Rechsteiner, Patrick Rüegg, Velt Model No. 15 Shoe, Silently Still Collection, Spring/Summer 2014. Photo: Andreas Zimmermann

820 Qwstion, Weekender Raw Canvas Bag, 2011 Photo: Urs Bigler, Matthias Graf

821 ECAL/Aurélie Mathieu & Philippe Karrer, Brick Vase, 2012, Master of Advanced Studies, Design for Luxury and Craftsmanship in collaboration with the Cristallerie Baccarat (FR). Photo: Nicolas Genta

822 Kueng Caputo – Sarah Kueng, Lovis Caputo, *Never Too Much* exhibition, Salon 94 Freemans, 2014, New York. Photo: Salon 94

823 Atelier Oï, Objets Nomades: Le Tabouret, 2012, Louis Vuitton, Paris. Foldable leather stool for deluxe travel. Photo: Atelier Oï

824

Big-Game, *Overview,* 2008. The Swiss-French-Belgian trio Big-Game (Grégoire Jeanmonod, Augustin Scott de Martinville, and Elric Petit), based in Lausanne, attracted attention for the first time at the 2005 Milan Furniture Fair with hunting trophies made of wood. The three graduates of ECAL Lausanne like to put an ironic twist on their status as designers, for example in their exhibition at the Musée du Grand-Hornu, Boussu (BE), in 2008.

Photo: Milo Keller, Big-Game

2000 — Perspectives

As authors or service providers, designers now work either under contract for different companies or build their own label or brand. Sometimes they switch roles and present themselves as the authors of works that are exhibited in galleries like art pieces, or they support brand communication. If they work for industry, they act as mediators between users, brand, and producer, while often remaining anonymous. Design is taught in diverse study programs at Swiss universities and is researched in state-funded projects. Design theory as a discipline establishes itself as an academic field without a direct connection to design practice, which continues to be determined by the market. The mediation of design criticism as a critique of consumer society, along with the detailed description and study of design objects, puts design in the spotlight again with exhibitions, publications, and events. In everyday life, however, good and useful Swiss products are no longer taken for granted.

825

Atelier Oï, *Oïphorique,* 2011. As a gift to themselves to celebrate twenty years of Atelier Oï, Aurel Aebi, Patrick Reymond, and Armand Louis designed a room installation that debuted at their *Moïtel,* La Neuveville. Two rows of movable columns that get their shape from lightweight pleated filter inserts are propelled by engines that otherwise drive Fiat windshield wipers. Accompanied by French accordion music, the columns, lit from within, appear to breath rhythmically.

Photo: Atelier Oï

826

INCH Furniture, Shanghai furniture collection. INCH Furniture was commissioned to design all the furnishings for the Swiss pavilion at the Expo 2010 in Shanghai. Inspired by the pavilion's architecture, Thomas Wüthrich and Yves Raschle conceived in collaboration with the architects Buchner Bründler solid wooden furniture for the restaurant and VIP lounge.

Photo: Mark Niedermann

827

Hannes Wettstein, SF News Studio, Tagesschau and 10vor10, 2005–07, SF Schweizer Fernsehen, Zurich. The development of the new corporate design for the Swiss television network also involved devising a dynamic and credible set for the daily news shows. Reduced to the four elements of pedestal, news desk, band of lights, and panorama wall, the studio manifests clean-lined architecture featuring elements inspired by Stanley Kubrick as well as Ken Adams's sets for the James Bond films.

Photo: SRF, Markus Bertschi, Oscar Alessio

828

Amy Winehouse with En Soie silk foulard, London, 2008. Amy Jade Winehouse (1983–2011) was a British singer and songwriter. In her eight-year career she sold over 25 million records and won five Grammy Awards. One of her trademarks was her beehive hairdo, which she and her 1960s-inspired music helped to revive. Winehouse often used scarves as a hair accessory, for example here with a silk foulard from En Soie. The silk shop, founded in Zurich in 1976 by Monique Meier Leuthold, also has a branch in Los Angeles run by her daughters Eleonore, Sophie, and Anna Meier.

Photo: Keith Hewitt

829

Alfredo Häberli, Origo bowl, littala, 1999, littala Oy AB, Helsinki. When Alfredo Häberli designed a white, freely combinable tableware service for Europe's second-oldest porcelain manufacturer, Rörstrand (which changed its name to littala Oy AB in 2000), he only gave the smallest item, the egg cup, a colorful striped pattern, similar to something by the British fashion designer Paul Smith. But when the striped egg cups became a bestseller, littala decided to produce an entire striped porcelain service. In the meantime, several colorways are on the market.

Photo: Walter Pfeiffer

830

Studio Achermann, Atelier Pfister, Presentation of the 2013 collection. With the promise "Pfister is devoted to contemporary design," the furniture-industry giant launched the new label Atelier Pfister in 2009. The first collection of thirty-seven furniture pieces and lamps, all bearing the names of Swiss towns, are supplemented yearly with new items by Swiss designers. Meinrad Fleischmann, Pfister's CEO, is personally committed to the collection, which is curated by Alfredo Häberli and visually communicated by Studio Achermann.

Photo: François Halard

Chaise
2010

831 A C E, Chaise, 2011
Ashwood, 85 × 72 × 57 cm, A C E Design, Lausanne
Museum für Gestaltung

832 Stacked Chaises, 2011
Photo: Nicolas Haeni

833 Chaise in use, 2011
Photo: Nicolas Haeni

831

832

833

Arnault Weber (b. 1985), Camille Blin (b. 1985), and Emmanuel Mbessé (b. 1983) completed their studies as industrial and product designers together at ECAL in 2009 and founded the furniture label A C E in Lausanne. Their Collection 1 appeared in 2011. It consists of six simple pieces of wooden furniture: a chair, a self-supporting bookshelf, a table, a coffee table, and two lamps. The *Chaise* is vastly different from traditional models because it breaks with formal ergonomic conventions. Its overlong, straight backrest is reminiscent of a park bench, but it can also be used for sitting at a table and is stackable. *Chaise* was the first product in which A C E consistently implemented its concept of reduction, and it best embodies the team's programmatic approach. Because all A C E furniture is produced in small cabinetmaking shops in Piedmont, Italy and French-speaking Switzerland, their designs are created with the machines available in mind. The products are fabricated using the simplest of tools—band and circular saws, planing and milling machines, a lathe—but the focus is on precision. The result is a restricted formal repertoire, which A C E then takes to humorous extremes. Playing with proportions, details, transitions, connections, and the clunky look of the furniture is all part of the program. European ash is used for all products. While the first collection was designed by the A C E team itself, the second collection also includes products by designer friends. RM

Young Swiss Design Kaleidoscope 2013, p. 8; www.a-c-e.ch.

834 Big-Game, Bold Chair, 2007, prototype
Steel, polyurethane foam with polyamide cover, 77.5 × 39 × 53 cm, Big-Game, Lausanne
Museum für Gestaltung*

835 Illustration of the "From light to bold" concept, 2008
Photo: Big-Game

836 Bold Chairs, produced by Moustache (FR), staged at the Musée des Arts Décoratifs, Paris, 2009
Photo: Tania & Vincent, Les Arts Décoratifs, Paris

834

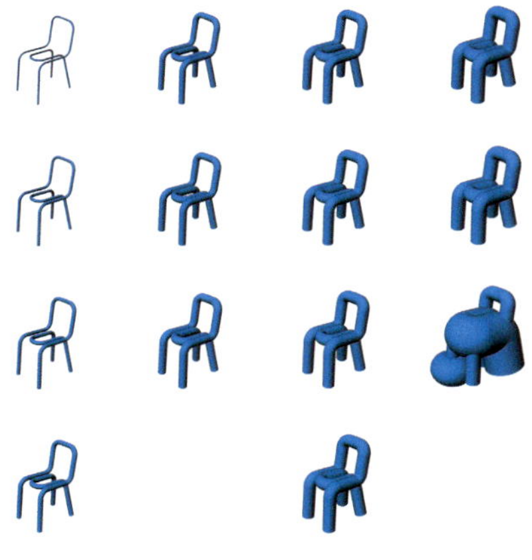

835

836

The designers Grégoire Jeanmonod (b. 1978), Elric Petit (b. 1978), and Augustin Scott de Martinville (b. 1980) come from Switzerland, Belgium, and France, respectively. They have been working together since 2004, when they founded the studio Big-Game, based in Lausanne. Their diverse cultural origins and penchant for exploring collaborative authorship are evident in their designs. In their very first collection, *Heritage in Progress,* the hybrid nature of the design collective's work and their relish for turning rules on their head were already evident: the hunting trophies assembled from wooden components were created for the 2005 Milan Furniture Fair. Each collection consists of an ensemble of simple furniture, lamps, or utilitarian objects that share a common theme or method of production. The titles are catchy and the photographs pointedly convey the idea (in the typical fashion of ECAL, where the members of Big-Game teach). The collection *Plus is more* alludes to the motto "less is more" coined by Braun designer Dieter Rams. These functional products are composed of basic shapes. The *Bold* chair gets its name from the bold font of the Univers typeface (Adrian Frutiger, 1957). Based on that concept, Big-Game developed a chair made of two U-shaped steel tubes and upholstered with polyurethane foam. Covered in nylon mesh, it recalls stocking-clad legs and gives the user the sensation of sitting on someone's lap. RM

Big-Game 2008; www.big-game.ch.

837

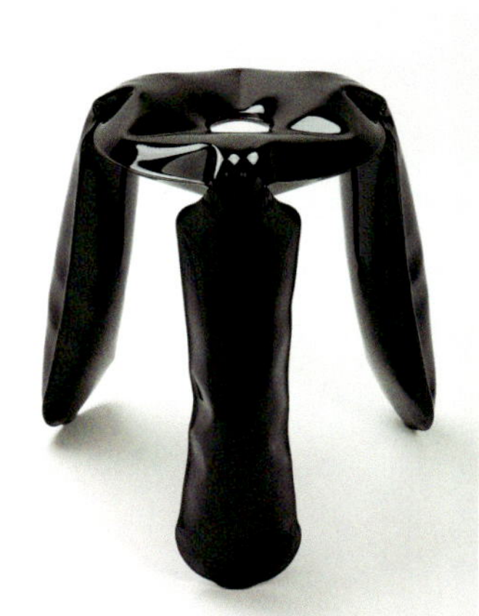

838

839

837 Oskar Zieta, Plopp Stool, 2007
838 Steel, various colors, spray-painted or chrome-plated, 50 × 55 cm, Hay/BS Studio, Horsens (DK), developed at the ETH Zürich, Chair for Computer-Aided Architectural Design (CAAD)
Museum für Gestaltung*

839 Oskar Zieta, Production process using the FIDU method: free inner pressure deformation.

In 2003, as a PhD student at the ETH Zürich, the Polish architect Oskar Zieta (b. 1975) began to explore new production technologies in architectural practice. His experiments with digital tools and computer-aided design led to the invention of a new production method. The so-called FIDU process—free inner pressure deformation—turns thin metal sheets into hollow volumes in a seamless digital chain. Two silhouettes cut out with CNC-controlled laser beams are welded together at the edges by a robot into a waterproof volume. The cavity formed between them can then be inflated under pressure with a liquid or air. The metal bulges outward and forms a stable hollow body. This lightweight construction method allows for customized mass production at a low cost using a common manufacturing process. The first application clearly showed the potential of the FIDU technology. The *Plopp* stool held up under 2.5 tons of weight in the testing hall at the ETH. The formally striking stool betrays a further characteristic of FIDU: in contrast to conventional CNC-manufactured products, *Plopp* looks like a handmade original and has a sculptural quality about it. Available in a choice of vibrant colors, it recalls the inflatable furniture of the seventies yet is as tough and durable as steel. RM

Sonna 2008; von Ascheraden 2009; www.blech.arch.ethz.ch; www.zieta.pl.

Yves Béhar

One Laptop per Child
2007

840

841

842

840 Yves Béhar, One Laptop per Child XO Laptop, 2005–07
Metal, plastic, rubber keyboard,
24.2 × 22.8 × 3 cm, weight: 1.4 kg, One Laptop per Child Association – OLPC, Somerville, Massachusetts (US); Quanta Computer Inc. (TW)
Museum für Gestaltung*

841 Laptop closed, showing personalizable XO logo

842 Learning how to learn project, Peru
Photo: OLPC

One Laptop per Child (OLPC) is a nonprofit initiative, developed at MIT in Cambridge, Massachusetts by Nicholas Negroponte, with the objective of giving all children access to a computer and thus improving educational opportunities in emerging and developing nations. In 2005, the Swiss industrial designer Yves Béhar (b. 1967) with his studio Fuseproject in San Francisco and New York was commissioned to do the final design of the laptop. Education policy, technology, and economy were the factors determining its technical features. The laptop had to be robust and the battery rechargeable either mechanically or via solar panel. The software allows the devices to be networked, so that children can communicate with each other without Internet access. Equally important is the design quality: despite low production costs, the computer has an upscale and refined look and feel. Carefully chosen textures and materials protect it from scratches and splashes while also making it fun to use. The foldout antennas conceal the cable sockets and support the screen when it is folded upward. At the same time, they turn the laptop into a friendly creature with green ears. Since the laptop was to be used throughout the world, Béhar also took into account cultural differences. In order to allow the attachment of a range of straps and different carrying options, he added holes next to the handle. The XO logo was conceived for universal appeal to children, but with a choice of 400 color combinations it can also be geared to a specific child. RM

Haffner 2001; Tobler 2010; www.one.laptop.org.

Allegro Assai Pendant Lamp
2007

843 Atelier Oï, Allegro Assai Hanging Pendant Lamp, 2007
Anodized aluminum, 136 × 136 cm, Foscarini S.r.l., Marcon (IT); Atelier Oï, La Neuveville
Museum für Gestaltung/
Donation of Foscarini S.r.l.

844 Atelier Oï, *A Composition of Cords* exhibition, 2006, Centro Culturale Svizzero, Milan
Photo: Barbara Werren

845 Atelier Oï, De l'objet à l'espace wallpaper collection, 2011, Hurter Tapeten AG, Winterthur
Photo: Barbara Werren

843

844

845

Atelier Oï in La Neuveville is known for experimenting with scale and with the cross-fertilization of architecture, scenography, and furniture and product design. An interdisciplinary team recruited by the founders Aurel Aebi (b. 1966), Patrick Reymond (b. 1962), and Armand Louis (b. 1966) pushes ahead with creative design development independent of client demand. Material samples, prototypes, and models are carefully archived and serve as DNA and a visual reservoir of ideas for the western (French-speaking) Swiss studio. This has made it possible for the team to realize spectacular projects even under time pressure, such as the 2006 exhibition *A Composition for Cords* at the Centro Culturale Svizzero in Milan. Taking as their point of departure the materials provided for the scenography of the 2005 Swiss Federal Design Awards, in only fifteen days Atelier Oï created an atmospherically charged space that attracted countless visitors. Oversized spools of rope served as both inviting seats and bell ropes, while freely suspended aluminum rods became ringing chandeliers. Further exhibitions followed the same year, at the Instituto Svizzero di Roma and the Designers' Saturday in Langenthal. The Italian lighting manufacturer Foscarini developed products with Atelier Oï based on the sound-producing elements in their shows, while the Swiss watchmaker Breguet had the studio design its worldwide corporate architecture out of gleaming aluminum rods. The *Allegro* family of lamps now includes six models. In 2011, Atelier Oï translated the lamp into a graphic pattern and used it on wallpaper for Vescom. RM

Schindler 2006a; Menzi 2007b; Atelier Oï 2012; www.atelier-oi.ch.

846 Michel Charlot, U-Turn Lamp, 2011–12
Die-cast aluminum, wet-painted, LED bulb, lamp head diameter: 14.3 cm, floor lamp height: 115 cm, clamping lamp height: 17.6 cm, table lamp height: 33 cm, Belux AG, Birsfelden
Museum für Gestaltung/Donation of Belux

847 The magnetic lamp head can be freely swiveled atop the ball joint.
Photo: Felix Wey

848 The angle of light can be adjusted at the lamp head and the light can also be dimmed.

846

847

848

Without LED technology, the *U-Turn* lamp would not have been possible. In 2011, the lighting manufacturer Belux handed the young industrial designer Michel Charlot (b. 1985) a prototype. It was a good idea, a good starting point for a new product. Charlot began to sketch, to call into question the development up to that point, and to evaluate new technical solutions. The engineers' main innovation was a steel ball on which a magnetic lamp head could be docked at any angle. By contrast with a restricted ball joint—like the one in Sigfried Giedion's *Indi Lamp*—the cylindrical lamp head can be removed from the stand completely and turned upside down, so that the beam is directed upward. This playful interaction with the light turns the user into a designer. Because the LED plate does not heat up, the lamp head needs no handles and the whole thing can be picked up in the hands. All components must to fit precisely together and consist of a minimum of parts to make the bulb easy to replace. *U-Turn* is available as a table lamp, floor lamp, or clamping lamp, and the product family is still being optimized and expanded. Charlot designs from the inside out so that the product is a result, not a goal. Designing products that are aesthetically and technologically contemporary is important to him. A product should be useful, efficient, simple, and durable. Design is thus a small but essential part of each project. RM

Ernst 2012; www.belux.ch; www.michelcharlot.com; Information provided by Michel Charlot.

849

850

851

849 Thilo Alex Brunner, On Running Shoe, 2010
3-D mesh, perforated nubuk PU, reflective material, PU-reinforced nylon, On AG, Zurich
Museum für Gestaltung/Donation of On AG

850 The logo on the shoe was designed in collaboration with Raffinerie AG, Zurich.

851 "Tegla Loroupe loves to run on clouds," world champion marathoner and world record holder in Zurich, 2011
Photo: Braschler/Fischer, On AG

The driving force behind the technical innovations in the CloudTec cushioning system is the former professional athlete Olivier Bernhard. After an injury, he set out in search of running shoes that would be easy on his feet, and when he failed to find what he was looking for, he began to experiment himself in 2005. An ETH engineer friend assisted him in the four-year process, from the initial prototype (with garden hose sections superglued to the sole) to devising effective three-dimensional cushioning. The flexible tubes ensure a soft landing; pressed flat on the ground, however, they provide a solid foundation for pushing off. Today, the so-called CloudTec system is protected by worldwide patents. In 2010, Oliver Bernhard partnered with the marketing experts David Allemann (production and marketing) and Caspar Coppetti (international market development) to establish the company On. They commissioned designer Thilo Alex Brunner (b. 1977) to design the new running shoe. Unlike traditional sporting goods brands, On makes its innovative cushioning visible as a central design element (the slogan is "Run on Clouds") and with its formal reduction stands out from shoes by Nike, Adidas, or Puma. The shoe's "playful post-minimalism" (Brunner) gives credibility to its functional promise, helped along by the enthusiasm of the running community. RM

Manz 2012; *Young Swiss Design Kaleidoscope* 2013, p. 58; www.on-running.com.

852

852 Jörg Boner, Wogg 42 Armchair, 2007–09
Varnished ash frame, wool cushion,
79 × 57 × 52 cm, Wogg AG, Baden. Photo: Wogg

853 Seat folded from two sheets of paper, 1:5 scale model from the design process, 2007–09
Museum für Gestaltung*

854 The two-part cushion is laid across the backrest and armrests and fixed in place with snaps.
Photo: Wogg

855 In the first version the cushions were made of polyamide mesh, and later in wool in a range of colors, produced by Kvadrat A/S, Ebeltoft (DK).
Photo: Wogg

853

854

855

Product designer Jörg Boner (b. 1968) took a head-on approach to the task of designing an upholstered chair for the Swiss company Wogg. Over the course of four years, he developed a whole new type of chair, which reached series maturity in 2009 and was then presented for the first time at the Milan Furniture Fair. As is customary for Boner, who apprenticed as a carpenter and then trained as a product designer at the Schule für Gestaltung Basel, he started from a visual image: a chair that would be able to change its clothes. Two pads are draped across a filigree wood framework and secured with snap fasteners. To do so, the user has to fold the two-dimensional quilted pads, which then form a seat shell with a "collar" that folds over the chair back. Boner searched a long time before finding the right industrial technology to manufacture his chair upholstery. As with a waffle iron, the padding is welded between the two textile surfaces (polyamide) at high frequency and punched out, all in a single step. The welded seams aid in folding the padding, and the segmentation they provide contributes to the look of the design. This technological innovation gives a new twist to a simple wooden chair, lending it a synthetic, high-tech appeal. With the exchangeable padding, Boner was responding not only to a customer request but even more so to the current trend in the furniture industry to work with diverse suppliers. The padding is manufactured in Sweden and Finland and the wooden frame by the furniture factory Muotathal Paul von Rickenbach AG in Switzerland. RM

Menzi 2007a; *Young Swiss Design Kaleidoscope* 2013, p. 40; www.joergboner.ch; www.wogg.ch.

SDS 011 Sunglasses
2006

856

857

858

856 Sandra Kaufmann, SDS 011 Sunglasses, 2006
Spring steel frame, etched and bent into shape, Strada del Sole GmbH, Zurich
Museum für Gestaltung/
Donation of Sandra Kaufmann

857 Detail of SDS 011 Sunglasses, silicon nose pads

858 Sandra Kaufmann and Monika Fink, Sunglasses produced under the label Solsolito, founded in 2013
Museum für Gestaltung*

Eyeglasses are quite straightforward from a technical standpoint. As long as the correct ergonomic dimensions are given, such as the width of the head, the interpupillary distance, and the nose width, their form has few constraints and tends to be oriented on the idiom and laws of fashion. "We work with tenths of a millimeter," says Sandra Kaufmann. Kaufmann (b. 1972) trained as an industrial designer in Zurich and then worked in various design studios, including with eyewear designer Alain Mikli in Paris. After returning to Switzerland, she boldly launched her own eyewear collection. The market seemed saturated, and yet she and her business partner, the master optician Markus Dudli, managed to build a successful eyewear brand. The first collection brought out by Strada del Sole provided functional added value. Kaufmann and Dudli experimented for over a year and a half to produce sunglasses with maximum flatness. With spring tension temples, the glasses can be folded like a hair clip in two places. Closed and with the temples folded in, they are so flat that they can be carried in a back pocket. The bar above the bridge is easy to open and the lenses easy to change. Laser-cut from thin sheet metal, the rim, temples, bridge, and bar are all made in Switzerland, where numerous companies still produce precision mechanical parts for the watch industry. RM

Ernst 2007a; www.stradadelsole.ch.

859 Frédéric Dedelley, Vanitas Mirror, 2009
Glass, 9 × 13 × 0.5 cm, edition of 100 mirrors (numbered, signed), Memento Mori object group, Helmrinderknecht, Berlin
Museum für Gestaltung*

860 Frédéric Dedelley, Reliquary, 2009, glass container in various sizes, part of the Memento Mori object group
Photo: Imagekontainer/Inga Knölke

861 Frédéric Dedelley, Altar Shelf, 2009, walnut, 46 × 18 × 2 cm, part of the Memento Mori object group
Photo: Imagekontainer/Inga Knölke

859

860

861

In 2009, the industrial designer Frédéric Dedelley (b. 1964), who trained at the ECAL and the Art Center College of Design in Pasadena, devised a series of limited-edition objects on the theme of transcendence and memory for the contemporary design gallery Helmrinderknecht. Inspired by Baroque vanitas paintings and religious relics, they were meant to fulfill both symbolic and utilitarian functions: glass soap bubbles seem frozen in black glass, an urn made of walnut wood and glass reveals its contents at all times, an altar top is used to display memorabilia, and glass relic vessels both conserve and expose what they hold. Finally, Dedelley's vanity mirror is a *Memento Mori* ("Remember that you will die") exhibiting an unsparing directness. All earthly things are mere vanity: fleeting, insignificant, ephemeral, unstable, empty semblance, futile. Influenced by Judeo-Christian notions of transience, Baroque vanitas motifs such as the hourglass or the skull admonish viewers that their life is finite. Those looking at themselves in Dedelley's mirror can partially see through it. The transparent lattice of lines in the form of a skull, designed by the illustrator Bastien Aubry (b. 1970), becomes part of the observer's own likeness. At once playful, beautiful, and eerie, the mirror optically casts the memento mori back at the viewer's face. RM

Horisberger 2010; www.fredericdedelley.ch; www.helmrinderknecht.de.

Renate Menzi

Fields of Activity in Current Design Practice: When Does Good Design Pay Off?

The economic situation for furniture design in Switzerland has changed since the turn of the millennium. Many of the small and medium-sized enterprises (SMEs) that took the plunge and invested in New Simplicity in the nineties are now instead turning out proven classics or relying on so-called cash cows, that is, risk-free products that are a sure sell.[1] Furniture exports declined after the economic crisis of 2008 as the strong Swiss franc made products from Switzerland too expensive, especially for European buyers.[2] Changes in management, mergers, and takeovers influenced corporate strategies and companies' appreciation for the cultural added value of design. The development of new products is not, after all, an investment guaranteeing a quick profit for shareholders. It has therefore become more difficult for designers to find good producers for their furniture designs in Switzerland. For freelance industrial designers as well, the road to realizing a product is long. Nevertheless, new products continue to be developed by many Swiss SMEs: bathroom fixtures, lamps, pans, knives, bicycles, scarves, watches, kitchen appliances, climbing hooks. But what about their design quality? Although "Swissness" is still in demand all over the world, it is associated more with production quality and utility value than with good design. Whether this is due to the lack of coordination in presenting Designland Switzerland or in the still common Swiss reserve—both formally and in terms of communication—is not the issue here.[3] Instead, our attention will be focused here on a selection of successful projects that can point the way to promising routes for the future of Swiss design.

Global Brand Auteur or Art Scene Figure

Because professional design costs money, it must either be sold in large numbers or create high added value. The Swiss luxury goods industry succeeds at the latter. After a brief dip around 2008, this industry is growing again, spearheaded by the watch industry.[4] Although watches as luxury items allow little in the way of innovative design approaches, they do effectively market the enduring values of "Swiss Made," such as functionality, precision, and reliability.[5]

Handcrafted in limited editions, design can be a luxury. A case in point is the leather furniture designed by the western (French-speaking) Swiss studio Atelier Oï for the 2012 *Objets Nomades* collection from Louis Vuitton fig. 823. The pieces are simple yet of the highest quality, almost like a luxury edition of the wooden folding furniture designed for postwar reconstruction in the forties. Their design is based on the artisanal tradition of the former saddlery and communicates the brand value.[6] Another leather project was realized by Adrien Rovero (b. 1981), a graduate of ECAL, for the Hermès brand in 2013. Simple animal silhouettes cut out of leather scraps can be snapped together into animal bodies. *Clickazoo* fig. 817 manifests the playful inventiveness of the designer, who slips into the role of the court fool here but is not mentioned by name on the Hermès homepage.[7] Such collaborations nonetheless offer an international platform for original design. ECAL introduced a master's degree in Design for Luxury and Craftsmanship in 2009 fig. 821. Product designers learn to design for large firms and come into contact with companies such as Hublot, Christofle, or Baccarat.[8] Their experiments in form and materials are geared not to industrial mass production but to limited, handcrafted editions with a media impact.

Turning to craftsmanship and the art world opens up new sales opportunities for designers even without financially strong clients backing them. Design galleries such as Helmrinderknecht, temporary spaces for contemporary design such as Depot Basel, local producers like Dadadum, or shops that host exhibitions present themselves as design mediators and take charge of communication and sales. This enables freelance designers to present their work to a select audience. Rather than everyday products, this field focuses on "design art" objects, which are sold as works but understood and evaluated as experiments.[9] For example, the Zurich duo Kueng-Caputo—Lovis Caputo (b. 1979) and Sarah Kueng (b. 1981), who both trained as industrial designers at the Zurich University of the Arts—sells its objects straddling art and design at the New York gallery Salon 94.

Serving the User or Consulting for Industry

In 2006, Jasper Morrison (b. 1958) curated in conjunction with Naoto Fukasawa (b. 1956) the exhibition *Super Normal* in London and Tokyo. They share the approach that design quality should be conveyed less through form than through function. The two curators published a design manifesto that can be read as a tribute to *Gute Form,* which should by no means be confused with minimalism: "Minimalism has always seemed to me to be an inappropriate term in relation to design which, as the best Swiss design clearly shows, is all about simplification of form, expression of purpose, economy of material and suitability for purpose. All of which may have links with Minimalism in Art but are in fact natural elements of good design. Reversing or ignoring any of these requirements doesn't lead to a viable alternative, as it can in Art. Super Normal is a simple extension of these ingredients to include a longer term assessment of design quality in relation to use."[10] Like many designers from Le Corbusier to Bruno Munari and onward to Franco Clivio (b. 1942) and Alfredo Häberli, Morrison collects "good design" that fits this description and intuitively derives his motifs from these models.[11] He runs a design shop in London where, in addition to his own designs, he also sells anonymous household appliances from all over the world. Morrison also speaks of "democratic design" that does not use any "dumb" materials, keeps prices low, and seeks an aesthetic that makes it possible in the long term to incorporate the object into diverse settings. Among the items on offer at his shop is the *Rex* peeler produced in Switzerland, which was also on view in the exhibition *Super Normal.* But new products can, in this sense, also be democratic. The normal, according to the curators, is no longer available as a matter of course. So it must be discovered or designed anew.

Many product designers in Switzerland see themselves as service providers and strive to design functional and sustainable forms. If they work for industry, they usually work anonymously. On the websites of Nose Design, Process Brand Evolution, Vetica, Walser Design, Milani Design, Studio Hannes Wettstein, and Les Ateliers du Nord, consumer products are shown that fulfill their everyday functions while at the same time evoking the client's visual identity. Whether working for Mammut, Betty Bossi, Logitech, or Nestlé, the designer as auteur stays in the background and, in the relationship with the brand, instead becomes an advocate for the user. The brand must not be allowed to dominate the design, because everyday products especially are based on existing types and should exhibit what they can do. The designer can be a valuable mediator between producer and user if he or she is able to bring together form, function, and corporate design. With a well-designed product, the designer exerts a long-lasting positive influence on the perception of the brand.

Corporate Creatives, Globally Networked

If young designers are forced to share the financial risk of their designs for industry due to licensing agreements, they may as well invest in building their own label and setting up their own production facilities—whether with their own factory in Switzerland like Freitag, or in Indonesia as in the case of INCH Furniture fig. 826, or in China as with Qwstion fig. 820. To monitor design quality in the production process from A to Z, designers work closely with the craftsmen and technicians, or sometimes produce certain parts or carry out production steps themselves.[12] The goal is ultimately to build a network that allows for economically and ecologically sound production. Many of these young labels publicize their production history and thus create added value through individuality, quality, and revealing a smart production process.[13] When design, production, and communication are interdependent and in the hands of the designers, this constellation often results in good products. Particularly in the fields of accessories and wooden furniture, there are new Swiss brands that have established an international reputation. Shoes by Velt fig. 819, bags by Freitag or Qwstion, eyewear from Sire or VIU, furniture from A C E or INCH Furniture: all demonstrate the designers' entrepreneurial skills and inspire the younger generation to take things into their own hands. Design thinking need not limit itself to the design of material products but may also include services, systems, processes, organizations, social work, teaching, and research. The Senior Design Factory fig. 315, for instance, offers in its barrier-free shop not only attractive products for seniors but also consulting for companies, or intergenerational do-it-yourself workshops.[14] In the niche markets of postindustrial Switzerland—traditional product and furniture design projects aside—locally rooted approaches and business models, which are somewhat summarily grouped together as creative industries, are leaving an indelible mark on the local culture.

In addition to the top achievements of a few Swiss design authors for international producers, which are still rather pertinent to the history of design, design-conscious Swiss SMEs, the art market, global luxury brands, and social projects have also provided a breeding ground for good design. Meanwhile, though, normal, everyday Swiss design threatens to disappear and should be duly recognized for its quality or continually reinvented.

1 The designer Frédéric Dedelley vividly describes his problem with "cash cows" using the example of the Swiss furniture maker De Sede. Frédéric Dedelley, cited in "Design versus Marketing" 2013.
2 The furniture market in Switzerland generates a sales volume of around four billion francs per year. It encompasses 1,500 furniture retailers and 50 industrial furniture makers. In the last two years, market share has been lost both domestically and on the export market. Prices for imported furniture dropped an average of 20 percent, while Swiss furniture went up in price in the euro export markets. Info from www.moebelschweiz.ch, cited in Meret Ernst 2014a.
3 In an interview on Swiss design, the product designer Jörg Boner said he wished for more state-funded presentations of Swiss design. Ott 2014, p. 46.
4 In 2013, an estimated 16 percent of the European production volume of personal luxury goods, at a value of around 25 billion euros, was produced in Switzerland, representing about 4 percent of gross domestic product. Müller-Stewens 2013. The Swiss watch industry has doubled its exports since the turn of the millennium; see Feubli et al. 2013.
5 Willy Gläser, CEO of Wogg, comparing the watch with the furniture industry, calls for more quality, including in design: "If the goal is to systematically promote product exports, this requires strict triage. Quality is everything—quality in terms of form, function, and durability, not only quality of production. You have to build relationships over the years, convey our understanding of design, plan compelling presentations. In short, it takes concepts." Cited in Ernst 2014a.
6 For example, they reinforce Louis Vuitton's themes of travel and adventure, the furniture pieces becoming so-called touch points of the brand and supporting brand communication.
7 Rovero originally developed the leather animals for the 2009 Vienna Design Week. Hermès sells the elephant (520 euros), the crocodile (520 euros), and the rabbit (320 euros) online.
8 François Hollande gave Michelle and Barack Obama Baccarat's *Brick* vase, which was designed by Aurélie Mathieu as an ECAL project.
9 See the anthology Huber et al. 2011.
10 Naoto Fukasawa and Jasper Morrison, excerpt from e-mail correspondence with the Design Collecion, March 6, 2013. See also Fukasawa / Morrison 2007.
11 "For me, this attitude comes from observing other objects I think are good. For me, it is a natural goal to simplify things. And it should certainly also be the goal of design. Because complicated things are not really made for everyday use." Jasper Morrison, in Imwinkelried 2013.
12 To build their eyewear label Strada del Sole, Sandra Kaufmann and Markus Dudli had to make certain parts of their glasses by hand because they were unable to find producers. See Ernst 2007a.
13 The Freitag story—from truck tarpaulin to individually designed bag—is thus the core of the company's brand communication and is drummed into every vendor in Zurich, New York, and Tokyo.
14 This is also the context of catch phrases such as "universal design," "social design," and "design for all": they all refer to socially engaged design with the aspiration to make the world a better place.

"Design means relating thinking and doing."
—Otl Aicher

How much of the academy can design take? Upheavals in education policy—rather than the designers themselves—brought this question to the fore in Switzerland. In 1995, the Swiss universities of applied sciences law, the Fachhochschulgesetz, was revised. Three years later, seven public universities of applied sciences were founded on the basis of this reform.[1] Today, design is among the courses offered in each of these universities. With the reform, the commercially oriented education at the schools of arts and crafts, which had sprung up all over Europe in the latter third of the nineteenth century, was first assumed by specialized classes at schools and secondary schools of design. Eventually this led to courses of study in design being offered at applied science universities, allowing them to permeate the realms of tertiary education—on an equal footing with the regular universities, yet different, as laid down in the law on the support and coordination of higher education.[2]

Universities instead of the lovingly nicknamed "Kunsti" art schools? Interdisciplinary modules instead of specialized classes? The commotion surrounding "academization" belied the fact that the reform in design training led to a real wave of development. This was reflected in the new courses that emerged: interaction design, game design, or animation design provide training in the design of virtual worlds, also tied into fields such as industrial design or visual communication. The Bologna education reform, enacted in 1999 and established across the European Union by 2011, was implemented in Switzerland at the same time.[3] Since then, traditional universities, institutes of technology, and universities of applied sciences have all formed part of a standardized education system. After three years of undergraduate study at a university of applied sciences, students receive their bachelor's degree and are "generally" considered to be fully prepared for employment according to the law. They can then further specialize during one-and-a-half to two-year master's degrees.[4] This master's degree is equivalent to the former *Diplom* or *Lizentiat* degree. A doctorate is necessary in order to qualify within scientific fields. This possibility is not yet within reach for Swiss universities of art and design. Resistance within the politics of the education system, coupled with uncertainty surrounding definitions, prevent the implementation of a third cycle of training in art and design.

While many may see this as nothing more than further evidence of excessive bureaucracy in education, the European Bologna reform and the Swiss universities of applied sciences reform not only shook up the educational institutions but also how design conceives of itself. Being the youngest course on offer at universities, it had to be subjected to a thorough examination. Old certainties and educational practices were challenged and new courses were developed and accredited. What is design? How does one define the specific knowledge required in professional practice, and how can it best be taught? As a craft or as a science? In the workshop or in the studio? Should it be understood as vocational or academic, serve industry or the creative industries? In practice, the crucial aim was to reduce the five-year *Diplom* course of studies—which included an integrated foundation course and an internship—to a three-year bachelor's degree without an internship in such a way that it would still ensure the same level of professional qualification. The Swiss Federal Office for Professional Education and Technology (OPET) only granted the subsequent master's degree a length of one-and-a-half years instead of the two years requested. Naturally, competition is also an issue, because the authorization and financing of these courses by the federal government and the respective cantons depend on student numbers. The education market is limited, however. Today, the overall number of students doing courses in all areas of design is around three thousand.[5] These concrete problems have always raised the crucial question: What functions must universities of applied sciences fulfill in order to be considered equal in value to regular universities?

Design Research at Universities of Applied Sciences

Aside from teaching, research is also an important factor. While universities of applied sciences are more focused on practical research, basic research is the reserve of the regular universities. As a result, the subject of design is faced with at least two problems: What does research mean to design as a discipline that was introduced in the sixties and was not initially established within the university system? And on what basis should applied research be undertaken when it still remains unclear who should do fundamental research and how?[6]

Universities of applied sciences have been successful at putting in place the structures necessary for the investigation of practice-oriented questions. Design research in Switzerland has been overseen by the Swiss Design Network since 2003. Projects that stay within their respective university's research budget can be financed through federal funds, which are administrated by the Commission for Technology and Innovation (CTI) and the Swiss National Science Foundation (SNSF).[7] The Science Foundation offered specialized programs to assist in the establishment of universities of applied sciences from the ground up. Nowadays, however, design research must compete with traditional research disciplines for funding and support programs.

The CTI, on the other hand, aligns itself more closely to business interests. As a promotional agency for federal innovation,[8] the independent commission provides support to all disciplines in which applied research is undertaken for the purpose of establishing systems, disciplines, finished and semifinished products, as well as services with the goal of reaching market maturity. The CTI's funds are reserved solely for regular universities and are limited to the amount the business partners are willing to match. Among the key funding criteria are points such as "innovative content" and "economic impact."

Driven by state funding for research, innovation, and economic development,[9] design research takes place in a variety of contexts including industry, universities, and research conferences. However, a disciplinary consensus on design science and research and its academic legitimization has yet to be reached. Discussions surround the question of whether research into, for, or through design can claim to be constitutive of a newly emerging discipline of design science.[10] Research into design regards design from the outside in an analytical manner, looking, for example, at design from historical, sociological, cultural, and technological perspectives.[11] Research for design, on the other hand, aids the design process by using the study of product semantics to explore the meaning behind the display function of artifacts. The third area of study, research through design—design's own method of research—investigates the possibility of creating an entirely design-based form of knowledge.

In a summary of the situation written in 2008, the authors identified at least two different trends being followed in Switzerland. Aside from the first, which has been committed to its methodological paradigm since the sixties and pursues discourses surrounding planning and engineering, a second trend emerged in the eighties from the growing interest in the implicit knowledge made available by creative practice. This latter trend is more closely connected to the traditions of training and education in art.[12] It remains to be seen which direction design research in Switzerland will take in its future development. Both of these schools of thought are related to local traditions, whether to an engineering-driven design approach—as exhibited by the USM Haller modular furniture system—or the more user-centered, research-based approach fostered by Willy Guhl in his design teaching. The question that must always be raised is whether these research efforts are focused on the knowledge that design practice can derive from other disciplines, or the knowledge that design itself provides.[13]

What was the effect of this reform in education on design practice? In the autumn of 2008, the first graduates of bachelor's degree programs in design entered the job market, and the first master's degree courses in design began. Is the bachelor's degree now accepted in design practice as a standard qualification? Or do graduates end up doing underpaid internships? No systematic surveys are available on where they gather their first work experience. This much is certain: the universities report that very few bachelor's degree holders go straight on to do master's degrees at the same institution. It appears that many of them would rather gain some practical experience before specializing.

General complaints about the "academization of design" are heard much more rarely these days. Instead, employers criticize the younger generation's lack of skills in model building, rendering, presentations, and economical thinking and planning. The Swiss Design Association agrees that education should be theory-based.[14] Every field is experiencing a rampant increase in knowledge. Why should a professional field as complex as design be an exception? The challenge, in the eyes of practicing designers, is to master the balancing act between reflecting and actively designing, between thinking and doing. In this sense, it is not an unreasonable requirement for every designer to be able to summarize their own thought and practice in three short sentences.

Promotion

Designers who seek to find a path into practice following their training can rely on receiving support. By state decree, the promotion of design, which has existed in Switzerland since 1917, is set up as cultural funding.[15] However, this does not entirely do justice to the origin of design, which is not only culturally relevant but also economically motivated. Design is systematically promoted on a federal level by the Federal Office of Culture (BAK) and the cultural foundation Pro Helvetia. By contrast, hardly any cantons consider design funding part of their models for cultural promotion. A few exceptions, such as the Berner Design Stiftung as well as individual canton-administered grants for design, do not make much of a difference. The exact reason why the promotion of design should be treated as an entirely federal matter, unlike fine art or film, remains unclear.

The new law for the promotion of culture that has been enforced since 2012 reallocated the functions performed by the BAK and Pro Helvetia between the two institutions. While the Federal Office of Culture awards prizes, the cultural foundation is now responsible for funding creative projects and up-and-coming designers. The law established a systemic change: every four years, the Federal Council presents a dispatch on the financing and focal points of cultural promotion.[16] In the first of these dispatches, the council acknowledged the economic importance of culture as well as the significance of the creative industries as it has been researched and statistically ascertained over the last ten years in Switzerland.[17] Today, design promotion that seeks to bridge the gap between itself and business funding does in fact react to the specific needs of young designers.[18]

After all, young designers face dual challenges.[19] On one hand, they are expected to set up their own business and find affordable work spaces—which are in ever shorter supply in urban centers—so that they can begin work. On the other hand, they must prove their competence as designers in order to kick-start their careers. Design processes cannot be fully rationalized as they are usually a question of personal expression. Effective design funding should therefore ideally be tied in with business and innovation funding, which mainly promotes inventions on the basis of their originality, their potential market success, or the novelty of the idea behind them.[20] Through these efforts, and in step with the newly introduced professionalization of education and establishment of design research, it is likely that Switzerland will remain fertile soil for design in the long run. Subsequent generations of designers will then be better able to master the many tasks that have since arisen in the global context.

1 In late 2003, the Swiss Federal Council granted them an indefinite permit. Since 2005, higher education courses in health, social work, and art have no longer been regulated by individual cantons but also fall under the rule of the Federal Council. In 2008, the council reviewed the accredited universities of applied sciences and commented positively on their development. On the development of Swiss universities of applied sciences, see http://www.sbfi.admin.ch/fh/index.html?lang=en (accessed June 16, 2014). Design is taught in seven different university schools in Basel, Bern, Geneva, Lausanne, Lugano, Lucerne, and Zurich.

2 According to Article 63a of the Federal Constitution, both the federal government and the cantons are responsible for creating a competitively viable higher education system (consisting of federal institutes of technology, cantonal universities, universities of applied sciences, pedagogical universities, and other institutions of higher education). The fundamental principles are laid down in the Federal Act for the Support and Coordination of Higher Education Institutions (HFKG). The law's regulations were in hearing in 2014. http://www.sbfi.admin.ch/themen/hochschulen/01640/index.html?lang=en (accessed June 16, 2014).

3 For a summary, see von Matt 2010.

4 On the introduction of the master's degree in design, see Ernst 2007b.

5 In 2012–13, the Federal Statistical Office counted 2,938 design students, 62.5 percent of whom were women and 27.4 percent international students. http://www.bfs.admin.ch/bfs/portal/en/index.html (accessed June 16, 2014).

6 Alois Martin Müller describes how the constraints on applied research suggest a questionable level of freedom in research: "Furthermore, this case is unique in the history of science in that we have a field that aspires to become a science but is not one yet. At the same time, however, the panel of the Federal Commission for Technological Innovation (CTI), which is responsible for judging the merit of this not yet constituted science, naturally also wants to have a say in defining what is scientific in this area and what is not." Cited in Ernst 2004, p. 4.

7 For eight years (2004 to 2011), the SNSF promoted practice-oriented research in social sciences and the arts at universities of applied sciences and pedagogical universities through the DORE program; of the 302 projects that received support, 62 were in the field of art and design. Altogether, the National Fund dispensed 46 million francs in total to universities of applied sciences and pedagogical universities, and over 15 million francs went to the schools of practice partners. Since August 2011, applicants can designate their research projects as "practice-oriented." SNFinfo print no. 13/ June 2011, pdf; SNF Annual Report 2011, pp. 18–19.

8 See The Commission for Technology and Innovation (CTI), http://www.kti.admin.ch/index.html?lang=en (accessed June 16, 2014).

9 An overview of a number of research projects that came out of universities of applied sciences over the past years can be found here: http://www.swissdesignnetwork.org (accessed June 16, 2014).

10 Brandes 2008, pp. 334–35.

11 In Switzerland, design historiography is taught neither at traditional universities nor at universities of applied sciences. Ernst 2014b.

12 Hugentobler et al. 2008, p. 5.

13 Mareis 2008, p. 20.

14 See http://www.swiss-design-association.ch (accessed June 16, 2014).

15 On the history of state programs for culture, see Münch 1997.

16 The Dispatch on Culture Promotion (Kulturbotschaft) is the central instrument with which the federal government determines national cultural politics. It has 632.7 million francs available until 2015. On the effects of this instrument on the promotion of culture, see Ernst 2013b.

17 Weckerle 2008.

18 Currently, Pro Helvetia is striving for the creation of a coherent system of support for design, which would be coordinated with state funding programs for business (the CTI, specifically). Annoni 2014.

19 The regulations for the Dispatch on Culture Promotion specify an age of thirty-five and five years of professional practice as the upper limits.

20 In Switzerland there is an abundance of well-established funding models. They range from competitions and scholarships to project, work, or financing contributions and what actually amounts to product promotion, and also include the supply of affordable infrastructure, coaching offers, and crowd funding, which is rather new to Switzerland. There is still, however, a lack of studies that explore the effectiveness of these funding models.

In 2014, the collections of the Museum für Gestaltung Zürich were relocated to shared premises in Zürich West. A center for the history of design of European significance is thus being developed in the area known as the Toni-Areal and as part of the Zurich University of the Arts (ZHdK). This project gives design history, particularly in Switzerland, enhanced visibility and makes it come alive for teaching and research as well as for a broader public. A review of the museum's collection history reveals that many ideas and concepts developed in the past are now becoming a reality with the new Schaudepot (Open Collections), along with the website eMuseum.ch and the planned permanent exhibition.

Models for Education and Industry

The Arts and Crafts Collection and the Graphics Collection were already established in 1875 as a model collection for professional education. The "South Kensington System" in which the later Victoria and Albert Museum and the Royal College of Art in London closely complemented each other and were mutually dependent provided a vital international reference point for this collection strategy.[1]

Most of these partnerships were dissolved in Europe over time, not least because the cultural-historical approach of many museums seemed incompatible with the future-oriented mission of the applied arts schools and universities. The tension between collecting for the preservation of cultural heritage and collecting for educational purposes also arose due to the different ways in which the objects were perceived and evaluated.

A Collection of Product Design

In 1979, Hansjörg Budliger (1925–2009), the director of the Kunstgewerbeschule and Kunstgewerbemuseum Zürich, outlined a pilot program for the development of design documentation. While the "free and the applied arts and communication design are already covered," he noted, Switzerland still clearly displayed "a gap in terms of product design."[2] Budliger ultimately saw the museum as a link between the educational institution and the public. The museum should therefore foster "stimulating lessons" as well as fulfilling economic and educational tasks with "accessible collections and exhibitions."

The 1983 exhibition *Design – Formgebung für jedermann* (Design for Everyone) presented for the first time object groups arranged typologically in a broader framework. Major acquisitions were carried out specifically for this project and donations accepted, including from the faculty. From the beginning, the design documentation was established as a model collection for the design department and it also regularly included exhibitions and, starting in 1984, the publication series Schweizer Design-Pioniere (Swiss Design Pioneers).[3]

Collection Concept

Claude Lichtenstein (b. 1949), who worked as exhibition curator from 1985 to 2001, devised a concrete concept for the design collection in 1986 in cooperation with Hansjörg Budliger and the art historian Lotte Schilder Bär (b. 1953). The main focus would be "serially produced objects" from 1900 onward that have a clear reference to Switzerland, in order to "prevent superficial, unsystematic collecting activity."[4]

With his concept, Lichtenstein laid important cornerstones for the museum's future collecting activity. From the same object category, different design variants were to be collected to "draw attention to the changes in the course of a long production period."[5] An indispensable complement is formed by a virtual "apparatus" of documents, photographs, brochures, and plans that points far beyond the object itself. Lichtenstein conceived the substantive orientation of the design collection following an extensive analysis of comparable institutions. The cross-comparison culminated in the ambition to establish Switzerland's first public collection for design at the Schule und Museum für Gestaltung Zürich.

862

863

862 Museum für Gestaltung Schaudepot (Open Collections), 2014
863 Jörg Hamburger and Georg Staehelin, The Design Collection of the Museum für Gestaltung, 1990. Poster

From Schaulager to Schaudepot

From the beginning, the collection strategy of the Museum für Gestaltung has been derived from its close association with design production and its relationship with the university. The museum's collections are to be understood as a thematic reference system and a reference work rather than a historically accurate time capsule. Design attitudes, trends, and exclusivity are of greater interest than mainstream tastes. The focal point is the design itself and thus the production aesthetic more than its historic significance.

Taking into consideration that the collection objects should be quickly accessible at all times for educational purposes, Lichtenstein suggested providing the archive cabinets with glass doors, and even in certain instances illuminating them. From the onset, the collection was therefore thought of as a viewing collection, described by Lichtenstein as a study collection or "institute."[6]

In the following years, exhibitions of industrial design were carried out on a regular basis. The collection grew continuously, partly thanks to the consistent support of the Swiss Confederation, and was rapidly acknowledged as a center of excellence for design in Switzerland.[7] Following Lichtenstein's concept, a part of the collection was indeed made accessible to the general public in 1999. The new showroom was called the Schaulager der Designsammlung (Display Storeroom of the Design Collection) and opened its doors on Zurich's Sihlquai with a smaller exhibition of objects organized in thematic groups based on variations in form, material, and utilization.[8]

With the new Schaudepot, literally "display depot," of the Museum für Gestaltung on the grounds of the former Toni dairy in the Areal, the collections are now being transported into the twenty-first century.[9] Closely related are several projects dealing with content and conservation as well as a novel and multifaceted approach to communicating the collection. The Schaudepot is essentially divided into four parts: the walk-in archive; the operational structures with offices, research space, workshops, and photo labs; several exhibition spaces; and the virtual space.

Graphics and Poster Design, Applied Art and Industrial Design

The most obvious new feature in 2015 is the physical relocation of all archives to a common site. As a result, it will become easier to place the objects into changing, thematic relationships with one another. This is one of the great attributes of the Zurich collection: in addition to the actual objects and the "apparatus" described above, complementary objects can often be found in the partner collections, which tremendously enhances the possibilities of a discursive reference. For this purpose, a portion of the Design Collection's holdings is stored in typological series to make cross-references and development stages visible. This is meant to provide the archive itself with added value, which can only partially be achieved by way of temporary exhibits. The dormant substantive energy, the latent potential of the collection, will thus become more comprehensible and grow. Through this new arrangement of objects within a category—rather than according to the year acquired for the collection, as was previously the case—thematic references can be established. During the preparations for the move, all collection objects were provided, either individually or in groups, with a bar code that allows the corresponding data record in the system to be accessed via a hand scanner. All sites were also provided with a code so that the storage system can always be newly arranged and linked in a logical way following a thematically based approach.

The idea of reading the collection's holdings in many different interpretive contexts is also reflected in the digital archive accessible online. Through eMuseum.ch, more than 80,000 data records are accessible and cross-linked.[10] To also represent the objects visually, they are photographed in the collections' own studio. In addition, the electronic museum offers the possibility of arranging one's own object and media compositions by way of the "my collection" tool. Here, too, the museum seeks debate and discussion with an extended circle of specialists, students, and interested parties, since it is the multilayered and open-ended embedding of one and the same object in different contexts that accounts for its value.

Exhibiting and Communicating

The new Schaudepot of the Museum für Gestaltung is meant to be accessible to an interested public. Several routes through the different areas of the archive—the high-bay storage, the textile warehouse, the exhibition niches designed as period rooms, and also the areas for the Applied Art, Poster, and Graphics Collections—are designed as curated tours that provide insight into the variety and abundance of the holdings.

In addition to the walk-in archive, the museum will be showing its collection holdings in the medium term in a constantly evolving permanent presentation. The style and concern of these exhibitions are diverse and encompass thematic, monographic, functional, narrative, as well as aesthetic approaches.

Ultimately, all these innovations reflect an open attitude toward the objects and their interpretation and contextualization. This concept is broadened by the mediation work taking place concurrently. With its approach to education and communication, the Museum für Gestaltung Zürich wants to create a platform for discursive engagement. Critiques of design and consumerism, theory and design research, are as much part of this as is a reflection on our society's product culture. Design is not understood as a hermetically sealed notion but rather as an intangible aspect of our society. For this discussion, the inventoried collection objects play an indispensable role as the protagonists of design history. Those collections that were created as a tool for cultivating taste education when the museum was founded[11] once again provide a platform for in-depth engagement with complex questions surrounding the concept of design. Prospective collecting,[12] not least for a forward-looking design practice, remains a central objective of the collecting activity.

1 It also had to do with design training in industry: "Parliamentary frugality was overcome by the argument that the manufacturing population needed training in design, so that Britain would thereby be better equipped to outdistance her international rivals." Cardoso Denis 1997, p. 107.
2 Budliger 1979.
3 The series was published from 1984 to 1992 and comprises six volumes.
4 Lichtenstein et al. 1986.
5 Sigfried Giedion presented developmental series like these in a book that became a seminal work for industrial culture. See Giedion 1948.
6 Lichtenstein et al. 1986.
7 Since 1987, the Federal Department of the Interior, Federal Office of Culture (BAK), has been financing the acquisitions of the works of living designers within the framework of its design promotion. They are selected and inventoried for the Design Collection as permanent loans from the BAK.
8 The term "Schaulager" (Display Storeroom) was adopted in 2003 by the Laurenz Foundation in Basel, which, ignoring other uses of the name, registered it as a trademark.
9 On the term "Schaudepot," see Natter et al. 2010.
10 The eMuseum also includes the Zurich University of the Art's archive of projects from studies there and documentation dating back to the founding of the university and the museum.
11 Beginning in 1908 with the interior design exhibitions all the way up to the Swiss Werkbund's interior design consulting starting in 1956, the Museum für Gestaltung was dedicated in its exhibitions to imparting quality design. On the interior design shows, see Schiller 2012. On interior design consulting in connection with the Swiss Werkbund, see Gnägi et al. 2013, pp. 377–79.
12 Lichtenstein 2006, p. 132.

Bibliography

The following consists of the full references for all works cited in shortened form in the essays as well as suggested further reading.

Aebersold/Meyer 1969
Alfred Aebersold and Heinz Meyer, "Papiermöbel aus der Klasse für Innenarchitektur und Produktgestaltung," in *Magie des Papiers,* exh. cat. Kunstgewerbemuseum Zürich, Wegleitung 277, Zurich 1969, pp. 67–70.

Afuhs/Reble 2007
Eva Afuhs and Christina Reble, *Sophie Taeuber-Arp. Gestalterin Architektin Tänzerin,* exh. cat. Museum Bellerive, Zurich 2007.

AIAG 1890
Aluminium-Industrie-Aktien-Gesellschaft, ed., *Die Anlagen der Aluminium-Industrie-Actien-Gesellschaft, ihre Produkte, deren Behandlung und Verwendung,* Schaffhausen 1890.

AIAG 1942–43
Geschichte der Aluminium-Industrie-Aktien-Gesellschaft Neuhausen 1888–1938, 2 vols., Chippis 1942–43.

Alexander 1977
Christopher Alexander, *A Pattern Language,* New York 1977.

Annoni 2014
Mario Annoni, "Das Ziel. Eine kreative und innovative Schweiz," in *Jahresbericht 2013,* ed. Pro Helvetia, Zurich 2014, pp. 2–3.

Antonio Vitali 1994
Antonio Vitali. Spielzeugdesigner/Creator of Toys, Weingarten 1994.

Arp 1955
Hans Arp, *Unsern täglichen Traum. Erinnerungen, Dichtungen und Betrachtungen aus den Jahren 1914–1954,* Zurich 1955.

Atelier Oï 2012
Atelier Oï, ed., *Workshop Guide,* Ludwigsburg 2012.

Baer 1927
Casimir (Heinrich) Baer, "Ein Doppel-Einfamilienhaus in Zürich," *Das ideale Heim* 1, no. 2 (1927), pp. 46–49.

BAG 1933
BAG Turgi, cat. no. 644 6737, 1933 edition.

Bähler 1997
Liana Bähler, "Kunststoff made in Switzerland," in "Design," *Kunst und Architektur in der Schweiz* 48, no. 3 (1997), pp. 30–39.

Baltensweiler AG 1996
Baltensweiler AG – Atelier für Beleuchtungstechnik, company brochure, 1996.

Banham 1953
Reyner Banham, "Vehicles of Desire," *Art,* no. 1 (September 1953), p. 3.

Banham 1955
Reyner Banham, "Machine Aesthetic," *Architectural Review,* no. 117 (April 1955), pp. 225–28.

Barsac 2005
Jacques Barsac, *Charlotte Perriand. Un art d'habiter,* Paris 2005.

Barten 1997
Sigrid Barten, ed., *Kreuz und quer der Farben,* exh. cat. Museum Bellerive, Zurich 1997.

Bärtschi 2011
Hans-Peter Bärtschi, *Die industrielle Schweiz vom 18. ins 21. Jahrhundert. Aufgebaut und ausverkauft,* Baden 2011.

Baudin 2002
Antoine Baudin, "Office Furniture, 1924/1925. Henri-Robert Von der Mühll," in *Swiss Furniture and Interiors in the 20th Century,* ed. Arthur Rüegg, Basel et al. 2002, pp. 90–91.

Baudrillard (1968) 1996
Jean Baudrillard, *The System of Objects,* London and New York 1996. Originally published in French in 1968 as *Le système des objets. La consommation des signes.*

Bauer et al. 1989
Tobias Bauer et al., *Silbersonne am Horizont. Alusuisse – eine Schweizer Kolonialgeschichte,* Zurich 1989.

Beaujon 1961
Edmond Beaujon, *L'art du potier. Paul A. Bonifas,* preface by Amédée Ozenfant, Neuchâtel 1961.

Benjamin (1982) 1999
Walter Benjamin, *The Arcades Project,* trans. Howard Eiland and Kevin McLaughlin, Cambridge, Mass. and London 1999. Originally published in German in 1982 as *Das Passegen-Werk.*

Berger 1983
Susi and Ueli Berger, "TISCH – Tisch," in "Einrichtungen/Aménagements," *Archithese* 13, no. 1 (1983), p. 49.

Berselli 2013
Silvia Berselli, "'Clarté' ou 'clair-obscur ornemental'?," *Kunst und Architektur in der Schweiz* 64, no. 3 (2013), pp. 20–27.

BIG-GAME 2008
BIG-GAME: Design Overview, exh. cat. Grand Hornu Images, Oostkamp 2008.

Bignens 1987
Christoph Bignens, "'Hermes-Baby' Schreibmaschine," in *Unbekannt – Vertraut. "Anonymes" Design im Schweizer Gebrauchsgerät seit 1920,* ed. Museum für Gestaltung Zürich, Reihe Schweizer Design-Pioniere 4, Zurich 1987, pp. 25–33.

Bignens 2001
Christoph Bignens, "'Die gute Form' gegen Amerika," *Zeitschrift für schweizerische Archäologie und Kunstgeschichte* 58, no. 3 (2001), pp. 185–94.

Bignens 2002a
Christoph Bignens, "EA-616 Sofa Bed. Ernst Ambühler," in *Swiss Furniture and Interiors in the 20th Century,* ed. Arthur Rüegg, Basel et al. 2002, pp. 202–3.

Bignens 2002b
Christoph Bignens, "Escape from 'Liberated Living,'" in *Swiss Furniture and Interiors in the 20th Century,* ed. Arthur Rüegg, Basel et al. 2002, pp. 177–95.

Bignens 2002c
Christoph Bignens, "S.T. Pull-up Table, 1954. Jürg Bally," in *Swiss Furniture and Interiors in the 20th Century,* ed. Arthur Rüegg, Basel et al. 2002, pp. 198–99.

Bignens 2003
Christoph Bignens, *The American Way of Life,* Sulgen 2003.

Bignens 2013
Christoph Bignens, "Die Zürcher Lichtwoche 1932," *Kunst und Architektur in der Schweiz* 64, no. 3 (2013), pp. 14–19.

Bill 1939
Max Bill, ed., *Le Corbusier et Pierre Jeanneret. Œuvre complète,* vol. 3: *1934–38,* Zurich 1939.

Bill 1945
Max Bill, *Wiederaufbau. Dokumente über Zerstörungen, Planungen, Konstruktionen,* Erlenbach-Zürich 1945.

Bill (1948) 2010.
Max Bill, "Schönheit aus Funktion und als Funktion," lecture in Basel, October 1948. Translated as "Beauty from Function and as Function," in *Form, Function, Beauty = Gestalt,* trans. Pamela Johnston London 2010, pp. 32–41. First published in German in *Werk* 36, no. 8 (1949), pp. 272–74.

Bill 1952
Max Bill, *Form. Eine Bilanz über die Formentwicklung um die Mitte des XX. Jahrhunderts/A Balance Sheet of Mid-Twentieth-Century Trends in Design/Un bilan de l'évolution de la forme au milieu du XXe siècle,* Basel 1952.

Bill 1954a
Max Bill, "Caro Ernesto (Lettere al direttore)," *Casabella-continuità* 18, no. 201 (1954), p. ii.

Bill 1954b
Max Bill, "Report on Brazil: 4," *Architectural Review* (October 1954), pp. 238–39.

Bill 1957
Max Bill, *Die gute Form: 6 Jahre Auszeichnung "Die gute Form" an der Schweizer Mustermesse in Basel,* Direktion der Schweizer Mustermesse in Basel, Winterthur 1957.

Bill 1988
Max Bill, "Die Funktion der gestalteten Objekte/The Function of Created Objects," *Zeitschrift für schweizerische Archäologie und Kunstgeschichte* 45, no. 1 (1988), pp. 50–52.

Bill/Rothebühler 1994
"Ein Bild von mir ist eine Organisation," Max Bill in an interview with Peter Rothenbühler, *Schweizer Illustrierte,* no. 51 (1994), pp. 34–36.

Binkley 2007
Stuart Binkley, *Getting Loose: Lifestyle Consumption in the 1970s,* Durham, NC and London 2007.

Birkner 1969
Othmar Birkner, ed., *Der Weg ins 20. Jahrhundert. Aus der Entwicklungsgeschichte der modernen Raumgestaltung in der Schweiz 1870–1918,* exh. cat. Gewerbemuseum Winterthur 1969.

Boesiger (1946) 1995
Willy Boesiger, *Le Corbusier. Œuvre complète/Complete Works,* vol. 4: *1938–1946,* Zurich 1946; repr. 1995.

Boesiger/Stonorov (1929) 1995
Willy Boesiger and Oscar Stonorov, *Le Corbusier. Œuvre complète/Complete Works,* vol. 1: *1910–1929,* Zurich 1929; repr. 1995.

Böhme (2006) 2014
Hartmut Böhme, *Fetishism and Culture: A Different Theory of Modernity,* Berlin and Boston 2014. Originally published in German in 2006 as *Fetischismus und Kultur. Eine andere Theorie der Moderne.*

Bonsiepe 1967
Gui Bonsiepe, "Arabesken der Rationalität. Anmerkungen zur Methodologie des Design/Arabesques of Rationality: Notes on the Methodology of Design," *Ulm* 19/20 (1967), pp. 9–23.

Bony 2004
Anne Bony, *Furniture and Interiors of the 1960s,* Paris 2004.

Borngräber 1989
Christian Borngräber, "Rezeptionsästhetik beweglicher Güter," *Kunstforum International* 99 (March/April 1989), pp. 59–60.

Bourban 2009
Elisabeth Bourban, "La brique de verre Falconnier," *Clins d'œil de l'Histoire – le blog de l'archiviste,* December 16, 2009, http://blogs.nyon.ch/archivistes/la-brique-de-verre-falconnier (accessed February 2, 2014).

Bourdieu (1979) 1984
Pierre Bourdieu, *Distinction: A Social Critique of the Judgement of Taste,* trans. Richard Nice, London and New York 1984. Originally published in French in 1979 as *La distinction. Critique sociale du jugement.*

Bracht et al. 2011
Uwe Bracht, Dieter Geckler, and Sigrid Wenzel, *Digitale Fabrik. Methoden und Praxisbeispiele,* Berlin 2011.

Brandes 2008
Uta Brandes, "Research," in *Design Dictionary: Perspectives on Design Terminology,* ed. Michael Erlhoff and Tim Marshall, Basel et al. 2008. pp. 332–36.

Breuer 1928
Marcel Breuer, “Metallmöbel,” in Werner Graeff, *Innenräume,* Stuttgart 1928, pp. 133–34.

Breuer 2001
Gerda Breuer, *Die Erfindung des Modernen Klassikers. Avantgarde und ewige Aktualität,* Ostfildern-Ruit 2001.

Breuer 2004
Gerda Breuer, “Überzeugen, nicht überreden. Das Wechselspiel von Sachfotografie und Produktdesign,” in *Im Rausch der Dinge,* ed. Thomas Seelig and Urs Stahel/Fotomuseum Winterthur and Fotostiftung Schweiz, Göttingen 2004, pp. 88–95.

Breuer et al. 1999
Gerda Breuer, Andrea Peters, and Kerstin Plüm, eds., *Die 60er. Positionen des Designs,* Cologne 1999.

Brillat-Savarin (1825) 2009
Jean Anthelme Brillat-Savarin, *The Physiology of Taste: Or Meditations on Transcendental Gastronomy,* New York 1949; repr. New York 2009. Originally published in French in 1825 as *Physiologie du goût, ou, Méditations de gastronomie transcendante.*

Brooks 1997
H. Allen Brooks, *Le Corbusier's Formative Years: Charles-Edouard Jeanneret at La Chaux-de-Fonds,* Chicago and London 1997.

Bruyère 1945
André Bruyère, “Légende,” in “Solutions d'urgence,” special issue, *L'Architecture d'aujourd'hui* 16, no. 2 (July/August 1945), p. 3.

Buck 1998
Alex Buck, *Dominanz der Oberfläche. Betrachtungen zu einer neuen Bedeutsamkeit der Gegenstände,* Frankfurt am Main 1998.

Buddensieg 1979
Tilmann Buddensieg, *Peter Behrens und die AEG, 1907–1914,* Berlin 1979.

Budliger 1979
Hansjörg Budliger, “Aufbau einer Design-Dokumentation,” internal concept paper, Kunstgewerbemuseum Zürich, March 21, 1979. Museum für Gestaltung.

Bundesamt für Kultur 1997
Bundesamt für Kultur/Patrizia Crivelli and Köbi Gantenbein, eds., *Made in Switzerland. 80 Jahre Förderung durch die Eidgenossenschaft,* Bern and Zurich 1997.

Bundesamt für Statistik 2004
Bundesamt für Statistik, ed., *Analysen zur Betriebszählung 2001. Standorte der Innovation,* Neuchâtel 2004.

Burckhardt 1980
Lucius Burckhardt, “Wo geht es weiter?,” *Werk, Bauen + Wohnen* 67, no. 1–2 (1980), pp. 19–20.

Burckhardt (1980) 2012
Lucius Burckhardt, “Design Is Invisible” (1980), in *Lucius Burckhardt Writings: Rethinking Man-Made Environments,* ed. Jesko Fezer and Martin Schmitz, trans. Jill Denton, Vienna and New York 2012, pp. 153–65. First published in German in 1981 as “Design ist unsichtbar.”

Bürdek 2005
Bernhard E. Bürdek, *Design: History, Theory and Practice of Product Design,* Basel et al. 2005.

Bürkle/Tropeano 1994
J. Christoph Bürkle and Ruggero Tropeano, *Ein Prototyp des Neuen Bauens in Zürich – Die Rotachhäuser,* Zurich 1994.

Bürkle/Tropeano 1995
J. Christoph Bürkle and Ruggero Tropeano, “Max Ernst Haefeli als Entwerfer: Werkbundgedanken, Schweizer Design und die Arbeit an der ‘Form ohne Ornament,'” in *Wege zur “Guten Form”. Neun Beiträge zur Geschichte der Schweizer Produktgestaltung,* ed. Arthur Rüegg and Ruggero Tropeano, Basel et al. 1995, pp. 40–48.

Cacciola 2008
Donatella Cacciola, *Moderne Klassiker. Die wiedergefundene Zeit. Die Reeditionen von Sitzmöbeln in Deutschland und in Italien und ihre Rezeption,* PhD diss., TU Delft, 2008.

Caplan 1976
Ralph Caplan, “Making Connections: The Work of Charles and Ray Eames,” in *Connections: The Work of Charles and Ray Eames,* exh. cat. Frederick S. Wright Art Gallery, University of California, Los Angeles, 1976, pp. 15–55.

Cardoso Denis 1997
Rafael Cardoso Denis, “Teaching by Example: Education and the Formation of South Kensington's Museums,” in *A Grand Design: The Art of the Victoria and Albert Museum,* ed. Malcolm Baker and Brenda Richardson, London 1997, pp. 107–16.

Changes – Chancen 2013
Depot Basel, ed., *Changes – Chancen,* Basel 2013.

Clark 1951
W. M. Clark, ed., *The Crystal Palace and Its Contents: An Illustrated Cyclopaedia of the Great Exhibition of the Industry of All Nations,* London 1851.

Claus et al. 2009
Sylvia Claus, Dorothee Huber, and Beate Schnitter, *Lux Guyer 1894–1955. Architektin,* Zurich 2009.

Clivio 2009
Franco Clivio, *Verborgene Gestaltung – Dinge sehen und begreifen,* Basel et al. 2009.

Cohen 2011
Jean-Louis Cohen, *Architecture en uniforme. Projeter et construire pour la Seconde Guerre mondiale,* Montreal 2011.

Colomina 2006
Beatriz Colomina, *Domesticity at War,* Barcelona 2006.

Corrodi 2006
Michelle Corrodi, “On the Kitchen and Vulgar Odors: The Path to a New Domestic Architecture between the Mid-Nineteenth Century and the Second World War,” in *The Kitchen: Life World, Usage, Perspectives,* ed. Klaus Spechtenhauser, Basel et al. 2006, pp. 21–42.

Craig-Martin/Hamilton (1990) 2010
Michael Craig-Martin, “Richard Hamilton in Conversation with Michael Craig Martin” (1990), in *Richard Hamilton,* ed. Hal Foster with Alex Bacon, October Files 10, Cambridge, Mass. 2010, pp. 1–13.

Crettaz-Stürzel 2013
Elisabeth Crettaz-Stürzel, “Heimatstil,” in *Historisches Lexikon der Schweiz,* article published December 16, 2013, http://www.hls-dhs-dss.ch/textes/d/D11186.php (accessed February 1, 2014).

Dalí 1933
Salvador Dalí, “De la beauté terrifiante et comestible de l'architecture Modern Style,” *Minotaure* 3–4 (1933), pp. 69–76.

Delfs 2007
Timm Delfs, “Die Zeit im Griff?,” in *On Time. Objekte der Zeitmessung/Objects of Chronometry,* ed. Norbert Wild/Museum für Gestaltung Zürich, Design Collection 2, Ludwigsburg 2007, pp. 6–13.

Der junge Le Corbusier 2002
Der junge Le Corbusier, exh. cat. Museum Langmatt, Baden 2002.

Design Preis Schweiz 2013
Design Preis Schweiz, ed., *Design Preis Schweiz. Edition 2013/2014,* Langenthal 2013.

“Design versus Marketing” 2013. “Design versus Marketing. Ein Einzelfall?,” *Form News,* December 19, 2013, http://www.form.de/news/design-versus-marketing-punkt-ein-einzelfall (accessed June 8, 2014).

Die Allgemeine Entwurfsanstalt 2011
Röthlisberger Kollektion, ed., *Die Allgemeine Entwurfsanstalt mit Trix und Robert Haussmann. Möbel für die Röthlisberger Kollektion,* Sulgen 2011.

Die praktische Küche 1930
Die praktische Küche, exh. cat. Gewerbemuseum Basel, 1930.

Die Schweiz im Spiegel 1940
Schweizerische Landesausstellung 1939, ed., *Die Schweiz im Spiegel der Landesausstellung 1939,* vol. 1, Zurich 1940.

Die Wohnung für das Existenzminimum 1930
Internationale Kongresse für Neues Bauen and Hochbauamt Frankfurt am Main, eds., *Die Wohnung für das Existenzminimum,* Frankfurt am Main 1930.

Donzé 2009
Pierre-Yves Donzé, *Historie de l'industrie horlogère suisse. De Jacques David à Nicolas Hayek,* Neuchâtel 2009.

Dosch 2002
Leza Dosch, “Kitchen/Living Room of a Working-Class House, Swiss Werkbund Exhibition, Zurich, 1918. Hans Bernoulli,” in *Swiss Furniture and Interiors in the 20th Century,* ed. Arthur Rüegg, Basel et al. 2002, p. 304.

Dürrenmatt 1956
Friedrich Dürrenmatt, *Die Panne,* Zurich 1956.

Eames 1943
Charles Eames, “Beauty Thrust Upon Us,” *Arts and Architecture* 60 (1943), p. 29.

Easy! Easy? 2011
Easy! Easy? Schweizer Industrie-design – das Büro M&E 1967–2002, exh. cat. Museum in der Burg Zug, Zug 2011.

Eckstein 1934
Hans Eckstein, ed., *Die schöne Wohnung,* Munich 1934.

Eimert 1994
Dorothee Eimert, ed., *Paper Art 5. Geschichte der Papierkunst,* exh. cat. Leopold-Hoesch-Museum, Düren, Cologne 1994.

Elsasser 1997
Killian Elsasser, “Die stärkste Lokomotive der Welt,” in “Eisenbahn,” *Kunst und Architektur in der Schweiz* 48, no. 1 (1997), pp. 62–67.

Erlhoff/Marshall 2008
Michael Erlhoff and Tim Marshall, “Design,” in *Design Dictionary: Perspectives on Design Terminology,* ed. Michael Erlhoff and Tim Marshall, Basel et al. 2008, pp. 104–9.

Erni 1983
Peter Erni, *Die gute Form. Eine Aktion des Schweizerischen Werkbundes. Dokumentation und Interpretationen,* Baden 1983.

Erni 1985
Peter Erni, “Neue Modernität – vage Zuversicht der Epoche: Willy Guhl und das Design der Nachkriegszeit,” in *Willy Guhl. Gestalter und Lehrer,* ed. Museum für Gestaltung Zürich, Reihe Schweizer Design-Pioniere 2, Zurich 1985, pp. 12–21.

Ernst 2004
Meret Ernst, “Forschen im Netzwerk,” supplement to *Hochparterre* 17, no. 5 (2004), pp. 4–7.

Ernst 2005
Meret Ernst, “Design Preise Schweiz,” *Hochparterre* 18, no. 11 (2005), pp. 28–30.

Ernst 2006
Meret Ernst, “Design aus der Schweiz,” in *DesignSuisse,* ed. Christian Eggenberger and Meret Ernst, Zurich 2006, pp. 318–50.

Ernst 2007a
Meret Ernst, “Selbst gebügelte Sonnenbrille,” *Hochparterre* 20, no. 8 (2007), pp. 26–27.

Ernst 2007b
Meret Ernst, “Die zweite Stufe zünden,” *Hochparterre* 20, no. 11 (2007), pp. 68–69.

Ernst 2011a
Meret Ernst, “Design – kühne Entwürfe,” in *Kultur digital. Begriffe, Hintergründe, Beispiele,* ed. Hedy Graber et al. Basel 2011, pp. 297–304.

Ernst 2011b
Meret Ernst, “Jetzt lernen die Ingenieure die Sprache der Gestalter,” *Hochparterre* 24, no. 3 (2011), pp. 18–27.

Ernst 2011c
Meret Ernst, “Peter Wirz, Designer. ‘Es braucht Einsatz,'” supplement to *Hochparterre* 24, no. 12 (2011), p. 11.

Ernst 2012
Meret Ernst, “Im Handumdrehen,” *Hochparterre* 25, no. 12 (2012), pp. 46–48.

Ernst 2013a
Meret Ernst, “Der prekäre Prozess des Entwerfens,” *Passagen* 61, no. 2 (2013), pp. 8–10.

Ernst 2013b
Meret Ernst, “Förderschub für junge Designer,” *Hochparterre* 26, no. 6–7 (2013), pp. 20–21.

Ernst 2014a
Meret Ernst, “Gut gemeint,” *Hochparterre* 27, no. 4 (2014), pp. 54–57.

Ernst 2014b
Meret Ernst, “Hallo Forschung (Kommentar),” *Hochparterre* 27, no. 5 (2014), p. 63.

Ernst/Hönig 2005
Meret Ernst and Roderick Hönig, “Materialtest,” *Hochparterre* 18, no. 1–2 (2005), pp. 16–24.

Eschbach 2006
Andrea Eschbach, “Hannes Wettstein,” in *Design Suisse,* ed. Meret Ernst and Christian Eggenberger, Zurich 2006, pp. 271–79.

Esposito 2004
Elena Esposito, *Die Verbindlichkeit des Vorübergehenden. Paradoxien der Mode,* Frankfurt am Main 2004.

Falk/Schwarz 1991
Susanne Falk and Roland Schwarz, “Aluminium – Metall der Moderne,” in *Aluminium: Das Metall der Moderne,* ed. Werner Schäfke, Cologne 1991, pp. 27–70.

Fallet/Veyrassat 2013
Estelle Fallet and Béatrice Veyrassat/ANS, “Uhrenindustrie,” in *Historisches Lexikon der Schweiz,* article published January 1, 2014, http://www.hls-dhs-dss.ch/textes/d/D13976.php (accessed April 11, 2014).

Fäth 2011
Reinhold J. Fäth, *Dornach Design. Möbelkunst 1910–2010,* Basel 2011.

Feubli et al. 2013
Patricia Feubli et al., “Swiss Watch Industry: Prospects and Challenges,” Credit Suisse Global Research, October 2013, https://www.credit-suisse.com/media/production/pb/docs/unternehmen/kmugrossunternehmen/uhrenstudie-en.pdf (accessed June 8, 2014).

Fischer 1971
Wend Fischer, "Die verborgene Vernunft. Funktionale Gestaltung im 19. Jahrhundert," in *Die verborgene Vernunft. Funktionale Gestaltung im 19. Jahrhundert*, Wegleitung 284, Kunstgewerbemuseum Zürich with Die Neue Sammlung, Munich, Zurich 1971, pp. 5–13.

Fischer/Hamilton 1999
Volker Fischer and Anne Hamilton, *Theorien der Gestaltung. Grundlagentexte zum Design*, vol. 1, Frankfurt am Main 1999.

Fischli 1958
Hans Fischli/Kunstgewerbemuseum Zürich, *neue metallmöbel. Arbeiten von Hans Eichenberger, Robert Haussmann, Kurt Thut*, Wegleitung 224, Zurich 1958.

Fischli 2012
Fredi Fischli and Niels Olsen, *Trix + Robert Haussmann*, Zurich 2012.

Flierl 1965
Bruno Flierl, ed., *Hans Schmidt. Beiträge zur Architektur 1924–1964*, Basel 1965.

Formlose Möbel 2008
Formlose Möbel, exh. cat. Museum für angewandte Kunst Wien, Vienna, MAK Studies 13, Ostfildern 2008.

Forty 1986
Adrian Forty, *Objects of Desire: Design and Society, 1750–1980*, London 1986.

Frampton 2005
Kenneth Frampton, "The Work of Architecture in the Age of Commodification," Introduction to *Commodification and Spectacle in Architecture*, ed. William Saunders, Minneapolis 2005, pp. ix–xviii.

Frey 1986
Gilbert Frey, *Mobilier suisse, creation de 1927 à 1984/Schweizer Möbeldesign 1927–1984*, Lausanne and Zurich 1986. With texts by Claude Lichtenstein and Arthur Rüegg.

Frisch 1979
Max Frisch, *Tagebuch 1966–1971*, Frankfurt am Main 1979.

Fukasawa/Morrison 2007
Naoto Fukasawa and Jasper Morrison, *Super Normal: Sensations of the Ordinary*, Baden 2007.

Gamboni 2007
Dario Gamboni, "Art Nouveau: The Shape of Life," in *Nature Design: From Inspiration to Innovation*, ed. Angeli Sachs, exh. cat. Museum für Gestaltung Zürich, Baden 2007, pp. 98–111.

Gantenbein 1989
Köbi Gantenbein, "Ludwig Walser: Gestalter für die Industrie," *Hochparterre* 2, no. 5 (1989), pp. 37–43.

Gantenbein 2013
Köbi Gantenbein, ed., *Die Bahnhofsuhr. Ein Mythos des Designs aus der Schweiz*, Zurich 2013.

Gantenbein/Hablützel 2001
Köbi Gantenbein and Alfred Hablützel, eds., *Thut-Möbel 1953 bis heute*, supplement, *Hochparterre* 14, no. 8 (2001).

Ganz/Rieker 2012
Robert Ganz and Roger Rieker, *Bau und Betrieb Schweizerischer Bergbahnen*, exh. cat. Schloss Hünegg, Hilterfingen 2012.

Garner 1996
Philippe Garner, *Sixties Design*, Cologne 1996.

Gasser 2007
Martin Gasser, ed., *Bilderstreit. Durchbruch der Moderne um 1930*, Fotostiftung Schweiz, Zurich 2007.

Gausemeier 2013
Jürgen Gausemeier, "Rechnerunterstützte Entwicklung (CAE/CAD)," in *Enzyklopädie der Wirtschaftsinformatik Online-Lexikon*, ed. Karl Kurbel et al., updated September 3, 2013, http://www.wi-lex.de (accessed May 30, 2014).

Georgiadis 1993
Sokratis Georgiadis, *Sigfried Giedion: An Intellectual Biography; Architectural Interfaces*, Edinburgh 1993.

Gianola 1983
Ivano Gianola, "Möbel," in "Einrichtungen/Aménagements," *Archithese* 13, no. 1 (1983), p. 47.

Giedion 1923
Giedion, Sigfried, "Bauhaus und Bauhauswoche zu Weimar," *Das Werk* 10, no. 9 (1923), pp. 232–34.

Giedion (1928) 1995
Sigfried Giedion, *Building in France, Building in Iron, Building in Ferroconcrete*, trans. J. Duncan Berry, Santa Monica, Calif. 1995. Originally published in German in 1928 as *Bauen in Frankreich, Bauen in Eisen, Bauen in Eisenbeton*.

Giedion 1929
Sigfried Giedion, *Befreites Wohnen*, Zurich and Leipzig 1929.

Giedion 1945
Sigfried Giedion, "Léger in America," *Art News*, 1945, pp. 295–99.

Giedion 1946
Sigfried Giedion, "Stromlinienstil und industrielles Entwerfen in USA," *Werk* 33, no. 5 (1946), pp. 155–62.

Giedion 1948
Sigfried Giedion, *Mechanization Takes Command: A Contribution to Anonymous History*, New York 1948.

Giedion (1948) 2013
Sigfried Giedion, *Mechanization Takes Command: A Contribution to Anonymous History*, New York 1948; repr. Minneapolis 2013, with an afterword by Stanislaus von Moos.

Giedion 1951
Sigfried Giedion, ed., *A Decade of New Architecture/Dix ans d'architecture contemporaine*, Zurich 1951, p. 151.

Ginsburger 1931
Roger Ginsburger, "Kunstgewerbe und Industrie in Frankreich," in *Französische Ausstellung*, Kunstgewerbemuseum Zürich, Wegleitung 98, Zurich 1931, pp. 3–16.

Glanzmann et al. 2003
Jutta Glanzmann et al., eds., "Autobiografie," in *Flora Steiger-Crawford, 1899–1991*, Dokumente zur modernen Schweizer Architektur, Zurich 2003, pp. 24–98.

Gnägi et al. 2013
Thomas Gnägi, Bernd Nicolai, and Jasmine Wohlwend Piai, eds., *Gestaltung Werk Gesellschaft. 100 Jahre Schweizerischer Werkbund SWB*, Zurich 2013.

Gomringer 1967
Eugen Gomringer, "Die Chance des Industrial Design in der Schweiz," *Interieur/Industrial Design* 13, no. 2 (1967), pp. 2–3.

Graber 1932
Rudolf Graber, "Vom Entwurf zum serienreifen Möbelstück," *Das Werk* 19, no. 11 (1932), pp. 335–37.

Graber 1956
Rudolf Graber, "Kleine Erinnerungen," in *1931–1956. 25 Jahre Wohnbedarf*, Zurich 1956, pp. 30–33.

Gropius (1935) 1965
Walter Gropius, *The New Architecture and the Bauhaus*, London 1935; repr. Cambridge, Mass. 1965.

GSK 1997
Gesellschaft für Schweizerische Kunstgeschichte, ed., *Kunst und Architektur in der Schweiz, Design* 48, no. 3 (1997).

Gtr. 1926
Gtr. [Josef Gantner], "Möbel und Innenräume," *Das Werk* 13, no. 12 (1926), pp. 365–66.

Gugelot 1987
Hans Gugelot, "beschreibung und anlayse des baukastensystems m 125," in *System-Design Bahnbrecher. Hans Gugelot, 1920–65*, ed. Hans Wichmann, Basel and Boston 1987.

Gugerli 1997
David Gugerli, "Wie die Jungfrau zu ihrer Bahn gekommen ist. Technische Naturbeherrschung an einer anthropomorphisierten Landschaft," in "Eisenbahn," *Kunst und Architektur in der Schweiz* 48, no. 1 (1997), pp. 42–55.

Guhl 1950
Willy Guhl, "Über Stuhl- und Sitzformen," *Werk* 37, no. 8 (1950), pp. 230–35.

Guyer/Kettiger 1946
Hans Guyer and Ernst Kettiger, eds., *Möbel und Wohnraum*, Erlenbach-Zürich 1946.

H. L. 1951
H. L., "Swiss Telephones Are Efficient and Unassuming in Design," *Design*, no. 25 (January 1951), pp. 20–21.

H. R. S. 1949
H .R. S., "Es klappt," *Neue Zürcher Zeitung*, Pentecost edition, June 5, 1949.

Haber 2011
Peter Haber, "Informatisierung," in *Historisches Lexikon der Schweiz*, article published March 3, 2011, at http://www.hls-dhs-dss.ch/textes/d/D13724.php (accessed May 30, 2014).

Hablützel 1963
Alfred Hablützel, "1960: das Polyesterbett von Andreas Christen," *Interieur* 9, no. 2 (1963), pp. 20–23.

Hablützel 1981
Alfred Hablützel, ed., *H-Design for Mira-X. Interieur-Textilien/Interior Textiles*, Mira-X AG, Suhr 1981.

Hablützel 1985
Alfred Hablützel, "Kurt Thut. Designer und Unternehmer," in *30 Jahre Teo Jakob. Kurt Thut (Katalog)*, ed. Teo Jakob and Roger Pfund, Bern 1985.

Hablützel 1989
Alfred Hablützel, ed., *Mobilier suisse*, exh. cat. Centre Pompidou, Paris 1989.

Hablützel 2002a
Alfred Hablützel, "Aluminium Tower Cupboard and Wardrobe, 1986. Kurt Thut," in *Swiss Furniture and Interiors in the 20th Century*, ed. Arthur Rüegg, Basel et al. 2002, pp. 280–81.

Hablützel 2002b
Alfred Hablützel, "Multifunctional Wogg 3 Bed, 1983. Ubald Klug," in *Swiss Furniture and Interiors in the 20th Century*, ed. Arthur Rüegg, Basel et al. 2002, pp. 240–41.

Hablützel et al. 1988
Alfred Hablützel et al., eds., *Möbelgestaltung. Möbelerfindung. Am Beispiel Schweiz*, exh. cat. Modus Möbel, Berlin 1988.

Hablützel/Huber 1993
Alfred Hablützel and Verena Huber, *Innenarchitektur in der Schweiz 1942–1992*, Sulgen 1993.

Haefeli 1936
Max Ernst Haefeli, "Entwurf für ein eingeschossiges Wohnhaus am rechten Zürichseeufer," *Weiterbauen*, no. 5 (1936), p. 38. Supplement to *Schweizerische Bauzeitung* 107, no. 1 (1936).

Haefeli 1950
Max Ernst Haefeli, "Die Formgebung von Industrieprodukten," in *SWB-Ausstellung 1950*, ed. Kunstgewerbemuseum Zürich, Wegleitung 184, Zurich 1950, n.p.

Haffner 2001
Peter Haffner, "Die Essenz der Dinge," *Das Magazin*, no. 6 (2001), pp. 23–27.

Häfliger et al. 1991
Lorenz Häfliger et al., Stiftung Eisenbibliothek, Georg Fischer AG, and SIG Schweizerische Industrie-Gesellschaft Holding AG, eds., *Beiträge der Schweiz zur Technik. Schweizerische Erfindungen von internationaler Bedeutung*, Festschrift celebrating 700 years of the Swiss Confederation, Oberbözberg 1991.

Haller 1965
Bruno and Fritz Haller, "Bürohaus Schärer's Söhne Münsingen," *Bauen+Wohnen* 19, no. 8 (1965), pp. 323–33.

Haller 1978
Fritz Haller, "USM Haller Möbelbausystem. Planung und Konstruktion eines Möbelsystems," 4th inaugural address at the Universität Karlsruhe, May 11, 1978, gta Archive, ETH Zürich, Estate of Fritz Haller.

Hamilton (1962) 1982
Richard Hamilton, *Collected Words, 1953–1982*, London 1982. Originally published as "An Exposition of $he," in *Architectural Design* 32, no. 10 (October 1962), pp. 485–86.

Hanak 2009
Michael Hanak, "Zum Eternit-Wettbewerb für die G59," *Faszination Faserzement, Design+Design*, no. 2 (2009), pp. 128–41.

Hanhart 1997
Rudolf Hanhart, "Traugott Stauss 1898–1952," in *Toggenburger Annalen*, Bazenheid 1997.

Hannes Wettstein 2012
Studio Hannes Wettstein, ed., *Hannes Wettstein: Seeking Archetypes*, Baden 2012.

Hans Coray 1986
Museum für Gestaltung Zürich, ed., *Hans Coray. Künstler und Entwerfer*, Reihe Schweizer Design-Pioniere 3, Zurich 1986. Concept and realization: Jürg Brühlmann and Peter Eberhard.

Hans Hilfiker 1984
Museum für Gestaltung Zürich, ed., *Hans Hilfiker – Ingenieur und Gestalter*, Reihe Schweizer Design-Pioniere 1, Zurich 1984. Concept and realization: Peter Eberhard.

Harbusch 2009
Gregor Harbusch, "Bestandesbeschrieb Sigfried Giedion," gta Archive, ETH Zürich, 2009, http://www.archiv.gta.arch.ethz.ch/nachlaesse-vorlaesse/giedion-sigfried (accessed October 2, 2013).

Haussmann 1958
Robert Haussmann, "Zur Ausstellung," in *Neue Metallmöbel*, exh. cat. Kunstgewerbemuseum Zürich, Zurich 1958, n.p.

Haussmann 1981
Trix and Robert Haussmann, "Zu unseren Arbeiten," in "Illusionismus," *Werk, Bauen + Wohnen* 68, no. 10 (1981), pp. 29–32.

Haussmann 2012
Robert Haussmann, "Manierismo Critico (1981)," in Fredi Fischli and Niels Olsen, *Trix + Robert Haussmann*, Zurich 2012, pp. 28–33.

Hegel (1822) 1970
Georg Wilhelm Friedrich Hegel, *Vorlesung über Ästhetik II* (1822), in *Werke in 20. Bänden*, vol. 14, Frankfurt am Main 1970.

Heller/Volk 1999
Martin Heller and Andreas Volk/Museum für Gestaltung Zürich, eds., *Die Schweizer Autobahn*, Zurich 1999.

Helvetica 2011
Helvetica. Porcelaine de Nyon à Mariemony, exh. cat. Musée royal de Mariemont, Morlanwelz 2011.

Hennessey/Papanek 1973
James Hennessey and Victor Papanek, *Nomadic Furniture: How to Build and Where to Buy Lightweight Furniture that Folds, Collapses, Stacks, Knocks Down, Inflates or Can Be Thrown Away and Recycled,* New York 1973.

Herding 2008
Klaus Herding, *1968. Kunst, Kunstgeschichte, Politik,* Frankfurt am Main 2008.

Herrmann 1989
Gabrielle Herrmann, "Ein Lehrstück wird Renner," *Hochparterre* 2, no. 5 (1989), pp. 18–39.

Herzog/de Meuron 1977
Jacques Herzog and Pierre de Meuron, "Rationale Architektur und historische Bezugnahme," *The Village Cry,* no. 3 (March/April 1977), pp. 18–39.

Herzog/de Meuron 1983
Jacques Herzog and Pierre de Meuron, "Haus-Bar," *Archithese* 13, no. 1 (1983), p. 50.

Hildebrand 2010
Sonja Hildebrand, "Kunsthaus Zürich," in *Karl Moser. Architektur für eine neue Zeit,* ed. Werner Oechslin and Sonja Hildebrand, Zurich 2010, pp. 132–41.

Hildebrand et al. 2007
Sonja Hildebrand, Bruno Maurer, and Werner Oechslin, eds., *Haefeli Moser Steiger. Die Architekten der Schweizer Moderne,* Zurich 2007.

Holenstein 1998
Peter Holenstein, *Die sprechenden Maschinen. Studer Revox. Das Lebenswerk des Audiopioniers Willi Studer,* Zurich 1998.

Horisberger 2010
Christina Horisberger, "Memento Mori," *Design+Design,* no. 3 (2010), pp. 20–27.

Hornbogen 2003
Knuth Hornbogen, "Vom Springen über die Dreiprozenthürde," *Hochparterre* 16, no. 9 (2003), pp. 26–27.

Hornung 2011
Annabelle Hornung, ed., *Do it yourself. Die Mitmach-Revolution,* exh. cat. Museum für Kommunikation Frankfurt, Mainz 2011.

Horowitz 2004
Daniel Horowitz, *The Anxieties of Affluence: Critiques of American Consumer Culture, 1939–1979,* Amherst and Boston 2004.

Huber 1973
Verena Huber, "Banken, Büros und Büromöbel," *Werk/Œuvre* 60, no. 8 (1973), pp. 921–27.

Huber 2009
Verena Huber, "'Basket' by Gian Franco Legler," *Raum und Wohnen* 35, no. 3 (2009), p. 27.

Huber et al. 2011
Jörg Huber, Burkhard Meltzer, Heike Munder, and Tido von Oppeln, *It's Not a Garden Table. Design und Kunst im erweiterten Feld,* Zurich 2011.

Huber/Rucki 1998
Dorothee Huber and Isabelle Rucki, eds., *Architekturlexikon der Schweiz,* Basel et al. 1998.

Hugentobler et al. 2008
Hans Kaspar Hugentobler, Claudia Mareis, Franziska Nyffenegger et al., *Designwissenschaft und Designforschung. Ein einführender Überblick,* Lucerne n.d. [2008].

Hummel 2004
Peter Hummel, "Glamouröser Klassiker wird neu aufgelegt," *Tages-Anzeiger,* January 3, 2004, p. 39.

Hünerwadel 2014
Jürg Hünerwadel, "Die gute Form. Der Schweizerische Werkbund und die Produktgestaltung nach dem Zweiten Weltkrieg," in *Gestaltung Werk Gesellschaft. 100 Jahre Schweizerischer Werkbund SWB,* ed. Thomas Gnägi et al. Zurich 2014, pp. 277–92.

Hüttinger 1977
Eduard Hüttinger, *Max Bill,* Zurich 1977.

Imwinkelried 2013
Rita Imwinkelried, "Jasper Morrison in London," *Interior Design Blog, Neue Zürcher Zeitung,* April 19, 2013, http://interiordesign.blog.nzz.ch/2013/04/19/jasper-morrison-in-london (accessed August 29, 2013).

Jackson 1998
Lesley Jackson, *The Sixties: Decade of Design Revolution,* Oxford 1998.

Jacob Müller 1988
Museum für Gestaltung Zürich, ed., *Jacob Müller. Handwerk, Technologie, Experiment,* Reihe Schweizer Design-Pioniere 5, Zurich 1988. Concept and realization: Jürg Brühlmann and Claude Lichtenstein.

Jakob 1987
Teo Jakob, *Schweizer Möbel Design. Kurt Thut, Hans Eichenberger, Trix und Robert Haussmann,* Bern 1987.

Janser/Rüegg 2001
Andres Janser and Arthur Rüegg, *Hans Richter. Die neue Wohnung,* Baden 2001.

Jeanneret 1912
Charles-Edouard Jeanneret, *Étude sur le mouvement d'art décoratif en Allemagne,* La Chaux-de-Fonds 1912.

Jencks 1978
Charles Jencks, *The Language of Postmodern Architecture,* London 1978.

Jollos 1918
Waldemar Jollos, "Vom Marionettentheater," *Das Werk* 5, no. 8 (1918), pp. 128–32.

Jollos 1955
Waldemar Jollos, *Arte tedesca fra le due guerre,* Milan 1955.

Joris 1990
Elisabeth Joris, "Die Schweizer Hausfrau. Genese eines Mythos," in *Schweiz im Wandel,* ed. Sebastian Brändli et al., Basel 1990, pp. 99–116.

Jørstian/Munk Nielsen 1994
Tina Jørstian and Poul Erik Munk Nielsen, eds., *Light Years Ahead: The Story of the PH Lamp,* Copenhagen 1994.

Jost 1981
Karl Jost, "Transportmittel: Leichtbauweise und Stromlinienform," in *Dreissiger Jahre Schweiz. Ein Jahrzehnt im Widerspruch,* exh. cat. Kunsthaus Zürich, Zurich 1981, pp. 157–65.

Just et al. 2007
Marcel Just, Christof Kübler, Matthias Noell, and Renzo Semadeni, eds., *Arosa. Die Moderne in den Bergen,* Zurich 2007.

Karbacher 2009
Ursula Karbacher, ed., *Sicherer, Schöner, Schneller. Textilien im Sport,* exh. cat. Textilmuseum St. Gallen, 2009.

Kaufmann 1950
Edgar Kaufmann, Jr., *Prize Designs for Modern Furniture [from the International Competition for Low-Cost Furniture Design],* New York 1950.

"Keramische Werkstätte" 1935.
"Keramische Werkstätte und Wohnhaus F. Haussmann, Uster (Kt. Zürich), erbaut von Max Ernst Haefeli, Arch. BSA, Zürich," *Werk* 22, no. 12 (1935) pp. 397–401.

Kettiger/Vetter 1957
Ernst Kettiger and Franz Vetter, *Möbel und Räume/Furniture and Interiors/Meubles et intérieurs,* Erlenbach-Zürich 1957.

Kienzle 1943
Hermann Kienzle, "'Heimatstil'? Über die Entwicklung und Beziehung von Funktion, Konstruktion, Form," *Werk* 30, no. 4 (1943), pp. 124–25.

Kieren 1990
Martin Kieren, *Hannes Meyer, Architekt. Dokumente zur Frühzeit,* Teufen 1990.

Kirchgraber et al. 1996
Jost Kirchgraber et al., *Raumkunst Lichtensteig. Ein Mann probt die Moderne. Traugott Stauss (1898–1952),* exh. cat. Historisches Museum, St. Gallen 1996.

Knöpfli 2010
Adrian Knöpfli, *Im Zeichen der Sonne. Licht und Schatten über der Alusuisse 1930–2010,* Baden 2010.

Koenig 2013
Gloria Koenig, *Charles & Ray Eames 1907–1978, 1912–1988. Vorreiter der Nachkriegsmoderne,* Cologne 2013.

Koenig/Gasser 2006
Thilo Koenig and Martin Gasser/Museum für Gestaltung Zürich, eds., *Hans Finsler und die Schweizer Fotokultur. Werk, Fotoklasse, moderne Gestaltung 1932–1960,* Zurich 2006.

Köhler 2002a
Bettina Köhler, "Between Industry and Craft—the Interior," in *Swiss Furniture and Interiors in the 20th Century,* ed. Arthur Rüegg, Basel et al. 2002, pp. 23–53.

Köhler 2002b
Bettina Köhler, "Bookcase circa 1850–1860. Guichard et Wanner," in *Swiss Furniture and Interiors in the 20th Century,* ed. Arthur Rüegg, Basel et al., pp. 60–61.

Kolb 1948
Otto Kolb, "Ideen zur Entwicklung von neuen Möbeln," *Bauen + Wohnen,* no. 2 (1948), pp. 54–55.

Koolhaas 1978
Rem Koolhaas, *Delirious New York: A Retroactive Manifesto for Manhattan,* New York 1978.

Kremp 2011
Matthias Kremp, "Apple-Spezialschrauben: Das iProblemchen," *Spiegel Online,* January 21, 2011, http://www.spiegel.de/netzwelt/gadgets/apple-spezialschrauben-das-iproblemchen-a-740801.html (accessed April 11, 2014).

Kries/Kugler 2013
Mateo Kries and Jolanthe Kugler, eds., *Lightopia,* vol. 2: *Masterpieces: 100 Lights,* trans. Brian Currid and Jeremy Gaines, Weil am Rhein 2013.

Kries/von Vegesack 2010
Mateo Kries and Alexander von Vegesack, *Rudolf Steiner: Alchemy of the Everyday,* exh. cat. Vitra Design Museum, Weil am Rhein 2010.

Kugler 2013
Jolanthe Kugler, "Bright Future?," in *Lightopia,* ed. Mateo Kries and Jolanthe Kugler, vol. 3: *Visions: Lighting Designs for Tomorrow,* Weil am Rhein 2013, pp. 7–16.

L'Eplattenier 1910
Charles L'Eplattenier, "Renouveau d'art," *L'Abeille. Supplément du National Suisse,* February 20, 1910, p. 1.

Lahuerta 1999
Juan José Lahuerta, *Decir ANTI es decir PRO. Escenas de la vanguardia en España,* Teruel 1999.

Landes 1983
David S. Landes, *Revolution in Time: Clocks and the Making of the Modern World,* Cambridge, Mass. 1983.

Landwehr 2001
Dominik Landwehr, "Computer 'Made in Switzerland' sind eine Seltenheit," in *Loading History. Computergeschichte(n) aus der Schweiz,* ed. Museum für Kommunikation Bern, Zurich 2001, pp. 42–57.

Lavisse (1902) 1906
Charles Emile Lavisse, *Comparative Studies of the Field Equipment of the Foot Solder of the French and Foreign Armies,* Washington, DC 1906. Originally published in French in 1902 as *Sac au dos. Études comparées de la tenue de campagne des fantassins, des armées française et étrangères.*

Le Corbusier (1912) 2008
Le Corbusier, *A Study of the Decorative Art Movement in Germany,* ed. Mateo Kries, trans. Alex Thomas Anderson, Weil am Rhein 2008. Originally published in French in 1912 as Charles-Edouard Jeanneret, *Étude sur le mouvement d'art décoratif en Allemagne.*

Le Corbusier (1930) 1991
Le Corbusier, *Precisions on the Present State of Architecture and City Planning,* trans. Edith Schreiber Aujame, Cambridge, Mass. 1991. Originally published in French in 1930 as *Précisions sur un état présent de l'architecture et de l'urbanisme.*

Le Corbusier 1931
Le Corbusier, "La guerre? Mieux vaut construire," *Plans,* no. 6 (1931), p. 66.

Le Corbusier 1933
Le Corbusier, "Das Haus aus Glas. Wohnhaus 'Clarté,'" special issue of *Die Kunst in der Schweiz,* no. 4–5 (1933).

Le Corbusier 1942
Le Corbusier, *Les Maisons "Murondins,"* Paris and Clermont-Ferrand 1942.

Le Corbusier 1960
Le Corbusier, "Tapisseries Muralnomad," *Zodiac,* no. 7 (1960), pp. 57–63.

Le Corbusier/Jeanneret 1930
Le Corbusier and Pierre Jeanneret, "Analyse des éléments fondamentaux du problème de la maison minimum," in *Die Wohnung für das Existenzminimum,* ed. Internationale Kongresse für Neues Bauen/Hochbauamt Frankfurt am Main Frankfurt am Main 1930, pp. 20–29.

Lepel 2009
Peter Lepel, *Über die Entstehung eines neuen Möbels. Marcel Breuer und der Wettbewerb der Alliance Aluminium Cie. Basel, 1933,* MAS thesis, ETH Zürich, 2009.

Lepel/Spies 2001
Peter Lepel and Oliver Spies, *Über Möbel. Ein Streifzug durch das Archiv der Embru-Werke, 1928–1943,* Rüti 2001.

Lepel/Spies 2011
Peter Lepel and Oliver Spies, *Über Reklame. Werbemittel der Embru-Werke,* Rüti 2011.

Leuschel 2009
Klaus Leuschel, ed., *Swissness. 43 helvetische Errungenschaften und 7 prägende Persönlichkeiten der Designgeschichte,* Sulgen and Zurich 2009.

Lichtenstein 1987
Claude Lichtenstein, "Das 'Patent Ochsner'. Die Entwicklung der Kehrichtabfuhr in technischen und ästhetischen Zusammenhängen," in *Unbekannt – Vertraut. "Anonymes" Design im Schweizer Gebrauchsgerät seit 1920,* ed. Museum für Gestaltung Zürich, Reihe Schweizer Design-Pioniere 4, Zurich 1987, pp. 49–65.

Lichtenstein 1988
Claude Lichtenstein, "Geschichte und Struktur der Werkgenossenschaft Wohnhilfe," in *Jacob Müller. Handwerk, Technologie, Experiment,* ed. Museum für Gestaltung Zürich, Reihe Schweizer Design-Pioniere 5, Zurich 1988, pp. 16–25.

Lichtenstein 1991
Claude Lichtenstein, "Typenmöbel," in *Wilhelm Kienzle,* ed. Museum für Gestaltung Zürich, Reihe Schweizer Design-Pioniere 6, Zurich 1991, pp. 55–69.

Lichtenstein 1995
Claude Lichtenstein, "Villen und Landhäuser 1911–1936," in *Otto Rudolf Salvisberg (1882–1940). Die andere Moderne,* ed. Martin Steinmann and Claude Lichtenstein, 2nd rev. ed., Zurich 1995, pp. 150–61.

Lichtenstein 1997
Claude Lichtenstein, "Der tiefe Ernst der guten Form," in *Made in Switzerland, 80 Jahre Förderung durch die Eidgenossenschaft,* ed. Bundesamt für Kultur, Bern and Zurich 1997, pp. 174–83.

Lichtenstein 1999
Claude Lichtenstein/Museum Bellerive, ed., *Schöne Möblierung. Möbelstücke aus 4 Sammlungen,* Zurich 1999.

Lichtenstein 2002
Claude Lichtenstein, "The End of Progress and the Rediscovery of History," in *Swiss Furniture and Interiors in the 20th Century,* ed. Arthur Rüegg, Basel et al. 2002, pp. 228–33.

Lichtenstein 2006
Claude Lichtenstein, "The Collection as Blueprint," in *Eidgenössische Förderpreise für Design/Bourses fédérales de design/Swiss Federal Design Grants 2006,* ed. Bundesamt für Kultur, Basel 2006, pp. 125–37.

Lichtenstein 2007
Claude Lichtenstein, *Playfully Rigid: Swiss Architecture, Graphic Design, Product Design, 1950–2006,* Baden 2007. Published in German as *Spielwitz und Klarheit. Schweizer Architektur, Grafik und Design 1950–2006.*

Lichtenstein 2014
Claude Lichtenstein, "The Rebirth of the Wristwatch from the Spirit of Graphic Design: On the Phenomenon of Swatch; Robert & Durrer," in *100 Years of Swiss Graphic Design,* ed. Christian Brändle/Museum für Gestaltung Zürich, Zurich 2014, pp. 206–7.

Lichtenstein et al. 1986
Claude Lichtenstein, Lotte Schilder Bär et al., "Konzept für die Design-Sammlung," internal paper, Museum für Gestaltung Zürich, June 9, 1986.

Lichtenstein/Engler 1992
Claude Lichtenstein and Franz Engler/Museum für Gestaltung Zürich, eds., *Stromlinienform,* exh. cat. Museum für Gestaltung Zürich et al., Baden 1992.

Lichtenstein/Engler 1993(?)
Claude Lichtenstein and Franz Engler/Museum für Gestaltung Zürich, eds., *Streamlined: A Metaphor for Progress; The Esthetics of Minimized Drag,* Baden and Zurich 1993(?). Based on the catalogue for the exhibition *Stromlinienform,* originally published in German in 1992.

Linder 1918
Hans Eduard Linder, "Basler Kunstgewerbe," *Das Werk* 5, no. 1 (1918), pp. 12–16.

Linke/Scharloth 2008
Angelika Linke and Joachim Scharloth, *Der Zürcher Sommer 1968. Zwischen Krawall, Utopie und Bürgersinn,* Zurich 2008.

Locher 1994a
Adalbert Locher, "CAD wird selbstverständlich," *Hochparterre* 7, no. 9 (1994), pp. 26–28.

Locher 1994b
Adalbert Locher, "Nose – Designer, Kaufmann, Ingenieur," *Hochparterre* 7, no. 12 (1994), pp. 16–21.

Locher 1995
Adalbert Locher, "Das Modell aus dem Computer," *Hochparterre* 8, no. 12 (1995), pp. 34–35.

Loderer 2002, Benedikt Loderer, "Seconda Chair, 1982. Mario Botta," in *Swiss Furniture and Interiors in the 20th Century,* ed. Arthur Rüegg, Basel et al. 2002, p. 250.

Lohse 1947
Richard Paul Lohse, "Surrealismus und Gebrauchsgrafik," *Chamäleon,* no. 9–10 (1947), pp. 60–61.

Lohse 1953
Richard Paul Lohse, *Neue Ausstellungsgestaltung/Nouvelles conceptions de l'exposition/New Design in Exhibitions,* Erlenbach-Zürich 1953.

Lohse Stiftung 1999
Richard Paul Lohse-Stiftung, ed., *Richard Paul Lohse. Konstruktive Gebrauchsgraphik,* Ostfildern-Ruit 1999.

Lucie-Smith 1969
Edward Lucie-Smith, *Movements in Art Since 1945,* London and New York 1969.

Lueg/Gantenbein 2001
Gabriele Lueg and Köbi Gantenbein, eds., *Swiss Made. Aktuelles Design aus der Schweiz,* Cologne 2001.

M. 1956
Alexandra M., "'Do it yourself – mach es selbst!' says America," in "Wohnen," special issue, *Zürcher Wochenblätter,* no. 7 (February 1956).

Máčel/van Geest 1980
Otakar Máčel and Jan van Geest, *Stühle aus Stahl,* Cologne 1980.

Mager 2008
Birgit Mager, "Customization," in *Design Dictionary: Perspectives on Design Terminology,* ed. Michael Erlhoff and Tim Marshall, Basel et al. 2008. p. 100.

Manz 2012
Ev Manz, "Der designte Schlauchschuh," *Tages-Anzeiger,* September 29, 2012, p. 32.

Marbach/Rüegg 1990
Ueli Marbach and Arthur Rüegg, *Werkbundsiedlung Neubühl 1928–1932,* Zurich 1990.

Mareis 2008
Claudia Mareis, "Einführung in die Designforschung," in Hans Kaspar Hugentobler et al., *Designwissenschaft und Designforschung. Ein einführender Überblick,* Lucerne n.d. [2008], pp. 19–22.

Mario Botta 1989
Mario Botta. Construire les objets. Œuvre design, 1982–89, exh. cat. Fondation Louis Moret, Martigny 1989.

Matter 1999
Hansuli Matter, "CAD, IT und andere Techniken im Designbüro," *Hochparterre* 12, no. 5 (1999), pp. 42–43.

Mauderli 2002
Laurence Mauderli, "Terre lustrée noire Bowl, 1931. Paul Ami Bonifas," in *Swiss Furniture and Interiors in the 20th Century,* ed. Arthur Rüegg, Basel et al. 2002, pp. 92–93.

Maurizio 1952
Julius Maurizio, *Der Siedlungsbau in der Schweiz,* Erlenbach-Zürich 1952.

Max Bill 1999
Max Bill. Typografie, Reklame, Buchgestaltung/Typography, Advertising, Book Design, with contributions by Gerd Fleischmann, Hans Rudolf Bosshard, and Christoph Bignens, Sulgen and Zurich 1999.

Mbiti et al. 1999
Kyeni Mbiti et al., *CAD Bericht 1999,* Hochschule für Gestaltung und Kunst Zürich, Zurich 1999.

McLuhan (1951) 2001
Marshall McLuhan, *The Mechanical Bride: Folklore of Industrial Man,* New York 1951, repr. Corte Madera, Calif. 2001.

McLuhan (1964) 1972
Marshall McLuhan, *Understanding Media: The Extensions of Man* (1964), 2nd ed., New York 1972.

Mebes 1908
Paul Mebes, *Um 1800. Architektur und Handwerk im letzten Jahrhundert ihrer traditionellen Entwicklung,* Munich 1908.

Medici-Mall 1998
Katharina Medici-Mall, *Im Durcheinandertal der Stile. Architektur und Kunst im Urteil von Peter Meyer (1894–1984),* Basel et al. 1998.

Mehlau-Wiebking et al. 1989
Friedrike Mehlau-Wiebking, Arthur Rüegg, and Ruggero Tropeano, *Schweizer Typenmöbel 1925–1935. Sigfried Giedion und die Wohnbedarf AG,* Zurich 1989.

Meier 2002a
Irene Meier, "Lifto Table Lamp, 1984. Benjamin Thut," in *Swiss Furniture and Interiors in the 20th Century,* ed. Arthur Rüegg, Basel et al. 2002, pp. 252–53.

Meier 2002b
Irene Meier, "Low-voltage Lighting System Metro, 1982/1984. Hannes Wettstein," in *Swiss Furniture and Interiors in the 20th Century,* ed. Arthur Rüegg, Basel et al. 2002, pp. 276–77.

Meier 2002c
Irene Meier, "Move-it Folding Table, 1996. Alfredo Häberli and Christophe Marchand," in *Swiss Furniture and Interiors in the 20th Century,* ed. Arthur Rüegg, Basel et al. 2002, pp. 292–93.

Mendini 1984
Alessandro Mendini, "Banal Design als Arbeitsmethode," in *Gestaltung zwischen "Good Design" und Kitsch,* Berlin 1984, pp. 27–39.

Menzi 2000
Renate Menzi, "Über die Schönheit des Turnschuhs," *Hochparterre* 13, no. 6–7 (2000), pp. 30–31.

Menzi 2007a
Renate Menzi, "Der bekleidete Stuhl," *Hochparterre* 20, no. 10 (2007), pp. 23–27.

Menzi 2007b
Renate Menzi, "Seiltanz und Glockenspiel," *Hochparterre* 20, no. 12 (2007), pp. 44–47.

Menzi 2010
Renate Menzi/Museum für Gestaltung Zürich, ed., *Make up. Design der Oberfläche/Designing Surfaces,* Ludwigsburg 2010.

Menzi 2012
Renate Menzi/Museum für Gestaltung Zürich, ed., *Freitag: Out of the Bag,* Zurich 2012.

Meyer (1926) 1965
Hannes Meyer, "Die neue Welt," *Das Werk* 13, no. 7 (1926), pp. 205–24. Translated in English as "The New World," in Claude Schnaidt, *Hannes Meyer. Bauten, Projekte und Schriften/Buildings, Projects, and Writings,* trans. D. Q. Stephenson, Teufen 1965, pp. 91–95.

Meyer (1926) 1994
Hannes Meyer, "Die neue Welt," *Das Werk* 13, no. 7 (1926), pp. 205–24. Translated in English as "The New World," in *The Weimar Republic Sourcebook,* ed. Anton Kaes et al. Berkeley and Los Angeles 1994, pp. 445–49.

Meyer (1938) 1965
Hannes Meyer, "Education of the Architect," lecture to San Carlos Academy, Mexico, September 30, 1938, in Claude Schnaidt, *Hannes Meyer: Bauten, Projekte und Schriften/Buildings, Projects, and Writings,* trans. D. Q. Stephenson, Teufen 1965, pp. 53–55.

Meyer 1928
Peter Meyer, *Moderne Schweizer Wohnhäuser,* Zurich 1928.

Meyer 1937
Peter Meyer, "Ornamentfragen. Ungegenständliche Kunst, Ornament und Architektur," *Das Werk* 24, no. 2 (1937), pp. 53–60.

Meyer 1943
Peter Meyer, "Heimatstil" (Werkbundfragen II), *Schweizerische Bauzeitung* 122, no. 7 (1943), p. 90.

Meyer 1944
Peter Meyer, *Das Ornament in der Kunstgeschichte,* Zurich 1944.

Meyer-Thoss 1996
Christiane Meyer-Thoss, *Meret Oppenheim – Buch der Ideen. Frühe Zeichnungen, Skizzen und Entwürfe für Mode, Schmuck und Design,* Bern 1996.

Minges 1996
Klaus Minges, "Staatsbildende Insekten," *Weltkunst,* November 11, 1996.

Moebius/Prinze 2012
Stephan Moebius and Sophia Prinze, eds., *Das Design der Gesellschaft. Zur Kultursoziologie des Designs,* Bielefeld 2012.

Molin 1904
Aloys de Molin, *Histoire documentaire de la manufacture de porcelaine de Nyon 1781–1813,* Lausanne 1904.

Molins 2003
Patricia Molins, ed., *Suiza constructiva,* exh. cat. Museo Nacional Centro de Arte Reina Sofía, Madrid 2003.

Monnerat 2003
Roger Monnerat, "Karl Zysset und sein Blitzhacker," *Work – Die Zeitung der Gewerkschaft Unia,* November 20, 2003.

Monnier/Vuille 1979
Pierre Monnier and Louis Vuille, *Les meubles d'Yverdon,* Yverdon 1979.

Müller 1949
Jacob Müller, *Neue Gebrauchsgeräte für Haus, Garten und Reise,* brochure, ca. 1949.

Müller 2001
Lars Müller, ed., *Freitag: Individual Recycled Freeway Bags,* Baden 2001.

Müller 2011
Esther Müller, "Das Geschenk des Mauritius Ehrlich," *Tachles. Das jüdische Wochenmagazin,* February 25, 2011, http://www.tachles.ch/news/print/das-geschenk-des-mauritius-ehrlich (accessed October 2, 2013).

Müller Horn 2012
Christine Müller Horn, *Bilder der Schweiz. Die Beiträge auf den Weltausstellungen 1851 bis 2010,* PhD diss., ETH Zürich, 2012.

Müller-Stewens 2013
Günter Müller-Stewens, "Disruptionen bei den Luxusgütern," *Neue Zürcher Zeitung,* June 4, 2013, http://www.nzz.ch/aktuell/wirtschaft/wirtschaftsnachrichten/disruptionen-bei-den-luxusguetern-1.18092264 (accessed June 8, 2014).

Munari 1944
Bruno Munari, "Ricerca della comodità in una poltrona scomada," *Domus,* no. 202 (October 1944), pp. 374–75. Available in English on the Internet as "Searching for Comfort in an Uncomfortable Chair," http://www.domusweb.it/en/from-the-archive/2012/03/31/searching-for-comfort-in-an-uncomfortable-chair.html.

Münch 1997
Andreas Münch, "Kunst oder Design? Der Bund und die Gestaltung," in *Made in Switzerland. Gestaltung – 80 Jahre Förderung durch die Eidgenossenschaft,* ed. Bundesamt für Kultur, Bern and Zurich 1997, pp. 88–107.

Muthesius (1904–05) 1979
Hermann Muthesius, *The English House,* ed. Dennis Sharp, trans. Janet Seligman, New York 1979. Originally published in German as three volumes in 1904–05 as *Das englische Haus. Entwicklung, Bedingungen, Anlage, Aufbau, Einrichtung und Innenraum.*

Muthesius 1908
Hermann Muthesius, *Wirtschaftsformen im Kunstgewerbe,* Volkswirtschaftlichen Gesellschaft 233, Berlin 1908.

Muthesius 1915
Hermann Muthesius, *Wie baue ich mein Haus,* Munich 1915.

Muthesius 1922
Hermann Muthesius, *Landhäuser von Hermann Muthesius. Ausgeführte Bauten mit Grundrissen, Gartenplänen und Erläuterungen,* 2nd expanded ed., Munich 1922.

N. N. 1991
N. N., "Wohnausstellung Neubühl," *Neue Zürcher Zeitung,* September 21, 1931.

Natter et al. 2010
Tobias G. Natter, Michael Fehr, and Bettina Habsburg-Lothringen, eds., *Das Schaudepot. Zwischen offenem Magazin und Inszenierung,* Bielefeld 2010.

Neuburg 1963
Hans Neuburg, "Einheit von Funktion, Konstruktion, Formgebung und Information/Unity of Function, Construction, Design and Information/Convergence de la fonction, de la construction, du dessin industrie et de l'information," *Neue Grafik/New Graphic Design/Graphisme actuel,* no. 15 (1963), pp. 2–38.

"Neue Möbel" 1956
"Neue Möbel," *Werk* 43, no. 9 (1956), p. 283.

Nichols 2001
Sarah C. Nichols, *Aluminum by Design,* exh. cat. Carnegie Museum of Art, Pittsburgh, New York 2001.

Nizon (1970) 1990
Paul Nizon, "Diskurs in der Enge" (1970), in *Diskurs in der Enge. Verweigerers Steckbrief. Schweizer Passagen,* Frankfurt am Main 1990, pp. 139–226.

Noseda 2997
Irma Noseda, "Von der guten Form zum unsichtbaren Design," in *Das gute Leben. Der Deutsche Werkbund nach 1945,* ed. Gerda Breuer, Tübingen 2007, pp. 176–85.

Oechslin 1988
Werner Oechslin, ed., *Le Corbusier & Pierre Jeanneret. Das Wettbewerbs-projekt für den Völkerbundspalast in Genf 1927,* Zurich 1988.

Oechslin/Hildebrand 2010
Werner Oechslin and Sonja Hildebrand, *Karl Moser. Architektur für eine neue Zeit, 1880 bis 1936,* Zurich 2010.

Oehler 2013
Felicitas Oehler, "Volkskunst," in *Historisches Lexikon der Schweiz,* article published September 26, 2013, http://www.hls-dhs-dss.ch/textes/d/D11192.php (accessed November 1, 2013).

Ott 2014
Stephan Ott, "Discussion on Swiss Design: Renate Menzi, Simon Husslein, Jörg Boner," *Form,* no. 253 (May/June 2014), pp. 40–47.

Ottilinger 2003
Eva B. Ottilinger, *Gebrüder Thonet. Möbel aus gebogenem Holz,* exh. cat. Museen des Mobiliendepots, Vienna 2003.

"Papiermöbel und Kinderspielsachen" 1970
"Papiermöbel und Kinderspielsachen aus Pappe," *Werk* 57, no. 5 (1970), pp. 320–24.

Paul Bonifas 1997
Paul Bonifas. Céramiste du purisme, exh. cat. Musée Ariana, Geneva 1997.

Perret/Cosandey 2013
Thomas Perret and Roland Cosandey, *Paillard – Bolex – Boolsky. La caméra de Paillard & Cie SA. Le cinéma de Jacques Boolsky,* Yverdon-les-Bains 2013.

Pivot 2002
Monique Pivot, *Maggi et la magie du bouillon Kub,* Paris 2002.

Pop Art Design 2012
Pop Art Design, exh. cat. Vitra Design Museum et al., Weil am Rhein 2012.

Porten 2008
Dirk Porten, "CAD/CAM/CIM/CNC," in *Design Dictionary: Perspectives on Design Terminology,* ed. Michael Erlhoff and Tim Marshall, Basel et al. 2008. p. 61.

Posener 1964
Julius Posener, ed., *Anfänge des Funktionalismus. Von Arts and Crafts zum Deutschen Werkbund,* Berlin et al. 1964.

Pradal 2006
Adriana Pradal, "Nose," in *Design Suisse,* ed. Meret Ernst and Christian Eggenberger, Zurich 2006, pp. 114–24.

Pradal 2008
Ariana Pradal, ed., *Frédéric Dedelley: Design Detective,* Baden 2008.

Probst 1968
Robert Probst, *The Office: A Facility Based on Change,* Elmhurst, Ill. 1968.

Rével 1982
Jean-François Rével, "Brillat-Savarin ou le style aimable," Introduction to Jean Anthelme Brillat-Savarin, *Physiologie du gout* (1825), Paris 1982, pp. 5–15.

Rey 1999
Patrick Rey, *Ramseier-Worblaufen. Schweizer Carrosseriebau der Spitzenklasse/Un carossier suisse haut de gamme, 1930–1960,* Geneva 1999.

Richter (1964) 1997
Hans Richter, *Dada: Art and Anti-Art,* trans. David Britt, London 1965; repr. London 1997. Originally published in German in 1964 as *Dada. Kunst und Anti-Kunst.*

Richtlinien Digitale Fabrik 2008
Verein Deutscher Ingenieure (VDI), ed., *Richtlinien Digitale Fabrik – Grundlagen,* VDI 4499, February 2008.

Rickert 1920
Heinrich Rickert, *Die Philosophie des Lebens. Darstellung und Kritik der philosophischen Modeströmungen unserer Zeit,* Tübingen 1920.

Robbins 1990
David Robbins, *The Independent Group: Postwar Britain ant the Aesthetics of Plenty,* Cambridge, Mass. and London 1990.

Rogers 1954
Ernesto N. Rogers, "Pretesti per una critica non formalistica," *Casabella-continuità,* no. 200 (February/March 1954), pp. 1–3.

Rotzler 1952
Willy Rotzler, "Neue Schweizer Pflanzenbehälter," *Werk* 39, no. 6 (1952), pp. 192–93.

Rüegg 1987
Arthur Rüegg, "Der Pavillon de L'Esprit Nouveau als Musée imaginaire," in *L'Esprit Nouveau. Le Corbusier und die Industrie,* ed. Stanislaus von Moos, exh. cat. Museum für Gestaltung Zürich et al., Berlin 1987, pp. 134–51.

Rüegg 1989
Arthur Rüegg, "Giedion in der Industrie: Programmierung und Entwurf," in *Sigfried Giedion. Der Entwurf einer modernen Tradition,* ed. Museum für Gestaltung Zürich, Zurich 1989, pp. 164–75.

Rüegg 1995a
Arthur Rüegg, "Hans Bellmann 1911–1990," in *Wege zur "Guten Form". Neun Beiträge zur Geschichte der Schweizer Produktgestaltung,* ed. Arthur Rüegg and Ruggero Tropeano, Basel et al. 1995, pp. 50–57.

Rüegg 1995b
Arthur Rüegg, "Möbel und Innenraum," in *Otto Rudolf Salvisberg (1882–1940). Die andere Moderne,* ed. Martin Steinmann and Claude Lichtenstein, 2nd rev. ed., Zurich 1995, pp. 162–69.

Rüegg 1996
Arthur Rüegg, *Ein Hauptwerk des Neuen Bauens in Zürich. Die Doldertalhäuser 1932–1936,* Zurich 1996.

Rüegg 1997
Arthur Rüegg, ed., *Das Atelierhaus Max Bill 1932/33,* Sulgen 1997.

Rüegg 1998
Arthur Rüegg, "Nemo propheta in patria," in *"Der Magus des Nordens". Alvar Aalto und die Schweiz,* ed. Teppo Jokinen and Bruno Maurer, Zurich 1998, pp. 119–33.

Rüegg 1999
Arthur Rüegg, "Couleurs, meubles et photographie: 'J'ai deux amours,'" in *La construction de l'immeuble Clarté,* ed. Luca Bellinelli, exh. cat. Museo d'Arte, Mendrisio 1999, pp. 50–71.

Rüegg 2002a
Arthur Rüegg, ed., *Swiss Furniture and Interiors in the 20th Century,* trans. Robin Benson et al., Basel et al. 2002. Published in German as *Schweizer Möbel und Interieurs im 20. Jahrhundert.*

Rüegg 2002b
Arthur Rüegg, "From Utopia to Concrete Fall," in *Swiss Furniture and Interiors in the 20th Century,* ed. Arthur Rüegg, Basel et al. 2002, pp. 96–98.

Rüegg 2002c
Arthur Rüegg, "H-Design for Mira-X, interior textiles, 1981. Trix and Robert Haussmann, Alfred Hablützel," in *Swiss Furniture and Interiors in the 20th Century,* ed. Arthur Rüegg, Basel et al. 2002, pp. 248–49.

Rüegg 2002d
Arthur Rüegg, "M 125, Furniture System, 1950/1956. Hans Gugelot," in *Swiss Furniture and Interiors in the 20th Century,* ed. Arthur Rüegg, Basel et al. 2002, pp. 164–65.

Rüegg 2002e
Arthur Rüegg, "One-point Chair (Bellmann Chair), 1951/52. Hans Bellmann," in *Swiss Furniture and Interiors in the 20th Century,* ed. Arthur Rüegg, Basel et al. 2002, pp. 170–71.

Rüegg 2002f
Arthur Rüegg, "Plan Cabinet, 1950. Benedikt Rohner," in *Swiss Furniture and Interiors in the 20th Century,* ed. Arthur Rüegg, Basel et al. 2002, pp. 196–97.

Rüegg 2002g
Arthur Rüegg, "Three-round Table, 1949/50. Max Bill," in *Swiss Furniture and Interiors in the 20th Century,* ed. Arthur Rüegg, Basel et al. 2002, pp. 162–63.

Rüegg 2003
Arthur Rüegg, "Die Möbel," in *Flora Steiger-Crawford, 1899–1991,* ed. Jutta Glanzmann, Zurich 2003, pp. 140–51.

Rüegg 2004
Arthur Rüegg, "Concrete Construction? Three Examples," in *Max Bill: Arquitecto/Architect,* ed. Karin Gimmi, Collection 2G, no. 29–30, Barcelona 2004.

Rüegg 2006
Arthur Rüegg, ed., *1931–2006. 75 Jahre Wohnbedarf,* Zurich 2006.

Rüegg 2008
Arthur Rüegg, "Die Erziehung zum 'bewussten menschen': Max Bill als Lehrer," in *Max Bill: Aspekte seines Werks,* exh. cat. Kunstmuseum und Gewerbemuseum Winterthur, Sulgen 2008, pp. 85–95.

Rüegg 2009a
Arthur Rüegg, "Kulturelle Kontinuität und Erfindung. Festes und Bewegliches in den Innenräumen Lux Guyers," in *Lux Guyer 1894–1955. Architektin,* ed. Sylvia Claus, Dorothee Huber, and Beate Reaper, Zurich 2009, pp. 106–19.

Rüegg 2009b
Arthur Rüegg, "Willy Guhl – die Schleife. Der 'Strandsitz' in Eternit, 1954," *Faszination Faserzement, Design+Design,* no. 2 (2009), pp. 32–39.

Rüegg 2010
Arthur Rüegg, "Notizen zur Glaslampe," *Wilhelm Wagenfeld, Design+Design,* no. 3 (2010), pp. 16–21.

Rüegg 2011
Arthur Rüegg, "Bedarf fürs Wohnen," in *Moderne Tradition – Der Wohnbedarf Zürich feiert seinen 80. Geburtstag,* supplement to *Hochparterre* 24, no. 10 (2011), pp. 4–7.

Rüegg 2012a
Arthur Rüegg, *Le Corbusier: Furniture and Interiors, 1905–1965,* Zurich 2012.

Rüegg 2012b
Arthur Rüegg, "Le Corbusier e la cultura del mobile in Italia," in *L'Italia di Le Corbusier,* ed. Marida Talamona, Rome 2012.

Rüegg 2013
Arthur Rüegg, "Lernen von Le Corbusier," in *Alfred Altherr junior. Protagonist der Schweizer Wohnkultur,* ed. Design+Design, Sulgen 2013, pp. 20–26.

Rüegg/Felder 2007
Arthur Rüegg and Lukas Felder, eds., *40 Europäische Wohnikonen neu gesehen,* Zurich 2007.

Rüegg/Mumenthaler-Grisard 2000
Arthur Rüegg and Stephanie Mumenthaler-Grisard, eds., *3m-Möbel von Ernst Mumenthaler und Otto Meier. Ein Basler Schrankprogramm, 1927–2000,* Basel 2000.

Rüegg/Steinmann 1983
Arthur Rüegg and Martin Steinmann, "Einrichtungen," in "Einrichtungen/Aménagements," *Archithese* 13, no. 1 (1983), pp. 34–53.

Rüegg/Tropeano 1982
Arthur Rüegg and Ruggero Tropeano, *Hermann Siegrist. Siedlung Leimenegg,* Zurich 1982.

Rüegg/Tropeano 1992
Arthur Rüegg and Ruggero Tropeano, "L'arredo svizzero 1925–1950," *Domus* 737, no. 4 (April 1992), pp. 238–53.

Rüegg/Wieser 1991
Arthur Rüegg and Ulrich P. Wieser, eds., *Hans Bellmann. Architekt und Industrie-Entwerfer, 1911–1990,* Basel 1991.

Ruppert 1998
Wolfgang Ruppert, ed., *Um 1968. Die Repräsentation der Dinge,* Marburg 1998.

Sachs 2007
Angeli Sachs/Museum für Gestaltung Zürich, ed., *Nature Design: From Inspiration to Innovation,* Baden 2007.

Scharf 2007
Armin Scharf, "Der Lotse hinter dem Ohr," *Hochparterre* 20, no. 12 (2007), pp. 52–53.

Scheerbart 1914
Paul Scheerbart, *Glass Architecture,* trans. James Palmes, in *Glass Architecture by Paul Scheerbart and Alpine Architecture by Bruno Taut,* ed. Dennis Sharp, New York and Washington 1972.

Schilder Bär 1987
Lotte Schilder Bär, "'Elna' Nähmaschine," in *Unbekannt – Vertraut. "Anonymes" Design im Schweizer Gebrauchsgerät seit 1920,* ed. Museum für Gestaltung Zürich, Reihe Schweizer Design-Pioniere 4, Zurich 1987, pp. 66–73.

Schilder Bär 1988
Lotte Schilder Bär, "Die Wohnhilfe-Moderne," in *Jacob Müller. Handwerk, Technologie, Experiment,* ed. Museum für Gestaltung Zürich, Reihe Schweizer Design-Pioniere 5, Zurich 1988, pp. 26–34.

Schilder Bär 1995
Lotte Schilder Bär, "Mechanische Objekte – Blickpunkte der Schweizer Präzisionsindustrie," in *Wege zur "Guten Form". Neun Beiträge zur Geschichte der Schweizer Produkt-gestaltung,* ed. Arthur Rüegg and Ruggero Tropeano, Basel et al. 1995, pp. 26–31.

Schilder Bär 1997
Lotte Schilder Bär, "Design Boom. Gesichtspunkte zum Schweizer Design," in "Design," *Kunst und Architektur in der Schweiz* 48, no. 3 (1997), pp. 51–59.

Schilder Bär 2002a
Lotte Schilder Bär, "Dietiker Chair, Model 3300, 1970. Bruno Rey," in *Swiss Furniture and Interiors in the 20th Century,* ed. Arthur Rüegg, Basel et al. 2002, pp. 238–39.

Schilder Bär 2002b
Lotte Schilder Bär, "From War-Time *Réduit* to Modern Living," in *Swiss Furniture and Interiors in the 20th Century,* ed. Arthur Rüegg, Basel et al. 2002, pp. 135–57.

Schilder Bär 2002c
Lotte Schilder Bär, "Illusionist Didactic Pieces—Manierismo critico, 1977–1986. Trix and Robert Haussmann," in *Swiss Furniture and Interiors in the 20th Century,* ed. Arthur Rüegg, Basel et al. 2002, pp. 246–47.

Schilder Bär 2002d
Lotte Schilder Bär, "Interior for the Exhibition 'Kunst und Wohnkultur' (Art and Interior Design), 1951. Hans Bellmann," in *Swiss Furniture and Interiors in the 20th Century,* ed. Arthur Rüegg, Basel et al. 2002, pp. 314–15.

Schilder Bär 2002e
Lotte Schilder Bär, "Squadra or Wing-door Cupboard, 1964. Klaus Vogt," in *Swiss Furniture and Interiors in the 20th Century,* ed. Arthur Rüegg, Basel et al. 2002, pp. 236–37.

Schilder Bär/Wild (2001) 2003
Lotte Schilder Bär and Norbert Wild, *Designland Switzerland: The Culture of Consumer Durables in the Twentieth Century,* trans. Eileen Walliser-Schwarzbart, Zurich 2003. Originally published in German in 2001 as *Designland Schweiz. Gebrauchs-güterkultur im 20. Jahrhundert.*

Schiller 2012
Janine Schiller, *Qualität, Einfachheit, Gediegenheit. Die Raumkunst-Ausstellungen in Zürich 1908–1912 zwischen Reform und Aufbruch in die Moderne,* master's thesis, ETH Zürich, Institute for the History and Theory of Architecture, 2012.

Schindler 2006a
Anna Schindler, "Atelier Oï," in *Design Suisse,* ed. Meret Ernst and Christian Eggenberger, Zurich 2006, pp. 174–84.

Schindler 2006b
Anna Schindler, "Freitag," in *Design Suisse,* ed. Meret Ernst and Christian Eggenberger, Zurich 2006, pp. 222–32.

Schivelbusch 1977
Wolfgang Schivelbusch, *Geschichte der Eisenbahnreise. Zur Industriali-sierung von Raum und Zeit im 19. Jahrhundert,* Munich and Vienna 1977.

Schmed/Rüegg 2012
Silvio Schmed and Arthur Rüegg, eds., *Evangelisch-Reformiertes Kirchenzentrum Altstetten,* Zurich 2012.

Schmidt 1927
Hans Schmidt, "Die Mietwohnungen der Schweizergruppe an der Ausstellung 'Die Wohnung' in Stuttgart," *Das Werk* 14, no. 9 (1927), pp. 272–78.

Schmidt 1932
Georg Schmidt, "Gebrauchsgerät," *Moderne Bauformen* 31, no. 9 (1932), pp. 489–99.

Schmidt (1965) 1993
Hans Schmidt, *Beiträge zur Architektur 1924–1964,* Berlin 1965; repr. Zurich 1993.

Scholz/Kaufmann 2003
Roland W. Scholz and Dieter Kaufmann, eds., *Zukunft der Schweizer Textilindustrie. Erkenntnisse einer gesamtschweizerischen Analyse aufbauend auf den Ergebnissen der ETH-UNS Fallstudie 2002 "Appenzell Ausserrhoden – Umwelt – Wirtschaft Region,"* Zurich 2003.

Scholz-Hänsel 2005
Michael Scholz-Hänsel, "'Avida Dollars' (Dollarjäger)? Breton's Parteidisziplin und Dalís kapitaler Aufstand," in *Salvador Dalí. Facetten eines Jahrhundertkünstlers,* ed. Lisa Puyplat, Adrian La Salvia, and Herbert Heinzelmann, Würzburg 2005, pp. 203–26.

Schulz 2000
Matthias Schulz, "Kult und Kamikaze," *Der Spiegel* 54, no. 28 (2000), pp. 162–63.

Schulze (2000) 2005
Gerhard Schulze, *Die Erlebnisgesellschaft: Kultursoziologie der Gegenwart,* Frankfurt am Main 2000; repr. 2005.

Schumacher 2012
Anne-Claire Schumacher, *Die Porzellanmanufaktur Langenthal, zwischen Industriedesign und Sonntagsgeschirr,* exh. cat. Musée Ariana, Geneva 2012.

Schwager 2004
Christian Schwager, *Falsche Chalets,* with a text by Gerold Kunz, Zurich et al. 2004.

Schwartz-Clauss/von Vegesack 2002
Mathias Schwartz-Clauss and Alexander von Vegesack, *Living in Motion,* exh. cat. Vitra Design Museum, Weil am Rhein 2002.

Schweizerische Carrosseriegeschichte 1994
Schweizerische Carrosseriegeschichte von 1914–1994: kreativ und sicher. 75 Jahre VSCI, Zofingen 1994.

Selden 1998
Brigitte Selden, "Toggenburger Avantgarde. Zum 100. Geburtstag des Gestalters Traugott Stauss," *Neue Zürcher Zeitung,* February 7–8, 1998.

Sembach 1971
Sembach, Klaus-Jürgen, "Das Jahr 1851 – Fixpunkt eines Wandels," in *Die verborgene Vernunft. Funktionale Gestaltung im 19. Jahrhundert,* Wegleitung 284, Kunstgewerbe-museum Zürich with Die Neue Sammlung, Munich, Zurich 1971, pp. 25–29.

Simonis 2001
Annette Simonis, *Gestalttheorie von Goethe bis Benjamin. Diskurs-geschichte einer deutschen Denkfigur,* Cologne et al. 2001.

Sonderegger 1995
Christina Sonderegger, "Norm und System: Hans Hilfiker und die Schweizer Einbauküche," in *Wege zur "Guten Form". Neun Beiträge zur Geschichte der Schweizer Produkt-gestaltung,* ed. Arthur Rüegg and Ruggero Tropeano, Basel et al. 1995, pp. 64–68.

Sonderegger 1997
Christina Sonderegger, "Aluminiumzeit," *Kunst und Architektur in der Schweiz* 48, no. 3 (1997), pp. 20–29.

Sonderegger 2002
Christina Sonderegger, "Model 600 Floor Lamp, 1950/1951. Rosmarie and Rico Baltensweiler," in *Swiss Furniture and Interiors in the 20th Century,* ed. Arthur Rüegg, Basel et al. 2002, pp. 168–69.

Sonderegger 2006
Christina Sonderegger, "Between Progress and Idling: The Standard Kitchen; Notes on the Development of Swiss Kitchen Norms," in *The Kitchen: Life World, Usage, Perspectives,* ed. Klaus Spechtenhauser, Basel et al. 2006, pp. 95–111.

Sonna 2008
Birgit Sonna, "Blecherne Provokation," *Art – Das Kunstmagazin,* no. 8 (2008), pp. 74–76.

Spechtenhauser 2006
Klaus Spechtenhauser, ed., *The Kitchen: Life World, Usage, Perspectives,* trans. Bill Martin with Laura Bruce, Basel et al. 2006. Published in German as *Die Küche. Lebenswelt, Nutzung, Perspektiven.*

Staber 1970
Margit Staber, *Sophie Taeuber Arp,* Geneva 1970.

Stalder 2013
Helmut Stalder, "Mission in Europas Mitte," in *Damit die Züge fahren. Mensch und Technik am Gotthard-Basistunnel,* Zurich 2013, pp. 121–31.

Steinmann/Lichtenstein 1995
Martin Steinmann and Claude Lichtenstein, eds., *Otto Rudolf Salvisberg (1882–1940). Die andere Moderne,* 2nd rev. ed., Zurich 1995.

Strauss 2008
Dorothea Strauss, ed., *Andreas Christen. Zwischen Malerei und Objekt,* exh. cat. Haus Konstruktiv, Heidelberg 2008.

Streiff 1943
Egidius Streiff, Forward to *Unsere Wohnung. Wanderausstellung des SWB,* ed. Kunstgewerbemuseum Zürich, Wegleitung 156, Zurich 1943, pp. 3–4.

Streitz 2006
Matthias Streitz, "Logitech-Gründer Borel im Interview: 'China ist brutal – aber es hat uns gerettet,'" *Spiegel Online,* May 24, 2006, http://www.spiegel.de/wirtschaft/logitech-gruender-borel-im-interview-china-ist-brutal-aber-es-hat-uns-gerettet-a-417607.html (accessed May 30, 2014).

Studio Marconi 1981
Studio Marconi, ed., *Trix & Robert Haussmann. Manierismo Critico. Progetti, Oggetti, Superfici,* exh. cat. Centro Culturale Studio Marconi, Documenti 5, Milan 1981.

Sturm 2013
Dominic Sturm, "Design im globalen Wettbewerb," *Passagen* 61, no. 2 (2013), pp. 12–13.

Susi und Ueli Berger 1999
Susi und Ueli Berger im Schaulager, exh. cat. Museum für Gestaltung Zürich, Design Collection, Zurich 1999.

Suter 1995
Ursula Suter, "Mumenthaler und Meiers '3m-Möbel': das erste Schweizer Schrankprogramm mit Typenmöbelcharakter," in *Wege zur "Guten Form". Neun Beiträge zur Geschichte der Schweizer Produkt-gestaltung,* ed. Arthur Rüegg and Ruggero Tropeano, Basel et al. 1995, pp. 58–63.

Swatch 2011
Phillips de Pury & Company, ed., *Swatch: The Blum Collection,* Hong Kong 2011.

Swatch Watch 2000
Swatch AG, ed., *Swatch Watch: A Definitive Guide, 1983–2000,* Biel 2000.

Teissonière 2012
Didier Teissonnière, *La lampe Gras,* Paris 2012.

Therma 1925–27
Therma – Fabrik für elektrische Heizung AG, Schwanden (Schweiz), Therma AG company catalogue, 1925–27.

Therma 1957
50 Jahre Therma Schwanden, 1907–1957, Festschrift, Schwanden 1957.

Thomson 2006
Helen Bieri Thomson, ed., *Une expérience Art Nouveau. Le Style sapin à La Chaux-de-Fonds,* Paris 2006.

Thut 1958
Kurt Thut, "Verschraubte Möbel," *Bauen+Wohnen* 12, no. 11 (1958), pp. 411–14.

Tobler 2001
Beatrice Tobler/Museum für Kommunikation Bern, ed., *Loading History. Computergeschichte(n) aus der Schweiz/Chronique(s) de l'informatique en Suisse,* vol. 1, Zurich 2001.

Tobler 2010
Beatrice Tobler, "Computers for All," in *Global Design: International Perspectives and Individual Concepts,* ed. Angeli Sachs/Museum für Gestaltung Zürich, Baden 2010, pp. 102–9.

Touchaleaume/Moreau 2010
Eric Touchaleaume and Gerald Moreau, *Le Corbusier, Pierre Jeanneret: L'aventure indienne / The Indian Adventure; Design, Art, Architecture,* Montreuil and Paris 2010.

Tropeano 2002a
Ruggero Tropeano, "Elektron Chair, 1927. Max Ernst Haefeli," in *Swiss Furniture and Interiors in the 20th Century,* ed. Arthur Rüegg, Basel et al. 2002, pp. 118–19.

Tropeano 2002b
Ruggero Tropeano, "Moser Latticed Garden Chair, 1931. Werner Max Moser," in *Swiss Furniture and Interiors in the 20th Century,* ed. Arthur Rüegg, Basel et al., 2002, pp. 124–25.

Troquet 2009
Daniel Troquet/EM, "Musikdosen," in *Historisches Lexikon der Schweiz,* article published June 19, 2009, http://www.hls-dhs-dss.ch/textes/d/D13978.php (accessed April 11, 2014).

Troy 1991
Nancy Troy, *Modernism and the Decorative Arts in France: Art Nouveau to Le Corbusier,* New Haven and London 1991.

Trueb 1999
Lucien F. Trueb, *Die Zeit der Uhren. Historie, Modelle, Hersteller,* Ulm 1999.

Trueb 2013
Lucien F. Trueb, "Die Quarzrevolution – Von der Mechanik zur Elektronik und zurück," in *Ingenieure bauen die Schweiz – Technikgeschichte aus erster Hand,* ed. Franz Betschon et al., Zurich 2013, pp. 354–74.

Unbekannt – Vertraut 1987
Museum für Gestaltung Zürich, ed., *Unbekannt – Vertraut. "Anonymes" Design im Schweizer Gebrauchsgerät seit 1920,* Reihe Schweizer Design-Pioniere 4, Zurich 1987. Realization: Christoph Bignens, Claude Lichtenstein, and Lotte Schilder Bär.

Ursprung 2011
Philip Ursprung, "Containers: The Backbone of Globalization," in *Global Design: International Perspectives and Individual Concepts,* ed. Angeli Sachs/Museum für Gestaltung Zürich, Baden 2011, pp. 122–33.

Vacchini 1983
Livio Vacchini, "Möbel," in "Einrichtungen/Aménagements," *Archithese* 13, no. 1 (1983), p. 46.

van de Velde (1914) 1975
Hermann Muthesius/Henry van de Velde, "Muthesius/Van de Velde: Werkbund Theses and Antithese" (1914), in *Programs and Manifestoes on 20th Century Architecture,* ed. Ulrich Conrads, Cambridge, Mass. 1975, pp. 28–31.

Vellay 1985
Marc Vellay, "Agli estremi del mattone Nevada," in "Microstorie di architettura," *Rassegna* 7, no. 24 (1985), pp. 6–17.

Venturi 1966
Robert Venturi, *Complexity and Contradiction in Architecture,* New York 1966.

Vinçon 1992
Hartmut Vinçon, *Frank Wedekinds Maggi-Zeit. Reklamen, Reiseberichte, Briefe,* Darmstadt 1992.

Virilio (1975) 1994
Paul Virilio, *Bunker Archeology,* New York 1994. First published in French in 1975 as *Bunker archéologie, étude sur l'espace militaire européen de la Seconde Guerre mondiale.*

von Ascheraden 2009
Alexandra von Ascheraden, "Blowing Metal," *ETH Life Online,* November 11, 2009, http://www.ethlife.ethz.ch/archive_articles/091106_blechblaeser_ava/index_EN, p. 2 (accessed April 18, 2014).

von Büren 2006
Charles von Büren, ed., *Kurt Naef. Der Spielzeugmacher/The Toymaker,* Basel et al. 2006.

Von der Mühll 1973
Henri-Robert Von der Mühll, *De l'architecture,* Lausanne and Paris 1973.

von Matt 2010
Hans-Kaspar von Matt, *Bologna Report Fachhochschulen 2010. Stand der Umsetzung der Bologna-Reform an Fachhochschulen,* study, drafted on behalf of the Rectors' Conference of the Swiss Universities of Applied Sciences (KFH), Lucerne 2010.

von Moos 1969
Stanislaus von Moos, "Anmerkungen zum Bauen in der Schweiz (2): Zum Stand der Dinge," *FSAI Bulletin,* no. 2 (1969), p. 4.

von Moos 1980
Stanislaus von Moos, "Wohnbedarf und Lebensform. Bruchstücke zur Schweizer Design-Geschichte der dreissiger Jahre," *Archithese* 10, no. 2 (1980), pp. 16–26.

von Moos (1982) 2013
Stanislaus von Moos, "The Second Discovery of America: Notes on the Prehistory of Mechanization Takes Command," afterword to Sigfried Giedion, *Mechanization Takes Command: A Contribution to Anonymous History,* Minneapolis 2013, pp. 725–58. Originally published in German in 1982 as "Die zweite Entdeckung Amerikas."

von Moos 1992
Stanislaus von Moos, *Industrieästhetik,* Ars Helvetica 11, Die visuelle Kultur der Schweiz, Disentis 1992.

von Moos 1995
Stanislaus von Moos, "Schönheit als Funktion. Anmerkung zu Max Bill," in *Wege zur "Guten Form". Neun Beiträge zur Geschichte der Schweizer Produktgestaltung,* ed. Arthur Rüegg and Ruggero Tropeano, Basel et al. 1995, pp. 69–72.

von Moos (2001) 2005
Stanislaus von Moos, "Villa Langmatt and the Dream of Country Life," in Eva-Maria Preiswerk-Lösl, *A Home for the Impressionists: The Langmatt Museum; The Sidney and Jenny Brown Foundation, Baden, Switzerland,* trans. Steven Lindberg, Ostfildern-Ruit 2005. Originally published in German in 2001 as "Die 'Langmatt' und der Traum vom Landleben."

von Moos 2002
Stanislaus von Moos, "Politische Architektur in der Schweiz. Die Macht des Ephemeren," in *Expo-Syndrom? Materialien zur Landesausstellung 1883–2002,* ed. Stanislaus von Moos and Georg Kohler, Zurich 2002, pp. 115–34.

von Moos 2004
Stanislaus von Moos, "Der 'nackte Mensch' im Stadtlabor. Pipilotti Rist, die Expo.02 und das Kulturganze," in *Nicht Disneyland. Und andere Aufsätze über Modernität und Nostalgie,* Zurich 2004, pp. 155–70.

von Moos 2006
Stanislaus von Moos, "Work Aesthetic and Mass Consumption: The Art of Max Huber," in *Max Huber,* ed. Stanislaus von Moos, Mara Campana, and Giampiero Bosoni, London and New York 2006, pp. 8–32.

von Moos 2007
Stanislaus von Moos, "Rules of Fasting and Desire Derailed: Notes on Architecture and Gastronomy," in *The Architect, the Cook, and Good Taste,* ed. Petra Hagen Hodgson and Rolf Toyka, Basel et al. 2007, pp. 60–71.

von Moos 2010
Stanislaus von Moos, "Karl Moser. Die Kunst. Das Haus," *Neue Zürcher Zeitung,* April 17, 2010.

Wagner 2006
Monika Wagner, "Vom Ende der materialgerechten Form. Kunst im Plastikzeitalter," in *Stoffe. Zur Geschichte der Materialität in Künsten und Wissenschaften,* ed. Barbara Naumann, Thomas Strässle, and Caroline Torra-Mattenklott, Zurich 2006, pp. 229–46.

Weckerle et al. 2008
Christoph Weckerle et al., eds., *Kreativwirtschaft Schweiz. Daten, Modelle, Szene,* Basel 2008.

Weinberg-Staber 1983
Margrit Weinberg-Staber, "Hocker (Sitze)," in *Design – Formgebung für jedermann. Typen und Prototypen,* ed. Margrit Weinberg-Staber/Kunstgewerbemuseum Zürich, Wegleitung 345, Zurich 1983, pp. 152–59.

Weisberg 2008
David E. Weisberg, *The Engineering Design Revolution: The People, Companies and Computer Systems That Changed Forever the Practice of Engineering,* 2008, at http://www.cadhistory.net (accessed May 30, 2014).

Werner Zemp 2010
Werner Zemp. Design, Objets, Reliefbilder, exh. cat. Technopark Zürich, Winterthur 2010.

Westheim 1925
Paul Westheim, "Schweizer Künstler im Ausland. Otto Rudolf Salvisberg-Berlin," *Das Werk* 12, no. 1 (1925), pp. 6–15.

When Attitudes Become Form 1969
Harald Szeemann, ed., *Live in Your Head: When Attitudes Become Form. Works, Concepts, Process, Situations, Information,* exh. cat. Kunsthalle Bern et. al., Bern 1969.

Whiteley 1987
Nigel Whiteley, "Toward a Throw-Away Culture: Consumerism, 'Style Obsolescence' and Cultural Theory in the 1950s and 1960s," *The Oxford Art Journal* 10, no. 2 (1987), pp. 3–27.

Wichmann 1987
Hans Wichmann, *System-Design. Bahnbrecher: Hans Gugelot 1920–65,* 2nd expanded ed., Basel and Boston 1987.

Wichmann 1989
Hans Wichmann, *System-Design. Fritz Haller: Bauten, Möbel, Forschung,* Basel et al. 1989.

Widmer 1999
Ruedi Widmer, "'Anna,' der Holztisch auf Verlangen," *Hochparterre* 12, no. 10 (1999), pp. 26–27.

Wild 2002a
Norbert Wild, "'New Simplicity' and Design in the Age of Microelectronics," in *Swiss Furniture and Interiors in the 20th Century,* ed. Arthur Rüegg, Basel et al. 2002, pp. 255–75.

Wild 2002b
Norbert Wild, "Shoe Tidy, 1984. Hanspeter Weidmann," in *Swiss Furniture and Interiors in the 20th Century,* ed. Arthur Rüegg, Basel et al. 2002, pp. 278–79.

Wilhelm Kienzle 1991
Museum für Gestaltung Zürich, ed., *Wilhelm Kienzle,* Reihe Schweizer Design-Pioniere 6, Zurich 1991. Concept, text, and realization: Claude Lichtenstein.

Willy Guhl 1985
Museum für Gestaltung Zürich, ed., *Willy Guhl. Gestalter und Lehrer,* Reihe Schweizer Design-Pioniere 2, Zurich 1985. Concept: Peter Eberhard; realization: Martin Steiner.

Wirth 1998
Hanspeter Wirth, "Auf der N2 nach Milano," *Hochparterre* 11, no. 4 (1998), p. 40.

Wiseman 2013
Raymond Wiseman, "Leuchtende Halbleiter," *NZZ am Sonntag,* February 10, 2013, p. 66.

Woba 1930
Woba, exh. cat. Schweizerische Wohnungsausstellung Basel, 1930.

Wohnen heute 1970a. *Wohnen heute 6,* product catalogue of the Swiss Werkbund, 1967–68, Niederteufen 1970.

Wohnen heute 1970b. *Wohnen heute 7,* product catalogue of the Swiss Werkbund, 1969/70, Niederteufen 1970, pp. 54–55.

Wyss 1997
Beat Wyss, *Die Welt als T-Shirt. Zur Ästhetik und Geschichte der Medien,* Cologne 1997.

Young Swiss Design Kaleidoscope 2013
Design Preis Schweiz, ed., *Young Swiss Design Kaleidoscope,* exh. cat. Beijing Design Week et al., Langenthal 2013.

Zeischegg/Auböck 1995
Walter Zeischegg and Carl Auböck, *Hand und Griff. Ausstellung Wien 1951,* Cologne 1995.

Zoelly 1992
Pierre Zoelly, *My Old Chairs,* Basel et al. 1992.

Zorn (1977) 1982
Fritz Zorn, *Mars,* trans. Robert and Rita Kimber, London 1982. Originally published in German in 1977.

Zwicky 2005
Stefan Zwicky/NR Neue Räume AG, *Schweizer Möbel Lexikon,* Zurich 2005.

Italics = Figure numbers

A. Sutter, Münchwilen 243
Aalto, Aino 157
Aalto, Alvar 109, 157, 207/*486, 732*
A C E, Lausanne 346/*831–833*
Adam, Ken *827*
Adler, Florian F. 191, 302
Advico Y & R, Zurich 241
Aebersold, Alfred (Alf) *503, 606*
Aebi & Co., Burgdorf *589, 590*
Aebi, Aurel 350/*825*
Aermo-Möbel GmbH, Zurich 163/*352, 353, 375*
AG Möbelfabrik Horgen-Glarus (ag möbelfabrik horgenglarus) 54, 92, 207, 327/*106, 135, 199, 306, 333, 410, 488, 490, 493, 786–790*
Agil, Stans *785*
Albert Stoll, Koblenz *556*
Alberti, Leon Battista 11
Alessi S.p.A., Omegna (IT) *722, 802*
Alessio, Oscar *827*
Alias S.r.l., Grumello del Monte, Bergamo (IT) 287/*684, 686*
Alliance Aluminium Compagnie (AAC) 108, 131/*299*
Altherr, Alfred 44, 47, 49, 104/*49, 83, 84, 99*
Altherr, Alfred, jun. 136, 195, 197/*313, 314, 408, 447, 463*
Altorfer, Huldreich 163
Aluminium-Industrie AG (AIAG), Neuhausen 125, 131/*278*
Aluminium-Schweisswerk AG, Schlieren *226, 227*
Amba, Basel 121
Ambühler, Ernst 186/*426*
Apothéloz, Christophe *601*
Apple 151
Arad, Ron 129
Arbeitsgemeinschaft Schweizer Werkbund *244*
Architektengemeinschaft Kantonsspital Zürich (AKZ) *322*
Arp, Hans *92*
Artaria, Paul 90, 163/*384*
Ascom Autophon AG, Bern *804*
Asplund, Gunnar 301
Atelier Eidenbenz, Basel *460*
Atelier Hablützel und Jaquet, Bern *532*
Atelier Halpern, Zurich *568, 569*
Atelier Oï, La Neuveville 335, 350, 356/*808, 823, 825, 843–845*
Atelier Pfister, Suhr 115/*257, 830*
Atelier Roger Pfund, Carouge *710*
Atelier S.r.l., Mariano Comense, Como (IT) *724*
Aubry, Bastien 355
Auchli, Peter *688*
BAG Turgi (BAG Bronzewarenfabrik, Turgi) 106, 119/*202, 203, 221, 260, 261, 268*
Bakker, Gijs 334
Baleri Italia S.p.A., Albano Sant'Alessandro (IT) 334/*681, 682*
Bally, Jürg 136, 184, 205, 209, 231, 263/*419–422, 501*
Baltensweiler AG, Ebikon 122, 136, 335/*269, 277, 423, 525, 526*
Baltensweiler Niederberger *277*
Baltensweiler, Rico 121, 185, 224, 243/*269, 409, 423, 500, 525, 526*
Baltensweiler, Rosmarie 121, 185, 224, 243/*269, 409, 423, 500, 525, 526*
Bandixen, Edlef 255
Bangerter, Rolf *579*
Banham, Reyner 16, 17
Baumgartner, Hans *206*
Bauwerkstatt, Zurich *207*
Bayer, Herbert 207/*230, 237, 301, 315, 484*
Bee, Anton G. *402*
Béhar, Yves 349/*840, 841*
Behrens, Peter 53/*103*
Belli, Robert *43*
Bellmann, Hans 152, 158, 184, 188, 190, 205, 207, 231/*342–345, 406, 410, 432, 438, 486, 488–490, 495, 732*
Belux, Wohlen/Birsfelden 121, 334/*276, 670, 692, 743, 846, 847*
Bemag Objekteinrichtungen AG, Sissach *701, 702*
Berger, Susi *659, 661*
Berger, Susi and Ueli 121, 218, 226, 259, 261, 265, 267, 271, 295/*506, 507, 604, 624, 629–631, 639, 643, 644, 649, 653, 660, 706, 714*
Berger, Ueli *642, 652*
Berliner Metallgewerbe Josef Müller *734*
Bernasconi, Elisabeth *122, 142*
Berndorf Luzern AG *473*
Bernhard, Olivier 352
Bernhardt, Jürg *351*
Bernheim, Erwin *175*
Bernina, Zurich 324
Bernoulli, Hans 47, 49, 53/*87–89*
Bertschi, Markus *827*
Besset, Lisa *471*
Bezzola, Leonardo *249, 499, 572*
Biella-Neher, Biel 231/*555*
Biffi, Debora *816*
Big-Game, Lausanne 347/*818, 824, 834–836*
Bigler + Ramuz *527*
Bill, Binia *22, 204, 378, 380*
Bill, Max 14, 16, 17–20, 109, 121, 136, 146, 149, 159, 163, 165, 167, 169, 172, 189, 193, 195, 197, 201, 203, 205, 207, 209, 241, 302/*22, 23, 169, 204, 205, 251, 258, 268, 303, 316, 322, 333–335, 378–380, 407, 435, 436, 444, 445, 461, 474, 491–493*
Binder, Walter *503*
Bircher, Rudolf *503*
Birkhäuser, Peter *224*
Blaser, Werner *734*
Blin, Camille 346
Blum, Kurt *461*
Blumer, Samuel 41/*66*
BMC Switzerland AG, Grenchen *797–799*
Bodmer, Albert 90
Bodum, Triengen 326/*785*
Boesch, Ines *764*
Boesiger, Willy 136
Bolex International S.A., Yverdon-les-Bains (Paillard-Bolex S.A., Yverdon-les-Bains / Sainte-Croix) 81, 262/*168, 632*
Boner, Jörg 327, 334, 353/*812, 852–855*
Bonifas, Paul Ami 47, 72, 113/*86, 148, 149, 252*
Bonnard Woeffray, Monthey *575*
Bonsiepe, Gui 325
Borde, Carlo *291*
Borel, Daniel 288, 324
Borer, Carlo *803*
Botta, Mario 287, 299/*684, 685*
Bottoni Porzellan, Bern *253*
Brac AG, Breitenbach *647*
Brand, Stewart *654*
Brassaï (Gyula Halász) 11/*2*
Brechbühler, Hans 205/*486*
Bredendieck, Hin 106, 119/*202, 203, 221, 222, 260, 267*
Breuer, Marcel 54, 68, 75, 108, 129, 131, 136, 169, 207, 301, 302/*300, 309, 731, 732*
Breuning, Olaf *257*
Brillat-Savarin, Jean Anthelme 13, 16
Brooks, H. Allen *264*
Brown Boveri, Baden *359*
Brüco Swiss AG, Rümlang *594*
Brunner, Thilo Alex 352/*849*
Bucher Leichtbau AG, Fällanden *618, 619*
Bucher, Heinrich 257/*618*
Bücherstube Girsberger & Co., Zurich 133/*115, 305, 307, 308*
Buchmann, Max *349, 410, 439*
Buchner Bründler, Basel *826*
Budliger, Hansjörg 360
Bührer, Werner 153/*346*
Buquet, Edouard-Wilfrid 121
Burckhardt, Ernst F. 54, 90/*460*
Burckhardt, Lucius 295
Bureau central de construction (BCC), Grenoble (FR) 163/*372*
Bürgi, Xaver *597*
Burkhard, Balthasar *642*
Burri Public Elements AG, Glattbrugg 321/*593, 773*
Buser, Hans *121*
Busnelli, Misinto (IT) 275
Butz, Fritz *201*
Butz, Gret *201*
Cafiero, Vincent 229
Caminada, Gion A. *482*
Cappellini, Meda (IT) 334
Caputo, Lovis 356/*822*
Caran d'Ache, Geneva 133
Carrosserie Hermann Graber, Wichtrach *361, 362*
Casas Robert, Ramon 147/*328*
Cassina S.p.A., Milan 302/*719*
Cassina, Franco 302
Charles Stendig, New York 186
Charlot, Michel 122, 351/*846–848*
Christen, Andreas 121, 129, 136, 205, 225, 231, 241, 267/*272, 295, 502, 528, 530, 585, 586*
Christofle, Charles *279*
ClassiCon, Munich (DE) 327
Colani, Luigi 229, 267
Cole, Henry 12
Collas, Jean *563*
Collischonn, Hermann *559*
Colombo, Joe 218, 265, 267
Comforto, Ahlen (DE) 229
Concava S.A., Lugano *166*
Contina Bureaux- & Rechenmaschinen-fabrik AG, Mauren (LI) *170*
Coppetti, Caspar 352
Coray, Hans 129, 169, 207, 321/*206, 293, 404*
Création Baumann, Langenthal *743*
Cristallerie Baccarat (FR) *821*
Curjel, Hans *201*
Curjel, Jella *201*
Curtiss, Glenn *485*
D'Urbino, Donato 265/*645*
Daepp, Margareta 115
Dalí, Salvador 11, 12, 13, 17, 19/*2, 26*
De Koninck, Louis Herman 235/*564*
de Meuron, Pierre 18, 122, 299/*275, 276*
De Pas, Jonathan 265/*645*
de Praetere, Jules 25
De Sede AG, Klingnau 260, 305/*626, 627, 648, 652, 733, 735, 738*
Dedelley, Frédéric 321, 355/*773–775, 806, 859–861*
Deplazes, Andrea *480*
Depot Basel *817*
Derungs, Beat *616*
Deuber, Christian 327
Deutsche Krankenversicherungs AG, Berlin *539*
Devico Design. Dübendorf 241, 243, 245/*477, 600*
Dictaphone International AG, Killwangen *585*
Didier des Gachons & Ravel, Paris 119
Dietiker AG, Stein am Rhein 255/*612*
Dietlicher, Christoph 285, 335
Diller + Scofidio. New York *25*
Dino Gavina, Bologna (IT) 302/*719, 731*
DOG Cie., Geneva 241
Dornier, Claude *287*
Dornier-Metallbauten, Friedrichshafen (DE) *287*
Dortu, Jacques 36
Drack, Thomas 285, 335
Drechsel, Martin *341*
Driade, Fossadello di Caorso (IT) 327, 334
Droog Design, Amsterdam 334
Dubacher, Hans Peter *295*
Dudli, Markus 354
Dupraz, Claude 241/*580, 581*
E. J. van de Ven, Brussels *564*
Eames, Charles 15, 169, 188, 225, 275, 330/*206, 390, 404, 453*
Eames, Ray 169, 188/*206, 390, 404, 453*
Ebéniste Frédéric Ballenegger, Lausanne *132, 133*
Ebénisterie Jean Egger, La Chaux-de-Fonds (?) *70*
Eberth & Co., Zurich *266*
Ebnöther, Yves *667*
ECAL (École cantonale d'art de Lausanne) *821*
Eckert & Gasser *560*
Eckstein, Hans 74, 135
Edra Mazzei, Pisa (IT) *640*
Eduard Binder, Brienz *42*
Egger, Hector 47
Egloff, Thomas 122
Ehrlich, Mauritius 163/*374, 375*
Eichenberger, Hans 121, 295, 302/*461, 665, 715, 738*
Eichenberger, Walter 241, 289/*598, 599, 602, 689, 690*
Eidenbenz, Hermann *286*
Elna International Sewing Machine Company Inc., New York *329*
Embru-Werke AG, Rüti 105, 108, 109, 131, 207, 231, 301/*158, 159, 208–210, 217, 218, 229, 235, 236, 299, 300, 302, 313, 314*
Emch, Peter *638*
En Soie, Zurich 318/*764, 828*
Engesser, Friedrich *567, 568, 571*
Ericsson *600*
Erni, Hans 207/*289*
Ernst Schweizer Metallbau, Hedingen 241
Erwitt, Elliott *862*
ETA AG, Grenchen *696*
Eternit AG, Niederurnen 172, 191, 334/*399, 402, 412, 441*
F. Blumer & Cie. Schwanden 37/*54*
FabLab, Zurich *667*
Fabric Frontline, Zurich 318/*762*
Falconnier, Gustave 43/*72*
Fehlbaum, Rolf 334
Fellmann Design, Zurich *601*
Fellmann, Hansruedi 241, 243/*477, 600, 601*
Fink, Monika *858*
Fink, Peter 115
Finsler, Hans 12, 13, 111, 119, 195/*202, 203, 222, 231, 237, 241, 243, 244, 247, 248, 258–261, 267, 334, 345, 411, 488, 489*
Finsler, Lita *201*
Fischer, Carl 44, 47/*75*
Fischer, Georg 245
Fischli, Hans 158/*318, 363*
Fischli, Peter *671*
Fischli / Weiss 19/*671*
Fleischmann, Meinrad *830*
Flug- und Fahrzeugwerke AG, Altenrhein (FFA) 201
Fopp, Damian *793*
Forster, Cornelia 89, 110
Forty, Adrian 17
Foscarini S.r.l., Marcon (IT) 350/*843*
Frampton, Kenneth 12
Franz, Edi 121/*270*
Fregoso, Luca *574*
Freitag lab. AG, Zurich 335, 357/*751, 752*
Freitag, Daniel and Markus 314/*745, 750, 814*
Frey, Walter 254/*366, 610*
Fröhlich Brunnschweiler & Cie., Ennenda 37/*30, 31*
Fukasawa, Naoto 357/*801*
Fuller, Richard Buckminster *580*
Fuseproject Inc., New York *840*
Gaab, Marcus *573*
Gafner, Daniel 271
Gallay, Jacques 241
Gangloff Cabins AG, Bern *475*
Gantenbein, Köbi *704*
Gassmann, Estelle 115
Gaston Blanpain Massonet, Brussels *43*
Gatti, Piero 265
Gauchat, Pierre 207/*219, 302*
Gaudí, Antoni 11, 301
Gebrüder Ouboter AG, Küsnacht *769*
Gebrüder Thonet GmbH, Frankenberg (DE) *776*
Gebrüder Vadàsz, Geneva 66/*128–130, 161*
Gehry, Frank 271
General Motors 14/*12*
Gerstner + Kutter, Basel *347*
Gessner, Robert 207
Gfeller Communications, Bümpliz *600*
Gianola, Ivano 299
Giauque, Elsi *411, 503*
Giedion, Sigfried 12, 13, 14, 17, 18, 75, 106, 108, 119, 121, 131, 135, 195, 205, 207, 243, 301, 351/*160, 202, 203, 221, 222, 260, 261, 267, 484*
Giesserei Oederlin & Co. AG, Baden *135*
Gigon, Annette *744*
Giroflex Entwicklungs-AG, Koblenz (DE) 229, *557*
Girsberger AG, Bützberg *501*
Girsberger, Elsa *201*
Girsberger, Hans 133/*308*
Giupponi, Andreas 285, 335
Glaeser, Willi and Otto 334
Glas Trösch AG, Bützberg *813*
Glashütte Bülach AG 107/*223–225, 577*
Gnamm, Sophie-Renate *737*
Gold Medal Mfg. Co., Racine (US) *483*
Goldsholl, Morton *452*
Good, Amata 110
Graber, Rudolf 135, 185, 207, 301/*237, 316*

Gradmann, Alfred 90
Gramazio & Kohler, Zurich *794–796*
Grandjean, Etienne 243
Grands Magasins Jelmoli, Zurich *355*
Gras, Bernard-Albin 106
Grässli, Tina 323/*779*
Gretag AG, Regensdorf *591*
Greutmann, Carmen and Urs *746*
Gropius, Walter 12, 86, 108, 131, 169, 195, 201
Gruppo DAM 275
Gugelot, Hans 183, 189, 207, 231/*416–418, 435*
Guhl, Emil 148, 169/*393, 397*
Guhl, Willy 148, 165, 169, 172, 173, 191, 205, 225, 241, 275/*206, 330–332, 368, 385–387, 389, 393, 395, 397, 398, 399, 401, 402, 412, 441–443, 503, 590, 662*
Guichard et Wanner, Yverdon 39/*60–62*
Guichard, Jean-Pierre Moïse 23, 39/*33, 34, 62*
Guignard, André 288
Guimard, Hector 11, 43
Gull, Gustav 43/*74*
Guyer, Claire 89
Guyer, Lux 47, 53, 89/*97, 187, 192–195*
Guyer, Mike *744*
H. P. Spengler, Rümlang 225, 241, 267/*502, 528, 635, 637*
Häberli, Alfredo 115, 239, 322, 334/*298, 573, 746, 776, 809, 829, 830*
Hablützel, Alfred 121, 260, 284, 305/*271, 414, 502, 530, 532–535, 546, 605, 675, 677, 699, 700, 710, 717, 718, 725, 729, 739, 740, 742, 753, 810, 811*
Haechler, Richard *121*
Haefeli, Max, sen. 90
Haefeli, Max Ernst 54, 68, 69, 70, 75, 92, 110, 119, 129, 133, 158, 163/*113–115, 117, 118, 120, 121, 123, 124, 135, 136, 138–140, 158–160, 196–200, 266, 305–307, 319, 322*
Haeni, Nicolas *832, 833*
Halard, François *830*
Haller, Fritz 223, 229/*522, 523, 543*
Hamburger, Jörg *863*
Hamilton, Richard 12, 15/*4, 16, 17*
Hans Kaufeld. Brake 305/*733, 735, 738–740*
Harbeke, Christian 317, 326
Hassenpflug, Gustav *110*
Haussmann und Haussmann, Zurich *742*
Haussmann, Fritz 110, 113/*205, 232*
Haussmann, Helene 110/*205, 232, 251*
Haussmann, Robert 121, 265, 267, 302/*274, 733, 735, 742*
Haussmann, Trix and Robert 226, 258, 284, 295, 299/*254, 532–535, 621–623, 669, 673, 675–677, 707–713, 716–718, 720*
Hay / BS Studio, Horsens (DK) *837, 838*
Hayward Brothers & Eckstein Ltd., London *72*
Heierling, Davos 203
Heinrich Kihm, Dübendorf *733*
Heinrich Kuhn Metallwaren, Rikon *286*
Helfenstein, Heinrich *75, 195, 674, 716, 744*
Helmrinderknecht, Berlin 355/*859–861*
Henne, Walter 199/*468*
Hennessey, James 275/*655*
Henz, Hannes *575*
Herdeg, Hugo *23*
Herman Miller 169
Hermès, Paris 356
Hero, Lenzburg *578*
Herzog & de Meuron, Basel 11, 18/*1, 276*
Herzog, Jacques 18, 122, 299/*275, 276*
Hesse-Honegger, Cornelia 318/*762, 763*
Heusser AG, Obermurgenthal 172
Hewitt, Keith *828*
Hiestand, Ernst *541*
Hildinger, Paul 189/*435*
Hilfiker, Hans 151, 235, 241/*324, 339, 340, 582, 673*
Hirst, Damien 289/*691*
Hoffmann, Josef 301
Hofkunst, Alfred *697*
Hofmann, Armin *446, 449*
Hofmann, Hans 90/*303*
Horisberger, Christian 275/*666*
hp Weidmann, Basel *701, 702*
Hubacher & Steiger, Zurich 109
Huber, Max *273*
Huber, Uli 245
Hübsch, Heinrich 23
Hugelshofer, Alice *160*
Hungerbühler, Willy 219
Hurter Tapeten AG, Winterthur *845*
Huygens, Christian 77
Ibico AG, Zurich *587*
Ibsen, Henrik 16
Iittala Oy AB, Helsinki 115/*829*
Ikea *808*
INCH Furniture, Basel 357/*815, 826*
Injecta AG Spritzgusswerke und Apparatefabrik, Teufenthal 125/*357*
Irion, Stefan 136
Itten, Arnold *124*
Itten, Johannes 146
J. F. Pfeiffer AG, Zurich *531*
J. Lüber AG, Basel 262/*629, 631, 644*
J. Ochsner AG, Zurich 71/*127, 145–147*
Jacomet, Simon 203
Jaray, Paul 199/*125, 465*
Jeanmonod, Grégoire 347/*824*
Jeanneret, Charles-Edouard (Le Corbusier) 38, 42, 53, 113, 119/*46, 58, 59, 69, 70, 101, 102, 104, 264*
Jeanneret, Pierre 75, 106, 136, 163, 207, 301, 302/*3, 109, 186, 372, 373, 563, 668, 726–728, 730, 736, 737*
Jegrt, Rudolph *397*
Jehle, Siegfried *406*
Jeker, Martin *799*
Jenaer Glaswerke Schott & Gen., Jena (DE) *138*
Jomini, Valérie *297*
Jones, Allen 269
Jucker, Carl 119
Judd, Donald 295
Junghans Uhren GmbH, Schramberg (DE) *492*
Jura AG, Niederbuchsiten 324/*599*
Kadler-Vögeli, Ernst 54/*306*
Kahn, Albert 163
Kandahar Schumanufaktur AG, Thun-Gwatt 150/*336–338, 481*
Karrer, Philippe *821*
Kartell, Milan 334
Kaufmann Jr., Edgar 169/*394*
Kaufmann, Sandra 354/*856–858*
Keramik-Manufaktur Potsfink, Ependes-Fribourg *257*
Keramische Werkstätte Alsiko, Uster 110/*232, 234, 251*
Kera-Werke AG, Laufenburg *438*
Kern & Co., Aarau *634*
Kestler, Sonnhild *765*
Kienzle, Hermann 157
Kienzle, Wilhelm 54, 89, 105, 133, 136, 165, 187, 205, 207, 209, 231, 241, 293, 301/*90, 91, 98, 100, 121, 217, 218, 304, 321, 369, 381, 429, 430, 486, 487, 489*
Kim, Dübendorf 169
Klose, Odo 229/*557*
Klug, Ubald 260, 265/*605, 626, 648, 746*
Knaus & Knaus, Volketswil *175*
Knell-Weber, Kristin *503*
Knoll Associates Inc., New York 207, 301
Knoll International 184, 185, 186, 225, 229, 301/*406, 540, 541, 732*
Knoll, Florence 229
Koch, Carl Hans 153
Kolb, Otto 169/*391*
Kollektivgruppe SWB 9, 69, 92, 133/*137*
Konstruktionswerkstätten Wartmann AG, Brugg *324*
Kramer, Ferdinand 275
Krohn, Michael *782–784*
Kubrick, Stanley 265/*641, 827*
Kudelski S.A., Cheseaux-sur-Lausanne *163*
Kudelski, Stefan *163*
Kueng Caputo, Zurich 271, 356/*822*
Kueng, Sarah 356/*822*
Kuhn, Charles *173*
Künzel, August *156*
Kuster Design, Freienbach 319/*766*
Kvadrat A / S, Ebeltoft (DK) *855*
L. Kellenberger & Co., St. Gallen 243/*602*
Laguet, Bertille *817*
Lalanne, François-Xavier 269
Lang, Fritz 290/*670*
Lange, Fritz *811*
Lasius, Georg 25
Lavater, Warja *304*
Le Corbusier (Jeanneret, Charles-Edouard) 12, 14, 19, 38, 53, 54, 72, 75, 89, 90, 106, 108, 119, 131, 136, 163, 185, 205, 301, 302/*3, 109, 186, 371, 563, 668, 726–728, 730, 736–738*
Léger, Fernand 16
Legler, Gian Franco 121, 169, 172/*273, 405*
Lehni AG, Dübendorf 241, 295/*272, 295, 703, 743*
Leisinger, Elsbeth *603*
Lenco, Burgdorf 241
L'Eplattenier, Charles 25, 38, 53/*46*
Lexon Design, Boulogne-Billancourt (FR) *818*
Libiszewski, Serge *726*
Licht + Raum, Bern 122
Linck Keramik, Zollikofen *250*
Linck, Ernst *76*
Linck, Margit 113/*250*
Linck, Philipp and Ernst *79, 80*
Lista Office S.A., Degersheim *791–793*
Loewy, Raymond 14, 207
Logitech Europe S.A., Romanel-sur-Morges 288, 324/*687*
Löhndorf, Alfred *406*
Lohse, Richard Paul 183, 195, 197, 207/*397, 418, 455, 462, 487, 494, 547*
Lomazzi, Paolo 265/*645*
Loos, Adolf 205, 258
Loroupe, Tegla *851*
Lotti, Martin *805*
Louis Vuitton, Paris *823*
Louis, Armand 350/*825*
Lübbens, Rudolf 267
Lüdeke, Christine 201/*473*
Luginbühl, Bernhard 265/*499, 572*
Luginbühl, Ursula 113/*249, 499*
Lumière – Société pour la fabrication de lustrerie d'art, La Chaux-de-Fonds 119
M&E Design, Zug/Cham 241, 243, 245, 289/*598, 599, 602, 690*
Mackintosh, Charles Rennie 301
Maggi AG, Kemptthal 23, 25, 40/*63–65, 126*
Maggi, Julius Michael Johannes 40/*126*
Magnin, Michel *503*
Maillart, Robert *466*
Mammut Sport Group AG, Seon 317/*479, 759, 761*
Man Ray 11
Manufacture de porcelaine, Nyon 36/*52, 53, 238, 255*
Manufacture Jaeger-Le Coultre S.A., Le Sentier 81/*167*
Marchand, Christophe 322, 334/*746, 776*
Mari, Enzo 275/*666, 817*
Mascarin, Mario 113
Mathieu, Aurélie *821*
Mathsson, Bruno 169/*392*
Matter, Herbert 199, 201, 203/*11, 464, 476*
Mbessé, Emmanuel 346
McLuhan, Marshall 16
Meier Leuthold, Monique *828*
Meier, Otto 70/*122, 141, 142, 562*
Meinzer, Manfred 243/*595, 596*
Mendini, Alessandro 301
MEWA Blattmann, Wädenswil 169, 187/*206, 296, 429, 431*
Meyer, Erna 74, 235/*558*
Meyer, Hannes 50, 54, 83, 86, 157, 205/*108, 110, 112, 180–184, 360*
Meyer, Peter 89, 92/*185, 186*
Meyer-Strasser, Luise 89, 110, 113
Micro Mobility Systems, Küsnacht 320/*771*
Mies van der Rohe, Ludwig 68, 75, 131, 136, 169, 301/*732, 734, 738*
Migros-Genossenschafts-Bund (MGB), Zurich 245, 289, 333/*576, 598, 689*
Milani Design & Consulting AG, Thalwil 81/*767, 768*
Milani, Arturo 187
Milani, Francesco 241/*591*
Mira-X AG, Suhr 284/*675–677, 717, 718*
Möbelfabrik Horgen-Glarus, see AG Möbelfabrik Horgen-Glarus
Möbelfabrik Muotathal Paul von Rickenbach AG 353
Möbel Pfister, Suhr *643*
Moeschlin + Baur, Basel *408, 447, 463*
Moholy-Nagy, László 136
Molitor, Karl 203
Mondaine Watch Ltd., Zurich *175, 341*
Moosbrugger, Bernhard 13/*9, 398*
Morach, Otto 47/*81, 93–95*
Moroso, Cavalicco (IT) 334
Morris, William 25
Morrison, Jasper 334, 357/*800, 801*
Morscher, Reinhart *351*
Moser, Benjamin *816*
Moser, Karl 45, 90, 133/*78–80*
Moser, Werner Max 102, 106, 111, 133, 136, 163, 207, 322/*208, 209, 235, 236, 315, 317, 723*
Mourgue, Olivier 265, 267/*641*
Moustache, Paris *836*
MS Modules Sanitaires S.A., Saint-Sulpice *586*
Muchenberger, Ernst 241, 289/*598, 599, 602, 689*
Müller, Alois Martin *704*
Müller, Erich 207
Müller, Ferdinand 36
Müller, Jacob 155, 158, 165, 187, 205/*207, 318, 352–354, 364, 365, 367, 383*
Müller, Ruedi Alexander 317
Müller, Ruedi August 317
Müller, Werner 263, 267/*635, 636*
Mumenthaler & Meier, Basel 133
Mumenthaler, Ernst 70/*122, 141, 142, 144, 562*
Mumenthaler, Marco 115
Mumford, Lewis 15
Munari, Bruno 169/*388*
Munz, Johann 316/*756, 757*
Muthesius, Hermann 47, 53
MWB Metallwerke Buchs *516*
N2 Design, Lucerne 327, 334
Naef Spiele AG, Zofingen *413–415*
Naef, Kurt 182/*413, 414*
Nagasawa, Setsuko 115
Namco, Thessaloniki (GR) *581*
Negroponte, Nicholas 349
Neisse, Hans *197*
Nestlé 23/*37*
Neue Werkstatt, Winterthur 285/*678–680*
Neweczerzal, Alfred 324/*290*
Newson, Marc 129
Nicoud, Jean-Daniel 288
Niedermann & Staehli *263*
Niedermann, Mark *826*
Niemeyer, Jo 122
Niemeyer, Oscar 18/*24*
Niesz, Anita *318, 364*
Nigg, Jürg 122
Nike Inc., Beaverton (US) *805*
Nils Holger Moormann, Aschau (DE) *812*
Noguchi, Isamu *433*
Nose Design, Zurich 317, 326/*469, 759–761, 799*
Noyes, Eliot 193/*450, 451*
Oberson, André 241
Obrecht, Otto Walter *607*
Ochsner AG, see J. Ochsner AG
Ochsner, Jakob, jun. 71
Ochsner, Kurt (Koni) 229, 295
Ochs-Walde *561, 562*
Omega S.A., Biel *27, 173, 174*
On AG, Zurich *849, 851*
One Laptop per Child Association – OLPC, Somerville (US) 349/*840, 842*
Oppenheim, Meret 226, 265, 301/*505, 719*
Osswald, Paul *83*
Ouboter, Wim 320/*769–771*
OWO-Presswerke, Mümliswil *607*
Ozenfant, Amédée 72, 113
Paillard & Cie. S.A., Sainte-Croix (Ernest Paillard & Cie. S.A., Yverdon) 66, 103, 262/*128–131, 161, 171, 172, 211–213, 634*
Pallucco Italia, Castagnole (IT) 327
Panton, Verner 265, 267, 284/*650, 651*
Paolini, Cesare 265
Papanek, Victor 275/*655*
Patek Philippe & Co., Geneva 77/*178*
Patek, Antoine Norbert de 77
Paul Eberth & Co., Zurich *138, 266*
Paul Matzinger, Basel *111*
Paulin, Pierre 265
Peduzzi Riva, Elena *652*
Pemberton-Billing, Noel 81/*167*
Pensi, Jorge 129
Peraves AG, Uster 201/*472*
Perret, Auguste and Gustave 43
Perriand, Charlotte 75, 136, 163, 231, 301, 302/*563, 668, 726–728, 730, 736, 737*
Perrin, Léon *43*
Perroud, Eveline *104*
Petit, Elric 347/*824*
Pfändler, Jörg 245
Pfeiffer, Walter *829*
Pfister, Otto and Heinrich *323*
Ph. Oswald Schreinerei und Innenausbau AG, Oberglatt 231/*435, 550–552*
Philippe, Jean Adrien 77

Phonak AG, Stäfa 81/*766–768*
Picasso, Pablo 25/*39*
Pilatuswerke (Pilatus Aircraft), Stans 201
Ponti, Gio 302
Porzellanfabrik Langenthal 113, 222/*242–248, 519, 521, 743*
Porzellanmanufaktur Kilchberg-Schooren, Zurich 36, 113/*36*
Postfossil, Zurich 271
Pozzati, Severo (Sepo) *38*
Precisa AG, Zurich *588*
Preisig, Dölf *503*
Preiswerk, Rudolf *561*
Prezioso, Giuseppe 103/*211*
Probst, Robert 229/*542*
Process Design 327
Prouvé, Jean 163, 241/*372*
Pugin, Augustus Welby 12
Quanta Computer Inc. (TW) *840*
Qwstion, Seefeld (AT) 357/*820*
Raacke, Peter 271, 275
Raffainer, Dumeng *623*
Raffinerie AG, Zurich *850*
Raichle, Kreuzlingen 203, 245/*477, 478*
Ramakers, Renny 334
Raschle, Yves *815, 826*
Rasser, Alois 227, 271/*536, 538*
Rechsteiner, Stefan Roland *819*
Remo Trunz AG, Arbon *298*
Renfer, Fernand 113, 222
Renfer, Pierre 113, 222/*519*
Reuge, Sainte-Croix 79
Rey, Bruno 255/*612–614*
Reymond, Patrick 350/*825*
Ribcap AG, Münsingen *471*
Richter, Hans 47, 49, 111/*123, 311*
Rietveld, Gerrit 53, 275, 295, 301/*107*
Rimpl, Herbert 163
Risom, Jens *406*
Riwisa AG, Hägglingen *597*
Robert & Durrer, Zurich 291/*698*
Robert Mey, Paris *106*
Robert Strub, Zurich *321*
Robert, Henri *43*
Roesch, Carl *83*
Roesch, Urs 165
Rohner, Benedikt 231/*550–555, 609*
Rosselli, Alberto 267
Rossi, Aldo 301/*722*
Rössler Porzellanfabrik AG, Ersigen *239*
Roth, Alfred 108, 131, 169, 301/*226–228*
Roth, Dieter 19
Roth, Emil 108
Röthlisberger Schreinerei AG, Gümligen 229, 295/*621, 665, 706, 714, 715, 743, 746*
Rovero, Adrien 356/*817*
Ruckstuhl AG, Langenthal *743*
Ruder, Emil *449*
Rüegg, Barbara 323
Rüegg, Patrick *819*
Ruf, Fred 207
Ruskin, John 25
Rütschi, Rudolf S. *96*
Ryffel, Heinz 322
Saarinen, Eero 207, 229, 269
Sabez Sanitär-Bedarf AG, Zurich 190/*438, 440*
Salvisberg, Otto Rudolf 89, 90/*188–191, 539*
Samsonite AG, Muttenz *771*
Sapper, Richard 122, 267
Schälli, Colin 271
Schärer, Paul 223, 229/*522, 523*
Scheerbart, Paul 43
Scheidegger, Ernst *335, 407, 437, 444*
Schiess, Hanspeter *216*
Schiffini, Ceparana (IT) 239/*573, 574*
Schilder Bär, Lotte 360
Schläpfer-Thöne, Erika 209
Schlossfabrik Heusser, Obermurgenthal 172/*401*
Schmed, Silvio 122
Schmid, Eric *767, 768*
Schmid, Martin 199
Schmid, Moritz *813*
Schmidt, Carl 133
Schmidt, Georg 159, 193/*459*
Schmidt, Hans 90, 133/*664*
Schneck, Adolf 89/*190*
Schobinger, Bernhard *672*
Schoeller Textil, Sevelen 317
Schott, Jena (DE) 119
Schreibmaschinenfabrik Aug. Birchmeier's Söhne, Murgenthal *169*
Schroeter, Rolf *569*
Schuler, Fridolin 40
Schultz, Richard 229
Schütte-Lihotzky, Margarete 235/*559*
Schütz, Josef *280, 281*
Schweizer, Martin 275/*656*
Schweizerische Bundesbahnen (SBB), Bern 151, 199, 333/*323, 324, 339, 340, 467–469, 582, 583*
Schweizerische Gemeinnützige Gesellschaft (SGG), Zurich 40/*126*
Schweizerische Industriegesellschaft, Neuhausen *359*
Schweizerische Post-, Telefon- und Telegrafenbetriebe (PTT), Bern 133/*804*
Schweizerische Unfallversicherungsanstalt (SUVA), Lucerne 73/*151–154*
Schweizerische Verkehrszentrale (SVZ) *464*
Schweizerische Wagons- und Aufzügefabrik AG, Schlieren-Zürich 245/*323, 583, 584*
Schweizerischer Werkbund (SWB) 17, 29, 47, 49, 50, 59, 74, 113, 119, 169, 193, 195, 197, 209, 218, 265, 334/*244, 455, 520, 527, 726*
Scolari, Carla 265/*645*
Scott Bader & Co., Zurich 148, 169/*330, 331, 396*
Scott de Martinville, Augustin 347/*824*
Seeger, Mia *457*
Seidendruckerei Mitlödi AG *764*
Semper, Gottfried 12, 25
Senior Design Factory, Zurich *816*
Shulman, Julius *404*
Siegrist & Co., Grenchen 81/*166*
Sigg AG, Frauenfeld 125, 241, 316, 324, 333/*284, 292, 756, 757*
Sigg, Ferdinand 316
Sinar AG, Feuerthalen *346*
Smith, Paul *829*
Solsolito, Zurich *858*
Sondor Willy Hungerbühler AG, Zollikon 219/*509, 510, 592*
Sottsass, Ettore 267
Spemot AG, Dulliken 245, 289/*598, 689*
Spichtig AG Kunststoffwerk, Steinen *554, 609*
Spillmann Echsle Architekten, Zurich *751*
Spoerri, Daniel 265
Stadler Rail AG, Bussnang *469*
Staehelin, Georg *863*
Stam, Mart 54, 68, 75, 131
Starck, Philippe 129/*802*
Stäubli, Nicola 271, 275
Stauss, Traugott 104/*214–216*
Steiger, Carl 133
Steiger, Rudolf 119/*117, 124*
Steiger-Crawford, Flora 109, 119, 133, 136/*117, 124, 229–231, 310, 312*
Stein, Pierre 207
Steinberg, Saul 295
Stella Werke AG, Bassecourt 199, 254, 267/*610, 611*
Stendig Inc. Contract Furniture, New York *741*
Stocker, Simone 115/*253*
Stoeckli, Sibylle 275/*666*
Stoll, Albert, jun. *556*
Strada del Sole GmbH, Zurich *856*
Straka, Sieghard 320
Strasser, Luise, see Meyer-Strasser, Luise
Strässle Söhne AG, Kirchberg 188, 265/*432–434, 643*
Strässle, Alex 188
Straumann, Marc *780*
Štrba, Annelies *672*
Streiff, Egidius *459*
Strub, Robert 165
Studer Revox, Regensdorf 243/*595, 596*
Studer, Walter *604, 647*
Studer, Willi 243
Studio 65, Turin *640*
Studio Achermann, Zurich *830*
Studio Hannes Wettstein, Zurich *790*
Stutz, Andi, Elsa and Maya 318
Suter-Strehler Söhne & Co. AG, Zurich *135*
Swatch AG, Biel 19, 79, 291, 333/*177, 696–698, 747*
Swid Powell Collection (US) *254*
Swiss Alpine Club (SAC), Bern *480, 482*
Swiss Federal Railways, see Schweizerische Bundesbahnen
Swiss National Accident Insurance Fund, see Schweizerische Unfallversicherungsanstalt
Swiss National Tourist Office, see Schweizerische Verkehrszentrale
Swiss Society for Public Utility, see Schweizerische Gemeinnützige Gesellschaft
Swiss Werkbund, see Schweizerischer Werkbund
Swissair-Fluggesellschaft, Kloten 201/*473, 620*
Swisslamps International, Zurich *270, 271, 274*
Taeuber, Erika *50*
Taeuber, Sophie 44, 47, 49, 104/*47, 50, 75, 77, 92*
Tania & Vincent, Paris *836*
Tännler, Roland *751*
Tappolet, Berta 47, 89, 110, 113/*233, 243, 256*
Tara Möbel, Burgdorf *297*
Tati, Jacques 185/*409*
Tavaro S.A., Geneva 147/*328*
Taylor, Frederick Winslow 235
Teo Jakob, Bern 186, 305, 315/*426*
Teodoro, Franco 265
Tettamanti, Joël *752*
Thalmann, Ueli 245
Therma AG, Schwanden 41, 74, 235, 237/*66–68, 155, 356, 565–571*
Theus, Tilla *674*
Thonet, Boppard (DE) 322, 334/*32, 116*
Thonet, Michael *32*
Thorens S.A., Sainte-Croix 66, 159/*162, 164, 370*
Thut Möbel AG, Möriken 335/*294, 512, 513, 699, 743, 753–755, 807, 810*
Thut, Benjamin 122/*755*
Thut, Kurt 129, 292, 295, 302, 315/*294, 699, 739–741, 753, 754, 807, 810*
Tinguely, Jean *406*
Tissot S.A., Le Locle *176, 179*
Trachsel, Thomas *662*
Troehler, Alexander *551*
Trüdinger, Reni 121, 209/*497*
Twike AG, Gelterkinden *470*
U. Schärer Söhne AG, Münsingen (USM) 223, 229/*522–524, 543, 544, 608*
Uhrenfabrik W. Moser-Baer AG, Sumiswald *339, 340*
Ulrich, Heinz *652*
V-Zug 324
Vacchini, Livio 299, 301
Vallotton, Félix 25
van de Velde, Henry 53
van Eyck, Aldo *320*
Vaucher, Eric 324
Velt, Uster *819*
Verband der Schweizer Möbelindustrie (SEM), Lotzwil *659, 661*
Vescom, Deurne (NL) 350
Victoria-Werke AG, Baar 218, 265/*506–508, 639*
Vitali Spielzeug GmbH, Zurich *400*
Vitali, Antonio 146, 172, 182/*325–327, 400*
Vitra, Birsfelden 330, 334
Vivarelli, Carlo L. *570*
Vogler, Felix 115
Vogt, Klaus 220, 265, 267/*512–515, 603, 652*
Vogt, Rosmarie *603*
von Allmen, Fritz 150, 203
Von der Mühll, Henri-Robert 67/*132–134*
von Rumohr, Carl Friedrich 13
von Steiger, Alexander *347, 405*
Wagenfeld, Wilhelm 110, 119
Wagner, Otto 301
Waldvogel, Zurich *270*
Walser, Ludwig 219, 241, 243, 245/*402, 403, 509, 587–590, 592*
Walti, Ruedi *701, 702*
Wanner, Edmond 121
Wanner, Edouard 23, 39/*60, 61*
Wanner, Jakob Friedrich *673*
Warhol, Andy 12
wb form, Zurich 334/*541*
Weber, Arnault 346
Weber, Heidi *725, 727–729*
Weidmann, Hanspeter 293/*701–703*
Weiss, David *671*
Wenger, Käthi *411*
Werkgenossenschaft Wohnhilfe, Zurich 158, 187/*318, 364, 365, 367, 369, 419, 420, 430, 436, 496, 497*
Werren, Barbara *844, 845*
Wettly, M. L. *28*
Wettstein, Hannes 121, 199, 286, 290, 327, 334/*670, 681–683, 692, 789, 827*
Wey, Felix *847*
Wieser, Ulrich Paul 205, 231, 293, 301/*721, 731*
Wilhelm Bofinger, Ilsfeld (DE) 183, 225, 231/*547–549*
Wilhelm Egloff & Co., Zurich 119/*262, 265*
Willimann, Alfred *258*
Willy Gläser Innenausbau *605*
Windlin, Cornel *781*
Winehouse, Amy *828*
Wireworks, London *806*
Wirth, Armin 322
Wirth, Joachim *749*
Wirth, Niklaus *688*
Wirz, Peter 326, 327
Wisa-Gloria AG, Lenzburg 256/*615–617*
Wittwer, Hans 83, 86/*181*
Witzig AG, Zurich/Frauenfeld 231/*553*
Witzig, Uli 245/*583, 584*
Wogg AG, Baden 334, 353/*743, 811, 852*
Wohnbedarf AG, Zurich/Basel 17, 106, 111, 133, 135, 136, 152, 183, 185, 186, 188, 189, 207, 209, 225, 231, 263, 301/*208, 209, 235–237, 261, 301, 315–317, 333, 342, 343, 406, 410, 416, 417, 488, 490, 493–495, 502, 528, 547, 721, 726, 730–732*
Wolf-Bender, Heinrich *121*
Wolgensinger, Michael *271, 412*
Wright, Frank Lloyd 229, 301
Wülser, Tobias 201/*472*
Würmli, Peter *749*
Wüthrich, Thomas *471, 815, 826*
Xess + Baba, Zurich 323/*779, 780*
Zahn, Willy *308*
Zai, Disentis 203
Zanini, Robert *749*
Zanotta S.p.A., Nova Milanese (IT) 189, 322/*645*
Zanuso, Marco 267
Zappacosta, Pierluigi 288
Zeller, Eugen 12/*7*
Zellweger Uster AG *646*
Zemp, Werner 245/*593, 594*
Zena AG, Affoltern am Albis *290*
Zeppelin-Werke, Friedrichshafen (DE) 199/*288*
Zieta, Oskar 348/*837–839*
Zimmerli, Kurt 316/*756–758*
Zimmermann, Stanislas *297*
Zingg, Siegfried *259, 269, 425*
Zoelly, Pierre 221, 267/*504, 516*
Zwicky, Stefan *668*
Zyliss AG, Lyss 154, 324/*351*
Zysset, Karl 154/*349*
Zysset & Co., Lyss *349, 350*

Author Biographies

Christian Brändle
Born in 1969. Architect ETH/SIA (Swiss Federal Institute of Technology / Swiss Society of Engineers and Architects). Director of the Museum für Gestaltung Zürich since 2003. Prior to that worked at the Kunsthalle Basel and Art Basel as well as the Opernhaus Zürich and the Swiss National Exhibition Expo.02. Lecturer at Zurich University of the Arts (ZHdK). Chevalier dans l'Ordre des Arts et des Lettres of the French Republic. Publications: *Wohnträume – Wohnräume,* Zurich 2003; *Sportdesign – zwischen Style und Engineering,* Zurich 2004; *Every Thing Design,* Zurich 2009 (with Verena Formanek).

Claude Enderle
Born in 1956. Lecturer and mentor at Basel School of Design/University of Applied Sciences and Arts Northwestern Switzerland. Freelance consulting and coaching (CAS) with a focus on creative industries. Trained as a mechanical draftsman and product/interior designer (Interior Design Course, Basel), followed by several years of professional practice. Studied art history, economic and social history, and philosophy at the University of Zurich. Senior lecturer in the Department of Architecture at ETH Zürich from 2006 to 2008, in connection with a guest lectureship by Jasmin Grego. Research sabbatical in the United States from 2010 to 2011. Publications on design and architecture: "Die Co-op-Werke von Hannes Meyer und ein Lehrbuch von Anny Eichhorn," in *Schaufensterkultur – inszenierte Warenwelt in Basel,* Basel 2013; "Analog, digital, hyperreal und mehr – Zur Visualisierung innenarchitektonischer Konzepte," in *Boden Wand Decke. Zur Poesie der Innenarchitektur,* Sulgen 2014.

Meret Ernst
Born in 1966. Art historian. Head of the culture and design department at the magazine *Hochparterre* in Zurich since 2003. Teaches design theory (University of Applied Sciences and Arts Northwestern Switzerland, Lucerne University of Applied Sciences and Arts, Zurich University of the Arts). Vice president of the Swiss Design Association (SDA) and design expert on numerous juries. Member of the Zurich Fachhochschule Council since 2010. Project manager for young talent promotion at Design Pro Helvetia from 2012 to 2013. Received the Greulich Stiftung Kulturpreis for design journalism in 2014. Curated exhibitions at the Museum für Gestaltung Zürich and Expo.02. Publications: *City in Space: The Inside and the Outside of Barcelona Architecture and Design from the 50ies to the 70ies,* Zurich 2001; *Design Suisse* (with Christian Eggenberger), Zurich 2006.

Fredi Fischli
Born in 1986. Curator. Studied art history at the University of Zurich. Has been co-director of the gta exhibitions at ETH Zürich with Niels Olsen since 2013. After each had gained experience at museums and galleries in Switzerland and abroad, Fischli and Olsen founded the exhibition space Studiolo in Zurich in 2011. In recognition of their curatorial work they received the Premio Lorenzo Bonaldi EnterPrize from the museum GAMeC in Bergamo. In collaboration with Edition Patrick Frey, they edit their own series of publications on international artists and architects, including *Trix + Robert Haussmann,* Zurich 2012.

Hans Ulrich Gumbrecht
Albert Guérard Professor in Literature at Stanford University [and also teaches regularly at the Collège de France, the Universidade de Lisboa, and Zeppelin University in Friedrichshafen, Germany]. Although he lives in Silicon Valley, the electronic world gets on his nerves, but he nonetheless writes a blog for the *Frankfurter Allgemeine Zeitung;* although he was born in Germany [sixty-six years ago, in Würzburg], he tries to be a patriotic American and therefore loves Switzerland; although, like the designers of the twenties, he rhapsodizes about elegant emptiness, his offices at the university and his study at home are filling up with far too many books [not least those he writes himself].

Sebastian Hackenschmidt
Born in 1971. Art historian. Curator of the Furniture and Woodwork Collection at the MAK – Austrian Museum of Applied Arts/Contemporary Art in Vienna since 2005. Studied art history and German language and literature at the Universität Hamburg, thereafter research associate in the Art History Department. Publications: co-editor of the *Lexikon des künstlerischen Materials,* Munich 2002/2010, and the book *Möbel als Media,* Bielefeld 2011; most recently published *Hot 95,* Salzburg 2013.

Christof Kübler
Born in 1956. Art historian. Studied art history at the University of Zurich and was assistant to the chair of the Department of Modern and Contemporary Art there. Publications in the fields of art, architecture, and cultural history: *Inventar der neueren Schweizer Architektur* (INSA), vol. 8: *Schwyz,* Bern 1996. Cofounder of the publishing house Zurich InterPublishers (Zip). Curator at the Swiss National Museum from 1996, chief curator and member of the executive board from 2006 to 2008. Freelance work for the company Querverweise since 2009.

Peter Lepel
Born in 1944. A member of senior management at the traditional Swiss company Embru-Werke AG, Rüti, for many years, in charge of design and marketing. Postgraduate studies (MAS) in History and Theory of Architecture at ETH Zürich from 2007 to 2009. After retiring, curator of the Embru Collection and responsible for the company archives. Publications: *Furniture: A Journey through the Embru-Werke Archives, 1928–1943,* Rüti 2001, as well as *Über Reklame* and *Pierre Gauchat und die Embru-Werbung,* both Rüti 2011.

Claude Lichtenstein
Born in 1949. Architect ETH (Swiss Federal Institute of Technology). Curator at the Museum für Gestaltung Zürich from 1985 to 2001. Lecturer at various Swiss universities since 2002 (Zurich University of the Arts, University of Applied Sciences and Arts Northwestern Switzerland, Zurich University of Applied Sciences), curator, author, and (co)editor of various publications: *Your Private Sky: R. Buckminster Fuller; Art of Design Science* (with Joachim Krausse), Baden 1999; *Playfully Rigid: Swiss Architecture, Graphic Design, Product Design, 1950–2006,* Baden 2007. Also various texts on exponents of graphic design, including Bruno Monguzzi and Reinhart Morscher.

Renate Menzi
Born in 1968. Curator of the Design Collection of the Museum für Gestaltung Zürich since 2008. Trained as a graphic designer and in cultural studies at the Zurich University of the Arts, the Bezalel Academy of Art and Design in Jerusalem, and Humboldt University in Berlin. Teaches, publishes, and conducts research in the field of design, and has been a member of the Swiss Federal Design Commission since 2009. Exhibitions and publications: *Formfächer,* Ludwigsburg 2009; *Make up,* Ludwigsburg 2010; *Freitag: Out of the Bag,* Zurich 2012.

Arthur Rüegg
Born in 1942. A practicing architect since 1971. Professor of architecture and design at ETH Zürich from 1991 to 2007. Research, publications, and exhibitions on construction, color, photography, and design, as well as on modern living and interiors. Editor and author of numerous monographs: *Le Corbusier – Polychromie architecturale,* Basel et al. 1997/2006; *Swiss Furniture and Interiors in the 20th Century,* Basel et al. 2002; *Charlotte Perriand – Livre de bord, 1928–1933,* Basel et al. 2004; *Le Corbusier – Möbel und Interieurs, 1910–1960,* Zurich 2013. Curator of several exhibitions: *Le Corbusier before Le Corbusier* (with Stanislaus von Moos), 2002; *Le Corbusier: The Art of Architecture* (with Stanislaus von Moos and Mateo Kries), 2007; *Sophie Taeuber-Arp. Gestalterin – Architektin – Tänzerin* (with Eva Afhus), 2007; *Charlotte Perriand. Designerin – Fotografin – Aktivistin* (with Andres Janser), 2010, *100 Years of Swiss Design* (with Renate Menzi), 2014.

Anne-Claire Schumacher
Born in 1960. Senior conservator and curator at the Musée Ariana in Geneva. Studied art history and Chinese studies at the University of Geneva. Various publications on the Musée Ariana's collection of ancient and contemporary ceramics, including *La manufacture de porcelaine de Langenthal, entre design industriel et vaisselle du dimanche,* Milan 2012.

Christina Sonderegger
Born in 1961. Art historian. Curator of Furniture and Interiors at the Swiss National Museum since 2005 and head of the Department of Cultural History there since 2013. Studied art history as well as social and economic history at the University of Zurich. Research associate for the Design Collection of the Museum für Gestaltung Zürich from 1994 to 2000. Interior design and product design editor at the magazine *Werk, Bauen + Wohnen* from 1998 to 2005, as well as working as a freelance journalist and author in the areas of design, vernacular culture, and interior design.

Klaus Spechtenhauser
Born in 1969. Art historian and Slavist. Scientific editor in the Department of Monument Preservation for the canton of Basel-Stadt since 2011. Previously assistant to Professor Arthur Rüegg in the Department of Architecture at ETH Zürich, research associate/project manager for the ETH Wohnforum – ETH CASE, as well as being employed in the Department of Monument Preservation of the canton of Grisons. Regular publications on the architectural and cultural history of the twentieth century: *The Kitchen: Life World, Usage, Perspectives,* Basel et al. 2006; *Maison Blanche—Charles-Edouard Jeanneret, Le Corbusier: History and Restoration of the Villa Jeanneret-Perret, 1912–2005* (with Arthur Rüegg), Basel et al. 2007.

Stanislaus von Moos
Born in 1940. Art historian. Professor of Modern and Contemporary Art at the University of Zurich from 1983 to 2005. Visiting professor at the Yale School of Architecture from 2010 to 2014. Author of monographs on *Le Corbusier* (1968–), Italian Renaissance architecture (*Turm und Bollwerk,* Zurich/Freiburg 1974), the architecture of Venturi, Scott Brown & Associates (1987, 1999), and Swiss design history (*Industrieästhetik,* Disentis 1992). Curator of several exhibitions, including *Le Corbusier before Le Corbusier* (with Arthur Rüegg), 2002.

Sophie Wirth Brentini
Born in 1973. Art historian. Project manager at Musée Ariana in Geneva since 2012. Master of Arts degrees in art history from the University of Geneva and art education from the University of British Columbia (Vancouver). Curator of the Parcours Céramique Carougeois (from 2009 to 2013). Author of diverse articles on contemporary ceramics, including "La céramique genevoise de 1950 à nos jours," in *Musée d'Art et d'Histoire, Décor, design et industrie. Les arts appliqués à Genève,* Geneva 2010.

Illustration Credits

If not otherwise indicated, all objects illustrated in this volume are from the collections of the Museum für Gestaltung Zürich. Their authors are named in the captions. All key objects from the museum's collections have been newly photographed for this publication. Names of companies and institutions refer to the names and locations at the respective date of design, if researchable.

The data in the captions are broken down into the following sections:

- Design
- Object description, title
- Year of design
- in the case of key objects: Material, dimensions
- Producer and/or manufacturer

The information supplied reflects the current state of research. Any further references or additional material are welcome.

Works and images from sources other than the collections of the Museum für Gestaltung Zürich.

Agil Stans, Photo: Martin Stollenwerck: *785*
Archives d'Architecture Moderne, Brussels: *564*
Alfredo Häberli Design Development: *573, 574,*
Archiv Porzellanfabrik Langenthal: *242*
Atelier für Architekturfotografie, Heinrich Helfenstein: 75, 195, 674, 716, 744
Atelier Oï: *823, 825, 844, 845*
Atelier Pfister: *257, 830*
Audiovisuelles Archiv MGB: *576*
Baltensweiler, Ebikon: *423, 500*
Baugeschichtliches Archiv der Stadt Zürich (BAZ): *189–191, 263*
Belux, Birsfelden: *276, 847*
Big-Game: *818, 824, 835*
Bill-Stiftung, Zurich: *378, 380, 491*
BMC Switzerland AG, Photo: Martin Jeker: *797–799*
Damien Hirst and Science Ltd.: *691*
Degonda, Lucia (photo): *482*
Depot Basel, Photo: Thierry Bruder: *666*
Design+Design: *365, 403;* Photo: Michael Lio: *314, 317;* Photo: Andreas Zimmermann: *367*
Dornier Stiftung für Luft- und Raumfahrt, Friedrichshafen, Photo: Karl Alfred Ziegler: *287*
Dupraz, Claude: *580, 581*
ECAL, Nicolas Genta: *821*
Ecole d'arts apppliqués, La Chaux-de-Fonds: 46, Photo: Musée des beaux-arts, La Chaux-de-Fonds: *57*
Eames Office LLC (eamesoffice.com): *390*
ETH-Bibliothek Zürich, Bildarchiv: *125, 281, 465, 466*
ETH-Bibliothek Zürich, Bildarchiv / Stiftung Luftbild Schweiz: *620*
ETH-Studio Monte Rosa, Photo: Tonatiuh Ambrosetti: *480*
Fablab Zürich, Photo: Yves Ebnöther: *667*
Fellmann, Hansruedi *477, 478*
Fondation Le Corbusier, Paris: *59, 109, 371*
Fondazione Marguerite Arp, Locarno: *47, 50*
Fotostiftung Schweiz: *572*
Film still from: Jacques Tati (dir.), *Mon oncle,* 1958: *409*
Film still from: Stanley Kubrick (dir.), *2001: A Space Odyssey,* 1968: *641*
FREITAG lab.ag: *745,* 751, 752, 814
Gasser, Martin/Christoph Eckert (photo): *196*
Glas Trösch, Photo: Simon B. Opladen: *813*
Good, Desirée (photo): 792
Gramazio & Kohler: *794* (staff: T. Bonwetsch (project manager), D. Kobel, M. Lyrenmann. Industry partner: Keller AG, Ziegeleien), *795* (staff: S. Oesterle (project manager), R. Bärtschi, M. Lyrenmann. In cooperation with: Jürgen Strauss (Strauss Elektroakustik). Selected experts: Kurt Eggenschwiler (Empa). Industry partners: Cellform AG, PU-Technik Meyer), *796* (staff: S. Oesterle (project manager), R. Bärtschi, M. Lyrenmann. Industry partners: Häring Holz- und Systembau, Isoflock AG, Schilliger Holz AG)
Grands Magasins Jelmoli S.A.: *355*
gta-Archiv, ETH Zürich: *80, 108, 117, 118, 124, 160, 182, 187, 201, 202, 228, 231, 267, 307, 308, 319, 539, 543, 561*
Hablützel, Alfred: *605, 741*
Haussmann, Trix and Robert: *669, 673, 707* (photo), *708, 709* (photo), *711* (photo), *712, 713, 739, 742*
Herman Miller Inc.: *542*
Herzog & de Meuron: *1*
Hitsch, Kaspar (photo): *475*
horgenglarus: *786–787*
INCH Furniture: *815, 826*
Jasper Morrison Ltd.: *800*
J. Paul Getty Trust © (photo). Used with permission. Julius Shulman Photography Archive, Research Library at the Getty Research Institute (2004.R.10): *404*
Krohn, Michael: *782–784*
Kunstmuseum Bern, Meret Oppenheim Estate: *505*
Lepel, Peter: *299, 303*
Lichtenstein, Claude: *308, 358–360, 370*
Linck Keramik, Zollikofen, Photo: David Willen: 250
Li Wan Po, Derek (photo): *791*
M&E Design: *599, 602*
MAK Vienna, Photo: Georg Mayer: *640, 654–656*
Mattson, Bruno: *392*
Maurizio Corraini S.r.l.: *388*
Milani Design & Consulting, Photo: Eric Schmid: 767, 768
Mondaine Watch Ltd., Zurich: *175, 341*
Musée des beaux-arts, La Chaux-de-Fonds: *104*
Museum für Gestaltung Zürich, © ZHdK (photo): *51, 52, 55, 56, 74, 105, 106, 115, 148, 275, 305, 596*
Museum für Kommunikation, Bern, Photo: Peter Auchli: *688*
The Museum of Modern Art, Architecture and Design archive: *452*
Naef Spiele AG: *415*
Nike Inc., New York: *805*
Nose Design: *469*
Okolo/Depot Basel: *817*
On AG, Photo: Braschler/Fischer: 851
Panton Design, Basel: *650, 651*
Patek Philipp & Co.: *178*
Ph. Oswald, Photo: Alexander Troehler: *551*
Plakatsammlung, Schule für Gestaltung Basel: *579*
Private collection, France: *719*
Private collection, Switzerland: *558–560, 563, 565–568, 577, 578, 628*
Qwstion, Photo: Urs Bigler, Matthias Graf: *820*
Richard Paul Lohse-Stiftung: *487, 547* (scans)
Rist, Pipilotti: *748*
RMN-Grand Palais (domaine de Compiègne): *279*
Röthlisberger Kollektion: *665*
Rüegg, Arthur: *53, 68, 71, 79, 122, 185, 186, 192, 206, 207, 261, 264, 265, 273, 309, 316, 335, 345, 379, 382, 407, 410, 440, 457, 484, 486, 487, 494–496, 524, 540, 544, 554, 555, 608, 721, 723–725, 730, 732–738, 740*
Schweizerisches Nationalmuseum: *54* (DIG-11463), *289* (DIG-23572)
Schweizerisches Wirtschaftsarchiv CH SWA H+I Bc 272 Firmendossier zur Porzellanfabrik Langenthal: *247*
SBB Historic: *323, 467, 468*
Senior Design Factory, Photo: Florian Kalotay*: 817*
SRF, Photo: Markus Bertschi, Oscar Alessio: *827*
Staatsarchiv Schaffhausen, Fotosammlung Alusuisse: *283, 284, 288*
Stadtarchiv Zürich: *82, 85,*
starcknetwork: *802*
Stiftung Hans Arp und Sophie Taeuber-Arp e.V., Rolandseck: *76, 92*
Stiftung Moritzburg, Halle/Saale: *241, 248, 258, 260, 488, 489*
Stoll Giroflex: *556*
Studer, Walter: *604*
Studio Hannes Wettstein: *682, 683, 694, 695, 789, 790*
Swatch Ltd.: *697*
Swiss Car Register, Photo: H.-J. Bendel: *361, 362*
Tissot S.A.: *179*
Treichler, Hans Peter, Richterswil: *128*
Velt, Photo: Andreas Zimmermann, Basel: *819*
Walser, Ludwig: *589*
Wisa-Gloria AG: *617*
Witzig, Uli: *583, 584*
Wogg AG: *854, 855*
Wohngenossenschaft Eglisee, Basel: *562*
Wolf-Bender, Heinrich (photo): *121, 197, 199, 200, 266*
ZHdK-MIZ: *49, 83, 84, 90, 91, 96–100, 119, 203, 222, 240, 251, 327, 334, 395, 405, 503, 537, 538, 633, 662, 726*
Zieta Process Design: *839*
Zwicky, Stefan: *668*
from: *a+u,* 2000/3, p. 103: *107*
from: ALA-Mitteilungen 24–26/1958: *278*
from: *L'Architecture d'aujourd'hui* ("Solutions d'urgence" special issue) 16 (July–August 1945): *372*
from: *Bauen+Wohnen* 1, no. 2 (1948), p. 53: *320*
from: *Bauhaus* 1928/4, p. 97: *110*
from: Bill 1952: *322*
from: Gordon Bruce, *Eliot Noyes: A Pioneer of Design and Architecture in the Age of American Modernism,* London 2006: *450, 451*
from: *Sapperlot. Der Eisenplastiker Bernhard Luginbühl. 111 Fotos von Leonardo Bezzola,* Bern 1967: *499*
from: Bourban 2009: *73*
from: Tilmann Buddensieg, *Peter Behrens und die AEG, 1907–1914,* Berlin 1979: *103*
from: Bryan Burkhart and David Hunt, *Airstream: The History of the Land Yacht,* San Francisco 2000: *485*
from: *Das goldene Buch der LA 1939,* Zurich 1939: *280*
from: *Das ideale Heim,* 1929/9: *143*
from: *Die Schweiz im Spiegel* 1940: *282*
from: Guyer/Kettiger 1946: *366, 381, 384, 385, 387*
from: *Interieur* 1965/4: *497*
from: *Interieur* 1966/3, Photo: Helmut Bauer: *274*
from: *Jacob Müller* 1988: *383*
from: Janser/Rüegg 2001: *123, 311*
from: Kieren 1990: *180, 181*
from: Lavisse 1902: *285*
from: *Le Corbusier. Sitzmöbel/Sièges/Chairs,* exh. cat. Galerie Mezzanin (Heidi Weber), Zurich 1959: *727, 729*
from: Lohse 1953: *453, 462*
from: *Maggi Revue,* Jubiläumsausgabe 3, 100 Jahre Maggi, 1983: *65*
from: *Mario Botta* 1989: *685*
from: Maurizio 1952: *363*
from: Molin 1904: *238*
from: Monnier/Vuille 1979*: 62*
from: von Moos 1992: *27*
from: *Neue Grafik,* no. 15 (1963): 569–571
from: Ottilinger 2003: *116*
from: *Paul Bonifas* 1997: *150*
from: Rüegg 2002a: *89, 134, 427, 515*
from: *SBZ,* vol. 90, 1927, table 20: *137*
from: *SBZ,* vol. 99, 1932: *188*
from: Schwartz-Clauss/von Vegesack 2002: *483*
from: René Spitz, *Hfg Ulm: The View behind the Foreground,* Stuttgart/London 2002 (Photo: Ernst Scheidegger, Archiv hfg, 56/00851): *437*
from: Touchaleaume/Moreau 2010: *373*
from: *Das Werk* 5, no. 10 (1918), pp. 149–157: *87, 88*
from: *Das Werk* 14, no. 2 (1927), p. 34: *198*
from: *Das Werk* 22, no. 12 (1935): *234*
from: *Werk* 60, no. 8 (1973): *553, 557*
from: Wichmann 1987: *548, 549*
from: Wichmann 1989: *545*
from: *Wilhelm Kienzle* 1991: *304*
from: *Werner Zemp* 2010: *593, 594*

We also would like to thank all the designers, photographers, firms, collections, and private owners for kindly entrusting us with their works for photographs or providing us with illustrative material.

Donors

The captions provide the sources of all donated design objects and drawings shown in the illustrations. The origins of the archival materials in the Design Collection were initially not recorded systematically. The donors of documents accompanying the object collection—such as project studies, manuals, catalogues, brochures, photographs, etc.—are listed alphabetically below, together with the respective illustration number.

Bally, Jürg: *421, 422, 501*
Bastianello-Menzi, Maja: *396, 498*
Berger, Susi and Ueli: *508, 630, 642, 643, 649, 653, 654, 657, 658, 660, 706*
Dubacher, Hans Peter: *295*
Embru-Werke AG, Rüti: *210, 219, 302*
Eternit AG, Niederurnen: *402, 412*
Erni, Peter: *434*
Fässli, Kurt: *65*
H. P. Spengler Erben, Rümlang: *529*
Kandahar Schuhmanufaktur, Thun-Gwatt: *338*
Klug, Ubald: *627, 648*
M&E Design: *599, 690*
J. Ochsner AG, Zurich: *127, 146, 147*
Rohner, Fabian: *531, 552, 609*
Therma AG, Schwanden: *156*
Schilliger, Eliane: *676*
Schweizerische Bundesbahnen (SBB), Bern: *340*
Schweizerischer Werkbund (SWB): *204, 205, 400, 408, 411, 444, 447, 448, 449, 454, 458, 461, 463, 490, 520, 664*
von Steiger, Margrit: *347*
Walser, Ludwig: *510, 588, 592*
Weidmann, Hanspeter: *703*
Wieser, Ulrich P.: *230, 237, 315, 406, 418, 431, 731*
Zuber, Max: *634*

Objects designated with an asterisk * are on permanent loan from the Swiss Confederation, Federal Office of Culture, Bern:
389, 393, 502, 591, 652, 659, 661, 677, 693, 699, 700, 701, 717, 753, 810, 811

Permanent loan of the Schweizerischen Werkbunds (SWB), Zurich office:
SWB Photo collection/
Picture collection of well-designed consumer projects:
111, 121, 140, 243, 244, 249, 266, 268, 350, 357, 368, 391, 401, 705

Copyrights

Acknowledgments

Publication projects are always welcome occasions to examine and work with our own extensive holdings, and also to update it with targeted acquisitions.

We would like to thank all of the designers and firms, collectors, firms, and friends of the museum who have generously and consistently made their objects and documents—as well as their knowledge and expertise—available to the museum's collections.

www.eMuseum.ch

Short biographies or brief company portraits for selected Swiss designers, producers, and manufacturers that are represented with exemplary works in the collections of the Museum für Gestaltung can be found in the museum's online database.

100 Years of Swiss Design

Editors	Museum für Gestaltung Zürich Christian Brändle, Renate Menzi, Arthur Rüegg
Graphic design	NORM/Dimitri Bruni, Manuel Krebs, Ludovic Varone
Editing	Renate Menzi, Christina Reble, Arthur Rüegg
Image editing	Renate Menzi, Arthur Rüegg
Editorial assistance	Sabine Träger (text), Nava Sutter (text, image), Louis Schlegel (image)
Image acquisition	Kaltrina Ahmetaj, Christine Kessler
Object documentation	Liana Bähler, Nico Lazúla Baur
Photography	(unless otherwise stated) Franz Xaver Jaggy, Umberto Romito
Scans/Data coordination	Louis Schlegel
Copyediting	Jonathan Fox
Index	Sandra Leitte
Translations	Jennifer Taylor
Production	Martina Mullis, Integral Lars Müller
Repro	Ast & Fischer, Wabern
Printing and binding	Kösel, Altusried-Krugzell, Germany
Typefaces	Akzidenz-Grotesk Next, Berthold Akzidenz Grotesk
Paper	Profisilk, 150 g/m²
Cover	Designed with Willy Guhl's Garden Chair (1954) and Feller AG's Light Switch (ca. 1948)
Endpapers	© 2015 swisstopo

Z hdk
Zürcher Hochschule der Künste
Zurich University of the Arts

Museum für Gestaltung Zürich
Ausstellungsstrasse 60
CH-8005 Zürich/Switzerland
www.museum-gestaltung.ch
www.eMuseum.ch

Lars Müller Publishers
CH-8005 Zürich
books@lars-muller.ch
www.lars-mueller-publishers.com

100 Years of Swiss Design ISBN 978-3-03778-441-9
100 Jahre Schweizer Design ISBN 978-3-03778-440-2

First Edition
Printed in Germany

For their support we thank

Kanton Zürich Lotteriefonds

Freundeskreis Museum für Gestaltung Zürich

Stadt Zürich Kultur

UBS Kulturstiftung

USM Möbelbausysteme

Museum für Gestaltung Zürich

A Publication of the Museum für Gestaltung Zürich
Christian Brändle, Director

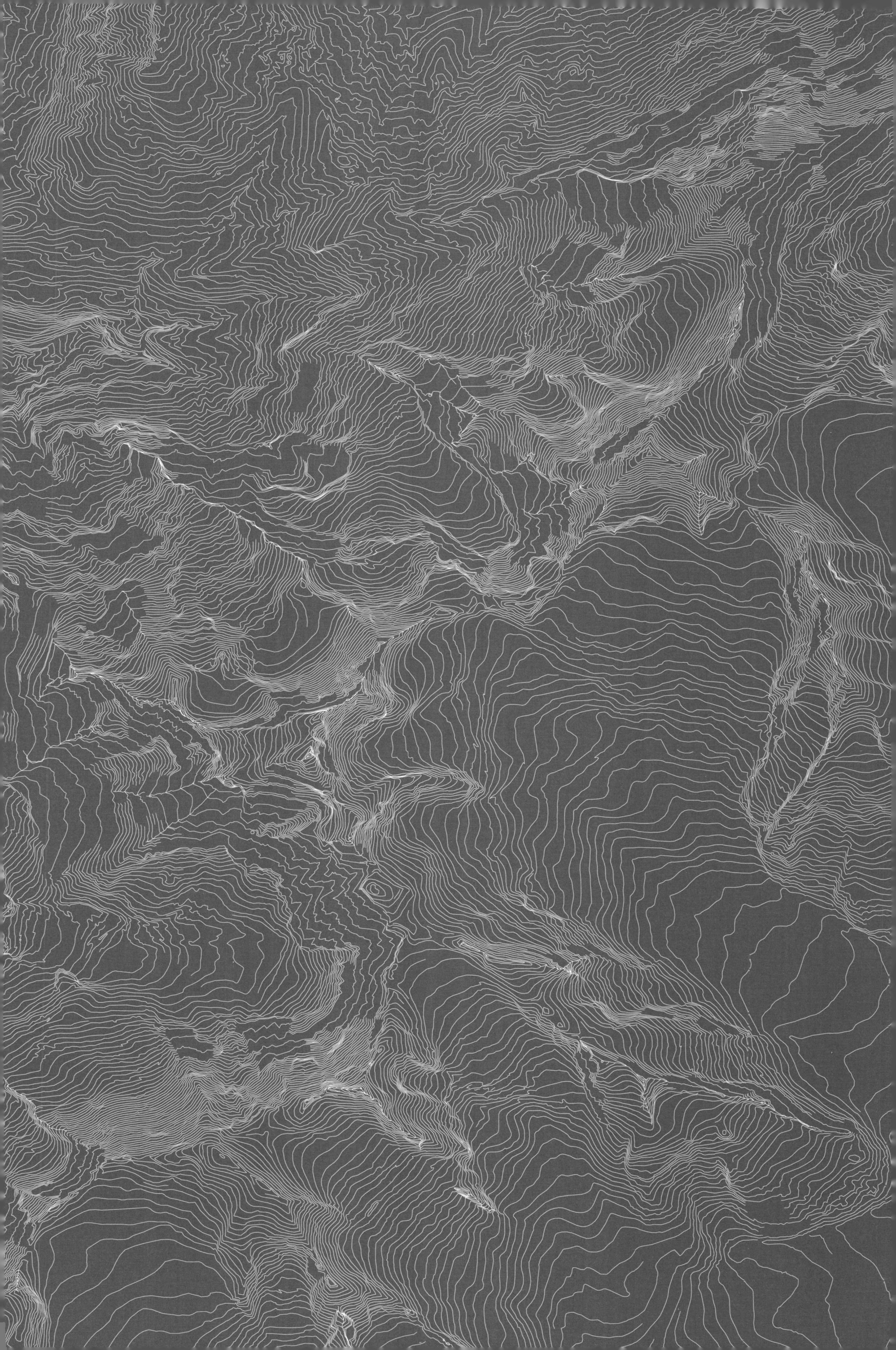

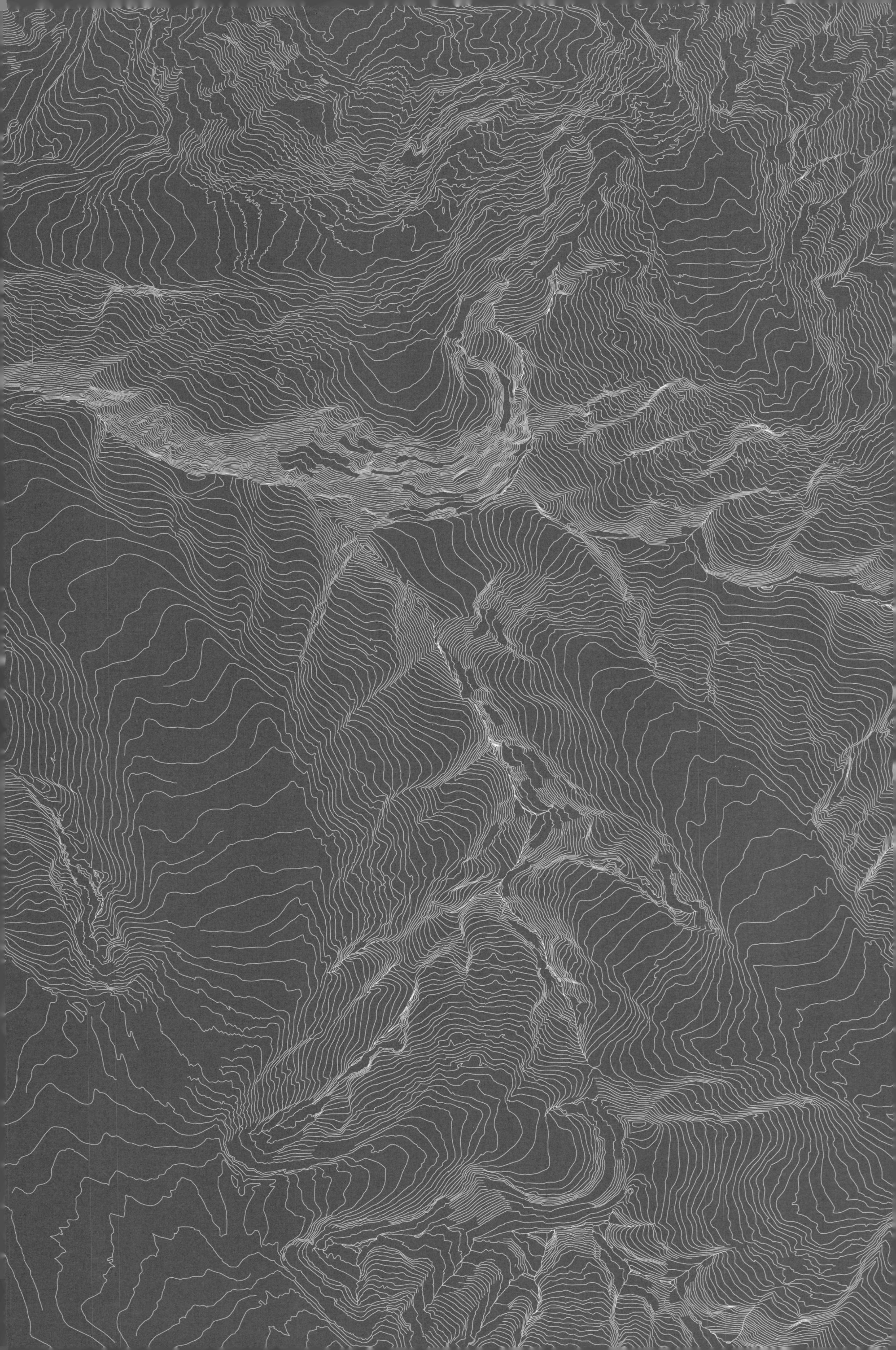